SYSTEMS ARCHITECTURE

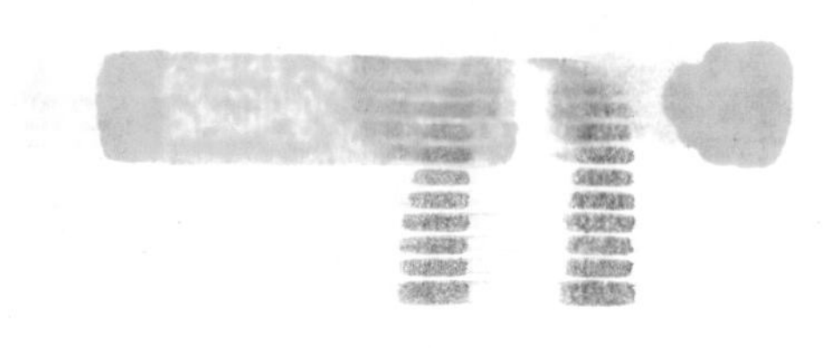

SYSTEMS ARCHITECTURE

Sixth Edition

Stephen D. Burd

University of New Mexico

Australia • Brazil • Japan • Korea • Mexico • Singapore • Spain • United Kingdom • United States

Systems Architecture, Sixth Edition
Stephen D. Burd

Executive Vice President and Publisher: Jonathan Hulbert

Executive Vice President of Editorial, Business: Jack Calhoun

Publisher: Joe Sabatino

Senior Acquisitions Editor: Charles McCormick, Jr.

Senior Product Manager: Kate Mason

Development Editor: Lisa M. Lord

Editorial Assistant: Nora Heink

Marketing Director: Keri Witman

Marketing Manager: Adam Marsh

Senior Marketing Communications Manager: Libby Shipp

Marketing Coordinator: Suellen Ruttkay

Content Project Manager: PreMediaGlobal

Media Editor: Chris Valentine

Senior Art Director: Stacy Jenkins Shirley

Cover Designer: Craig Ramsdell

Cover Image: ©Getty Images

Manufacturing Coordinator: Julio Esperas

Compositor: PreMediaGlobal

Library of Congress Control Number: 2010927431

Student Edition:

ISBN-13: 978-0-538-47533-4
ISBN-10: 0-538-47533-1

Instructor's Edition:

ISBN-13: 978-0-538-47536-5
ISBN-10: 0-538-47536-6

Course Technology
20 Channel Center Street
Boston, MA 02210
USA

Course Technology, a part of Cengage Learning, reserves the right to revise this publication and make changes from time to time in its content without notice.

Cengage Learning is a leading provider of customized learning solutions with office locations around the globe, including Singapore, the United Kingdom, Australia, Mexico, Brazil, and Japan. Locate your local office at: **www.cengage.com/global**

Cengage Learning products are represented in Canada by Nelson Education, Ltd.

To learn more about Course Technology, visit **www.cengage.com/coursetechnology**

Purchase any of our products at your local college store or at our preferred online store **www.cengagebrain.com**

Printed in the United States of America
2 3 4 5 6 7 16 15 14 13 12 11

To William I. Bullers, Jr., friend and colleague.

CONTENTS

PREFACE

INTENDED AUDIENCE

This book is intended for undergraduate students majoring or concentrating in information systems (IS) or information technology (IT) and as a reference for IS/IT professionals. It provides a technical foundation for systems design, systems implementation, hardware and software procurement, and computing resource management. Computer hardware and system software topics that are most useful to IS/IT students and professionals are described at an appropriate level of detail. For some topics, readers gain enough knowledge to solve technical problems. For other topics, they gain knowledge on communicating effectively with technical specialists.

Computer science students are exposed to computer hardware and system software technology in many undergraduate courses. Computer science books usually focus on a subset of the topics in this book. However, coverage of hardware and system software in an IS/IT curriculum is usually limited. A brief overview of hardware and system software might be provided in an introductory course, and some specific technical topics are often covered in other courses, but there's at most one course devoted to hardware and system software.

At this writing (May 2010), the latest curricula recommendations in IS and IT are IS 2010 and IT 2008. Many schools are still using curricula modeled on IS 2002 (see *www.acm.org* for details on these curricula). The topics covered in this book are mapped to all three curricula as follows:

- *IS 2002*—This book covers a superset of the requirements for IS 2002.4, Information Technology Hardware and System Software. Additional topics beyond those in IS 2002.4 include networks, application development software, and system administration.
- *IT 2008*—This book covers topics in four of the body of knowledge components: Integrative Programming and Technologies—Intersystems Communications and Overview of Programming Languages; Networking—all topics; Platform Technologies—all topics except Enterprise Deployment Software; and Systems Administration and Maintenance—Operating Systems and portions of Applications and Administrative Activities.
- *IT 2010*—This book covers the topics and learning objectives of the IT 2010.4 core course, IT Infrastructure.

This book can also serve as a supplement in courses on system design and computer resource management. For system design, it covers many technical topics to address when selecting and configuring hardware and system software. For computer resource management, it offers the broad technical foundation needed to manage resources effectively.

READERS' BACKGROUND KNOWLEDGE

Because target courses for this book are typically placed early in the recommended curricula, few assumptions are made about readers' background knowledge. Unlike many computer science books, readers aren't assumed to have an extensive background in mathematics, physics, or engineering. When necessary, background information in these areas is given in suitable depth.

In addition, readers aren't assumed to know any particular programming language. However, classroom or practical experience with at least one language is helpful to comprehend the discussions of CPU instruction sets, operating systems, and application development software. Programming examples are given in several programming languages and in pseudocode.

Detailed knowledge of a particular operating system isn't required. However, as with programming experience, practical experience with at least one operating system is helpful. Lengthy examples from operating systems are purposely avoided, but there are some short examples from MS-DOS, UNIX/Linux, and recent Windows versions.

Finally, knowledge of low-level machine instructions or assembly-language programming isn't assumed. Assembly-language instructions are described in several chapters, but a generic assembly language is used, and no detailed coverage of assembly-language program organization is included.

CHANGES IN THIS EDITION

The fifth edition was first published in 2005. Updates were needed throughout the book to address changes since that time. The following sections summarize major updates and additions, although most chapters include many additional minor changes, such as updates to screen captures, hardware specifications, and standards.

- *Chapter 1*—Updated the discussion of periodical literature and online sources of technology information.
- *Chapter 2*—Updated typical computer specifications; revised definitions of computer classes; expanded the discussion of multicomputer, distributed, and cloud computer architectures; added a Technology Focus on distributed simulation applications; modernized the Business Focus case; and updated the Technology Focus features on IBM POWER processors and the parallel evolution of Intel CPUs and Microsoft operating systems.
- *Chapter 3*—Updated the coverage of floating-point formats and Unicode standards to the latest standards.

- *Chapter 4*—Updated the discussion of RISC and the Pentium Technology Focus, added a Technology Focus on SPEC and TPC benchmarks, and updated several sections (including the Technology Focus features) to reflect current CPU clock rates, word sizes, fabrication technology, and multicore architecture.
- *Chapter 5*—Moved the discussion of memory allocation and addressing to Chapter 11, added details of solid-state drives, updated the coverage of memory packaging and nonvolatile memory technologies, and modernized the coverage of magnetic tapes, magnetic disks, and optical discs.
- *Chapter 6*—Expanded the discussion of buses to include internal/external subsidiary buses and serial bus technology, replaced the SCSI Technology Focus with one on PCI, modernized coverage and examples of buffering and caching, updated the Technology Focus about on-chip memory cache to current Intel multicore CPUs, and expanded the coverage of compression, including an updated Technology Focus.
- *Chapter 7*—Reduced the coverage of older display technologies and expanded the discussion of current display types and video adapters.
- *Chapter 8*—Expanded the discussion of bandwidth and S/N ratio, updated and expanded the coverage of copper and optical cabling, expanded the coverage of wireless transmission, and updated the Technology Focus and Business Focus.
- *Chapter 9*—Reduced the coverage of bus and ring topologies, expanded the discussion of switching and routing and sharpened the distinction between them, expanded wireless network coverage (including a new Technology Focus on WiMAX), updated the Ethernet coverage, and updated the Business Focus.
- *Chapter 10*—Updated the coverage of application development tools, including related Technology Focus and Business Focus features.
- *Chapter 11*—Revised the introductory material extensively, incorporated and updated memory allocation and addressing material moved from Chapter 5, and added material on hypervisors and a Technology Focus on VMware.
- *Chapter 12*—Updated and expanded the coverage on RAID, added material on backup and recovery procedures, and added a Technology Focus on the Google File System.
- *Chapter 13*—Reorganized the chapter for improved flow of topics, expanded the discussion of distributed architectures to include peer-to-peer architectures, revised the discussion of protocol stacks to be consistent with changes in Chapter 9, updated the Technology Focus to cover Java Platform, Extended Edition, and added new material on cloud computing architectures, including a Business Focus.
- *Chapter 14*—Streamlined to eliminate repetition with expanded coverage in other chapters, updated the Technology Focus on Windows monitoring tools, and updated screen captures to current Windows versions.

RESOURCES FOR INSTRUCTORS

Systems Architecture, Sixth Edition includes the following resources to support instructors in the classroom. All the teaching tools available with this book are provided to the instructor on a single CD. They can also be accessed with your single sign-on (SSO) account at Cengage.com.

- *Instructor's Manual*—The Instructor's Manual provides materials to help instructors make their classes informative and interesting. It includes teaching tips, discussion topics, and solutions to end-of-chapter materials.
- *Classroom presentations*—Microsoft PowerPoint presentations are available for each chapter to assist instructors in classroom lectures or to make available to students.
- *ExamView®*—ExamView is a powerful testing software package that enables instructors to create and administer printed, computer (LAN-based), and Internet exams. It includes hundreds of questions corresponding to the topics covered in this book so that students can generate detailed study guides with page references for further review. The computer-based and Internet testing components allow students to take exams at their computers and save instructors time by grading each exam automatically.
- *Distance learning content*—Course Technology is proud to present online content in WebCT and Blackboard to provide the most complete and dynamic learning experience possible. For more information on how to bring distance learning to your course, contact your local Cengage sales representative.

WORLD WIDE WEB SITES

Two support sites for this book (instructor and student), located at *www.cengage.com/mis/burd*, offer the following:

- The Instructor's Manual
- Figure files
- End-of-chapter questions and answers
- Web resource links for most book topics and research problems
- Additional content on virtualization
- Text updates and errata
- Glossary

ORGANIZATION

This book's chapters are organized into four groups. The first group contains two chapters with overviews of computer hardware, software, and networks and describes sources of technology information. The second group consists of five chapters covering hardware technology. The third group contains two chapters on data communication and computer networks. The fourth group includes five chapters covering software technology and system administration.

The chapters are intended for sequential coverage, although other orderings are possible. The prerequisites for each chapter are described in the following section. Other chapter orders can be constructed based on these prerequisites. Chapter 2 should always be covered before other chapters, although some sections can be skipped without loss of continuity, depending on which subsequent chapters are included or skipped.

There should be time to cover between 9 and 12 chapters in a three-credit-hour undergraduate course. This book contains 14 chapters to offer flexibility in course content. Topics in some chapters can be covered in other courses in a specific curriculum. For example, Chapters 8 and 9 are often covered in a separate networking course, and Chapter 14 is often included in a separate system administration course. Instructors can choose specific chapters to best match the overall curriculum design and teaching preferences.

CHAPTER DESCRIPTIONS

Chapter 1, "Computer Technology: Your Need to Know," briefly describes how knowledge of computer technology is used in the systems development life cycle. It also covers sources for hardware and system software information and lists recommended periodicals and Web sites. It can be skipped entirely or assigned only as background reading.

Chapter 2, "Introduction to Systems Architecture," provides an overview of hardware, system and application software, and networks. It describes main classes of hardware components and computer systems and describes the differences between application and system software. This chapter introduces many key terms and concepts used throughout the book.

Chapter 3, "Data Representation," describes primitive CPU data types and common coding methods for each type. Binary, octal, and hexadecimal numbering systems and common data structures are also discussed. Chapter 2 is a recommended prerequisite.

Chapter 4, "Processor Technology and Architecture," covers CPU architecture and operation, including instruction sets and assembly-language programming. It describes traditional architectural features, including fetch and execution cycles, instruction formats, clock rate, registers, and word size. It also discusses methods for enhancing processor performance as well as semiconductor and microprocessor fabrication technology. Chapters 2 and 3 are necessary prerequisites.

Chapter 5, "Data Storage Technology," describes implementing primary and secondary storage with semiconductor, magnetic, and optical technologies. Principles of each storage technology are described first, followed by factors affecting each technology and guidelines for choosing secondary storage technologies. Chapters 3 and 4 are necessary prerequisites, and Chapter 2 is a recommended prerequisite.

Chapter 6, "System Integration and Performance," explains communication between computer components and performance enhancement methods. It starts with a discussion of system bus and subsidiary bus protocols, followed by coverage of device controllers, mainframe channels, and interrupt processing. Performance enhancement methods include buffering, caching, parallel and multiprocessing, and compression. Chapters 4 and 5 are required prerequisites, and Chapters 2 and 3 are recommended prerequisites.

Chapter 7, "Input/Output Technology," describes I/O devices, including keyboards, pointing devices, printers and plotters, video controllers and monitors, optical input devices, and audio I/O devices. It also covers fonts, image and color representation, and image description languages. Chapter 3 is a necessary prerequisite, and Chapters 2, 5, and 6 are recommended prerequisites.

Chapter 8, "Data and Network Communication Technology," covers data communication technology, beginning with communication protocols, analog and digital signals, transmission media, and bit-encoding methods. This chapter also explains serial and parallel transmission, synchronous and asynchronous transmission, wired and wireless transmission, channel-sharing methods, clock synchronization, and error detection and correction. Chapters 2 and 3 are recommended prerequisites.

Chapter 9, "Computer Networks," describes network architecture and hardware. It starts with network topology and message forwarding, and then explains media access control and network hardware devices, such as routers and switches. This chapter also covers IEEE and OSI networking standards and includes an in-depth look at Internet architecture and TCP/IP. Chapters 3, 4, and 8 are necessary prerequisites, and Chapter 2 is a recommended prerequisite.

Chapter 10, "Application Development," begins with a brief overview of the application development process and development methodologies and tools, and then discusses programming languages, compilation, link editing, interpretation, and symbolic debugging. The final section describes application development tools, including CASE tools and integrated development environments (IDEs). Chapters 2, 3, and 4 are necessary prerequisites.

Chapter 11, "Operating Systems," describes the functions and layers of an operating system, explains resource allocation, and describes how an operating system manages the CPU, processes, threads, and memory. Chapters 2, 4, and 5 are necessary prerequisites, and Chapter 10 is a recommended prerequisite.

Chapter 12, "File and Secondary Storage Management," gives an overview of file management components and functions, including differences between logical and physical secondary storage access, and describes file content and structure and directories. Next, this chapter describes storage allocation, file manipulation, and access controls and ends with file migration, backup, recovery, fault tolerance, and storage consolidation methods. Chapters 5 and 11 are necessary prerequisites, and Chapter 10 is a recommended prerequisite.

Chapter 13, "Internet and Distributed Application Services," begins by discussing distributed computing and network resource access, network protocol stacks, and directory services. This chapter also explains interprocess communication, Internet protocols for accessing distributed resources, and component-based application development. Finally, it describes cloud computing models. Chapters 2, 8, and 9 are necessary prerequisites. Chapters 4, 11, and 12 are recommended prerequisites.

Chapter 14, "System Administration," gives an overview of system administration tasks and the strategic role of hardware and software resources in an organization. It then describes the hardware and software acquisition process. Next, the chapter discusses methods for determining requirements and monitoring performance. The next section describes measures for ensuring system security, including access controls, auditing, virus protection, software updates, and firewalls. The last section discusses physical environment factors affecting computer operation. Chapters 2, 4, 8, 11, and 12 are recommended prerequisites.

Appendix, "Measurement Units," summarizes common measurement units, abbreviations, and usage conventions for time intervals, data storage capacities, and data transfer rates.

ACKNOWLEDGMENTS

The first edition of this book was a revision of another book, *Systems Architecture: Software and Hardware Concepts*, by Leigh and Ali. Some of their original work has endured through all subsequent editions. I am indebted to Leigh, Ali, and Course Technology for providing the starting point for all editions of this book.

I thank everyone who contributed to this edition and helped make previous editions a success. Jim Edwards took a chance on me as an untested author for the first edition. Kristen Duerr, Jennifer Locke, Maureen Martin, and Kate Mason shepherded the text through later editions. Kate Mason oversaw this edition, and Lisa Lord helped make the text much more readable. Thanks to all past and present development and production team members. Thanks also to the peer reviewers of this edition: Angela Clark, University of South Alabama; Raymond Hansen, Purdue University; Jim Mussulman, Southern Illinois University Edwardsville; Barbara Ozog, Benedictine University; John Reynolds, Grand Valley State University; and Jeffrey Sprankle, Purdue University.

I thank students in the undergraduate MIS concentration at the Anderson Schools of Management, University of New Mexico, who have used manuscripts and editions of this book over the past two decades. Student comments have contributed significantly to improving the text. I also thank my department chair, Steven Yourstone, and my faculty colleagues—Ranjit Bose, Nick Flor, Peter Jurkat, Xin Luo, Laurie Schatzberg, Josh Saiz, and Alex Seazzu—for their continued support and encouragement of my textbook-writing activities.

Finally, I'd like to thank Dee, my wife, and Alex and Amelia, my children. Developing this book as a sole author through multiple editions has been a time-consuming process that often impinged on family time and activities. My success as an author would not be possible without my family's love and support.

CHAPTER 1

COMPUTER TECHNOLOGY: YOUR NEED TO KNOW

CHAPTER GOALS

- Describe the activities of information systems professionals
- Describe the technical knowledge of computer hardware and system software needed to develop and manage information systems
- Identify additional sources of information for continuing education in computer hardware and system software

The words you're reading now result, in part, from computer-based information systems. The author, editors, production team, and distribution team all relied on computer systems to organize, produce, and convey this information. Although many different kinds of computer information systems were involved, they share similar technology: Each consists of computer hardware, system and application software, data, and communication capabilities. In this chapter, you learn why you need to study hardware and software technology if you work, or plan to work, in information systems. You also learn about additional sources of information that can help expand and update your knowledge of hardware and software.

TECHNOLOGY AND KNOWLEDGE

The world is filled with complex technical devices that ordinary people use every day. Fortunately, you don't need a detailed understanding of how these devices work to use them. Imagine needing three months of training just to use a refrigerator or needing a detailed understanding of mechanics and electronics to drive a car. Using the earliest computers in the 1940s took years of training, but today, even though computers are increasingly complex and powerful, they're also easier to use. As a result, computers have proliferated beyond their original scientific applications into businesses, classrooms, and homes. If computers have become so easy to use, why do you need to know anything about their inner technology?

Acquiring and Configuring Technological Devices

The knowledge required to purchase and configure technically complex devices is far greater than the knowledge required to use them effectively. Many people can use complex devices, such as cars, home theater systems, and computers, but few people feel comfortable purchasing or configuring them. Why is this so?

When you walk into a store or visit a Web site to purchase a computer, you're confronted with a wide range of choices, including processor type and speed, hard disk speed and capacity, memory capacity, and operating system. To make an informed choice, you must know your preferences and requirements, such as the application software you plan to use and whether you plan to discard or upgrade the computer in a year or two. To evaluate the alternatives and determine their compatibility with your preferences and requirements, you must be able to comprehend technical terms (for example, gigahertz, gigabyte, DDR, and USB), technical and sales documentation, product and technology reviews, and the advice of friends, experts, and salespeople.

An information systems (IS) professional faces computer acquisition, upgrade, and configuration choices that are far more complex. Large computer systems and the software that runs on them use more complex technology than smaller ones do. There are many more components and, therefore, more complex configuration, compatibility, and administrative issues. Of course, the stakes are higher. Employers and users rely on the expertise of IS professionals, and companies invest substantial sums of money based on their recommendations. Are you (or will you be) able to meet the challenge?

INFORMATION SYSTEM DEVELOPMENT

When developing an information system, IS professionals follow a series of steps called a **systems development life cycle (SDLC)**. Figure 1.1 shows a modern SDLC called the **Unified Process (UP)**. Under the UP, an information system is built in a series of 4- to 6-week repeated steps called **iterations** (the vertical columns separated by dashed lines). Although Figure 1.1 shows six iterations, the number of iterations is tailored to each development project's specifications. Typically, the first iteration or two produces documentation and a prototype (model) system that's refined and expanded in subsequent iterations until it becomes the final system.

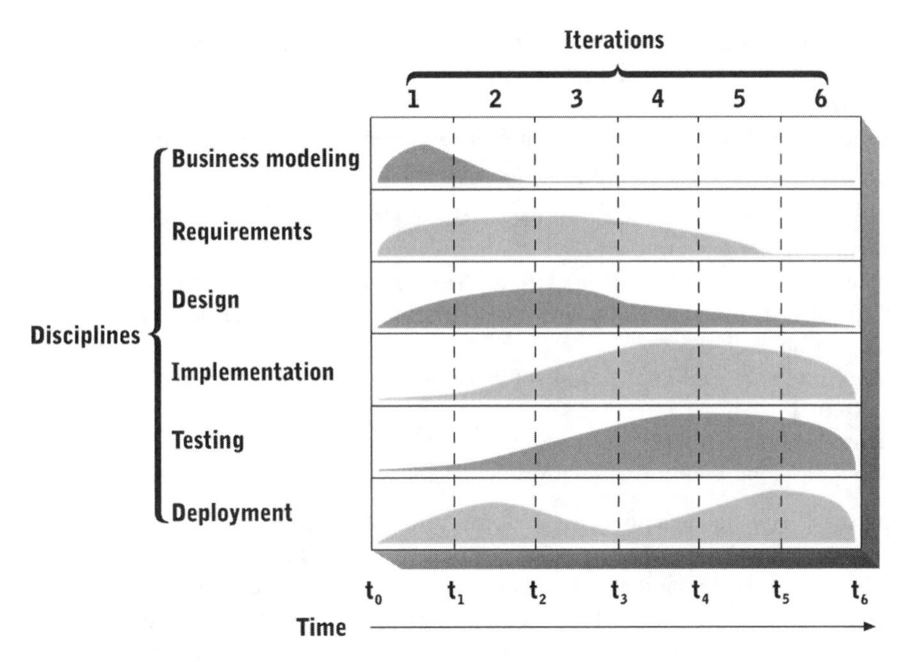

FIGURE 1.1 Disciplines and iterations in the Unified Process
Courtesy of Course Technology/Cengage Learning

Each iteration includes whatever activities are needed to produce testable models or working software. Related activities are grouped into UP **disciplines**. For example, the testing discipline includes activities such as creating test data, conducting tests, and evaluating test results. Activities and efforts in each discipline vary across iterations, as shown by the shaded curves in Figure 1.1. For example, in this figure, activities in iteration 1 are drawn primarily from the business modeling, requirements, design, and deployment disciplines, and activities in iteration 6 are drawn primarily from the implementation, testing, and deployment disciplines. As with the number of project iterations, the mix of activities in each iteration is tailored to each development project. Therefore, efforts in each discipline aren't always distributed across iterations exactly as shown in Figure 1.1.

The following sections explore the UP disciplines in more detail and describe the knowledge of computer hardware and system software each one requires.

Business Modeling and Requirements Disciplines

Activities in the **business modeling discipline** and the **requirements discipline** are primarily concerned with building models of the organization that will own and operate the system, models of the system's environment, and models of system and user requirements. The models can include narratives, organizational charts, workflow diagrams, network diagrams, class diagrams, and interaction diagrams. The purpose of building business and requirements models is to understand the environment in which the system will function and the tasks the system must perform or assist users to perform.

While building business and requirements models, developers ask many questions about the organization's needs, users, and other constituents and the extent to which these needs are (or aren't) being met and how they'll be addressed by a new system.

Technical knowledge of computer hardware and system software is required to assess the degree to which users' needs are being met and to estimate the resources required to address unmet needs.

For example, an analyst surveying a point-of-sale system in a retail store might pose questions about the current system, such as the following:

- How much time is required to process a sale?
- Is the system easy to use?
- Is enough information being gathered (for example, for marketing purposes)?
- Can the hardware and network handle periods of peak sales volume (such as during holidays)?
- Can hardware and application software be supported by a different operating system?
- Are cheaper hardware alternatives available?
- Could a cloud computing environment support the application software?

Answering these questions requires an in-depth understanding of the underlying hardware and software technologies. For example, determining whether a system can respond to periods of peak demand requires detailed knowledge of processing and storage capabilities, operating systems, networks, and application software. Determining whether cheaper alternatives exist requires technical knowledge of a wide range of hardware and software options. Determining whether cloud computing could be used requires a detailed understanding of the software environment and whether it's compatible with various cloud computing environments.

Design Discipline

The **design discipline** is the set of activities for determining the structure of a specific information system that fulfills the system requirements. The first set of design activities, called **architectural design**, selects and describes the exact configuration of all hardware, network, system software, and application development tools to support system development and operations (see Figure 1.2). These selections affect all other design decisions and serve as a blueprint for implementing other systems.

Specific systems design tasks include selecting the following:

- Computer hardware (processing, storage, input/output [I/O], and network components)
- Network hardware (transmission lines, routers, and firewalls)
- System software (operating system, database management system, Web server software, network services, and security software and protocols)
- Application development tools (programming languages, component libraries, and integrated development environments)

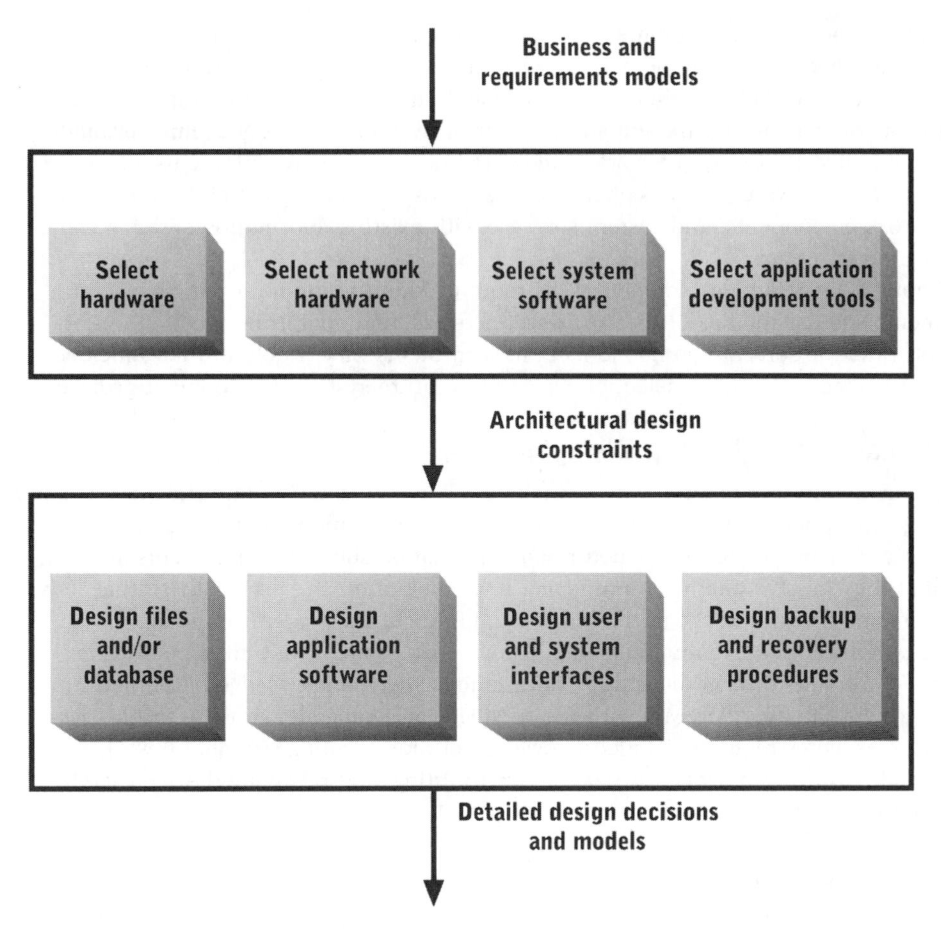

FIGURE 1.2 Design activities in the Unified Process
Courtesy of Course Technology/Cengage Learning

Collectively, these choices define an **information architecture**—requirements and constraints that define important characteristics of information-processing resources and how these resources interact with one another. When actual hardware, network, and system software components are acquired and installed, they make up an information technology infrastructure for one or more information systems. Operating and maintaining the infrastructure is a complex and costly endeavor in most organizations.

The remaining design activities, called **detailed design**, are narrower in scope and constrained by the information architecture. Detailed design activities include the following:

- File or database design (such as grouping data elements into records and files, indexing, and sorting)
- Application software design
- User and external system interface design (input screen formats, report formats, and protocols to interact with external services and systems)
- Design of system backup and recovery mechanisms

Technical knowledge of computer hardware and system software is most important for performing architectural design activities. Selecting hardware and network components requires detailed knowledge of their capabilities and limitations. When multiple hardware and network components are integrated into a single system, the designer must evaluate their compatibility. Hardware, network, and overall performance requirements affect the choice of system software. The designer must also consider the compatibility of new hardware, network components, and system software with existing information systems and computing infrastructure.

Selecting appropriate development tools requires knowing the information system requirements and capabilities of the hardware, network, and operating system. Development tools (and the software components built with them) vary widely in their efficiency, power, and compatibility. Tool selection also affects future system development projects.

Implementation and Testing Disciplines

The **implementation discipline** of the UP includes all activities for building, acquiring, and integrating application software components. The **testing discipline** includes activities that verify correct functioning of infrastructure and application software components and ensure that they satisfy system requirements. Implementation and especially testing activities require specific knowledge of the hardware, network, and system software.

For example, developing an application software component that interacts with an external Web service to schedule a shipment requires specific knowledge of the network protocols used to find and interact with the service. Diagnosing an error that occurs when the software executes requires detailed knowledge of the operating system, network services, and network protocols for creating, transmitting, and receiving the Web service request and response.

Deployment Discipline

The **deployment discipline** is the set of activities for installing and configuring infrastructure and application software components and bringing them into operation. Questions addressed by deployment discipline activities include the following:

- Who should be involved in and responsible for deploying each part of the system?
- In what order should parts of the system be deployed?
- Will any parts of the new system operate in parallel with the previous system?

Technical knowledge of computer hardware and system software is needed to perform many deployment tasks. Installing and configuring hardware, networks, and system software is a specialized task that requires a thorough understanding of the components being installed and the purposes for which they'll be used. Tasks such as formatting storage devices, setting up system security, installing and configuring network services, and establishing accounting and auditing controls require considerable technical expertise.

Systems Evaluation and Maintenance

Although not a formal UP discipline, systems evaluation and maintenance is an important group of activities that accounts for much of the long-range system cost. Over time, problems with the system can and do happen. Errors that escaped detection during testing and deployment might show up. For example, a Web-based order-entry system might become overloaded because of inadequate estimates of processing volume, network congestion, or capacity limits in underlying hardware or database services. Information needs can and do change, necessitating changes to collect, process, and store additional data.

Minor system changes, such as correcting application software errors or minor processing adjustments, are normally handled as maintenance changes. Maintenance changes can require extensive technical knowledge, and some technical knowledge might be needed to classify a proposed change as major or minor. Will new processing requirements be the "straw that breaks the camel's back" in terms of hardware, network, or software capacity? Do proposed changes require application development tools that aren't compatible with the current system's design or configuration? The answers to these questions determine whether the existing system will be modified or replaced by a new system.

If the existing system is to be modified, the application software components and files to be changed are identified, modified, tested, and deployed. The technical knowledge requirements depend heavily on the specific hardware, network, and software components affected by the change. If a new system is required, a new systems development life cycle is initiated.

MANAGING COMPUTER RESOURCES

So far, the need for technological knowledge has been discussed in the context of developing a single information system. However, think about the complexities and knowledge needed to manage the thousands of computer resources in a large organization, where many new development projects or system upgrades can be in progress at once.

In this type of environment, you must pay more attention to two critical technological issues—compatibility and future trends. Both issues are important because of the integration of computing technology in and across every function of modern organizations. For example, accounts payable and accounts receivable programs usually share a common hardware platform and operating system. Data from both systems is shared by a financial reporting system, which might be a different software system running on an entirely different computer. Data from many sources in the organization is often stored in a common database and accessed via an internal network or the Internet.

Managers of integrated collections of information systems and supporting infrastructure must contend with a great deal of technical complexity. They must ensure that each new system not only operates correctly by itself, but also operates smoothly with all the other systems in the organization. They must also make sure hardware and software acquisitions are a good foundation for both current and future systems.

Given the rapid pace of change in computer technology, a manager must have a broad understanding of current technology and future technology trends. Will the computer purchased today be compatible with the hardware available three years from now? Can the organization's communication network be expanded easily to meet future needs? Should the organization invest only in tried-and-true technologies, or should it acquire cutting-edge technologies in hopes of improving performance or gaining a competitive advantage? Should the organization invest in buying enterprise-level software packages to use on local hardware, or should applications be housed in a rented server farm?

The answers to these questions require in-depth technical knowledge—far more knowledge than any one person has. Typically, managers confronted by these questions rely on the advice of experts and other sources of information. Even so, they must have an adequate base of technical knowledge to understand this information and advice.

ROLES AND JOB TITLES

Many people in many different jobs are called computer professionals. There's a bewildering array of job titles, specializations, and professional certifications to describe the wide range of computer-related roles. To add to the confusion, responsibilities associated with specific job titles vary considerably from organization to organization. The following sections attempt to classify computer professionals into groups, explain some of their common characteristics, and describe the computer hardware and system software knowledge each group needs.

Software Developers

Software can be categorized loosely into two types: application software and system software. End users use application software to perform specific tasks, such as processing customer orders or developing and formatting documents and financial analyses. System software generally hides in the background, unnoticed or barely noticed by most end users. Examples include many parts of an operating system, database management systems, and software that protects networks against intruders. Of course, some software doesn't fit neatly into either category.

Many **software developers** create application software for specific processing needs. They have many different job titles, including programmer, systems analyst, and systems designer. Each role contributes to a different part of the systems development life cycle. A **systems analyst** performs activities in the business modeling and requirements disciplines. A **systems designer** performs activities in the design discipline and sometimes the deployment discipline. A **programmer** builds and tests software.

Many software developers have responsibilities that match these job descriptions precisely. For example, a systems analyst is often responsible for business modeling, requirements, design, and management tasks for a development project. Programmers often perform some requirements and design discipline tasks in addition to building and testing software.

The wide variety of application software causes further confusion about job titles, activities, responsibilities, and required education and training. For example, developers of application software that processes business transactions or provides information to managers usually have college or technical degrees in management or business with a specialization in information processing.

NOTE

Other names for the information processing field include management information systems, data processing, and business computer systems.

Developers of application software for scientific areas, such as astronomy, meteorology, and physics, typically have degrees in mathematics or **computer science**. Developers of application software for technical areas, such as robotics, flight navigation, and scientific instrumentation, typically have degrees in computer science or some branch of engineering.

All application software developers need technical knowledge of computer hardware and system software, as described earlier in "Information System Development." Software developers in technical areas typically need in-depth hardware knowledge because of the hardware control characteristics of many of these applications. Software developers in scientific areas must also have in-depth hardware knowledge if the applications they develop push the boundaries of hardware capabilities, such as simulating weather patterns on a global scale with supercomputers.

Systems programmers develop system software, such as operating systems, compilers, database management systems, Web servers, and network security monitors. These programmers typically have degrees in computer science or computer engineering. Organizations using a lot of computer equipment and software employ systems programmers to perform tasks such as hardware troubleshooting and software installation and configuration. Organizations that develop and market system and application development software, such as Microsoft, Oracle, and Cisco, employ many systems programmers.

Systems programmers must have in-depth knowledge of system software as well as computer hardware and networks because many types of system software interact directly with computer or network hardware. For this reason, computer science and computer engineering programs usually require students to take several different courses to learn the subjects discussed in this book.

Hardware Personnel

Computer hardware vendors employ a variety of people for design, installation, and maintenance. Lower-level personnel usually have technical degrees and/or vendor-specific training, and higher-level personnel usually have degrees in computer science or computer engineering. Employees must have extensive knowledge of computer hardware, including processing, data storage, I/O, and networking devices. Hardware designers need the most in-depth knowledge, far exceeding the scope of this book.

System Managers

The proliferation of computer hardware, software, and networks in today's organizations has created a need for many computer-related managers and administrators. The job descriptions vary widely because of differences in organizational structure and the nature of the organization's information systems and infrastructure. Common job titles include computer operations manager, network administrator, database administrator, and chief information officer.

A **computer operations manager** oversees the operation of a large information processing facility. These facilities usually house one or more large computer systems and all related peripheral equipment in a central location. They often have large databases, thousands of application programs, and dozens to hundreds of employees and perform a lot of batch processing. Organizations requiring this type of computing facility include large banks, credit reporting bureaus, the Social Security Administration, and the Internal Revenue Service. Organizations such as Google and Electronic Data Systems also operate this type of computing facility for themselves and for clients. Management of day-to-day operations in these facilities is extremely complex. Scheduling, staffing, security, system backups, maintenance, and upgrades are some important responsibilities of a computer operations manager.

A computer operations manager can have many technical specialists on staff. Staff members usually have considerable technical knowledge in narrow specialties, such as data storage hardware, network configuration and security, mainframe operating systems, and performance tuning. A computer operations manager needs a broad base of technical knowledge to understand the organization's information systems and infrastructure and must be capable of understanding the advice of technical staff.

Typically, the title of **network administrator** is applied to one of two roles. The first is responsibility for an organization's network infrastructure, such as for an Internet service provider or a large multinational corporation. Designing, operating, and maintaining a large network require substantial technical expertise in computer hardware, telecommunications, and system software. The role of network administrator in this environment is an important high-level position. Technical knowledge requirements are similar to those for a computer operations manager, although the emphasis is on network and data communication technologies.

In a smaller organization, the title of network administrator is used for the manager of a local area network. These networks connect anywhere from a half dozen to a few hundred computers (mostly desktop and portable computers) and provide access to shared databases. The network administrator can be responsible for many tasks other than operating and maintaining the network, including installing and maintaining end-user software, installing and configuring hardware, training users, and assisting management in selecting and acquiring software and hardware. This position is one of the most demanding in breadth and depth of required skills and technical knowledge.

The technology for managing and accessing large collections of data, called databases, is specialized and highly complex. This complexity, combined with managerial recognition of the importance of data resources, has resulted in creating many positions with the title of **database administrator**. This role requires both technical expertise and the ability to help the organization make optimal use of its data resources for tasks such as market research.

A large organization with a substantial investment in computer, network, and software technology usually has one high-level manager with the title **chief information officer (CIO)**. Many of the previously defined positions (database administrator, network administrator, and computer operations manager) report to the CIO. The CIO is responsible for the organization's computers, networks, software, and data as well as for strategic planning and the effective use of information and computing technology.

CIOs can't possibly be experts in every aspect of computer technology related to their organizations, but they must have a broad enough base of technical knowledge to interact effectively with all the organization's technical specialists. They must also be aware of how technology is changing and how best to respond to these changes to support an organization's mission and objectives.

COMPUTER TECHNOLOGY INFORMATION SOURCES

This book gives you a foundation of technical knowledge for a career in information system development or management. Unfortunately, this foundation can erode quickly because computer and information technologies change rapidly. How will you keep up with the changes?

You can use many resources to keep your knowledge current. Periodicals and Web sites offer a wealth of information on current trends and technologies. Training courses from hardware and software vendors can teach you the specifics of current products. Additional coursework and self-study can keep you up to date on technologies and trends not geared toward specific products. By far, the most important of these activities is reading periodical literature.

Periodical Literature

The volume of literature on computer topics is huge, so IS professionals face the difficult task of determining which sources are the most important and relevant. Because of differences in training and focus among computer professionals, sources of information oriented toward one specialty are difficult reading for professionals in other specialties. For example, a detailed discussion of user interaction and validation of requirement models targeted to IS professionals might be beyond a computer scientist or engineer's training. Similarly, detailed descriptions of optical telecommunication theory might be beyond an IS professional's training. These differences pose a problem for IS professionals who need current information on hardware and system software technology. The following periodicals are good information sources for IS professionals:

- *ACM Computing Surveys* (*http://surveys.acm.org*)—An excellent source of information on the latest research trends in computer software and hardware. Contains in-depth summaries of technologies or trends geared toward a readership with moderate to high familiarity with computer hardware and software.
- *Computerworld* (*www.computerworld.com*)—A weekly magazine focusing primarily on computer news items. Covers product releases, trade shows, and occasional reports of technologies and trends.
- *Communications of the ACM* (*http://cacm.acm.org*)—A widely used source of information about research topics in computer science. Many of the articles are highly technical and specialized, but some are targeted to a less research-oriented audience.
- *Computer* (*www.computer.org/computer*)—A widely used source of information on computer hardware and software. Many of the articles are research-oriented, but occasionally they cover technologies and trends for a less technical audience.
- *InformationWeek* (*www.informationweek.com*)—An online magazine focusing mainly on computer news items, covering a wide range of computer-related organizations and technologies.

These periodicals are only a small sample of the available literature. A complete list of today's recommended periodicals would become out of date quickly because of rapid changes in computer technology and the computer publishing industry. You can find a current list of recommended periodicals in the Research Links section of this book's Web site (*www.cengage.com/mis/burd*).

Most periodical publishers have a Web site with content and services that augment what's available in printed periodicals, such as the following:

- Content from back issues
- Additional current content that's not included in the printed periodical
- Search engines

Figure 1.3 shows the Web site for *InformationWeek*. Web-based periodicals often include Web links to article references and other related material. For example, an article referring to another article in a back issue might contain a link to the back issue. Links to reference lists and bibliographies can also be included, so the task of accessing background and reference material is as easy as clicking the mouse. Similarly, reviews of software or hardware might contain links to product information and specifications on vendor or manufacturer Web sites. Articles and other postings often include follow-up comments from readers and links to related blogs.

FIGURE 1.3 The InformationWeek.com home page
Courtesy of InformationWeek.com

Technology-Oriented Web Sites

Computer technology is a big business; there are millions of computer professionals worldwide and many Web sites devoted to serving their needs. Table 1.1 lists some of the most widely used sites. Check this book's Web site for updates to this table.

TABLE 1.1 Technology-related Web sites

Organization	URL	Description
CNET	*www.cnet.com*	Oriented toward consumers of a broad range of electronics devices but contains some computer content of interest to IS professionals
Earthweb	*www.earthweb.com*	Offers a broad range of information for IS professionals
Gartner Group	*www.gartnergroup.com*	A consulting and research company specializing in services for CIOs and other executive decision makers
Internet.com	*www.internet.com*	Contains a broad range of information for IS professionals, with an emphasis on Internet technology
ITworld	*www.itworld.com*	Provides a broad range of information for IS professionals, with an emphasis on reader-contributed content
NetworkWorld	*www.networkworld.com*	Contains a broad range of information for IS professionals, with an emphasis on network-related content
TechRepublic	*www.techrepublic.com*	Offers a broad range of information for IS professionals
Techweb	*www.techweb.com*	Contains a broad range of information for IS professionals
Tom's Hardware	*www.tomshardware.com*	Includes detailed articles covering hardware and software technologies and product reviews

Consolidation in periodical publishers has created large corporate families of technology-related Web sites and publications. These Web sites are owned by, or associated with, periodical publishers. For example, Computerworld.com is owned by International Data Group, Inc., which also owns JavaWorld, LinuxWorld, and many other Web sites and their affiliated print publications. Technology-oriented Web sites serve as a common interface to these publication families. They also enable publishers to provide content and services that transcend a single paper publication, such as cross-referencing publications and sites and offering online discussion groups, blogs, RSS and Twitter newsfeeds, employment services, and Web-based interfaces to hardware, software, and technology service vendors.

Technology Web sites can make money in several ways. A few companies, such as the Gartner Group, charge customers for Web-based information and services, but most companies earn revenue in other ways, including the following:

- Advertising
- Direct sales of goods and services
- Commissions on goods and services sold by advertisers and partners

Any site that generates revenue from advertising, referrals, commissions, or preferred partner arrangements might have biased content. Some examples of bias are as follows:

- Ordering content, links, or search results to favor organizations that have paid a fee to the Web site owner
- RSS and Twitter newsfeeds emphasizing organizations that have paid a fee to the Web site owner
- Omitting information from organizations that haven't paid a fee to the search provider
- Omitting information that's against the interests of organizations that have paid a fee to the search provider

CAUTION

These same biases might be reflected in general Internet search engines. It's not always obvious to readers which, if any, of these biases are present in Web sites.

Unbiased information exists on the Web, although it's not always easy to find. The old saying "You get what you pay for" applies. High-quality, unbiased information is the product of intensive research. Some computer professional societies and government agencies offer this information as a public service, but much of what you find on the Web is produced for a profit. Expect to pay for unbiased information. When dealing with publicly accessible information sources, be sure to use information from several unrelated sources to balance the biases of each source.

Vendor and Manufacturer Web Sites

Most vendors and manufacturers of computer hardware and software have an extensive Web presence (see Figure 1.4). Vendor Web sites are oriented toward sales, but they usually contain detailed information on specific products or links to manufacturer Web sites. Manufacturer Web sites have detailed information on their products and offer technical and customer support services.

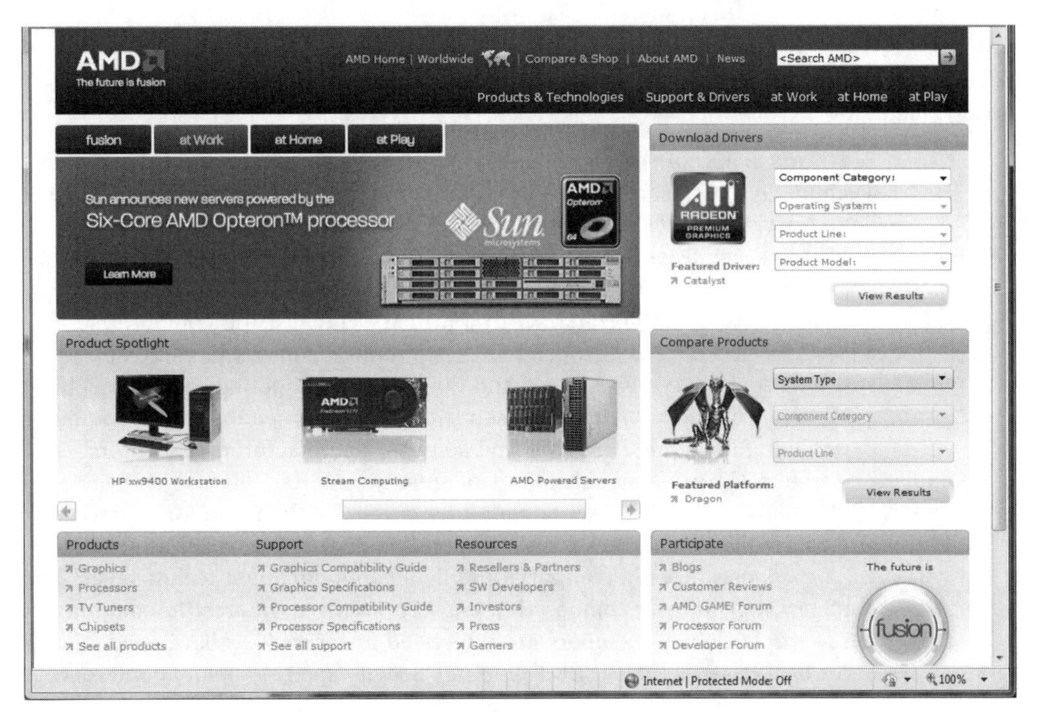

FIGURE 1.4 A typical computer hardware manufacturer's Web page

Manufacturer Web sites are mainly marketing and customer support tools. They supply technical product information that's far more detailed and current than what's available in printed brochures and technical documents. The breadth and depth of this information can help IS professionals make faster, better, and more informed choices. The downside is that this information is often biased in favor of vendors' products and the technologies they're based on. As with any source of information, you must consider the provider's motives and objectivity. Hardware and software manufacturers aren't in the business of providing unbiased information; they're in the business of selling what they produce. You should expect the content of manufacturer and vendor Web sites to be biased toward their products.

In the best case, biased content might consist of marketing hype that readers must wade through or filter out to get to the real information. In the worst case, the information might be biased purposefully by content, omissions, or both. Many sites include technology overviews, often called white papers, and similar information that's presented as independent research, even though it's not. In addition, some sites have links to biased supporting reviews or research. It's the reader's responsibility to balance potentially biased vendor and manufacturer information with information from unbiased sources.

Professional Societies

Several professional societies are excellent sources of information about computer technology, including the following:

- *AITP (www.aitp.org)*—The membership of the **Association for Information Technology Professionals (AITP)** consists mainly of IS managers and application developers. AITP has local chapters throughout the country and publishes several periodicals, including *Information Executive*.
- *ACM (www.acm.org)*—The **Association for Computing Machinery (ACM)** is a well-established organization with a primary emphasis on computer science. ACM has dozens of special-interest groups, sponsors hundreds of research conferences each year, and publishes many periodicals and technical books. The membership represents a broad cross-section of the computer community, including hardware and software manufacturers, educators, researchers, IT professionals, and students.
- *IEEE Computer Society (www.computer.org)*—The **Institute for Electrical and Electronics Engineers (IEEE) Computer Society** is a subgroup of the IEEE that specializes in computer and data communication technologies. The membership is largely composed of engineers with an interest in computer hardware, but many members are interested in software or have academic or other backgrounds. The IEEE Computer Society sponsors many conferences (some jointly with the ACM). Its publications include several periodicals, such as *IEEE Computer*, and a large collection of technical books and standards.

Summary

- Developing information systems requires technical knowledge of computer hardware, networks, and system software. The type and depth of required knowledge differ among disciplines of the Unified Process (UP). A broad base of knowledge is needed for activities in the business modeling and requirements disciplines. More in-depth knowledge is required for activities in the implementation, testing, and deployment disciplines.
- Technical knowledge is also needed to manage an organization's information systems and infrastructure, with particular attention to compatibility issues and future trends. Compatibility is important because organizational units and subsystems typically share computer hardware and system software. Future trends must be considered in acquisitions because of the long-term nature of hardware and software investments.
- With rapid changes in hardware and software technologies, technical knowledge must be updated constantly. IS professionals must engage in continuing education and study to keep pace with these changes. You can get training from vendors, educational organizations, and self-study, which relies heavily on reading periodical literature and Web resources. Articles in periodical literature can vary widely in the intended audience and the background and training you need to understand the material, so be careful when selecting periodical literature sources.
- Information on computer hardware and software is readily available on the Web, including sites maintained by publishers, vendors, manufacturers, and professional organizations. You can find a wealth of information, but be cautious because it might be biased or incomplete.

In this chapter, you've learned why you need to understand computer technology and how to keep this knowledge current. In the next chapter, you see an overview of hardware, software, and networking technology and examine concepts and terms that are explored in detail in the rest of the book. Your journey through the inner workings of modern information systems is about to begin.

Key Terms

architectural design
Association for Computing Machinery (ACM)
Association for Information Technology Professionals (AITP)
business modeling discipline
chief information officer (CIO)
computer operations manager
computer science
database administrator
deployment discipline
design discipline
detailed design
discipline
implementation discipline
information architecture
Institute for Electrical and Electronics Engineers (IEEE) Computer Society
iteration
network administrator
programmer
requirements discipline
software developers
systems analyst
systems designer

systems development life cycle (SDLC)
systems programmer
testing discipline
Unified Process (UP)

Vocabulary Exercises

1. Students of information systems generally focus on application software. Students of ________ generally focus on system software.
2. Configuring hardware and system software is an activity of the UP ________ discipline.
3. IS students and professionals should be familiar with professional societies, such as ________, ________, and ________.
4. Selecting hardware, network components, and system software is an activity of the UP ________ discipline.
5. Typically, a(n) ________ is responsible for a large computer center and all the software running in it.
6. The computer specialties most concerned with hardware and the hardware-software interface are ________ and computer engineering.
7. During the ________ UP disciplines, the business, its environment, and user requirements are defined and modeled.
8. The job titles of people responsible for developing application software include ________, ________, and ________.

Review Questions

1. How is the knowledge needed to operate complex devices different from the knowledge needed to acquire and configure them?
2. What knowledge of computer hardware and system software is necessary to perform activities in the UP business modeling and requirements disciplines?
3. What knowledge of computer hardware and system software is necessary to perform activities in the UP design and deployment disciplines?
4. What additional technical issues must be addressed when managing a computer center or campuswide network compared with developing a single information system?

Research Problems

1. The U.S. Bureau of Labor Statistics (BLS) compiles employment statistics for a variety of job categories and industries and predictions of employment trends by job category and industry. Most of the information is available on the Web. Go to the BLS Web site (*www.bls.gov*) and investigate current and expected employment prospects for IS professionals.
2. You're an IS professional, and your boss has asked you to prepare a briefing for senior staff on the comparative advantages and disadvantages of three competing tape drive technologies: Digital Linear Tape (DLT), Advanced Intelligent Tape (AIT), and Linear Tape

Open (LTO). Search the technology Web sites in Table 1.1 for source material to help you prepare the briefing. Which sites provide the most useful information? Which sites enable you to find useful information easily?

3. Read several dozen job listings at *http://itjobs.computerworld.com* or a similar site. Which jobs (and how many) require a broad knowledge of computer technology? At what level are these jobs? Try searching the listings based on a few specific keywords, such as "database," "developer," and "network." Examine the companies that are hiring and where their job postings show up in search results. Can you draw any conclusions about how the listings are ordered?

CHAPTER 2

INTRODUCTION TO SYSTEMS ARCHITECTURE

CHAPTER GOALS

- Discuss the development of automated computing
- Describe the general capabilities of a computer
- Describe computer hardware components and their functions
- List computer system classes and their distinguishing characteristics
- Define the roles, functions, and economics of application and system software
- Describe the components and functions of computer networks

Computer systems are complex combinations of hardware, software, and network components. The term **systems architecture** describes the structure, interaction, and technology of computer system components. The term "architecture" is a misnomer, however, because it implies a concern with only static structural characteristics. In this book, you also learn about the dynamic behavior of a computer system—that is, how its components interact as the computer operates.

This chapter lays the foundation for the book with a brief discussion of the major components and functions of hardware, software, and networks. Each component is described in detail in later chapters. Avoid the temptation to rush through this chapter because it's as important to know how all a computer system's components interrelate as it is to know their internal workings.

AUTOMATED COMPUTATION

A simple definition of a computer is any device that can do the following:

- Accept numeric inputs.
- Perform computational functions, such as addition and subtraction.
- Communicate results.

This definition captures a computer's basic functions, but it can also apply to people and simple devices, such as calculators. These functions can be implemented by many methods and devices. For example, a modern computer might perform computation electronically (using transistors in a microprocessor), store data optically (using a laser and an optical disc's reflective coating), and communicate with people by using a combination of electronics and mechanics, such as the mechanical and electrical components of a printer. Some experimental computers have even used quantum physics to perform data storage and computation.

Mechanical Implementation

Early mechanical computation devices were built to perform repetitive mathematical calculations. The most famous of these machines is the Difference Engine, built by Charles Babbage in 1821 (see Figure 2.1), which computed logarithms by moving gears and other mechanical components. Many other mechanical computation machines were developed well into the 20th century. Mechanical computers were used during World War II to compute trajectory tables for naval guns and torpedoes, and bookkeepers and accountants used mechanical adding machines as late as the 1970s.

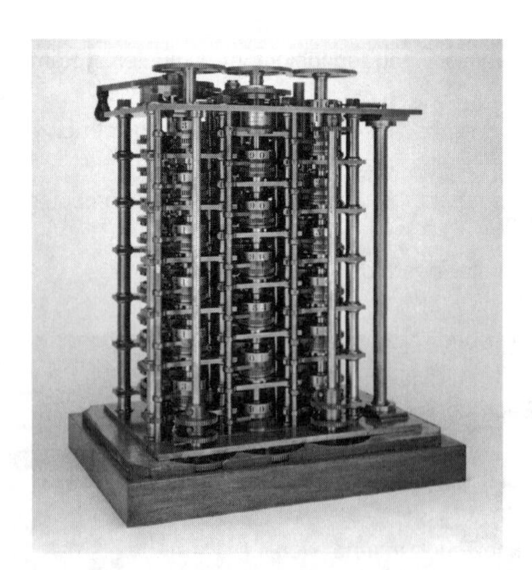

FIGURE 2.1 Charles Babbage's Difference Engine
Courtesy of Science Museum/Science & Society Picture Library

The common element in all these computation devices is a mechanical representation of a mathematical calculation. A mechanical clock is driven by a spring and pendulum, and each swing of the pendulum allows a gear to move one step under pressure from the spring. As the pendulum swings, the gears advance the clock's hands. The user inputs the current time by adjusting the hour and minute hands manually. Starting the pendulum in motion activates a calculation that repeatedly increments the current time and displays the result by using the clock hands and numbers printed on the clock face.

Mechanical computation devices can also perform more complex calculations. For example, whole numbers can be multiplied by repeated addition. A machine capable of addition can perform multiplication by executing the addition function multiple times. (For example, 6 times 3 can be calculated by adding 6 plus 6, storing the result temporarily, and then adding 6 a third time.) Combinations of moving parts can also be used to perform complex functions, such as logarithms and trigonometric functions.

Mechanical computation has some inherent limitations and shortcomings, such as the following:

- Complex design and construction
- Wear, breakdown, and maintenance of mechanical parts
- Limits on operating speed

Automated computation with gears and other mechanical parts requires a complex set of components that must be designed, manufactured, and assembled to exacting specifications. As the complexity of the computational function increases, the complexity of the mechanical device performing it also increases, exacerbating problems of design, construction, wear, and maintenance.

Electronic Implementation

Much as the era of mechanical clocks gave way to the era of electrical clocks, the era of mechanical computation eventually gave way to electronic computers. The biggest impetus for the change to electronic computing devices came during World War II. The military needed to solve many complex computational problems, such as navigation and breaking enemy communication codes. The mechanical devices of the time were simply too slow and unreliable.

In an electronic computing device, the movement of electrons performs essentially the same functions as gears and wheels in mechanical computers. Numerical values are stored as magnetic charges or by positioning electrical switches rather than gears and wheels. When necessary, electromechanical devices convert physical movement into electrical signals or vice versa. For example, a keyboard converts the mechanical motion of keystrokes into electrical signals, and ink pumps in an inkjet printer convert electrical signals into mechanical motion to force ink through a nozzle and onto paper.

Electronic computers addressed most shortcomings of mechanical computation. They were faster because of the high speed of moving electrons, and as electronic devices and fabrication technology improved, they became more reliable and easier to build than their mechanical counterparts. Electronic computers made it possible to perform complex calculations at speeds previously thought impossible. Larger and more complex problems could be addressed, and simple problems could be solved much faster.

Optical Implementation

Light can also be used as a basis for computation. A particle of light, called a photon, moves at a high rate of speed. As with electrons, a moving photon's energy can be harnessed to perform computational work. Light can be transmitted over conductors, such as laser light through a fiber-optic cable. Data can be represented as pulses of light and stored directly (such as storing an image as a hologram) or indirectly by reflective materials (such as the surface of a DVD).

Optical data communication is now common in computer networks that cover large distances. Optical discs are widely used for video and other data types. Some input/output devices (laser printers and optical scanners) are based on optical technologies and devices, and experimental optical computer processors have been developed. Optical and hybrid electro-optical devices connect system components in some experimental and high-performance computers.

Optics are expected to gradually supplant electronics during the 21st century, although the rate of change is unknown and will likely vary across computing applications. In theory, optics have clear advantages in each area of computing technology. Optical signals can carry more data than electrical signals, and optical storage media can store more data in a given physical space than magnetic or electrical media can. In addition, optical processors might be easier to fabricate than current processors and are better matched to optical communication technologies. However, turning the theoretical advantages of optical technology into more capable computer components will require time and resources. There's also the possibility that yet-to-be-developed technologies will eclipse optics in some or all areas of computing.

TECHNOLOGY FOCUS

Quantum Computing

Current computer technology is based on principles of classical physics developed during the 17th through 20th centuries, including electronics, magnetism, and optics. These principles are based on mathematical rules describing the behavior of matter at the level of atoms, molecules, and larger units. By manipulating matter at these levels, mechanical, electrical, and optical processors perform the mathematical functions underlying classical physics.

Quantum physics describes the behavior of matter differently—at a subatomic level. For example, in classical physics, an atom or a molecule has a specific position in space at any point in time. In quantum physics, a subatomic particle, such as a photon, can be in multiple places at one time. Larger particles can also be in multiple states at once, although only one state can be observed at any time.

As with classical physics, quantum physics describes subatomic behavior with mathematical rules. The rules differ from classical physics and are often more complex, but they're still rules based on specific computations. In theory, a processor that manipulates matter at the quantum level can perform quantum mathematics calculations.

In a modern digital computer, data is represented by groups of bits, each having a clearly defined binary physical state. For example, a bit value might be represented by a switch that's open or closed or an atom with a positive or negative electrical charge. Each bit has two possible states, representing the values 0 and 1. Each bit, obeying the rules of classical physics, must represent 0 or 1 at any specific time. A processor that

(*continued*)

manipulates a bit by using principles of classical physics "sees" only one physical state and produces only one computational result per bit.

At the quantum level, matter can be in multiple states at the same time. Just as a photon can be in two places at once, an atom can be both positively and negatively charged at the same time. In effect, the atom stores both 0 and 1 at the same time. This atom, or any other matter that stores data in multiple simultaneous quantum states, is called a **qubit**.

At first glance, a qubit might seem like little more than a physicist's toy, but the capability to store two data values at once has important advantages when tackling certain computational problems, such as cryptography. One advantage comes from increased storage capacity. In classical physics, a group of 3 bits can store only one of eight (2^3) possible values at a time. A group of 3 qubits can store all eight possible values at once, an eightfold increase in storage capacity.

Another advantage is computational efficiency. A computer that can manipulate 3 qubits at the quantum level can perform a calculation on all eight values at the same time, producing eight different results at once. A conventional computer must store these eight values separately and perform eight different computations to produce eight results. As the number of qubits increases, so does a quantum computer's comparative efficiency. For example, a 64-bit quantum computer is 2^{64} times more efficient than a 64-bit conventional computer.

Some prototype components for quantum computing have already been built, although a fully functional quantum computer has yet to be demonstrated publicly. Figure 2.2 shows two prototype quantum cryptography systems used in 2004 to encrypt messages sent across the Internet.

FIGURE 2.2 Prototype quantum cryptography systems
Courtesy of BBN Technologies

COMPUTER CAPABILITIES

All computers are automated computing devices, but not all automated computing devices are computers. The main characteristics distinguishing a computer from other automated computation devices include the following, discussed in more detail in the next sections:

- General-purpose processor capable of performing computation, data movement, comparison, and branching functions
- Enough storage capacity to hold large numbers of program instructions and data
- Flexible communication capability

Processor

A **processor** is a device that performs data manipulation and transformation functions, including the following:

- Computation (addition, subtraction, multiplication, and division)
- Comparison (less than, greater than, equal to, and not equal to)
- Data movement between memory, mass storage, and input/output (I/O) devices

An **instruction** is a signal or command to a processor to perform one of its functions. When a processor performs a function in response to an instruction, it's said to be **executing** that instruction.

Each instruction directs the processor to perform one simple task (for example, add two numbers). The processor performs complex functions by executing a sequence of instructions. For example, to add a list of 10 numbers, a processor is instructed to add the first number to the second and store the result temporarily. It's then instructed to add the stored result to the third number and store that result temporarily.

More instructions are issued and executed until all 10 numbers have been added. Most useful computational tasks, such as recalculating a spreadsheet, are accomplished by executing a long sequence of instructions called a program. A **program** is a stored set of instructions for performing a specific task, such as calculating payroll or generating paychecks and electronic fund transfers. Programs can be stored and reused over and over.

A processor can be classified as general purpose or special purpose. A **general-purpose processor** can execute many different instructions in many different sequences or combinations. By supplying it with a program, it can be instructed to perform a multitude of tasks, such as payroll calculation, text processing, or scientific calculation.

A **special-purpose processor** is designed to perform only one specific task. In essence, it's a processor with a single internal program. Many commonly used devices, such as automobiles, kitchen appliances, and MP3 players, contain special-purpose processors. Although these processors or the devices containing them can be called computers, the term "computer" usually refers to a device containing a general-purpose processor that can run a variety of programs.

Formulas and Algorithms

Some processing tasks require little more than a processor's computation instructions. For example, take a look at the following calculation:

GROSS_PROFIT = (QUANTITY_SOLD × SELLING_PRICE) - SELLING_EXPENSES

The only processor functions needed to compute GROSS_PROFIT are multiplying, subtracting, and storing and accessing intermediate results. In statement 30 (see Figure 2.3), QUANTITY_SOLD and SELLING_PRICE are multiplied, and the result is stored temporarily as INTERMEDIATE_RESULT. In statement 40, SELLING_EXPENSES is subtracted from INTERMEDIATE_RESULT. The GROSS_PROFIT calculation is a **formula**. A processor executes a sequence of computation and data movement instructions to solve a formula.

```
10    INPUT QUANTITY_SOLD
20    INPUT SELLING_PRICE
30    INTERMEDIATE_RESULT = QUANTITY_SOLD * SELLING_PRICE
40    GROSS_PROFIT = INTERMEDIATE_RESULT - SELLING_EXPENSES
50    OUTPUT GROSS_PROFIT
60    END
```

FIGURE 2.3 A BASIC program to compute gross profit
Courtesy of Course Technology/Cengage Learning

Computer processors can also perform a more complex type of processing task called an **algorithm**, a program in which different sets of instructions are applied to different data input values. The program must make decisions to determine what instructions to execute to produce correct data output values. Depending on the data input values, different subsets of instructions might be executed. In contrast, all the instructions that implement a formula are always executed in the same order, regardless of the data input.

The procedure for computing U.S. income tax is an example of an algorithm. Figure 2.4 shows income tax computation formulas. Note that different income values require different formulas to calculate the correct tax amount. A program that computes taxes based on this table must execute only the set of instructions that implements the correct formula for a particular income value.

Schedule X—If your filing status is **Single**

If your taxable income is:		The tax is:	
Over—	*But not over—*		*of the amount over—*
$0	$8,350	.... 10%	**$0**
8,350	33,950	**$835.00 + 15%**	**8,350**
33,950	82,250	**4,675.00 + 25%**	**33,950**
82,250	171,550	**16,750.00 + 28%**	**82,250**
171,550	372,950	**41,754.00 + 33%**	**171,550**
372,950	....	**108,216.00 + 35%**	**372,950**

FIGURE 2.4 A tax table
Courtesy of Course Technology/Cengage Learning

Comparisons and Branching

Decisions in a processing task are based on numerical comparisons. Each numerical comparison is called a **condition**, and the result of evaluating a condition is true or false. In the tax example, the income value is compared with the valid income range for each formula. Each formula is implemented as a separate instruction or set of instructions in the program. When a comparison condition is true, the program branches (jumps) to the first instruction that implements the corresponding formula.

In Figure 2.5, a BASIC program uses comparison and branching instructions to calculate income taxes. In statements 20, 50, 80, 110, and 140, INCOME is compared with the maximum income applicable to a particular tax calculation formula. The comparison result (either true or false) determines which instruction is executed next. If the comparison condition is false, the next program instruction is executed. If the comparison condition is true, the program jumps to a different point in the program by executing a GOTO, or branch, instruction.

```
10 INPUT INCOME
20 IF INCOME > 8350 THEN GOTO 50
30 TAX = INCOME * 0.10
40 GOTO 180
50 IF INCOME > 33950 GOTO 80
60 TAX = 835.00 + (INCOME - 8350) * 0.15
70 GOTO 180
80 IF INCOME > 82250 GOTO 110
90 TAX = 4675.00 + (INCOME - 33950) * 0.25
100 GOTO 180
110 IF INCOME > 171550 GOTO 140
120 TAX = 16750.00 + (INCOME - 82250) * 0.28
130 GOTO 180
140 IF INCOME > 372950 GOTO 170
150 TAX = 41754.00 + (INCOME - 171550) * 0.33
160 GOTO 180
170 TAX = 108216.00 + (INCOME - 372950) * 0.35
180 OUTPUT TAX
190 END
```

FIGURE 2.5 A BASIC program to calculate income taxes
Courtesy of Course Technology/Cengage Learning

Comparison instructions are part of a group of **logic instructions**, implying a relationship to intelligent decision-making behavior. A general-purpose processor in a computer is more restricted in comparative capabilities than a person is. A person can compare complex objects and phenomena and handle uncertainty in the resulting conclusions, whereas a computer can perform only simple comparisons (equality, less than, and greater than) with numeric data, in which the results are completely true or completely false. Despite these limitations, comparison and branching instructions are the building blocks of all computer intelligence. They're also the capabilities that distinguish a computer processor from processors in simpler automated computation devices, such as calculators.

Storage Capacity

A computer stores a variety of information, including the following:

- Intermediate processing results
- Data
- Programs

Because computers break complex processing tasks into smaller parts, a computer needs to store intermediate results (for example, the variable INTERMEDIATE_RESULT on line 30 in Figure 2.3). Programs that solve more complex, real-world problems might generate and access hundreds, thousands, or millions of intermediate results during execution.

Larger units of data, such as customer records, transactions, and student transcripts, must also be stored for current or future use. A user might need to store and access thousands or millions of data items, and a large organization might need to store and access trillions of data items. This data can be used by currently running programs, held for future processing needs, or held as historical records.

Programs must also be stored for current and future use. A simple program can contain thousands of instructions, and complex programs can contain millions or billions of instructions. A small computer might store thousands of programs, and a large computer might store millions of programs.

Storage devices vary widely in characteristics such as cost, access speed, and reliability. A computer uses a variety of storage devices because each device provides storage characteristics best suited to a particular type of data. For example, program execution speed is increased when the processor can access intermediate results, current data inputs, and instructions rapidly. For this reason, they're stored in devices with high access speeds but usually high costs, too. Programs and data held for future use can be stored on slower and less expensive storage devices. Data that must be transported physically is stored on removable media storage devices, such as DVDs or USB drives, which are slower and sometimes less reliable than fixed media devices.

Input/Output Capability

Processing and storage capabilities are of little use if a computer can't communicate with users or other computers. A computer's I/O devices must encompass a variety of communication modes—sound, text, and graphics for humans and electronic or optical communication for other computers. A typical small computer has up to a dozen I/O devices, including a video display, keyboard, mouse, printer, and modem or network interface. A large computer can have many more I/O devices (for example, multiple printers and network interfaces) and might use substantially more powerful and complex devices than a smaller computer does.

COMPUTER HARDWARE

Not surprisingly, computer hardware components parallel the processing, storage, and communication capabilities just outlined (see Figure 2.6). Computer hardware has four major functions:

- *Processing*—Executing computation, comparison, and other instructions to transform data inputs into data outputs

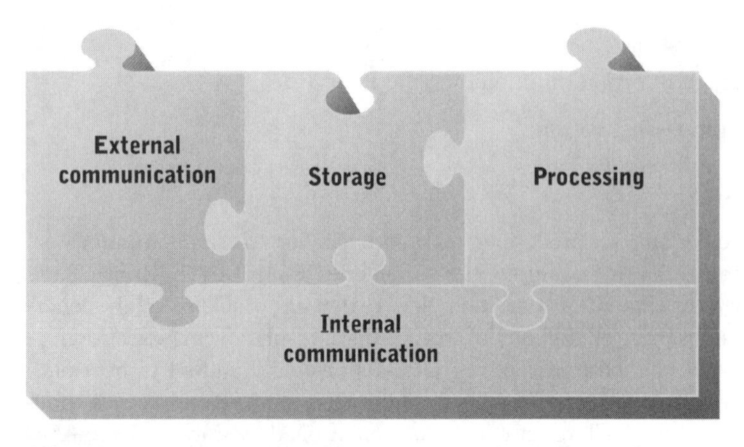

FIGURE 2.6 The major functions of computer hardware
Courtesy of Course Technology/Cengage Learning

- *Storage*—Storing program instructions and data for temporary, short-term, and long-term use
- *External communication*—Communicating with entities outside the computer system, including users, system administrators, and other computer systems
- *Internal communication*—Transporting data and instructions between internal and peripheral hardware components, such as processors, disk drives, video displays, and printers

All computer systems include hardware to perform each function. However, each function isn't necessarily implemented in a single device. For example, a computer processor performs processing and some storage functions, and external communication is handled by many hardware devices, such as keyboards, video displays, modems, network interface devices, sound cards, and speakers.

Figure 2.7 shows the hardware components of a computer system. The number, implementation, complexity, and power of these components can vary substantially from one computer system to another, but the functions performed are similar.

The **central processing unit (CPU)** is a general-purpose processor that executes all instructions and controls all data movement in the computer system. Instructions and data for currently running programs flow to and from **primary storage**. **Secondary storage** holds programs that aren't currently running as well as groups of data items that are too large to fit in primary storage. It can be composed of several different devices (for example, magnetic disk drives, optical disc drives, and solid-state drives), although only one device is shown in Figure 2.7.

Input/output (I/O) units perform external communication functions. Two I/O units are shown in Figure 2.7, although a typical computer has many more. The **system bus** is the internal communication channel connecting all hardware devices.

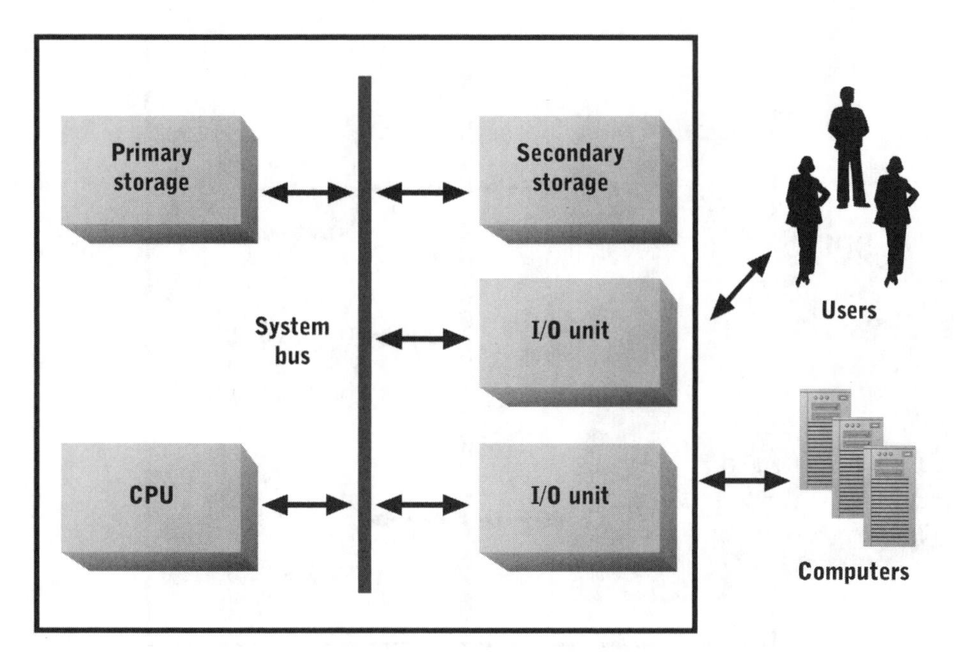

FIGURE 2.7 The hardware components of a computer system
Courtesy of Course Technology/Cengage Learning

Central Processing Unit

Figure 2.8 shows components of the CPU, which include the following:

- Arithmetic logic unit
- Registers
- Control unit

The **arithmetic logic unit (ALU)** contains electrical circuits that carry out each instruction. A CPU can execute dozens or hundreds of different instructions. Simple arithmetic instructions include addition, subtraction, multiplication, and division. More advanced computation instructions, such as exponentiation and logarithms, can also be implemented. Logic instructions include comparison (equal to, greater than, less than) and other instructions discussed in Chapter 4.

The CPU contains a few internal storage locations called **registers**, each capable of holding a single instruction or data item. Registers store data or instructions that are needed immediately or frequently. For example, each of two numbers to be added is stored in a register. The ALU reads these numbers from the registers and stores the sum in another register. Because registers are located in the CPU, other CPU components can access their contents quickly.

The **control unit** has two primary functions:

- Control movement of data to and from CPU registers and other hardware components.
- Access program instructions and issue appropriate commands to the ALU.

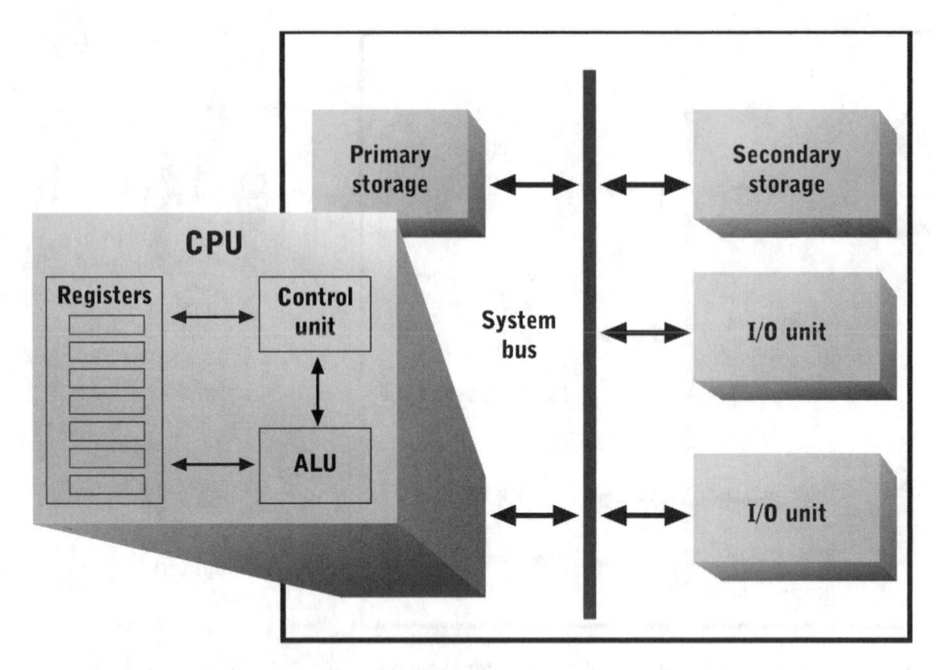

FIGURE 2.8 Components of the CPU
Courtesy of Course Technology/Cengage Learning

As program instructions and data are needed, the control unit moves them from primary storage to registers. The control unit examines incoming instructions to determine how they should be processed. It routes computation and logic instructions to the ALU for processing and executes data movement instructions to and from primary storage, secondary storage, and I/O devices.

System Bus

The system bus is the main channel for moving data and instructions to and from hardware components, and its capacity is a critical factor in computer performance. A powerful CPU needs a high-capacity system bus to keep it supplied with instructions and data from primary storage. Bus capacity is also important for secondary storage and I/O device performance.

Primary Storage

Primary storage contains millions or billions of storage locations that hold currently executing program instructions as well as data being processed by these instructions. Primary storage is also referred to as **main memory**, or simply memory. A running program moves instructions and data continually between main memory and the CPU. Because the CPU is a fast device, main memory devices must be capable of rapid access.

In current computer hardware, main memory is implemented with silicon-based semiconductor devices commonly called **random access memory (RAM)**. RAM provides the access speed the CPU requires and allows the CPU to read from or write to specific memory locations. Unfortunately, RAM can be expensive, which often limits the amount of main memory a computer system can include.

Another problem with RAM is that it doesn't provide permanent storage. When the power is turned off, RAM's contents are lost. This characteristic is called volatility. Any type of storage device that can't retain data values indefinitely is said to be volatile. In contrast, storage devices that retain data values permanently are said to be nonvolatile. Because of the volatility and limited capacity of primary storage, a computer system must have other devices to store data and programs for long periods.

Secondary Storage

Secondary storage is composed of high-capacity nonvolatile storage devices that hold the following:

- Programs not currently running
- Data not needed by currently running programs
- Data needed by currently running programs that doesn't fit in available primary storage

In a typical information system, the number of programs and amount of data are quite large. A typical computer system must have much more secondary storage capacity than primary storage capacity. For example, a microcomputer might have 4 billion primary storage locations and 500 billion secondary storage locations. Table 2.1 summarizes differences in the content and implementation of storage devices.

TABLE 2.1 Comparison of storage types

Storage type	Implementation	Content	Typical quantity
CPU registers	High-speed electrical devices in the CPU	Currently executing instructions and associated data inputs and outputs	Several dozen to a few hundred instructions and data items
Primary storage	High-speed electrical devices (RAM) outside but close to the CPU	Currently running programs and data needed immediately (if they fit in primary storage)	1 to 8 billion data items per CPU
Secondary storage	Low-speed electromagnetic and optical devices	Programs not currently running and data not currently being accessed by programs	Billions (gigabytes), trillions (terabytes), or quadrillions (petabytes) of data items

Secondary storage devices are constructed with slower and less expensive technology to keep total cost within acceptable limits. The most common secondary storage devices are magnetic disks, optical discs, and magnetic tape. Magnetic disks provide fast access compared with optical discs and magnetic tapes. Optical discs provide high-capacity portable storage at a low cost per gigabyte. Magnetic tape offers the slowest but cheapest storage method.

Input/Output Devices

The variety of I/O devices in modern computers addresses the many different forms of human-to-computer and computer-to-computer communication. From an architectural perspective, each I/O device is a separate hardware component attached to the system bus. All I/O devices interact similarly with other hardware devices, but their internal implementation varies depending on the exact form of communication they support.

I/O devices can be classified broadly into human-oriented and computer-oriented communication devices. Examples of input devices for human use include keyboards, pointing devices (such as a mouse or trackball), and voice-recognition devices. The purpose of these devices is to accept input from a person (voice, touch, or physical movement) and convert this input into something the computer can understand (electrical signals). Output devices for human use include video displays, printers, and devices for speech and sound output. All these devices convert electrical signals into a format that a person can understand, such as pictures, words, or sound. Computer-oriented I/O devices include modems and network interface devices, which handle communication between computer systems or between a computer and a distant I/O device.

COMPUTER SYSTEM CLASSES

Computer systems are available in many configurations that vary in CPU power, storage capacity, I/O capacity, number of simultaneous users, and intended application software. Computer systems can be grouped loosely into the following classes:

- Microcomputer
- Portable
- Midrange computer
- Mainframe
- Supercomputer

A **microcomputer** is a computer system designed to meet a single user's information-processing needs. It can also be called a **personal computer (PC)** or **workstation**. Portable computers, such as laptops, netbooks, and handheld computers, are also microcomputers. Examples of tasks performed with microcomputers include word processing, computer games, Web browsing, and small to medium application programs, such as programs to compute a person's income tax, calculate budgets for a home or small business, and compute payroll for a small business.

NOTE

To some people, the terms "personal computer" and "workstation" are interchangeable. To others, the term "workstation" implies a more powerful system than a typical PC, particularly one used in scientific and engineering organizations. Processing tasks in these organizations require more hardware power than typical business and home processing do. Examples of hardware-intensive tasks include complex mathematical computation, computer-aided design (CAD), and manipulation of high-resolution video images. The power of a workstation is often similar to that of a midrange computer (described later in this section), but a workstation's overall design is targeted toward a single-user operating environment.

Portable microcomputers have proliferated as the cost and size of computer components have decreased and battery technology has improved. Current portable microcomputer types include the following:

- A **laptop computer** is a full-featured microcomputer with an integrated display (typically 12 to 17 inches measured diagonally) and a battery; a laptop rivals traditional microcomputers in power and cost.
- A **netbook computer** is a laptop computer that emphasizes small size, reduced weight, low cost, and wireless networking and is capable of performing only light-duty tasks, such as Web browsing, e-mailing, and word processing.
- A **personal digital assistant (PDA)** is a handheld computer, usually integrated with a cell phone, that supports light-duty tasks.

A **midrange computer**, sometimes called a minicomputer, is designed to provide information processing for multiple users and run many application programs simultaneously. Supporting multiple users and programs requires fairly powerful processing, storage, and I/O subsystems and more sophisticated system software than is typically installed on microcomputers.

There are many ways of supporting multiple users. They can be connected to a computer system with simple I/O devices, such as video display terminals. In this case, processing, data storage, and network communication are handled by the shared computer system. Multiple users can also share resources (for example, printers, databases, and Web sites). A midrange computer can support several dozen people using video display terminals or respond to a few hundred simultaneous requests for shared resources.

A **mainframe** computer system handles the information-processing needs of a large number of users and applications. It can support hundreds of people using video display terminals, run hundreds or thousands of programs at one time, and respond to thousands of simultaneous requests for shared resources. Typical use might involve 250 users entering customer orders, several programs generating reports, dozens of users querying a large database's contents, and an operator making backup copies of disk files—all at the same time.

The feature that distinguishes mainframes from other computer classes is the capability to store large quantities of data and move it from one place to another quickly. Data can be moved between up to several dozen or a few hundred CPUs, hundreds or thousands of secondary storage devices, and hundreds or thousands of users connected via a network. Fast CPUs and large amounts of primary and secondary storage are also required, but a mainframe is optimized primarily for rapid and efficient data movement.

A **supercomputer** is designed for one purpose—rapid mathematical computation (billions, trillions, or more computations per second). Supercomputers are used for computation-intensive applications, such as simulations, 3D modeling, weather prediction, computer animation, and real-time analysis of large databases. These tasks require hundreds or thousands of CPUs with the highest possible computational speed. Storage and communication requirements are also extremely high but are secondary to computational speed and often delegated to other computers accessed via a high-speed network. Supercomputers use the most up-to-date (and expensive) computer technology.

The term **server** can describe computers as small as microcomputers and as large as supercomputers. It doesn't imply a minimum set of hardware capabilities; instead, it implies a specific mode of use. A server is a computer system that manages shared resources, such as file systems, databases, Web sites, printers, and high-speed CPUs, and allows users to access these resources over a local or wide area network.

Server hardware capabilities depend on the resources being shared and the number of simultaneous users. For example, an ordinary microcomputer might be more than adequate for a dozen users on a local area network sharing a file system and two printers. A mainframe server might be needed for sharing more demanding resources, such as large databases accessed by thousands of users. Supercomputer servers are sometimes used to augment the computational capabilities of workstations used for CAD and animation.

Table 2.2 summarizes the configuration and capabilities of each class of computer system. Each class is represented by a typical model available in 2009, but keep in mind that specifications and performance are in a constant state of flux. Rapid advances in computer technology lead to rapid performance improvements as well as redefinitions of computer classes and specifications in each class. There's also a technology pecking order—that is, the newest and most expensive hardware and software technology usually appear first in supercomputers. As experience is gained and costs decrease, these technologies move downward to the less powerful and less costly computer classes.

TABLE 2.2 Representative products in various computer classes (2009)

Class	Typical product	Typical specifications	Approximate cost	CPUs
Portable	Dell Latitude E6400	4 billion main memory cells 250 billion disk storage cells Rewritable DVD drive 14-inch display	$1150	2
Microcomputer	Dell Optiplex 760	4 billion main memory cells 500 billion disk storage cells Rewritable DVD drive	$1000	2
Workstation	Dell Precision T7500	12 billion main memory cells 1.5 trillion disk storage cells Rewritable high-capacity DVD drive Dual high-speed 3D graphics processors	$8350	8
Midrange	Dell PowerEdge T610	16 billion main memory cells 4 trillion high-speed disk storage cells High-speed fault-tolerant storage subsystem Tape backup	$15,050	8
Mainframe	IBM Z10 E64	512 billion main memory cells 100 trillion high-speed disk storage cells High-capacity tape archive system Four high-speed network interfaces	$500,000	64
Supercomputer	IBM Blue Gene/P	2 trillion main memory cells No internal disk storage	$1,300,000	4096

Several factors blur the lines between computer classes. One is the ability to configure subsystems, such as processors, primary and secondary storage, and I/O. For example, a microcomputer can be configured to support some multiuser applications simply by upgrading the capacity of a few subsystems and installing more powerful system software.

Another factor is virtualization, a technique that divides a single computer's capacity among multiple virtual machines. For example, an organization might buy a single mainframe and divide its capacity among several virtual servers: one for Web services, another for database management, another for background applications, and another for e-mail services. Each virtual machine is a collection of files on the physical server that define the virtual machine's configuration and the contents of its virtual disk drives (including the operating system and other software). When a virtual machine is powered on, it's connected to physical resources on the host machine, such as CPUs, memory, network interfaces, and disk drives.

Virtualization offers flexibility in server configuration and deployment, including the ability to "resize" virtual machines easily to match changing requirements. It also enables an organization to move a virtual machine from one physical computer to another simply by copying its files across a network. With this technique, an organization can also clone virtual servers on multiple physical servers to increase capacity or provide service redundancy.

TECHNOLOGY FOCUS

IBM POWER-Based Computer Systems

International Business Machines (IBM) introduced the POWER CPU architecture in 1990. With this CPU, IBM broke away from its previous designs and laid the foundation for two decades of future computer systems. POWER CPUs have been incorporated into systems ranging from workstations to supercomputers.

The POWER architecture has been through several generations and variations since 1990. The latest generation is called the POWER7 and incorporates IBM's latest microprocessor technologies, including the following:

- 45-nanometer fabrication technology
- Up to eight CPUs (cores) per microprocessor
- Multiple on-chip memory caches

As of this writing, POWER7-based computer systems aren't available yet. Table 2.3 summarizes sample computer systems based on POWER6 CPUs.

TABLE 2.3 IBM system configurations for POWER6 CPUs (2009)

Model	System type	Number of CPUs	CPU speed	Memory capacity
pSeries 520	Midrange	1–4	4.7 GHz	2–64 GB
pSeries 595	Mainframe	8–64	5.0 GHz	16–4096 GB
pSeries 575	Supercomputer node	32	4.7 GHz	32–256 GB

The pSeries 520 is a midrange computer based on a single POWER6 microprocessor with one, two, or four processing cores (microprocessors with multiple processing cores, discussed in detail in Chapter 6). A single POWER6 core can operate as two virtual

(*continued*)

processors via a technique IBM calls "simultaneous multithreading." (A similar technique in Intel CPUs is called "hyperthreading.") Internal disk capacity is up to 2.7 terabytes (TB), but the system can be expanded well beyond this limit with external storage units.

The pSeries 595 is a large-capacity mainframe with many possible uses. One use is as a large storage or database server accessing up to 2200 disk drives. Another use is as a virtualization platform hosting a few dozen high-capacity or many smaller-capacity virtual machines. The pSeries 595 can be configured for high I/O capacity, high storage capacity, high computational speed, or any combination of these capabilities.

The pSeries 575 is optimized for high computational performance and data transfer capacity between CPUs and between CPUs and memory. pSeries 575 computers are building blocks for supercomputer clusters. Multiple computers are linked by high-speed network connections to achieve very high computational capacity. As of this writing, the largest supercomputers (two identical clusters) based on the pSeries 575, containing 8384 CPU cores in each cluster, are used by the European Centre for Medium-Range Weather Forecasts (see Figure 2.9).

FIGURE 2.9 One of two supercomputer clusters at the European Centre for Medium-Range Weather Forecasts
Courtesy of European Centre for Medium-Range Weather Forecasts

Multicomputer Configurations

A final factor that blurs distinctions between computer classes is the trend toward multicomputer configurations. Simply put, a **multicomputer configuration** is any arrangement of multiple computers used to support specific services or applications. There are a variety of common multicomputer configurations, including clusters, blades, and grids.

A **cluster** is a group of similar or identical computers, connected by a high-speed network, that cooperate to provide services or run a single application. A Web server farm is one example of a cluster. Incoming Web service requests are routed to one of a set of Web servers, each with access to the same content. Any server can respond to any request, so the load is distributed among the servers. In effect, the cluster acts as a single large Web server. Modern supercomputers are another example of clustering, with dozens to thousands of identical computers operating in parallel on different portions of a large computational problem.

The advantages of a cluster include scalability and fault tolerance. Overall capacity can be expanded or reduced by adding or subtracting servers from the cluster, which is usually simpler than adding or subtracting capacity from a single larger computer. Also, if any one computer fails, the remaining computers continue to function, thus maintaining service availability at a reduced capacity. Clusters can be replicated in multiple locations for additional fault tolerance. The main disadvantages of clusters are complex configuration and administration.

A **blade** is a circuit board that contains most of a server. Typically, a blade has one or more CPUs, memory areas, and network interfaces. It lacks secondary storage, external I/O connections, and a power supply. Up to a dozen or so blades can be installed in a single cabinet, which also contains a shared power supply and external I/O connections. Secondary storage is typically provided by a storage server placed near the blade cabinet.

In essence, a blade is a specialized cluster. It has the same advantages and disadvantages as a cluster, although modifying a cluster of blades is usually simpler than modifying a cluster of stand-alone computers. Blades also concentrate more computing power in less space and with lower power requirements than a typical cluster needs.

A **grid** is a group of dissimilar computers, connected by a high-speed network, that cooperate to provide services or run a shared application. Besides dissimilarity of the component computers, grids have three other differences from clusters:

- Computers in a cluster are typically located close to one another (for example, in the same room or rack). Computers in a grid might be in separate rooms or buildings or even on different continents.
- Computers in a cluster are connected by dedicated high-speed networks that enable them to exchange information rapidly. Computers in a grid are generally connected by ordinary network connections that serve multiple purposes.
- Computers in a cluster work exclusively on the same services or applications. Computers in a grid work cooperatively at some times and independently at others.

Grids are typically implemented by installing software on each machine that accepts tasks from a central server and performs them when not busy doing other work. For example, all the desktop computers in a building could form a grid that performs computation-intensive tasks at night or on weekends. Grids of mainframes or supercomputers are sometimes used to tackle computation-intensive research problems.

A **cloud** is a set of computing resources with two components:

- *Front-end interfaces*—Typically Web sites or Web-based services that users interact with
- *Back-end resources*—A large collection of computing and data resources, typically organized with a combination of cluster- and grid-based architectures; used to provide sites or services to users through the front-end interface

A cloud isn't a specific multicomputer configuration. Rather, it's a specific way of organizing computing resources for maximum availability and accessibility with minimum complexity in the user or service interface. Users see only the front-end interface, not the myriad computing resources behind it. They're unaware of details such as computer system sizes, organization of stored data, geographical distribution, and redundancy. Clouds typically make use of both multicomputer configuration and virtualization, which make them fault tolerant (because of redundancy) and flexible enough to respond rapidly to changes in resource access patterns. The computing infrastructure and services of many service providers, such as Google and MSN, are organized as a cloud.

TECHNOLOGY FOCUS

Folding@Home Distributed Computing

Proteins are complex chains of amino acids that act as structural elements of living tissue, enzymes to initiate chemical reactions, and antibodies to fight disease. Although the structure of the human genome—and, therefore, the blueprint of all human proteins—has been mapped, chemical reactions in proteins still aren't understood well. A crucial aspect of protein chemistry is protein folding. In many chemical reactions, proteins must first assemble themselves into special shapes by folding parts of their structure (see Figure 2.10). Correct protein folding is necessary for biological processes, and folding errors are associated with many diseases, such as Alzheimer's and cystic fibrosis.

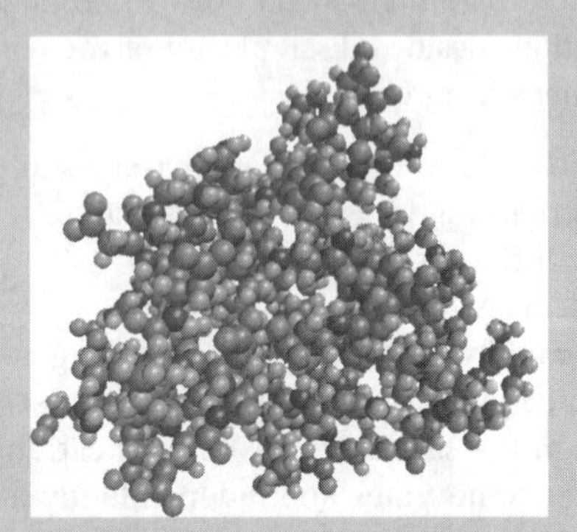

FIGURE 2.10 Visualization of a folded protein in the Folding@Home program
Courtesy of Course Technology/Cengage Learning

Biochemistry researchers at Stanford University wanted to use computers to simulate protein folding, but they faced a dilemma. Because proteins are highly complex molecules, simulating correct and incorrect folding requires many CPUs. Tackling the simulations these researchers envisioned would have required large, expensive supercomputers, and the funds to buy them weren't available. By cooperating with computer science researchers, they were able to address the problem by using grid computing.

The CPU in a typical PC has roughly the same power as a single CPU in a supercomputer. What differentiates a supercomputer from a PC is the number of CPUs and the ability to interconnect them. When a supercomputer runs a large simulation program, it divides the problem into many smaller pieces and assigns each piece to a CPU. In many simulation programs, such as weather-forecasting systems, CPUs must exchange data

(continued)

frequently. Therefore, execution speed is determined by the number of CPUs and the speed of data exchange between CPUs and dedicated primary storage areas.

However, pieces of a protein-folding simulation running on each CPU have little need to interact during execution. Therefore, the high-speed interconnection of CPUs in a typical supercomputer isn't a critical performance factor. A more important performance factor is the number of CPUs. Protein-folding simulations can be scaled to millions of CPUs—far more than any supercomputer has today.

Folding@Home is a distributed simulation program that uses servers at Stanford and a grid of hundreds of thousands of ordinary computers to work simultaneously on protein folding. Computer owners download and install a small application program that downloads simulation problems from the Stanford servers, executes them, and uploads the results. The application consumes idle CPU capacity that would otherwise be wasted on a system idle process (see Figure 2.11). The servers gather and assemble the results for use by researchers. The project has advanced knowledge of protein folding by assembling donated computing resources to create a virtual supercomputer more powerful than any current supercomputer or cluster.

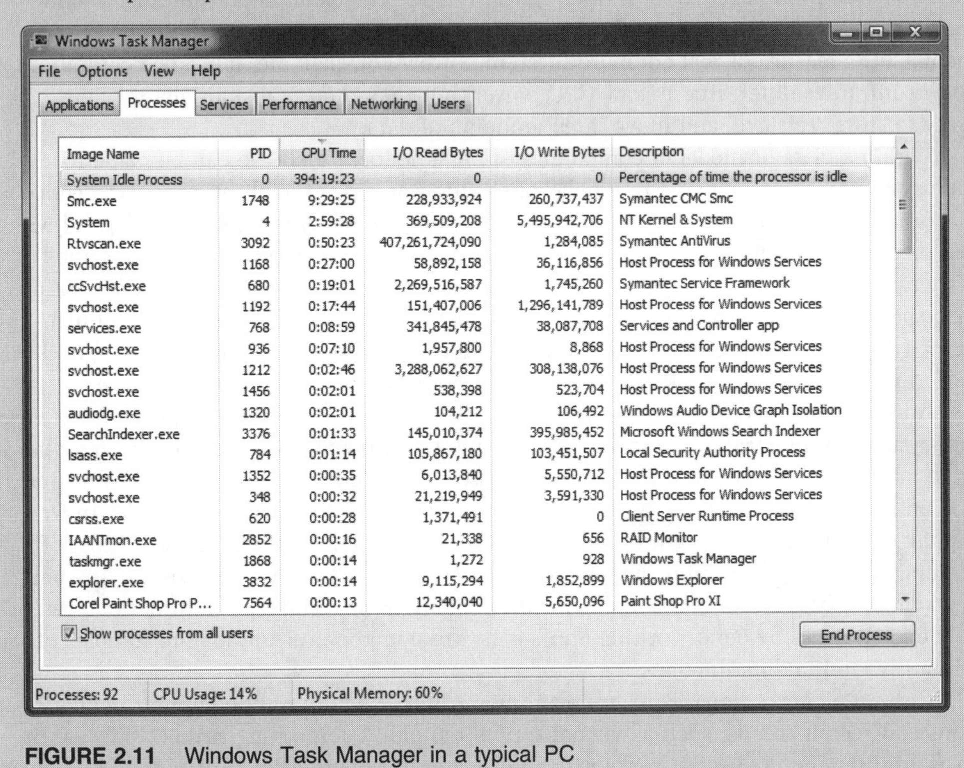

Image Name	PID	CPU Time	I/O Read Bytes	I/O Write Bytes	Description
System Idle Process	0	394:19:23	0	0	Percentage of time the processor is idle
Smc.exe	1748	9:29:25	228,933,924	260,737,437	Symantec CMC Smc
System	4	2:59:28	369,509,208	5,495,942,706	NT Kernel & System
Rtvscan.exe	3092	0:50:23	407,261,724,090	1,284,085	Symantec AntiVirus
svchost.exe	1168	0:27:00	58,892,158	36,116,856	Host Process for Windows Services
ccSvcHst.exe	680	0:19:01	2,269,516,587	1,745,260	Symantec Service Framework
svchost.exe	1192	0:17:44	151,407,006	1,296,141,789	Host Process for Windows Services
services.exe	768	0:08:59	341,845,478	38,087,708	Services and Controller app
svchost.exe	936	0:07:10	1,957,800	8,868	Host Process for Windows Services
svchost.exe	1212	0:02:46	3,288,062,627	308,138,076	Host Process for Windows Services
svchost.exe	1456	0:02:01	538,398	523,704	Host Process for Windows Services
audiodg.exe	1320	0:02:01	104,212	106,492	Windows Audio Device Graph Isolation
SearchIndexer.exe	3376	0:01:33	145,010,374	395,985,452	Microsoft Windows Search Indexer
lsass.exe	784	0:01:14	105,867,180	103,451,507	Local Security Authority Process
svchost.exe	1352	0:00:35	6,013,840	5,550,712	Host Process for Windows Services
svchost.exe	348	0:00:32	21,219,949	3,591,330	Host Process for Windows Services
csrss.exe	620	0:00:28	1,371,491	0	Client Server Runtime Process
IAANTmon.exe	2852	0:00:16	21,338	656	RAID Monitor
taskmgr.exe	1868	0:00:14	1,272	928	Windows Task Manager
explorer.exe	3832	0:00:14	9,115,294	1,852,899	Windows Explorer
Corel Paint Shop Pro P...	7564	0:00:13	12,340,040	5,650,096	Paint Shop Pro XI

FIGURE 2.11 Windows Task Manager in a typical PC
Courtesy of Course Technology/Cengage Learning

Bigger Isn't Always Better

In 1952, computer scientist H. A. Grosch asserted that computing power, measured by millions of instructions per second (MIPS), is proportional to the square of hardware cost. According to Grosch, large and powerful computers will always be more cost effective than smaller ones. The mathematical formula describing this relationship is called **Grosch's Law**. For many years, computer system managers pointed to this law as justification for investments in ever larger mainframe computers.

Many modern trends in computer design and use, such as the following, have combined to rewrite Grosch's Law:

- Multiple classes of computers
- Expanded abilities to configure computers for specific purposes
- Increases in software and system administration costs compared with hardware costs
- Multicomputer configurations

In Grosch's time, there was basically only one class of computer—the mainframe. Purchasing larger machines in a single class offers more CPU power for the money; therefore, Grosch's Law appears to hold. If all classes of computers are considered as a group, however, Grosch's Law doesn't hold. ***The cost of CPU power actually increases on a per-unit basis as the computer class increases.*** Despite this statement, midrange, mainframe, and supercomputers aren't necessarily cost inefficient because CPU power is an incomplete measure of real computer system power. For example, midrange computers and mainframes differ little in raw CPU power, but they differ substantially in their capability to store, retrieve, and move large amounts of data.

Within a class or model of computer system, a deliberate choice can be made to tune system performance to a particular application or type of applications cost effectively. For example, in a transaction-processing application, the system might be configured with fast secondary storage and I/O subsystems but sacrifices CPU performance. The same system might be tailored to a simulation application cost effectively by providing a lot of high-speed main memory, multiple CPUs with high computational capacity, and a high-capacity communication channel between memory and CPUs. Secondary storage and I/O performance would be far less important, and less costly alternatives could be used.

Another difficulty in applying Grosch's Law today is the reality of expensive software and system administration. Software development and management of software and hardware infrastructure are labor-intensive processes. Despite the pressures of globalization and outsourcing, labor costs haven't fallen substantially in the developed world. However, rapidly decreasing hardware costs make hardware a smaller factor in the total cost of building and operating an information system. Although hardware costs should be managed effectively, an IS professional is far more likely to maximize an information system's cost effectiveness by concentrating on its software components and long-term costs of administration and maintenance.

Cost-efficiency comparisons are even more difficult when multicomputer configurations and recent trends, such as virtualization and cloud computing, are considered. In Grosch's time, computer networks with multicomputer configurations and Web-based services didn't exist. Today, however, high-speed computer and storage networks enable many organizations to deliver large-scale computing power with distributed networks of smaller computers. At the same time, cloud computing and virtualization enable organizations to rent computing capacity and control these resources across the Internet. Cloud computing providers are able to achieve economies of scale by building large computer centers that support many different organizations.

NOTE

"Economies of scale" is the concept that the per-unit cost of producing goods or providing services decreases as the organization size increases. Economies of scale occur in many industries, including electronics, agriculture, and electrical power generation. They result from many factors, such as higher levels of automation, more efficient use of expensive resources, and more specialized and efficient production processes.

The bottom line is that in today's world, Grosch's Law still holds in some cases but not in others. Bigger is better when it lowers long-term software, maintenance, and administration costs and achieves economies of scale. Smaller is better, however, when an organization needs flexibility, redundancy, and local control of computing resources.

BUSINESS FOCUS

Does Your Company Need a Mainframe?

Bauer Industries (BI) is a manufacturer and wholesale distributor of jewelry-making machinery and components. The company has 200 employees and approximately 2500 products. The current inventory of automated systems is small for a company of its size. Applications include order entry, inventory control, and general accounting functions, such as accounts payable, accounts receivable, payroll, and financial reporting. BI's applications currently run on a Hewlett-Packard (HP) AlphaServer DS midrange computer purchased in 2000. Dedicated character-based video display terminals are used to interact with application programs.

BI has grown rapidly over the past decade. Increased transaction volume has strained existing computer hardware capability. Typical order-entry response time is now 30 to 60 seconds compared with 2 to 5 seconds a few years ago. The order-entry system consumes all memory and CPU capacity during daytime hours. To reduce order-entry response time, all nonessential processing has been moved to nighttime hours. The AlphaServer is configured with two CPUs, 2 GB of primary storage, and 500 GB of disk storage, and no further hardware enhancements are possible.

Bauer Industries management realizes the seriousness of the situation but isn't sure how to address it. Other factors, such as the following, complicate the decision:

- An expected fourfold increase in business over the next 10 years
- Relocation to new facilities in the next year
- Plans to move the Web-based catalog and online order system, which are now contracted out, in house
- Managers' goals to increase and modernize automated support of operating and management functions

Management has identified these viable options to address the problem:

- Purchase one or more used or factory-refurbished AlphaServer systems (which are no longer manufactured), connect them with a high-speed network, and partition the application software among these systems.
- Accept an offer from HP to upgrade to an Integrity mainframe computer.
- Develop a highly scalable hardware platform consisting of HP blade servers or a cluster of HP midrange computers.

(*continued*)

- Migrate some applications, such as accounting and payroll, to a cloud computing service.

Option 1 is least disruptive to the status quo but has two obvious disadvantages. First, because the AlphaServer is no longer manufactured, parts are in short supply. The second problem is data overlap among application programs. The current operating and application software won't support access to data stored on one machine by applications running on another machine. Therefore, data would have to be stored on both machines and synchronized periodically to reflect updates made by different application programs. Estimated two-year costs are $75,000 for hardware and $100,000 for labor.

Option 2 theoretically preserves existing application software and allows reusing some hardware, such as video display terminals. It also provides enough capacity to operate current software at many times the current transaction volume. However, the new hardware requires a more recent version of the UNIX operating system. This new UNIX version is designed to support applications running on older versions on AlphaServer hardware, but management has heard of many problems in migrating application software. Estimated two-year costs are $350,000 for hardware and $150,000 for labor.

Option 3 is the most expensive and the biggest departure from current operating methods. Migration problems are similar to Option 2. Existing video display terminals can be used initially and gradually replaced with workstations. This option seems to hold the most promise for implementing a new generation of application software, but the company has little experience in operating a multicomputer configuration. Estimated (guesstimated?) two-year costs are $150,000 for hardware and $400,000 for labor (including hiring or contracting three new IS staff members for two years).

Option 4 theoretically extends the life of the current system by reducing demand on its capacity from applications other than order entry. However, BI has no experience in outsourcing IT and business functions and is leery of locking itself into a long-term relationship with an outsourcing vendor. BI is also concerned that it might need to redesign its network to expand capacity and improve security so that on-site employees can interact with distant servers effectively to perform routine tasks. BI has received a preliminary estimate of $75,000 per year to host basic accounting and payroll functions on an external server farm. However, the cost and details of migrating BI's accounting functions to the new environment haven't been estimated yet.

Questions:

- What problems would you anticipate in two to five years if Option 1 is selected? Option 2? Option 3? Option 4?
- In the past, management has opted for the lowest-cost solutions. How can you justify the higher costs of options 2 and 3? Will choosing option 4 really save money over the long term?
- Which alternatives would you recommend to management? Why?

THE ROLE OF SOFTWARE

The primary role of software is to translate users' needs and requests into CPU instructions that, when executed, produce a result that satisfies the need or request. People usually ask computers to perform complex tasks, such as "generate my company's income statement," "spell-check my term paper," or "print a weekly sales report" (see Figure 2.12).

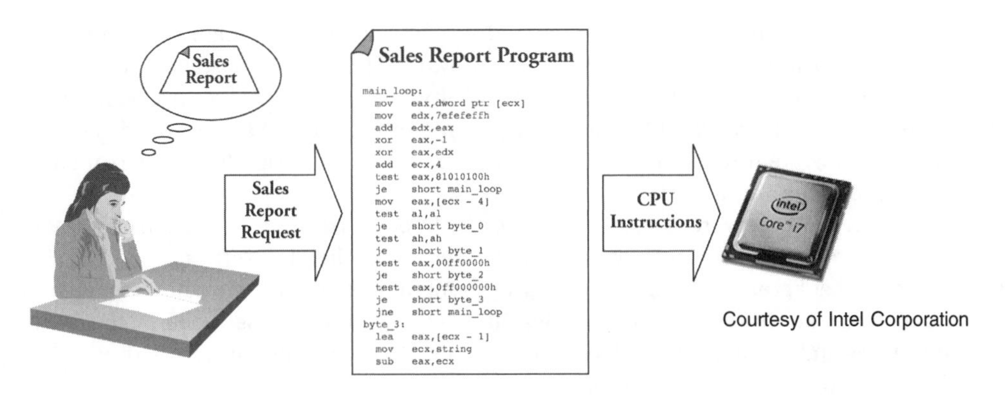

Courtesy of Intel Corporation

FIGURE 2.12 The role of software as a translator between user requests and CPU instructions

Software is complex because it performs a complex translation process that bridges two gaps:

- Human language to machine language
- High-level abstraction to low-level detail

People think, speak, and write in their native language, which has specific vocabulary, grammar, and semantics. Computers communicate in binary languages consisting of 0s and 1s, which are grouped by specific grammatical (syntax) rules to form meaningful statements or instructions.

The need or idea that motivates a request for computer processing is stated at a high (abstract) level. The tasks that are performed (instructions that are executed) to satisfy a request are specific ("low-level"). Because CPU instructions are very small units of work, the number of instructions that must be executed to satisfy a processing request is large.

Software Types

A program is a set of detailed instructions for directing a computer to perform a complex task. Much like a recipe, a program can be written down, stored, and retrieved when needed. Programs are usually stored in files on secondary storage devices. When needed, they're retrieved by name, copied into primary storage, and then executed by the CPU.

An application program, or **application software**, is a stored set of instructions for responding to a specific request, much as you might look up a recipe to prepare a particular dish. Examples of application programs in a system devoted to payroll processing include programs to print checks, enter new employee information, and produce annual tax reports. Application software is often purchased off the shelf. Examples include accounting packages, such as QuickBooks; project management software, such as Microsoft Project; and graphics software, such as Adobe Photoshop.

System software consists of programs for handling the following tasks:

- Perform utility functions needed by many application programs.
- Allocate computer resources to application programs.
- Manage computer resources.

Examples of system software include operating systems, database management systems, antivirus software, and network security software. The most important distinction between application and system software is specificity of purpose and use. Application software is targeted to specific information-processing tasks, such as generating customer credit card bills. System software is targeted to general-purpose tasks that support many application programs and users.

Most application software is used by end users. In contrast, most users don't interact with system software, and most utility programs used by application software operate invisibly in the background. Programs allocating computer resources shared by application software also operate in the background. System administrators can set policies governing allocation rules and methods, but allocation programs don't interact directly with users or system administrators during execution.

Utility programs that manage computer resources can be used directly by end users or system administrators. Examples of user-oriented management utilities include those that copy files and directories, install software on PCs or workstations, and create and manage connections to ISPs. Examples of management utilities intended for system administrators include those that create and modify user accounts, install hardware devices on servers, and carry out computer security policies.

Figure 2.13 shows the interaction between the user, application software, system software, and computer hardware. User input and program output are limited to direct communication with the application program. Application software, in turn, communicates with system software to request basic services (such as opening a file or a new window). System software translates a service request into a sequence of machine instructions, passes these instructions to hardware for execution, receives the results, and passes the results back to the application software.

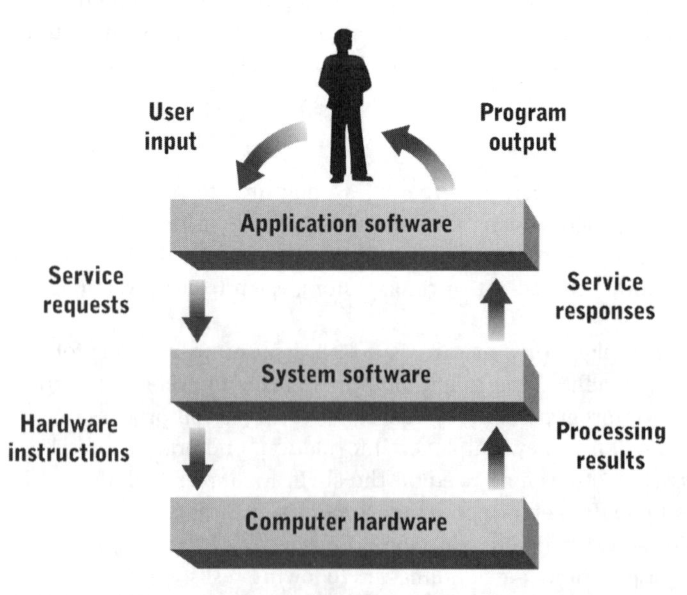

FIGURE 2.13 The interaction between the user, application software, system software, and computer hardware

Courtesy of Course Technology/Cengage Learning

Figure 2.13 also shows what's called a "layered approach" to software construction and operation. Application software is layered above system software, which is layered above hardware. A key advantage of this approach is that users and application programmers don't need to know the technical details of hardware. Instead, they interact with hardware via standardized service requests. Knowledge of the machine's physical details is embedded into system software and hidden from users and application programmers. This advantage is commonly referred to as **machine independence** or **hardware independence**.

System Software Layers

Figure 2.14 shows a more detailed view of software layers and their relationship to hardware functions. System software functions have been divided into four layers:

- *System management*—Utility programs used by end users and system administrators to manage and control computer resources
- *System services*—Utility programs used by system management and application programs to perform common functions
- *Resource allocation*—Utility programs that allocate hardware and other resources to multiple users and programs
- *Hardware interface*—Utility programs that control and interact with specific hardware devices

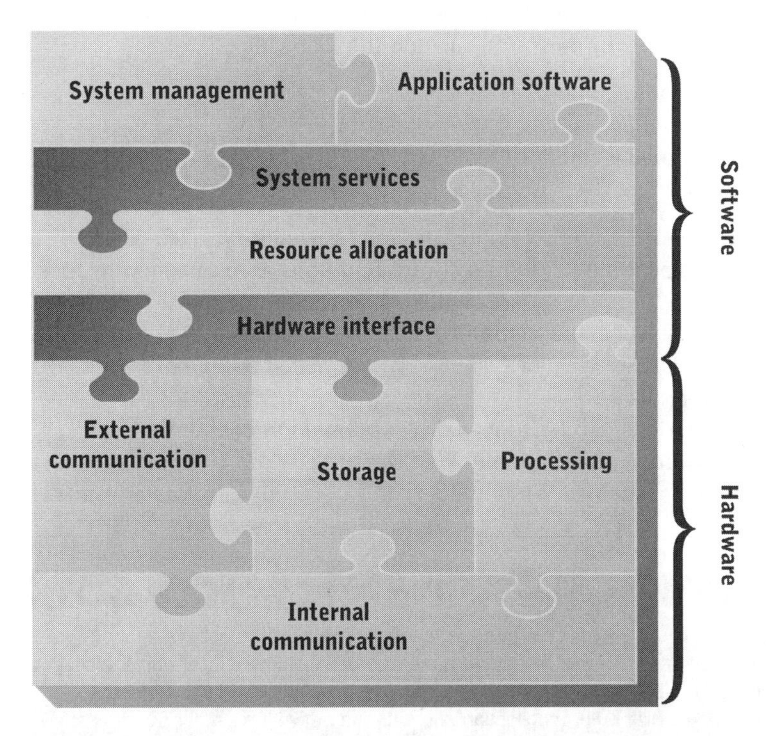

FIGURE 2.14 Software layers and their relationship to hardware
Courtesy of Course Technology/Cengage Learning

The principle of software layers applies not only to the relationship between application and system software, but also to the relationship between components of system software. Machine independence is achieved by placing all hardware interface functions in a single system software layer. In theory, system software can be modified to operate on new computer hardware by modifying only the hardware interface layer. In practice, machine independence isn't always that simple, as you learn in Chapter 11.

Resource allocation is an important, but mostly invisible, function of system software. A modern computer system can support multiple users and hundreds or thousands of programs running simultaneously. Under these conditions, resources such as files and I/O devices can't always be made available to a program or user immediately. System software balances competing resource demands of users and processes in such a way that users and programs don't interfere with one another, but all users and programs are assured of getting the resources they need.

Operating Systems

An **operating system (OS)** is a collection of utility programs for supporting users and application programs, allocating resources to multiple users and application programs, and controlling access to hardware. Note that this definition is similar to the definition of system software and includes a wide range of functions. OSs are the most common, but not the only, type of system software. Examples of OSs include Windows, UNIX, Linux, Mac OS X, and OpenVMS.

OSs include utility programs that meet the needs of most or all users, administrators, and application programs. Functions in most OSs include the following:

- Program storage, loading, and execution
- File manipulation and access
- Secondary storage management
- Network and interactive user interfaces

Although operating systems are the most important and prevalent type of system software, they aren't the only type. Some system software functions aren't needed by most users, so they aren't included as a standard part of all OSs. For example, most OSs don't include utility programs for database manipulation because few users and application programs need this capability. A separate type of system software called a database management system is used for this purpose.

Other system software functions can be provided by optional OS components or stand-alone system software utilities. For example, Web server software is included as an optional component of some OSs, such as Windows Server, but it's also available from other vendors, such as the Apache Software Foundation.

NOTE

Windows OSs tend toward an all-inclusive approach to system software, bundling most system software functions in the OS. UNIX and Linux tend toward a less inclusive approach, with basic resource allocation and system services embedded in the OS and third-party system software used for extended system services, such as Web services, distributed application support, and network security. Some users prefer the bundled approach of Windows, although this approach has occasionally resulted in antitrust litigation and claims of unfair competition.

Application Development Software

Early programmers developed application programs without assistance from computers. They wrote programs consisting entirely of binary CPU instructions, and there were no intermediaries, such as compilers, interpreters, programming languages, and OS services. As computers became more powerful, writing larger programs became possible, but these programs were too complex to develop without automated assistance.

Application development software describes programs used to develop other programs. Although the term implies otherwise, most application development programs can also be used to develop system software or other application development programs. Examples of application development software include many compilers and interpreters for programming languages, such as Java and Visual Basic, and integrated software development packages, such as Microsoft Visual Studio, IBM WebSphere, and Oracle JDeveloper.

Earlier in the chapter, software complexity was said to have resulted from the need to bridge the gaps between human-oriented and machine-oriented language and between abstract, high-level statements of needs and the low-level machine instructions that satisfy these needs. The same challenge of translating human intelligence into computer instructions applies to the process of developing programs.

A **program translator** is a program that translates instructions in a **programming language** into CPU instructions. Examples of programming languages include FORTRAN, Java, C++, and Visual Basic. A program translator doesn't just translate one language into another; it also translates high-level program instructions into detailed CPU instructions. With modern programming languages, programmers can express complex processing tasks in a single statement or instruction. For example, in the BASIC program shown earlier in Figure 2.3, hundreds or thousands of CPU instructions must be executed to perform that task. Programming languages free programmers from having to specify every CPU instruction.

Program translators aren't the only type of application development software. Other types include the following:

- **Program editors**—Writing tools similar to word-processing applications but customized for writing programs instead of documents
- **Debugging tools**—Tools that simulate program execution and help programmers trace errors
- **System development tools**—Tools that enable systems analysts and designers to develop models of information systems that are then used as the starting point for developing application programs

Like any tool, application development software increases programmers' productivity and the quality of what they produce. The complexity of modern software demands application development tools of at least equal complexity.

Economics of System and Application Development Software

The system and application development software used today bears little resemblance to software of the past. Programming languages, program development tools, and OSs weren't developed until the late 1950s and were primitive compared with their modern equivalents.

Why is modern software so advanced? The answer is derived from three economic facts of computer hardware and software:

- System software consumes hardware resources.
- The cost per unit of computing power has decreased rapidly.
- Software is more cost effective when it's reused many times.

These facts underlie a continually shifting tradeoff between the cost of developing and supporting application programs and the cost of hardware resources.

As discussed previously, in the early years of computers, application development was a manual process. Programmers wrote binary CPU instructions with pencil and paper, checked their code manually, and used punch cards or similar tools to convert the written 0s and 1s into something the computer could read. Then they tested their programs to see whether they worked as expected. If too many rounds of testing and correction were required, programmers were criticized for using too much valuable computer time. Hardware was simply too expensive to waste it on debugging software.

As the cost of hardware decreased, the economic balance between programming labor and computer hardware resources shifted. Tedious programming using binary CPU instructions and extensive manual code checks became more expensive than writing and running programs with automated assistance. The trend toward more automated support for application development and computer operation—as well as software reuse—had begun.

When programmers use a programming language, application development software, and a computer to develop programs, they're substituting software and hardware resources for the labor needed to write binary CPU instructions. As programming languages and development tools become more powerful, application development software becomes more complex and consumes more hardware resources. As hardware costs fall, the hardware resources used by application software become cheaper than the extra labor resources that would be needed without application development tools. As you can see in Figure 2.15, the balance of hardware and software cost has shifted over time because of the declining cost of hardware and the emergence of system software. Hardware is typically the cheapest component of current information systems, and system and application software are nearly equal components in the total cost.

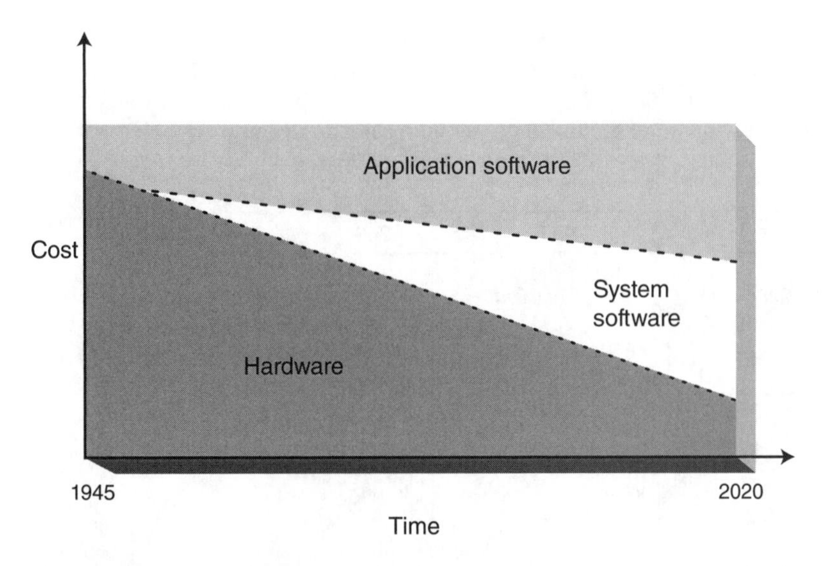

FIGURE 2.15 The change over time in related costs of hardware, application software, and system software in a typical information system
Courtesy of Course Technology/Cengage Learning

Both system and application development software also increase cost effectiveness by promoting software reuse. Each system service is a utility program, and application software reuses these programs each time it makes a service request. Each time a program translator translates a generic instruction, such as Print Customer_Name, into the equivalent CPU instructions, it's reusing the small utility program that implements a PRINT statement. Modern application development tools are even better examples of software reuse because they provide libraries of utility programs and generate program skeletons from templates.

TECHNOLOGY FOCUS

Intel CPUs and Microsoft Operating Systems

IBM introduced the first mass-market microcomputer in 1981. The original IBM PC platform, the Intel 8086, and MS-DOS 1.0 were primitive by modern standards. MS-DOS's primary functions were loading and executing application programs one at a time; providing minimal application support services, such as file I/O; and providing device drivers for the few PC hardware resources, such as the diskette drive, keyboard, and character-oriented video display. More sophisticated OS capabilities were available on more expensive hardware platforms using other CPUs. However, MS-DOS's capabilities were about all that could be provided, given the hardware capabilities of the 8086 and the original PC.

Figure 2.16 shows a timeline for Intel CPUs and Microsoft desktop OSs. Major advances in OS functions have typically lagged behind supporting CPU technology by several years because software development is a long and labor-intensive process. Also, OS development depends on hardware other than the CPU, such as video displays,

(*continued*)

network communication devices, primary storage, and secondary storage, which can be slower to develop. The evolution of Microsoft OSs is a good example of how software development depends on hardware technology. The increasing power of Intel CPUs made it possible for Microsoft to gradually improve the power and features of its OSs.

	1980s	1990s	2000s	2010s
Intel processors	8086 80286 80386 80486	Pentium Pentium Pro Pentium II Pentium III Pentium 4	Pentium D Core 2 Duo Core 2 Quad Core i7	
Microsoft operating systems	MS-DOS Windows 1.0 Windows 2.0	Windows NT 3.0 Windows NT 4.0 Windows 3.0 Windows 95 Windows 98	Windows 2000 Windows XP Windows Vista Windows 7 Windows Me	

FIGURE 2.16 The timing of Intel CPU and Microsoft OS releases
Courtesy of Course Technology/Cengage Learning

80x86 processors enhanced the capabilities of the original 8088/8086 processors, which allowed Microsoft OSs to develop beyond MS-DOS. In particular, the 80386 marked a major advance in CPU performance and capability by replacing earlier approaches to memory addressing that limited application programs to no more than 640 KB of memory. The 80386 also provided hardware support for running multiple programs simultaneously, simplified partitioning primary storage among programs, and provided mechanisms for preventing programs from interfering with one another. Microsoft developers used these processor advances to include OS functions, such as task switching and virtual memory management, in its first true multitasking OS—Windows 3.0. The 80486 provided integrated memory caches, enhanced computational capabilities, and increased raw CPU speed. Windows 95 was developed to take better advantage of this chip's capabilities.

Pentium processors improved memory access and raw CPU speeds and added features such as support for higher-speed system buses, pipelined instruction execution, and multimedia processing instructions. Microsoft OS development split into two distinct paths. The first path started with Windows 95, which evolved into Windows 98 and finally Windows Me. Multimedia instructions served as a foundation for improved high-resolution graphics and audio and video. The second path was a new family of OSs that began with Windows NT and continued through Windows 2000 and XP. Increased CPU speed and improved memory management enabled Microsoft to embed more sophisticated memory and hardware management capabilities in Windows NT than in other Windows OSs. These improvements also allowed Microsoft to develop server OSs, including Windows 2000 Server and Windows Server 2003.

Multiple-core CPUs resulted in only minor changes to CPU internals but substantially improved performance by integrating multiple processing cores and memory caches on a single chip and by increasing raw CPU speed. They also incorporated

(continued)

architectural changes introduced in earlier Pentium processors that expanded memory addressing capabilities beyond 4 GB. These improvements allowed incremental changes in Microsoft OSs, such as introducing more advanced user interface technologies and native OS support for virtualization. These improvements are included in Windows Vista, Windows 7, and Windows Server 2008.

In the push-pull relationship of hardware power and software capability, hardware is usually the driving force. Software developers have a wish list of capabilities that hardware manufacturers try to satisfy to gain a competitive advantage. Software development doesn't proceed in earnest until the necessary hardware is available to support development and testing. However, the situation is occasionally reversed. Software vendors might develop software that pushes past the limits of current hardware, confident that hardware manufacturers will soon provide more power. More demanding software in turn fuels users' drive to acquire new and more powerful machines, and the cycle continues. Competition between software developers guarantees continual and rapid assimilation of new hardware technology.

COMPUTER NETWORKS

In the past, computers were processing islands, and people had to carry data and programs to them, but computer networking and the Internet have changed the landscape of computing forever. Today, a computer isn't considered useful if it doesn't have the capability to interact with almost every other computer on the planet.

A **computer network** consists of hardware and software components that enable users and computer systems to share information, software, and hardware resources and make it possible to use many types of communication methods, such as e-mail, team collaboration, and social networking. The number and complexity of network functions and components have grown as network technology has matured and become more widespread.

NOTE

This section gives you an overview of networks. Chapters 8, 9, and 13 delve into more detail on this topic.

External Resources

Most network functions are extensions of the hardware and software functions discussed previously (see Figure 2.17) that allow a computer system to interface with a physical network and external resources. The complexity of modern networks arises from the huge quantity of distributed resources and the difficulties in finding, accessing, and managing these resources. In the early days of networks, the only important distributed resource was raw data. As networks have matured and expanded, so have the types of resources that can be accessed. Data is now available in many different forms, including text files, sound and video, databases, and Web pages. A computer can ask another computer to run a program and transmit the results, or it can retrieve a file containing the program and run it locally.

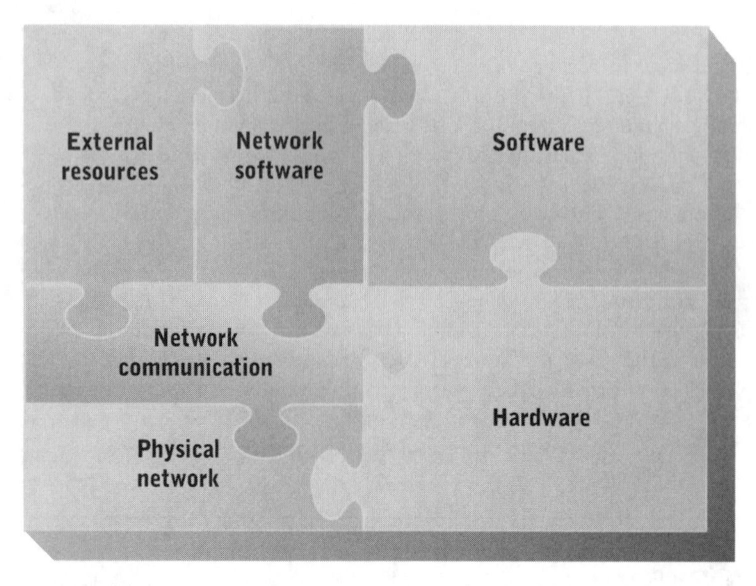

FIGURE 2.17 Computer network functions and their relationship to computer hardware and software
Courtesy of Course Technology/Cengage Learning

Hardware devices of all types can also be shared across a network. One computer can use the CPU, storage, or I/O devices of another. Devices such as printers and secondary storage arrays can be attached directly to the network. The capability to share data, programs, and hardware resources among computers gives organizations the flexibility to deploy and redeploy computing and information resources to satisfy rapidly changing needs.

Network Software

Recall that a key function of system software is allocating resources to users and programs. Allocation is simple when a single user and program access local resources, but it's more difficult when many users or programs compete for hardware and other resources on a single computer. Allocating and accessing resources are complex when a user or program can request resources that aren't on the local computer and aren't managed by locally installed system software. In this case, system software must do the following:

- Find requested resources on the network.
- Negotiate resource access with remote resource allocation software.
- Receive and deliver resources to the requesting user or program.

If a computer system makes its local resources available to other computers, system software must perform the following additional functions:

- Listen for resource requests.
- Validate resource requests.
- Deliver resources via the network.

In essence, system software plays two roles in each network resource access: request and response. Many computer systems fill both roles and must perform both system software functions.

System software has the intelligence needed to make and respond to external resource requests, and most operating systems include support for both functions. Specialized response functions are often provided as optional OS components or as stand-alone system software packages. Examples include Web server software, e-mail distribution software, and Internet security software.

Network Communication and the Physical Network

A computer system requires at least one hardware device to connect to a network. Network communication devices were once exotic and expensive but are now commonplace. They differ from I/O devices in two important ways. First, they're usually simpler because they don't need to convert electronic data into another form. Second, they must support communication at high speeds so that external resource access isn't far slower than access to local resources.

Each computer's network communication hardware is attached to a physical network. A physical network is a complex combination of communication protocols, methods of data transmission (cables and/or wireless), and network hardware devices (network interface cards, switches, routers, other devices to make physical data connections, and networked printers and other resources). A physical network can be implemented with a dizzying array of technologies and architectural approaches, which are discussed in more detail later in the book.

Summary

- A computer is an automated device for performing computational tasks. It accepts input data from the external world, performs one or more computations on the data, and then returns results to the external world. Early computers were mechanical devices with limited capabilities. Modern computers have more extensive capabilities and are implemented with a combination of electronic and optical devices. The advantages of electronic and optical implementation include speed, accuracy, and reliability.
- Computer capabilities include processing, storage, and communication. A computer system contains a general-purpose processor that can perform computations, compare numeric data, and move data between storage locations and I/O devices. A command to the processor to perform a function is called an instruction. Complex tasks are implemented as a sequence of instructions called a program. Changing a program changes a computer system's behavior.
- A computer system consists of a central processing unit (CPU), primary storage, secondary storage, and I/O devices. The CPU performs all computation and comparison functions and directs all data movement. Primary storage holds programs and data currently in use by the CPU. Primary storage is generally implemented by electronic devices called random access memory (RAM). Secondary storage consists of devices with high capacity and lower cost and access speed. Secondary storage holds programs and data not currently in use by the CPU. I/O devices allow the CPU to communicate with the external world, including users and other computers.
- A computer system can be classified as a microcomputer, midrange computer, mainframe, or supercomputer. Microcomputers are designed for use by a single user. Midrange computers and mainframes are designed to support many programs and users simultaneously. Mainframe computers are designed for large amounts of data storage and access. Supercomputers are designed to perform large amounts of numeric computations quickly. Clusters and grids can mimic the performance of mainframes and supercomputers with groups of smaller computers.
- The role of software is to translate user requests into machine instructions. The two primary types of software are application software and system software. Application software consists of programs that satisfy specific user processing needs, such as payroll calculation, accounts payable, and report generation. System software consists of utility programs designed to be general in nature and can be used many times by many different application programs or users. Examples of system software include operating systems, database management systems, antivirus software, and network security software.
- The operating system is the most important system software component in most computers. It provides administrative utilities, utility services to application programs, resource allocation functions, and direct control over hardware.
- System and application development software reduce the cost of developing and deploying application programs and allow organizations to substitute less expensive computer hardware for expensive labor.
- A computer network consists of hardware and software components that enable multiple users and computers to share information, software, and hardware resources. Network software allows a computer to find and retrieve resources on other computers or to respond

to requests from other computers for access to local resources. Network hardware implements direct communication between computer systems.

In this chapter, you've learned about the basic elements of systems architecture: hardware, software, and networks. You've been introduced to many new terms and concepts but haven't learned about them in detail. You'll get a chance to digest and expand your knowledge of these topics in later chapters. Figure 2.18 shows the key computer system functions covered so far and identifies the chapters that explore each function in depth.

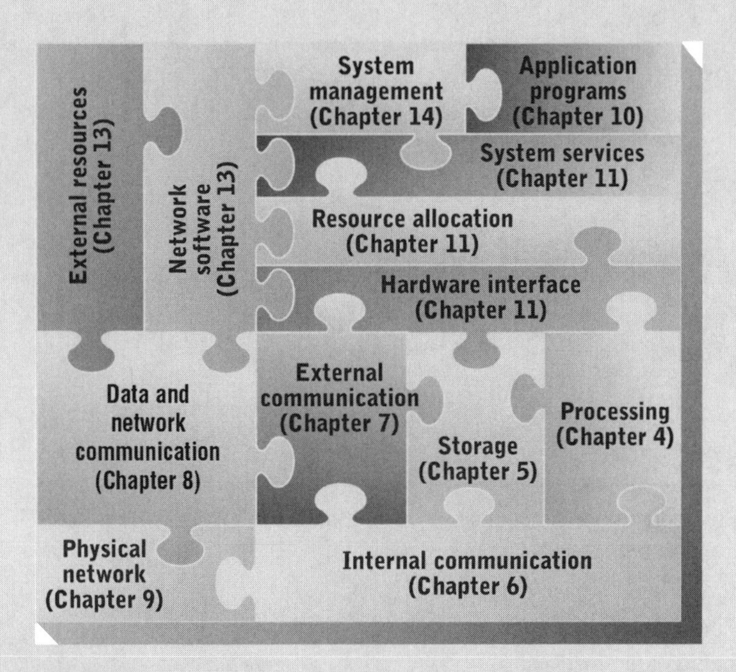

FIGURE 2.18 Topics covered in the rest of the book
Courtesy of Course Technology/Cengage Learning

Key Terms

80x86 processor
algorithm
application development software
application software
arithmetic logic unit (ALU)
blade
central processing unit (CPU)
cloud
cluster
computer network
condition
control unit
debugging tool
executing
formula
general-purpose processor
grid
Grosch's Law

hardware independence
Input/output (I/O) unit
instruction
laptop computer
logic instruction
machine independence
main memory
mainframe
microcomputer
midrange computer
multicomputer configuration
multiple-core CPU
netbook computer
operating system (OS)
Pentium processor
personal computer (PC)
personal digital assistant (PDA)
primary storage
processor
program
program editor
program translator
programming language
qubit
random access memory (RAM)
register
secondary storage
server
special-purpose processor
supercomputer
system bus
system development tool
system software
systems architecture
workstation

Vocabulary Exercises

1. A(n) ________ generally supports more simultaneous users than a(n) ________. Both are designed to support more than one user.
2. A(n) ________ is a storage location implemented in the CPU.
3. The term ________ refers to storage devices, not located in the CPU, that hold instructions and data of currently running programs.
4. A problem-solving procedure that requires executing one or more comparison and branch instructions is called a(n) ________.
5. A(n) ________ is a command to the CPU to perform one processing function on one or more data inputs.
6. The term ________ describes the collection of storage devices that hold large quantities of data for long periods.
7. A(n) ________ is a computer that manages shared resources and allows other computers to access them through a network.
8. A program that solves a(n) ________ requires no branching instructions.
9. The major components of a CPU are the ________, ________, and ________.
10. Primary storage can also be called ________ and is generally implemented with ________.
11. A set of instructions that's executed to solve a specific problem is called a(n) ________.
12. A(n) ________ typically uses the latest and most expensive technology.

13. A(n) ________ is a group of similar or identical computers, connected by a high-speed network, that cooperate to provide services or run an application.
14. A(n) ________ is a group of dissimilar computer systems, connected by a high-speed network, that cooperate to provide services or run a shared application.
15. A(n) ________ consists of computing resources with a Web-based front-end interface to a large collection of computing and data resources.
16. A(n) ________ is a hardware device that enables a computer to communicate with users or other computers.
17. A CPU is a(n) ________ processor capable of performing many different tasks simply by changing the program.
18. The ________ is the "plumbing" that connects all computer system components.
19. The CPU ________ program instructions one at a time.
20. The term ________ describes a computer system's components and their interactions.
21. Most programs are written in a(n) ________, such as C or Java, which is then translated into equivalent CPU instructions.
22. Resource allocation and direct hardware control are the responsibilities of a(n) ________.
23. ________ software is general-purpose software. ________ software is specialized for specific user needs.
24. A(n) ________ consists of hardware and software components that enable multiple users and computers to share information, software, and hardware resources.

Review Questions

1. What similarities exist in mechanical, electrical, and optical methods of computation?
2. What shortcomings of mechanical computation did the introduction of electronic computing devices address?
3. What shortcomings of electrical computation will optical computing devices address?
4. What is a CPU? What are its primary components?
5. What are registers? What are their functions?
6. What is main memory? How does it differ from registers?
7. What are the differences between primary and secondary storage?
8. How does a workstation differ from a PC?
9. How does a supercomputer differ from a mainframe computer?
10. Describe three types of multicomputer configurations. What are their comparative advantages and disadvantages?
11. What is virtualization? What are the advantages and disadvantages of using a single large computer to host multiple virtual machines compared with using multiple smaller computers without virtualization?
12. What classes of computer systems are normally used for servers?

13. What is Grosch's Law? Does it hold today? Why or why not?
14. How can a computer system be tuned to a particular application?
15. What characteristics differentiate application software from system software?
16. In what ways does system software make developing application software easier?
17. Why has the development of system software paralleled the development of computer hardware?
18. List at least five types of resources that computers on a local area network or wide area network can share.
19. Describe the dual roles most operating systems play in external resource access.
20. Describe the relationship between the resource allocation and management functions of system software and external resources accessible via a network. What system software functions must be provided to access external resources?

Research Problems

1. Find a catalog or visit the Web site of a major distributor of microcomputer equipment, such as Computer Discount Warehouse (*www.cdw.com*) or Dell (*www.dell.com*). You have a budget of $1000 to purchase workstation hardware (without peripheral devices, such as a scanner or printer). Select or configure a system that provides optimal performance for the following types of users:

   - A home user who uses word-processing software, such as Microsoft Office; a home accounting package, such as Quicken or TurboTax; children's games; and multimedia software for editing pictures and creating video DVDs
   - An accountant who uses office software, such as Microsoft Office, and statistical software, such as SPSS or SAS, and downloads large amounts of financial data from a corporate server for statistical and financial modeling
   - An architect who uses typical office software and CAD software, such as AutoCAD

   Pay particular attention to whether CPU power, disk space, and I/O capabilities are adequate.
2. You have been asked to recommend a computer system to be used as a server for a 250-person office. The server must provide shared access to a 10 TB file system, several printers, e-mail and Web services, and database services (for a 2 TB database) to a variety of application programs. Microsoft products, such as Windows Server and Microsoft SQL Server, will be the primary software used on the server. Gather product information from IBM (*www.ibm.com*), Hewlett-Packard (*www.hp.com*), and Dell Computer Corporation (*www.dell.com*). Determine which of their products best meet the server requirements.
3. Table 2.2 will probably be out of date by the time this book is published. Go to the manufacturers' Web sites (*www.ibm.com* and *www.dell.com*) or contact a sales representative, and update Table 2.2 with current representative models for each computer class. At the Web site for this book (*www.cengage.com/mis/burd*), download copies of Table 2.2 from previous editions. What is the rate of change in CPU speed, memory and disk capacity, and cost?

CHAPTER **3**

DATA REPRESENTATION

CHAPTER GOALS

- Describe numbering systems and their use in data representation
- Compare different data representation methods
- Summarize the CPU data types and explain how nonnumeric data is represented
- Describe common data structures and their uses

Computers manipulate and store a variety of data, such as numbers, text, sound, and pictures. This chapter describes how data is represented and stored in computer hardware. It also explains how simple data types are used as building blocks to create more complex data structures, such as arrays and records. Understanding data representation is key to understanding hardware and software technologies.

DATA REPRESENTATION AND PROCESSING

People can understand and manipulate data represented in a variety of forms. For example, they can understand numbers represented symbolically as Arabic numerals (such as 8714), Roman numerals (such as XVII), and simple lines or tick marks on paper (for example, ||| to represent the value 3). They can understand words and concepts represented with pictorial characters (电脑) or alphabetic characters ("computer" and компьютер, Cyrillic text of the Russian word for computer) and in the form of sound waves (spoken words). People also extract data from visual images (photos and movies) and from the senses of taste, smell, and touch. The human brain's processing power and flexibility are evident in the rich variety of data representations it can recognize and understand.

To be manipulated or processed by the brain, external data representations, such as printed text, must be converted to an internal format and transported to the brain's processing "circuitry." Sensory organs convert inputs, such as sight, smell, taste, sound, and skin sensations, into electrical impulses that are transported through the nervous system to the brain. Processing in the brain occurs as networks of neurons exchange data electrically.

Any data and information processor, whether organic, mechanical, electrical, or optical, must be capable of the following:

- Recognizing external data and converting it to an internal format
- Storing and retrieving data internally
- Transporting data between internal storage and processing components
- Manipulating data to produce results or decisions

Note that these capabilities correspond roughly to computer system components described in Chapter 2—I/O units, primary and secondary storage, the system bus, and the CPU.

Automated Data Processing

Computer systems represent data electrically and process it with electrical switches. Two-state (on and off) electrical switches are well suited for representing data that can be expressed in binary (1 or 0) format, as you see later in Chapter 4. Electrical switches are combined to form processing circuits, which are then combined to form processing subsystems and entire CPUs. You can see this processing as an equation:

$$A + B = C$$

In this equation, data inputs A and B, represented as electrical currents, are transported through processing circuits (see Figure 3.1). The electrical current emerging from the circuit represents a data output, C. Automated data processing, therefore, combines physics (electronics) and mathematics.

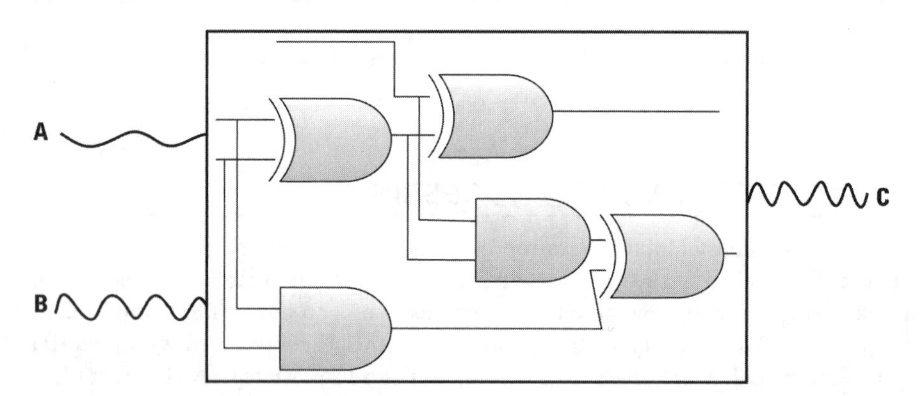

FIGURE 3.1 Two electrical inputs on the left flow through processing circuitry that generates their sum on the right

Courtesy of Course Technology/Cengage Learning

The physical laws of electricity, optics, and quantum mechanics are described by mathematical formulas. If a device's behavior is based on well-defined, mathematically described laws of physics, the device, in theory, can implement a processor to perform the equivalent mathematical function. This relationship between mathematics and physics underlies all automated computation devices, from mechanical clocks (using the mathematical ratios of gears) to electronic microprocessors (using the mathematics of electrical voltage and resistance). As you learned in Chapter 2, in quantum mechanics, the mathematical laws are understood but not how to build reliable and cost-effective computing devices based on these laws.

Basing computer processing on mathematics and physics has limits, however. Processing operations must be based on mathematical functions, such as addition and equality comparison; use numerical data inputs; and generate numerical outputs. These processing functions are sufficient when a computer performs numeric tasks, such as accounting or statistical analysis. When you ask a computer to perform tasks such as searching text documents and editing sound, pictures, and video, numeric-processing functions do have limitations, but ones that modern software has largely overcome. However, when you want to use a computer to manipulate data with no obvious numeric equivalent—for example, literary or philosophical analysis of concepts such as "mother," "friend," "love," and "hate"—numeric-processing functions have major shortcomings. As the data you want to process moves further away from numbers, applying computer technology to processing the data becomes increasingly difficult—and less successful.

Binary Data Representation

In a decimal (base 10) number, each digit can have 1 of 10 possible values: 0, 1, 2, 3, 4, 5, 6, 7, 8, or 9. In a **binary number**, each digit can have only one of two possible values: 0 or 1. Computers represent data with binary numbers for two reasons:

- Binary numbers represented as binary electrical signals can be transported reliably between computer systems and their components (discussed in detail in Chapter 8).
- Binary numbers represented as electrical signals can be processed by two-state electrical devices that are easy to design and fabricate (discussed in detail in Chapter 4).

For computer applications to produce accurate outputs, reliable data transport is important. Given current technology, binary signals and processing devices represent the most cost-efficient tradeoffs between capacity, accuracy, reliability, and cost.

Binary numbers are also well suited to computer processing because they correspond directly with values in **Boolean logic**. This form of logic is named for 19th-century mathematician George Boole, who developed methods of reasoning and logical proof that use sequences of statements that can be evaluated only as true or false. Similarly, a computer can perform logical comparisons of two binary data values to determine whether one data value is greater than, equal to, less than, less than or equal to, not equal to, or greater than or equal to another value. As discussed in Chapter 2, a computer uses this primitive logical capability to exhibit intelligent behavior.

Both computers and humans can combine digits to represent and manipulate larger numbers. Decimal and binary notations are alternative forms of a positional numbering system, in which numeric values are represented as groups, or strings, of digits. The symbol used to represent a digit and the digit's position in a string determine its value. The value of the entire string is the sum of the values of all digits in the string.

For example, in the decimal numbering system, the number 5689 is interpreted as follows:

$$(5 \times 1000) + (6 \times 100) + (8 \times 10) + 9 = 5000 + 600 + 80 + 9 = 5689$$

The same series of operations can be represented in columnar form, with positions of the same value aligned in columns:

$$\begin{array}{r} 5000 \\ 600 \\ 80 \\ +\ \ 9 \\ \hline 5689 \end{array}$$

For whole numbers, values are accumulated from right to left. In the preceding example, the digit 9 is in the first position, 8 is in the second position, 6 is in the third, and 5 is in the fourth.

The maximum value, or weight, of each position is a multiple of the weight of the position to its right. In the decimal numbering system, the first (rightmost) position is the ones (10^0), and the second position is 10 times the first position (10^1). The third position is 10 times the second position (10^2, or 100), the fourth is 10 times the third position (10^3, or 1000), and so on. In the binary numbering system, each position is 2 times the previous position, so position values for whole numbers are 1, 2, 4, 8, and so forth. The multiplier that describes the difference between one position and the next is the **base**, or **radix**, of the numbering system. In the decimal numbering system, it's 10, and in the binary numbering system, it's 2.

The fractional part of a numeric value is separated from the whole part by a period, although in some countries, a comma is used instead of a period. In the decimal numbering system, the period or comma is called a **decimal point**. In other numbering systems, the term **radix point** is used for the period or comma. Here's an example of a decimal value with a radix point:

5689.368

The fractional portion of this real number is .368, and its value is interpreted as follows:

$$(3 \times 10^{-1}) + (6 \times 10^{-2}) + (8 \times 10^{-3}) =$$
$$(3 \times .1) + (6 \times .01) + (8 \times .001) =$$
$$0.3 + 0.06 + 0.008 = 0.368$$

Proceeding toward the right from the radix point, the weight of each position is a fraction of the position to its left. In the decimal (base 10) numbering system, the first position to the right of the decimal point represents tenths (10^{-1}), the second position represents hundredths (10^{-2}), the third represents thousandths (10^{-3}), and so forth.

In the binary numbering system, the first position to the right of the radix point represents halves (2^{-1}), the second position represents quarters (2^{-2}), the third represents eighths (2^{-3}), and so forth. As with whole numbers, each position has a weight 10 (or 2) times the position to its right. Table 3.1 compares decimal and binary notations for the values 0 through 10.

TABLE 3.1 Binary and decimal notations for the values 0 through 10

	Binary system (base 2)					Decimal system (base 10)			
Place	2^3	2^2	2^1	2^0		10^3	10^2	10^1	10^0
Values	8	4	2	1		1000	100	10	1
	0	0	0	0	=	0	0	0	0
	0	0	0	1	=	0	0	0	1
	0	0	1	0	=	0	0	0	2
	0	0	1	1	=	0	0	0	3
	0	1	0	0	=	0	0	0	4
	0	1	0	1	=	0	0	0	5
	0	1	1	0	=	0	0	0	6
	0	1	1	1	=	0	0	0	7
	1	0	0	0	=	0	0	0	8
	1	0	0	1	=	0	0	0	9
	1	0	1	0	=	0	0	1	0

The number of digits needed to represent a value depends on the numbering system's base: The number of digits increases as the numbering system's base decreases. Therefore, values that can be represented in a compact format in decimal notation might require lengthy sequences of binary digits. For example, the decimal value 99 requires two decimal digits but seven binary digits. Table 3.2 summarizes the number of binary digits needed to represent decimal values up to 16 positions.

TABLE 3.2 Binary notations for decimal values up to 16 positions

Number of bits (n)	Number of values (2^n)	Numeric range (decimal)
1	2	0–1
2	4	0–3
3	8	0–7
4	16	0–15
5	32	0–31
6	64	0–63
7	128	0–127
8	256	0–255
9	512	0–511
10	1024	0–1023
11	2048	0–2047
12	4096	0–4095
13	8192	0–8191
14	16,384	0–16,383
15	32,768	0–32,767
16	65,536	0–65,535

To convert a binary value to its decimal equivalent, use the following procedure:

1. Determine each position weight by raising 2 to the number of positions left (+) or right (-) of the radix point.
2. Multiply each digit by its position weight.
3. Sum all the values calculated in Step 2.

NOTE

The standard Windows calculator can convert between binary, octal, decimal, and hexadecimal. To open the calculator in Windows 7, click Start, All Programs, Accessories, Calculator. To convert a binary number to decimal, click View, Programmer from the menu (View, Scientific for Vista and earlier Windows versions). Click the Bin (for binary) option button at the upper left, enter the binary number in the text box, and then click the Dec (for decimal) option button.

Figure 3.2 shows how the binary number 101101.101 is converted to its decimal equivalent, 45.625.

Position weights				Binary value		Decimal value
2^5	=	32	×	1	=	32.000
2^4	=	16	×	0	=	0.000
2^3	=	8	×	1	=	8.000
2^2	=	4	×	1	=	4.000
2^1	=	2	×	0	=	0.000
2^0	=	1	×	1	=	1.000
Radix point						
2^{-1}	=	½	×	1	=	0.500
2^{-2}	=	¼	×	0	=	0.000
2^{-3}	=	⅛	×	1	=	+ 0.125
						45.625

FIGURE 3.2 Computing the decimal equivalent of a binary number
Courtesy of Course Technology/Cengage Learning

In computer terminology, each digit of a binary number is called a **bit**. A group of bits that describe a single data value is called a **bit string**. The leftmost digit, which has the greatest weight, is called the **most significant digit**, or **high-order bit**. Conversely, the rightmost digit is the **least significant digit**, or **low-order bit**. A string of 8 bits is called a **byte**. Generally, a byte is the smallest unit of data that can be read from or written to a storage device.

The following mathematical rules define addition of positive binary digits:

$$
\begin{aligned}
0 + 0 &= 0 \\
1 + 0 &= 1 \\
0 + 1 &= 1 \\
1 + 1 &= 10
\end{aligned}
$$

To add two positive binary bit strings, you first must align their radix points as follows:

$$
\begin{array}{r}
101101.101 \\
\underline{+10100.0010}
\end{array}
$$

The values in each column are added separately, starting with the least significant, or rightmost, digit. If a column result exceeds 1, the excess value must be carried to the next column and added to the values in that column.

NOTE

The standard Windows calculator can add and subtract binary integers. To use this feature, click View, Programmer from the menu (View, Scientific for Vista and earlier Windows versions), and click the Bin option button. You can also click View, Digit grouping from the menu to place digits in groups of four for easier readability.

This is the result of adding the two preceding numbers:

```
  111    1
  101101.101
 +10100.0010
 -----------
1000001.1100
```

The result is the same as when adding the values in base-10 notation:

Binary		Real fractions		Real decimal
101101.101	=	$45\,^5\!/_8$	=	45.625
+10100.0010	=	$20\,^1\!/_8$	=	20.125
1000001.1100	=	$65\,^3\!/_4$	=	65.750

Binary numbers usually contain many digits and are difficult for people to remember and manipulate without error. Compilers and interpreters for high-level programming languages, such as C and Java, convert decimal numbers into binary numbers automatically when generating CPU instructions and data values. However, sometimes programmers must deal with binary numbers directly, such as when they program in machine language or for some operating system (OS) utilities. To minimize errors and make dealing with binary numbers easier, numbering systems based on even multiples of 2 are sometimes used. These numbering systems include hexadecimal and octal, discussed in the following sections.

Hexadecimal Notation

Hexadecimal numbering uses 16 as its base or radix ("hex" = 6 and "decimal" = 10). There aren't enough numeric symbols (Arabic numerals) to represent 16 different values, so English letters represent the larger values (see Table 3.3).

TABLE 3.3 Hexadecimal and decimal values

Base-16 digit	Decimal value	Base-16 digit	Decimal value
0	0	8	8
1	1	9	9
2	2	A	10
3	3	B	11
4	4	C	12
5	5	D	13
6	6	E	14
7	7	F	15

The primary advantage of **hexadecimal notation**, compared with binary notation, is its compactness. Large numeric values expressed in binary notation require four times as many digits as those expressed in hexadecimal notation. For example, the data content of a byte requires eight binary digits (such as 11110000) but only two hexadecimal digits (such as F0). This compact representation helps reduce programmer error.

Hexadecimal numbers often designate memory addresses. For example, a 64 KB memory region contains 65,536 bytes (64 × 1024 bytes/KB). Each byte is identified by a sequential numeric address. The first byte is always address 0. Therefore, the range of possible memory addresses is 0 to 65,535 in decimal numbers, 0 to 1111111111111111 in binary numbers, and 0 to FFFF in hexadecimal numbers. As you can see from this example, hexadecimal addresses are more compact than decimal or binary addresses because of the numbering system's higher radix.

When reading a numeric value in written text, the number's base might not be obvious. For example, when reading an OS error message or a hardware installation manual, should the number 1000 be interpreted in base 2, 10, 16, or something else? In mathematical expressions, the base is usually specified with a subscript, as in this example:

$$1001_2$$

The subscript 2 indicates that 1001 should be interpreted as a binary number. Similarly, in the following example, the subscript 16 indicates that 6044 should be interpreted as a hexadecimal number:

$$6044_{16}$$

The base of a written number can be made explicit by placing a letter at the end. For example, the letter B in this example indicates a binary number:

1001B

The letter H in this example indicates a hexadecimal number:

6044H

Normally, no letter is used to indicate a decimal (base 10) number. Some programming languages, such as Java and C++, use the prefix 0x to indicate a hexadecimal number. For example, 0x1001 is equivalent to 1001_{16}.

Unfortunately, these conventions aren't observed consistently. Often it's left to the reader to guess the correct base by the number's content or the context in which it appears. A value containing a numeral other than 0 or 1 can't be binary, for instance. Similarly, the use of letters A through F indicates that the contents are expressed in hexadecimal. Bit strings are usually expressed in binary, and memory addresses are usually expressed in hexadecimal.

Octal Notation

Some OSs and machine programming languages use **octal notation**. Octal notation uses the base-8 numbering system and has a range of digits from 0 to 7. Large numeric values expressed in octal notation are one-third the length of corresponding binary notation and double the length of corresponding hexadecimal notation.

GOALS OF COMPUTER DATA REPRESENTATION

Although all modern computers represent data internally with binary digits, they don't necessarily represent larger numeric values with positional bit strings. Positional numbering systems are convenient for people to interpret and manipulate because the sequential processing of digits in a string parallels the way the human brain functions and because people are taught to perform computations in a linear fashion. For example, positional numbering systems are well suited to adding and subtracting numbers in columns by using pencil and paper. Computer processors, however, operate differently from a human brain. Data representation tailored to human capabilities and limitations might not be best suited to computers.

Any representation format for numeric data represents a balance among several factors, including the following:

- Compactness
- Range
- Accuracy
- Ease of manipulation
- Standardization

As with many computer design decisions, alternatives that perform well in one factor often perform poorly in others. For example, a data format with a high degree of accuracy and a large range of representable values is usually difficult and expensive to manipulate because it's not compact.

Compactness and Range

The term "compactness" (or size) describes the number of bits used to represent a numeric value. Compact representation formats use fewer bits to represent a value, but they're limited in the range of values they can represent. For example, the largest binary integer that can be stored in 32 bits is 2^{32}, or $4,294,967,296_{10}$. Halving the number of bits to make the representation more compact decreases the largest possible value to 2^{16}, or $65{,}535_{10}$.

Computer users and programmers usually prefer a large numeric range. For example, would you be happy if your bank's computer limited your maximum checking account balance to 65,535 pennies? The extra bit positions required to increase the numeric range have a cost, however. Primary and secondary storage devices must be larger and, therefore, more expensive. Additional and more expensive capacity is required for data transmission between devices in a computer system or across computer networks. CPU processing circuitry becomes more complex and expensive as more bit positions are added. The more compact a data representation format, the less expensive it is to implement in computer hardware.

Accuracy

Although compact data formats can minimize hardware's complexity and cost, they do so at the expense of accurate data representation. The accuracy, or precision, of representation increases with the number of data bits used.

It's possible for routine calculations to generate quantities too large or too small to be contained in a machine's finite circuitry (that is, in a fixed number of bits). For example, the fraction 1/3 can't be represented accurately in a fixed number of bits because it's a nonterminating fractional quantity (0.333333333 … , with an infinite number of 3s). In these cases, the quantities must be manipulated and stored as approximations, and each approximation introduces a degree of error. If approximate results are used as inputs for other computations, errors can be compounded and even result in major errors. For this reason, a program can have no apparent logical flaws yet still produce inaccurate results.

If all data types were represented in the most compact form possible, approximations would introduce unacceptable margins of error. If a large number of bits were allocated to each data value instead, machine efficiency and performance would be sacrificed, and hardware cost would be increased. The best balance in performance and cost can be achieved by using an optimum coding method for each type of data or each type of operation to be performed. Striving for this balance is the main reason for the variety of data representation formats used in modern CPUs.

Ease of Manipulation

When discussing computer processing, **manipulation** refers to executing processor instructions, such as addition, subtraction, and equality comparisons, and "ease" refers to machine efficiency. A processor's efficiency depends on its complexity (the number of its primitive components and the complexity of the wiring that binds them together). Efficient processor circuits perform their functions quickly because of the small number of components and the short distance electricity must travel. More complex devices need more time to perform their functions.

Data representation formats vary in their capability to support efficient processing. For example, most people have more difficulty performing computations with fractions than with decimal numbers. People process the decimal format more efficiently than the fractional format.

Unfortunately, there's no best representation format for all types of computation operations. For example, representing large numeric values as logarithms simplifies multiplication and division for people and computers because $\log A + \log B = \log (A \times B)$ and $\log A - \log B = \log (A \div B)$. Logarithms complicate other operations, such as addition and subtraction, and they can increase a number's length substantially (for example, $\log 99 \approx 1.9956351945975499153402557777533$).

Standardization

Data must be communicated between devices in a single computer and to other computers via networks. To ensure correct and efficient data transmission, data formats must be suitable for a wide variety of devices and computers. For this reason, several organizations have created standard data-encoding methods (discussed later in the "Character Data" section). Adhering to these standards gives computer users the flexibility to combine hardware from different vendors with minimal data communication problems.

CPU DATA TYPES

The CPUs of most modern computers can represent and process at least the following primitive data types:

- Integer
- Real number
- Character
- Boolean
- Memory address

The arrangement and interpretation of bits in a bit string are usually different for each data type. The representation format for each data type balances compactness, range, accuracy, ease of manipulation, and standardization. A CPU can also implement multiple versions of each type to support different types of processing operations.

Integers

An **integer** is a whole number—a value that doesn't have a fractional part. For example, the values 2, 3, 9, and 129 are integers, but the value 12.34 is not. Integer data formats can be signed or unsigned. Most CPUs provide an **unsigned integer** data type, which stores positive integer values as ordinary binary numbers. An unsigned integer's value is always assumed to be positive.

A **signed integer** uses one bit to represent whether the value is positive or negative. The choice of bit value (0 or 1) to represent the sign (positive or negative) is arbitrary. The sign bit is normally the high-order bit in a numeric data format. In most data formats, it's 1 for a negative number and 0 for a nonnegative number. (Note that 0 is a nonnegative number.)

The sign bit occupies a bit position that would otherwise be available to store part of the data value. Therefore, using a sign bit reduces the largest positive value that can be stored in any fixed number of bit positions. For example, the largest positive value that can be stored in an 8-bit unsigned integer is 255, or 2^8 - 1. If a bit is used for the sign, the largest positive value that can be stored is 127, or 2^7 - 1.

With unsigned integers, the lowest value that can be represented is always 0. With signed integers, the lowest value that can be represented is the negative of the highest value that can be stored (for example, -127 for 8-bit signed binary).

Excess Notation

One format that can be used to represent signed integers is **excess notation**, which always uses a fixed number of bits, with the leftmost bit representing the sign. For example, the value 0 is represented by a bit string with 1 in the leftmost digit and 0s in all the other digits. As shown in Table 3.4, all nonnegative values have 1 as the high-order bit, and negative values have 0 in this position. In essence, excess notation divides a range of ordinary binary numbers in half and uses the lower half for negative values and the upper half for nonnegative values.

TABLE 3.4 Excess notation

Bit string	Decimal value	
1111	7	Nonnegative numbers
1110	6	
1101	5	
1100	4	
1011	3	
1010	2	
1001	1	
1000	0	
0111	-1	Negative numbers
0110	-2	
0101	-3	
0100	-4	
0011	-5	
0010	-6	
0001	-7	
0000	-8	

To represent a specific integer value in excess notation, you must know how many storage bits are to be used, whether the value fits within the numeric range of excess notation for that number of bits, and whether the value to be stored is positive or negative. For any number of bits, the largest and smallest values in excess notation are $2^{(n-1)} - 1$ and $-2^{(n-1)}$, where n is the number of available storage bits.

For example, consider storing a signed integer in 8 bits with excess notation. Because the leftmost bit is a sign bit, the largest positive value that can be stored is 2^7 - 1, or 127_{10}, and the smallest negative value that can be stored is -2^7, or -128_{10}. The range of positive values appears to be smaller than the range of negative values because 0 is considered a positive (nonnegative) number in excess notation. Attempting to represent larger positive or smaller negative values results in errors because the leftmost (sign) bit might be overwritten with an incorrect value.

Now consider how +9 and -9 are represented in 8-bit excess notation. Both values are well within the numeric range limits for 8-bit excess notation. The ordinary binary representation of $+9_{10}$ in 8 bits is 00001001. Recall that the excess notation representation of 0 is always a leading 1 bit followed by all 0 bits—10000000 for 8-bit excess notation. Because +9 is nine integer values greater than 0, you can calculate the representation of +9 by adding its ordinary binary representation to the excess notation representation of 0 as follows:

$$\begin{array}{r} 10000000 \\ +\,00001001 \\ \hline 10001001 \end{array}$$

To represent negative values, you use a similar method based on subtraction. Because -9 is nine integer values less than 0, you can calculate the representation of -9 by subtracting its ordinary binary representation from the excess notation representation of 0 as follows:

$$\begin{array}{r} 10000000 \\ -\,00001001 \\ \hline 01110111 \end{array}$$

Twos Complement Notation

In the binary numbering system, the complement of 0 is 1, and the complement of 1 is 0. The complement of a bit string is formed by substituting 0 for all values of 1 and 1 for all values of 0. For example, the complement of 1010 is 0101. This transformation is the basis of **twos complement notation**. In this notation, nonnegative integer values are represented as ordinary binary values. For example, a twos complement representation of 7_{10} using 4 bits is 0111.

Bit strings for negative integer values are determined by the following transformation:

$$\text{complement of positive value} + 1 = \text{negative representation}$$

Parentheses are a common mathematical notation for showing a value's complement; for example, if A is a numeric value, (A) represents its complement. In 4-bit twos complement representation, -7_{10} is calculated as follows:

$$\begin{aligned} (0111) + 0001 &= \\ 1000 + 0001 &= \\ 1001 &= -7_{10} \end{aligned}$$

As another example, take a look at the twos complement representation of $+35_{10}$ and -35_{10} in 8 bits. The ordinary binary equivalent of $+35_{10}$ is 00100011, which is also the twos complement representation. To determine the twos complement representation of -35_{10}, use the previous formula with 8-bit numbers:

$$(00100011) + 00000001 =$$
$$11011100 + 00000001 =$$
$$11011101 = -35_{10}$$

Twos complement notation is awkward for most people, but it's highly compatible with digital electronic circuitry for the following reasons:

- The leftmost bit represents the sign.
- A fixed number of bit positions are used.
- Only two logic circuits are required to perform addition on single-bit values.
- Subtraction can be performed as addition of a negative value.

The latter two reasons enable CPU manufacturers to build processors with fewer components than are needed for other integer data formats, which saves money and increases computational speed. For these reasons, all modern CPUs represent and manipulate signed integers by using twos complement format.

Range and Overflow

Most modern CPUs use 64 bits to represent a twos complement value and support 32-bit formats for backward compatibility with older software. A 32-bit format is used in the remainder of this book to simplify the discussion and examples. A small positive value, such as 1, occupies 32 bits even though, in theory, only 2 bits are required (one for the value and one for the sign). Although people can deal with numeric values of varying lengths, computer circuitry isn't nearly as flexible. Fixed-width formats enable more efficient processor and data communication circuitry. The additional CPU complexity required to process variable-length data formats results in unacceptably slow performance. Therefore, when small numeric values are stored, the "extra" bit positions are filled with leading 0s.

The **numeric range** of a twos complement value is $-(2^{n-1})$ to $(2^{n-1} - 1)$, where n is the number of bits used to store the value. The exponent is n-1 because 1 bit is used for the sign. For 32-bit twos complement format, the numeric range is $-2,147,483,648_{10}$ to $2,147,483,647_{10}$. With any fixed-width data storage format, it's possible that the result of a computation will be too large to fit in the format. For example, the Gross Domestic Product of each U.S. state was less than \$2 billion in 2005. Therefore, these values can be represented as 32-bit twos complement integers. Adding these numbers to calculate Gross National Product (GNP), however, yields a sum larger than \$2 billion. Therefore, a program that computes GNP by using 32-bit twos complement values will generate a value that exceeds the format's numeric range. This condition, referred to as **overflow**, is treated as an error by the CPU. Executing a subtraction instruction can also result in overflow—for example, $-(2^{31}) - 1$. Overflow occurs when the absolute value of a computational result contains too many bits to fit into a fixed-width data format.

As with most other aspects of CPU design, data format length is one design factor that needs to be balanced with others. Large formats reduce the chance of overflow by increasing the maximum absolute value that can be represented, but many bits are wasted (padded with leading 0s) when smaller values are stored. If bits were free, there would be no tradeoff. However, extra bits increase processor complexity and storage requirements, which increase computer system cost. A CPU designer chooses a data format width by balancing numeric range, the chance of overflow during program execution, and the complexity, cost, and speed of processing and storage devices.

To avoid overflow and increase accuracy, some computers and programming languages define additional numeric data types called **double-precision** data formats. A double-precision data format combines two adjacent fixed-length data items to hold a single value. Double-precision integers are sometimes called **long integers**.

Overflow can also be avoided by careful programming. If a programmer anticipates that overflow is possible, the units of measure for program variables can be made larger. For example, calculations on centimeters could be converted to meters or kilometers, as appropriate.

Real Numbers

A **real number** can contain both whole and fractional components. The fractional portion is represented by digits to the right of the radix point. For example, the following computation uses real number data inputs and generates a real number output:

$$18.0 \div 4.0 = 4.5$$

This is the equivalent computation in binary notation:

$$10010 \div 100 = 100.1$$

Representing a real number in computer circuitry requires some way to separate the value's whole and fractional components (that is, the computer equivalent of a written radix point). A simple way to accomplish this is to define a storage format in which a fixed-length portion of the bit string holds the whole portion and the remainder of the bit string holds the fractional portion. Figure 3.3 shows this format with a sign bit and fixed radix point.

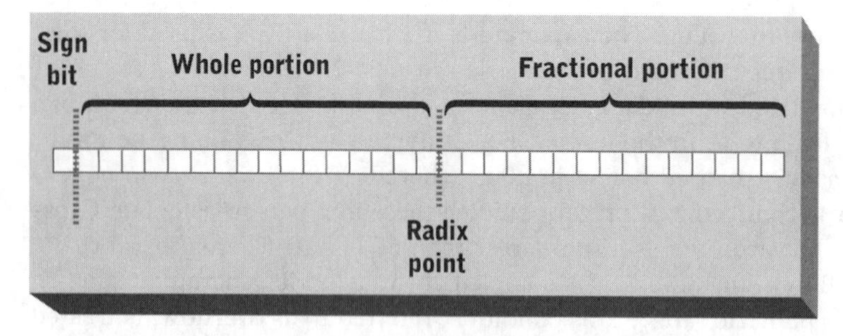

FIGURE 3.3 A 32-bit storage format for real numbers using a fixed radix point
Courtesy of Course Technology/Cengage Learning

The format in Figure 3.3 is structurally simple because of the radix point's fixed location. The advantage of this simplicity is simpler and faster CPU processing circuitry.

Unfortunately, processing efficiency is gained by limiting numeric range. Although the sample format uses 32 bits, its numeric range is substantially less than 32-bit twos complement. Only 16 bits are allocated to the whole portion of the value. Therefore, the largest possible whole value is 2^{16} - 1, or 65,535. The remaining bits store the fractional portion of the value, which can never be greater than or equal to 1.

You could increase the format's numeric range by allocating more bits to the whole portion (shifting the radix point in Figure 3.3 to the right). If the format's total size is fixed at 32 bits, however, the reallocation would reduce the number of bits used to store the fractional portion of the value. Reallocating bits from the fractional portion to the whole portion reduces the precision of fractional quantities, which reduces computational accuracy.

Floating-Point Notation

One way of dealing with the tradeoff between range and precision is to abandon the concept of a fixed radix point. To represent extremely small (precise) values, move the radix point far to the left. For example, the following value has only a single digit to the left of the radix point:

0.0000000013526473

Similarly, very large values can be represented by moving the radix point far to the right, as in this example:

1352647300000000.0

Note that both examples have the same number of digits. By floating the radix point left or right, the first example trades range of the whole portion for increased fractional precision, and the second example trades fractional precision for increased whole range. Values can be very large or very small (precise) but not both at the same time.

People tend to commit errors when manipulating long strings of digits. To minimize errors, they often write large numbers in a more compact format called scientific notation. In scientific notation, the two preceding numbers shown are represented as $13{,}526{,}473 \times 10^{-16}$ and $13{,}526{,}473 \times 10^{8}$. Note that the numbering system's base (10) is part of the multiplier. The exponent can be interpreted as the number and direction of positional moves of the radix point, as shown in Figure 3.4. Negative exponents indicate movement to the left, and positive exponents indicate movement to the right.

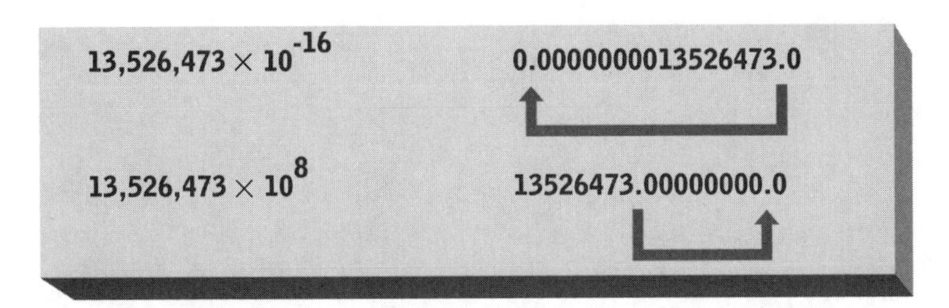

FIGURE 3.4 Conversion of scientific notation to decimal notation
Courtesy of Course Technology/Cengage Learning

Real numbers are represented in computers by using **floating-point notation**, which is similar to scientific notation except that 2 (rather than 10) is the base. A numeric value is derived from a floating-point bit string according to the following formula:

$$\text{value} = \text{mantissa} \times 2^{\text{exponent}}$$

The mantissa holds the bits that are interpreted to derive the real number's digits. By convention, the mantissa is assumed to be preceded by a radix point. The exponent value indicates the radix point's position.

Many CPU-specific implementations of floating-point notation are possible. Differences in these implementations can include the length and coding formats of the mantissa and exponent and the radix point's location in the mantissa. Although twos complement can be used to code the exponent, mantissa, or both, other coding formats might offer better design tradeoffs. Before the 1980s, there was little compatibility in floating-point format between different CPUs, which made transporting floating-point data between different computers difficult or impossible.

The Institute of Electrical and Electronics Engineers (IEEE) addressed this problem in standard 754, which defines the following formats for floating-point data:

- *binary32*—32-bit format for base 2 values
- *binary64*—64-bit format for base 2 values
- *binary128*—128-bit format for base 2 values
- *decimal64*—64-bit format for base 10 values
- *decimal128*—128-bit format for base 10 values

The binary32 and binary64 formats were specified in the standard's 1985 version and have been adopted by all computer and microprocessor manufacturers. The other three formats were defined in the 2008 version. Computer and microprocessor manufacturers are currently in the process of incorporating these formats into their products, and some products (such as the IBM POWER6 processor) already include some newer formats. For the remainder of this chapter, all references to floating-point representation refer to the binary32 format, unless otherwise specified.

Figure 3.5 shows the binary32 format. The leading sign bit applies to the mantissa, not the exponent, and is 1 if the mantissa is negative. The 8-bit exponent is coded in excess notation (meaning its first bit is a sign bit). The 23-bit mantissa is coded as an ordinary binary number. It's assumed to be preceded by a binary 1 and the radix point. This format extends the mantissa's precision to 24 bits, although only 23 are actually stored.

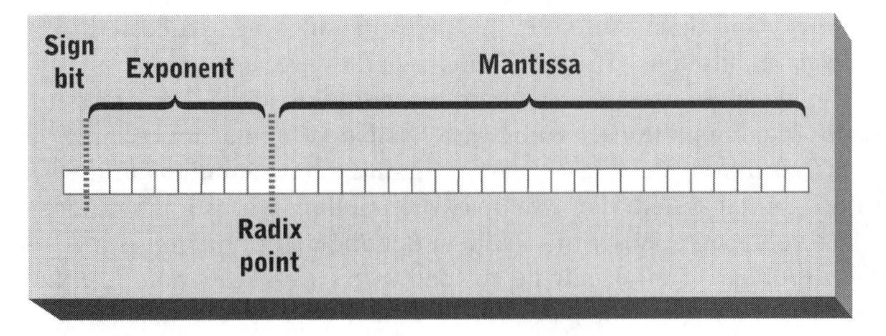

FIGURE 3.5 IEEE binary32 floating-point format
Courtesy of Course Technology/Cengage Learning

Range, Overflow, and Underflow

The number of bits in a floating-point string and the formats of the mantissa and exponent impose limits on the range of values that can be represented. The number of digits in the mantissa determines the number of significant (nonzero) digits in the largest and smallest values that can be represented. The number of digits in the exponent determines the number of possible bit positions to the right or left of the radix point.

Using the number of bits assigned to mantissa and exponent, the largest absolute value of a floating-point value appears to be the following:

$$1.11111111111111111111111 \times 2^{11111111}$$

Exponents containing all 0s and all 1s, however, represent special data values in the IEEE standards. Therefore, the usable exponent range is reduced, and the decimal range for the entire floating-point value is approximately 10^{-45} to 10^{38}.

Floating-point numbers with large absolute values have large positive exponents. When overflow occurs, it always occurs in the exponent. Floating-point representation is also subject to a related error condition called **underflow**. Very small numbers are represented by negative exponents. Underflow occurs when the absolute value of a negative exponent is too large to fit in the bits allocated to store it.

Precision and Truncation

Recall that scientific notation, including floating-point notation, trades numeric range for accuracy. Accuracy is reduced as the number of digits available to store the mantissa is reduced. The 23-bit mantissa used in the binary32 format represents approximately seven decimal digits of precision. However, many useful numbers contain more than seven nonzero decimal digits, such as the decimal equivalent of the fraction 1/3:

$$1/3 = 0.33333333 \ldots$$

The number of digits to the right of the decimal point is infinite. Only a limited number of mantissa digits are available, however.

Numbers such as 1/3 are stored in floating-point format by **truncation**. The numeric value is stored in the mantissa, starting with its most significant bit, until all available bits are used. The remaining bits are discarded. An error or approximation occurs any time a floating-point value is truncated. However, the truncated digits are insignificant compared with the significant, or large, value that's stored. Problems can result when truncated values are used as input to computations. The error introduced by truncation can be magnified when truncated values are used and generate inaccurate results. The error resulting from a long series of computations starting with truncated inputs can be large.

An added difficulty is that more values have nonterminating representations in the binary system than in the decimal system. For example, the fraction 1/10 is nonterminating in binary notation. The representation of this value in floating-point notation is a truncated value, but these problems can usually be avoided with careful programming. In general, programmers reserve binary floating-point calculations for quantities that can vary continuously over wide ranges, such as measurements made by scientific instruments.

When possible, programmers use data types other than binary32 to avoid or minimize the impact of truncation. Most current microprocessors can store and manipulate binary64 values, and support for binary128 is gradually being added. In addition, most programming languages can emulate binary128, decimal64, and decimal128 values, although processing these values is considerably slower than when the microprocessor supports them as hardware data types. Programmers seeking to minimize representation and computation errors should choose the largest floating-point format supported by hardware or software.

Monetary values are particularly sensitive to truncation errors. Most monetary systems have at least one fractional monetary unit, such as pennies—fractions of a U.S. dollar. Novice programmers sometimes assume that monetary amounts should be stored and manipulated as binary floating-point numbers. Inevitably, truncation errors caused by nonterminating representations of tenths and other fractions occur. Cumulative errors mount when truncated numbers, or approximations, are input in subsequent calculations.

One way to address the problem is to use integer arithmetic for accounting and financial applications. To do so, a programmer stores and manipulates monetary amounts in the smallest possible monetary unit—for example, U.S. pennies or Mexican pesos. Small denominations are converted to larger ones only when needed for output, such as printing a dollar amount on a check or account statement.

Although representing and manipulating monetary values as integers provides computational accuracy, this method has limitations. For example, complex formulas for computing interest on loans or investment balances include exponents and division. Intermediate calculation results for programs using these formulas can produce fractional quantities unless monetary amounts are scaled to very small units (for example, millionths of a penny).

The decimal64 and decimal128 bit formats defined in the 2008 version of IEEE standard 754 are intended to address the shortcomings of both binary floating-point and integer representation of monetary units. These formats provide accurate representation of decimal values, and the standard specifies rounding methods that can be used instead of truncation to improve computational accuracy. Both formats use the same basic approach as in binary formats—a mantissa and an exponent—but they encode three decimal digits in each 10-bit group.

Processing Complexity

The difficulty of learning to use scientific and floating-point notation is understandable. These formats are far more complex than integer data formats, and the complexity affects both people and computers. Although floating-point formats are optimized for processing efficiency, they still require complex processing circuitry. The simpler twos complement format used for integers requires much less complex circuitry.

The difference in processing circuitry complexity translates to a difference in speed of performing calculations. The magnitude of the difference depends on several factors, including the computation and the exact details of the processing circuitry. As a general rule, simple computational operations, such as addition and subtraction, take at least twice as long with floating-point numbers than with integers. The difference is even greater for operations such as division and exponentiation. For this reason and for reasons of accuracy, careful programmers never use a real number when an integer can be used, particularly for frequently updated data items.

Character Data

In their written form, English and many other languages use alphabetic letters, numerals, punctuation marks, and a variety of other special-purpose symbols, such as $ and &. Each symbol is a **character**. A sequence of characters that forms a meaningful word, phrase, or other useful group is a **string**. In most programming languages, single characters are surrounded by single quotation marks (`'c'`), and strings are surrounded by double quotation marks (`"computer"`).

Character data can't be represented or processed directly in a computer because computers are designed to process only digital data (bits). It can be represented indirectly by defining a table that assigns a numeric value to each character. For example, the integer values 0 through 9 can be used to represent the characters (numerals) `'0'` through `'9'`, the uppercase letters `'A'` through `'Z'` can be represented as the integer values 10 through 36, and so forth.

A table-based substitution of one set of symbols or values for another is one example of a coding method. All coding methods share several important characteristics, including the following:

- All users must use the same coding and decoding methods.
- The coded values must be capable of being stored or transmitted.
- A coding method represents a tradeoff among compactness, range, ease of manipulation, accuracy, and standardization.

The following sections describe some common coding methods for character data.

EBCDIC

Extended Binary Coded Decimal Interchange Code (EBCDIC) is a character-coding method developed by IBM in the 1960s and used in all IBM mainframes well into the 2000s. Recent IBM mainframes and mainframe OSs support more recent character-coding methods, but support for EBCDIC is still maintained for backward compatibility. EBCDIC characters are encoded as strings of 8 bits.

ASCII

The **American Standard Code for Information Interchange (ASCII)**, adopted in the United States in the 1970s, is a widely used coding method in data communication. The international equivalent of this coding method is **International Alphabet 5 (IA5)**, an **International Organization for Standardization (ISO)** standard. Almost all computers and OSs support ASCII, although a gradual migration is in progress to its newer relative, Unicode.

ASCII is a 7-bit format because most computers and peripheral devices transmit data in bytes and because parity checking was used widely in the 1960s to 1980s for detecting transmission errors. Chapter 8 discusses parity checking and other error detection and correction methods. For now, the important characteristic of parity checking is that it requires 1 extra bit per character. Therefore, 1 of every 8 bits isn't part of the data value, leaving only 7 bits for data representation.

The standard ASCII version used for data transfer is sometimes called ASCII-7 to emphasize its 7-bit format. This coding table has 128, or 2^7, defined characters. Computers that use 8-bit bytes are capable of representing 256, or 2^8, different characters. In most computers, the ASCII-7 characters are included in an 8-bit character coding table as the first, or lower, 128 table entries. The additional, or upper, 128 entries are defined by the computer manufacturer and typically used for graphical characters, such as line-drawing characters and multinational characters—for example, á, ñ, Ö, and Ω. This encoding method is sometimes called ASCII-8. The term is a misnomer, as it implies that the entire table (all 256 entries) is standardized. In fact, only the first 128 entries are defined by the ASCII standard. Table 3.5 shows portions of the ASCII and EBCDIC coding tables.

TABLE 3.5 Partial listing of ASCII and EBCDIC codes

Symbol	ASCII	EBCDIC
0	0110000	11110000
1	0110001	11110001
2	0110010	11110010
3	0110011	11110011
4	0110100	11110100
5	0110101	11110101
6	0110110	11110110
7	0110111	11110111
8	0111000	11111000
9	0111001	11111001
A	1000001	11000001
B	1000010	11000010
C	1000011	11000011
a	1100001	10000001
b	1100010	10000010
c	1100011	10000011

Device Control When text is printed or displayed on an output device, often it's formatted in a particular way. For example, text output to a printer is normally formatted in lines and paragraphs, and a customer record can be displayed onscreen so that it looks like a printed form. Certain text can be highlighted when printed or displayed by using methods such as underlining, bold font, or reversed background and foreground colors.

ASCII defines several device control codes (see Table 3.6) used for text formatting by sending them immediately before or after the characters they modify. Among the simpler codes are carriage return, which moves the print head or insertion point to the beginning of a line; line feed, which moves the print head or insertion point down one line; and bell, which generates a short sound, such as a beep or bell ring. In ASCII, each of these functions is assigned a numeric code and a short character name, such as CR for carriage return, LF for line feed, and BEL for bell. In addition, some ASCII device control codes are used to control data transfer. For example, ACK is sent to acknowledge correct receipt of data, and NAK is sent to indicate that an error has been detected.

NOTE

The 33 device control codes in the ASCII table occupy the first 32 entries (numbered 0 through 31) and the last entry (number 127).

TABLE 3.6 ASCII control codes

Decimal code	Control character	Description
000	NUL	Null
001	SOH	Start of heading
002	STX	Start of text
003	ETX	End of text
004	EOT	End of transmission
005	ENQ	Enquiry
006	ACK	Acknowledge
007	BEL	Bell
008	BS	Backspace
009	HT	Horizontal tabulation
010	LF	Line feed
011	VT	Vertical tabulation
012	FF	Form feed
013	CR	Carriage return
014	SO	Shift out
015	SI	Shift in

TABLE 3.6 ASCII control codes (*continued*)

Decimal code	Control character	Description
016	DLE	Data link escape
017	DC1	Device control 1
018	DC2	Device control 2
019	DC3	Device control 3
020	DC4	Device control 4
021	NAK	Negative acknowledge
022	SYN	Synchronous idle
023	ETB	End of transmission block
024	CAN	Cancel
025	EM	End of medium
026	SUB	Substitute
027	ESC	Escape
028	FS	File separator
029	GS	Group separator
030	RS	Record separator
031	US	Unit separator
127	DEL	Delete

Software and Hardware Support Because characters are usually represented in the CPU as unsigned integers, there's little or no need for special character-processing instructions. Instructions that move and copy unsigned integers behave the same whether the content being manipulated is an actual numeric value or an ASCII-encoded character. Similarly, an equality or inequality comparison instruction that works for unsigned integers also works for values representing characters.

The results of nonequality comparisons are less straightforward. The assignment of numeric codes to characters follows a specific order called a **collating sequence**. A greater-than comparison with two character inputs (for example, `'a'` less than `'z'`) returns a result based on the numeric comparison of the corresponding ASCII codes—that is, whether the numeric code for `'a'` is less than the numeric code for `'z'`. If the character set has an order and the coding method follows the order, less-than and greater-than comparisons usually produce expected results.

However, using numeric values to represent characters can produce some unexpected or unplanned results. For example, the collating sequence of letters and numerals in ASCII follows the standard alphabetic order for letters and numeric order for numerals, but uppercase and lowercase letters are represented by different codes. As a result, an equality comparison between uppercase and lowercase versions of the same letter returns false because the numeric codes aren't identical. For example, `'a'` doesn't equal `'A'`, as shown previously in Table 3.5. Punctuation symbols also have a specific order in the collating sequence, although there's no widely accepted ordering for them.

ASCII Limitations ASCII's designers couldn't foresee the code's long lifetime (almost 50 years) or the revolutions in I/O device technologies that would take place. They never envisioned modern I/O device characteristics, such as color, bitmapped graphics, and selectable fonts. Unfortunately, ASCII doesn't have the range to define enough control codes to account for all the formatting and display capabilities in modern I/O devices.

ASCII is also an English-based coding method. This isn't surprising, given that when it was defined, the United States accounted for most computer use and almost all computer production. ASCII has a heavy bias toward Western languages in general and American English in particular, which became a major limitation as computer use and production proliferated worldwide.

Recall that 7-bit ASCII has only 128 table entries, 33 of which are used for device control. Only 95 printable characters can be represented, which are enough for a usable subset of the characters commonly used in American English text. This subset, however, doesn't include any modified Latin characters, such as ç and á, or those from other alphabets, such as Σ.

The ISO partially addressed this problem by defining many different 256-entry tables based on the ASCII model. One, called **Latin-1**, contains the ASCII-7 characters in the lower 128 table entries and most of the characters used by Western European languages in the upper 128 table entries. The upper 128 entries are sometimes called **multinational characters**. The number of available character codes in a 256-entry table, however, is still much too small to represent the full range of printable characters in world languages.

Further complicating matters is that some printed languages aren't based on characters in the Western sense. Chinese, Japanese, and Korean written text consists of ideographs, which are pictorial representations of words or concepts. Ideographs are composed of graphical elements, sometimes called strokes, that number in the thousands. Other written languages, such as Arabic, present similar, although less severe, coding problems.

Unicode

The Unicode Consortium (*www.unicode.org*) was founded in 1991 to develop a multilingual character-encoding standard encompassing all written languages. The original members were Apple Computer Corporation and Xerox Corporation, but many computer companies soon joined. This effort has enabled software and data to cross international boundaries. Major Unicode standard releases are coordinated with ISO standard 10646. As of this writing, the latest standard is Unicode 5.2, published in October 2009.

Like ASCII, **Unicode** is a coding table that assigns nonnegative integers to represent printable characters. The ISO Latin-1 standard, which includes ASCII-7, is incorporated into Unicode as the first 256 table entries. Therefore, ASCII is a subset of Unicode. An important difference between ASCII and Unicode is the size of the coding table. Early versions of Unicode used 16-bit code, which provided 65,536 table entries numbered 0 through 65,535. As development efforts proceeded, the number of characters exceeded the capacity of a 16-bit code. Later Unicode versions use 16-bit or 32-bit codes, and the standard currently encompasses more than 100,000 characters.

The additional table entries are used primarily for characters, strokes, and ideographs of languages other than English and its Western European siblings. Unicode includes many other alphabets, such as Arabic, Cyrillic, and Hebrew, and thousands of Chinese,

Japanese, and Korean ideographs and characters. Some extensions to ASCII device control codes are also provided. As currently defined, Unicode can represent written text from all modern languages. Approximately 6000 characters are reserved for private use.

Unicode is widely supported in modern software, including most OSs and word-processing applications. Because most CPUs support characters encoded in 32-bit unsigned integers, there's no need to upgrade processor hardware to support Unicode. The main impact of Unicode on hardware is in storage and I/O devices. Because the numeric code's size is quadruple that of ASCII, pure text files are four times as large. This increased size seems to be a problem at first glance, but the impact is reduced with custom word-processing file formats and the ever-declining cost of secondary storage. In addition, most documents aren't stored as ASCII or Unicode text files. Instead, they're stored in a format that intermixes text and formatting commands. Because formatting commands also occupy file space, the file size increase caused by switching from ASCII to Unicode is generally less than implied by quadrupling per-character storage.

Before Unicode, devices designed for character I/O used ASCII by default and vendor-specific methods or older ISO standards to process character sets other than Latin-1. The typical method was to maintain an internal set of alternative coding tables, with each table containing a different alphabet or character set. Device-specific commands switched from one table to another. Unicode unifies and standardizes the content of these tables to provide a standardized method for processing international character sets. This standard has been widely adopted by I/O device manufacturers. In addition, backward compatibility with ASCII is ensured because Unicode includes ASCII.

Boolean Data

The **Boolean data type** has only two data values—true and false. Most people don't think of Boolean values as data items, but the primitive nature of CPU processing requires the capability to store and manipulate Boolean values. Recall that processing circuitry physically transforms input signals into output signals. If the input signals represent numbers and the processor is performing a computational function, such as addition, the output signal represents the numerical result.

When the processing function is a comparison operation, such as greater than or equal to, the output signal represents a Boolean result of true or false. This result is stored in a register (as is any other processing result) and can be used by other instructions as input (for example, a conditional or an unconditional branch in a program). The Boolean data type is potentially the most concise coding format because only a single bit is required. For example, binary 1 can represent true, and 0 can represent false. To simplify processor design and implementation, most CPU designers seek to minimize the number of different coding formats used. CPUs generally use an integer coding format for integers to represent Boolean values. When coded in this manner, the integer value zero corresponds to false, and all nonzero values correspond to true.

To conserve memory and storage, sometimes programmers "pack" many Boolean values into a single integer by using each bit to represent a separate Boolean value. Although this method conserves memory, generally it slows program execution because of the complicated instructions required to extract and interpret separate bits from an integer data item.

Memory Addresses

As described in Chapter 2, primary storage is a series of contiguous bytes of storage. Conceptually, each memory byte has a unique identifying or serial number. The identifying numbers are called addresses, and their values start with zero and continue in sequence (with no gaps) until the last byte of memory is addressed.

In some CPUs, this conceptual view is also the physical method of identifying and accessing memory locations. That is, memory bytes are identified by a series of nonnegative integers. This approach to assigning memory addresses is called a **flat memory model**. In CPUs using a flat memory model, using twos complement or unsigned binary as the coding format for memory addresses is logical and typical. The advantage of this approach is that it minimizes the number of different data types and the complexity of processor circuitry.

To maintain backward compatibility with earlier generations of CPUs that don't use a flat memory model, some CPU designers choose not to use simple integers as memory addresses. For example, Intel Core CPUs maintain backward compatibility with the 8086 CPU used in the first IBM PC. Earlier generations of processors typically used a different approach to memory addressing called a **segmented memory model**, in which primary storage is divided into equal-sized (for example, 64 KB) segments called pages. Pages are identified by sequential nonnegative integers. Each byte in a page is also identified by a sequentially assigned nonnegative integer. Each byte of memory has a two-part address: The first part identifies the page, and the second part identifies the byte in the page.

Because a segmented memory address contains two integer values, the data-coding method for single signed or unsigned integers can't be used. Instead, a specific coding format for memory addresses must be defined and used. For this reason, CPUs with segmented memory models have an extra data type, or coding format, for storing memory addresses.

TECHNOLOGY FOCUS

Intel Memory Address Formats

Intel microprocessors have been used in PCs since 1981. The original IBM PC used an Intel 8086 microprocessor. All later generations of Intel processors (the 80x86, Pentium, and Core processor families) have maintained backward compatibility with the 8086 and can execute CPU instructions designed for the 8086. Backward compatibility has been considered an essential ingredient in the success of PCs based on Intel CPUs.

Early Intel microprocessors were designed and implemented with speed of processor operation as a primary goal. Because a general-purpose CPU uses memory addresses in nearly every instruction execution cycle, making memory access as efficient as possible is important. Processor complexity rises with the number of bits that must be processed simultaneously. Large coding formats for memory addresses and other data types require complicated and slow processing circuitry. Intel designers wanted to minimize the complexity of processor circuitry associated with memory addresses. They also needed to balance the desire for efficient memory processing with the need to maintain a large

(*continued*)

range of possible memory addresses. To achieve this balance, they made two major design decisions for the 8086:

- A 20-bit size for memory addresses
- A segmented memory model

The 20-bit address format limited usable (addressable) memory to 1 MB (2^{20} addressable bytes). This limitation didn't seem like a major one because few computers (including most mainframes) had that much memory at the time. The 20-bit format is divided into two parts: a 4-bit segment identifier and a 16-bit segment offset. The 16-bit segment offset identifies a specific byte in a 64 (2^{16}) KB memory segment. Because 4 bits are used to represent the segment, there are 16, or 2^4, possible segments.

Intel designers anticipated that most programs would fit in a single 64 KB memory segment. Further, they knew that 16-bit memory addresses could be processed more efficiently than larger addresses. Therefore, they defined two types of address-processing functions—those using the 4-bit segment portion of the address and those ignoring it. When the segment identifier was used, memory access was slow, but it was much faster when the processor ignored the segment identifier. Intel designers assumed that most memory accesses wouldn't require the segment identifier.

Both the 64 KB segment size and the 1 MB memory limit soon became significant constraints. Although early programs for the IBM PC were generally smaller than 64 KB, later versions of these programs were larger. Because the processor design imposed a performance penalty when the segment identifier was used, later versions ran more slowly than earlier versions did. In addition, both the OS and the computer hardware were becoming more complex and consuming more memory resources. As a result, computer designers, OS developers, and users chafed under the 1 MB memory limit.

Intel designers first addressed these constraints in the 80286 by increasing the segment identifier to 8 bits, thus increasing total addressable memory to 16 MB. The 80386 went a step further by providing two methods of addressing. The first, called real mode, was compatible with the 8086's addressing method. The second, called protected mode, was a new method based on 32-bit memory addresses. Protected-mode addressing eliminated the performance penalty for programs larger than 64 KB. Pentium and Core microprocessors continue to use both addressing methods.

The 32-bit protected-mode addresses are adequate for physical memory up to 4 GB. By the late 1990s and 2000s, larger computer classes were reaching this limit, and PCs and workstations were expected to follow in a decade. Intel expanded protected-mode memory addresses to 64 bits beginning in the mid-2000s. The change was far less disruptive than earlier changes because addresses of both sizes use the flat memory model. Nonetheless, users have encountered some bumps in the road as software manufacturers shifted from 32-bit to 64-bit addressing.

This discussion, and the earlier discussion of ASCII, illustrates a few pitfalls for computer designers in choosing data representation methods and formats. A fundamental characteristic of any data representation method is that it involves balancing several variables. For Intel memory addresses, designers had to balance processor cost, speed of program execution, and memory requirements of typical PC software. The balance might have been optimal, or at least reasonable, given the state of these variables in 1981, but neither time nor technology stands still. The outcome of any CPU design decision, including data representation formats, is affected quickly as technology changes. Attempts to maintain backward compatibility with previous CPU designs can be difficult and expensive and might severely limit future design choices.

DATA STRUCTURES

The previous sections outlined the data types that a CPU can manipulate directly: integer, real number, character, Boolean, and memory address. The data types a CPU supports are sometimes called **primitive data types** or **machine data types**. Computers can also process more complex data, such as strings, arrays, text files, databases, and digital representations of audio and image data, such as MP3, JPEG, and MPEG files. Chapter 7 discusses audio and image data representations and related hardware devices. The remainder of this chapter concentrates on other complex data formats commonly manipulated by system and application software.

If only primitive data types were available, developing programs of any kind would be difficult. Most application programs need to combine primitive data items to form useful aggregations. A common example is a character or text string. Most application programs define and use character strings, but few CPUs provide instructions that manipulate them directly. Application development is simplified if character strings can be defined and manipulated (that is, read, written, and compared) as a single unit instead of one character at a time.

A **data structure** is a related group of primitive data elements organized for some type of common processing and is defined and manipulated in software. Computer hardware can't manipulate data structures directly; it must deal with them in terms of their primitive components, such as integers, floating-point numbers, single characters, and so on. Software must translate operations on data structures into equivalent machine instructions that operate on each primitive data element. For example, Figure 3.6 shows a comparison of two strings decomposed into comparison operations on each character.

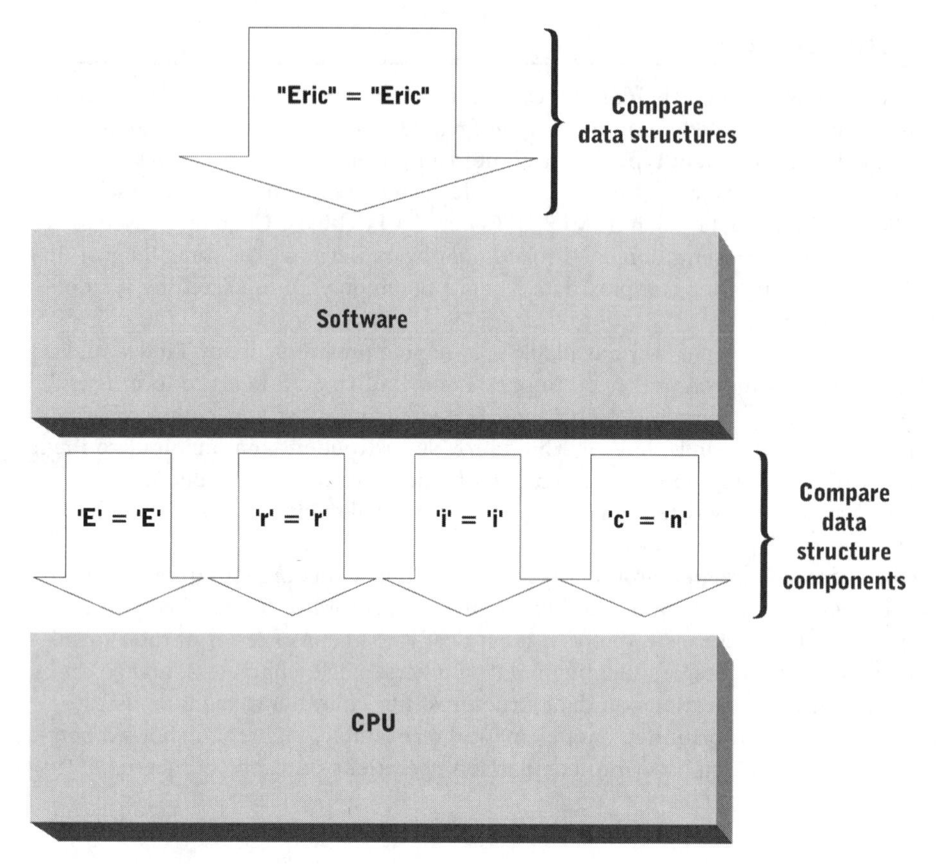

FIGURE 3.6 Software decomposes operations on data structures into operations on primitive data elements

Courtesy of Course Technology/Cengage Learning

The complexity of data structures is limited only by programmers' imagination and skill. As a practical matter, certain data structures are useful in a wide variety of situations and are used commonly, such as character strings or arrays, records, and files. System software often provides application services to manipulate these commonly used data structures. For example, an OS normally provides services for reading from and writing to files.

Other data structures are less commonly supported by system software. Examples include numeric arrays, indexed files, and complex database structures. Indexed files are supported in some, but not all, OSs. Numeric arrays are supported in most programming languages but not in most OSs. Database structures are normally supported by a database management system. Most programming languages support direct manipulation of character strings.

Data structures have an important role in system software development. For example, OSs commonly use linked lists to keep track of memory blocks allocated to programs and disk blocks allocated to files and directories. Indexes are widely used in database management systems to speed search and retrieval operations. Programmers who develop system software generally take an entire course in data structures and efficient methods for manipulating them.

Pointers and Addresses

Whether implemented in system or application software, almost all data structures make extensive use of pointers and addresses. A **pointer** is a data element containing the address of another data element. An **address** is the location of a data element in a storage device. Addresses vary in content and representation, depending on the storage device being addressed. Secondary storage devices are normally organized as a sequence of data blocks. A block is a group of bytes read or written as a unit. For example, disk drives usually read and write data in 512-byte blocks. For block-oriented storage devices, an address can be represented as an integer containing the block's sequential position. Integers can also be used to represent the address of a single byte in a block.

As discussed in the previous Technology Focus, memory addresses can be complex if a segmented memory model is used. For the purpose of discussing data structures, a flat memory model is used, and memory addresses are represented by nonnegative integers.

Arrays and Lists

Many types of data can be grouped into lists. A list is a set of related data values. In mathematics, a list is considered unordered, so no specific list element is designated as first, second, last, and so on. When writing software, a programmer usually prefers to impose some ordering on the list. For example, a list of days of the week can be ordered sequentially, starting with Monday.

An **array** is an ordered list in which each element can be referenced by an index to its position. Figure 3.7 shows an example of an array for the first five letters of the English alphabet. Note that the index values are numbered starting at 0, a common (although not universal) practice in programming. Although the index values are shown in Figure 3.7, they aren't stored. Instead, they're inferred from the data value's location in the storage allocated to the array. In a high-level programming language, array elements are normally referenced by the array name and the index value. For example, the third letter of the alphabet stored in an array might be referenced as follows:

alphabet[2]

In this example, "alphabet" is the array name, and 2 is the index value (with numbering starting from 0).

Index	Value
0	A
1	B
2	C
3	D
4	E

FIGURE 3.7 Array elements in contiguous storage locations
Courtesy of Course Technology/Cengage Learning

Figure 3.8 shows a character array, or string, stored sequentially in contiguous memory locations. In this example, each character of the name John Doe is stored in a single byte of memory, and the characters are ordered in sequential byte locations starting at byte 1000. An equivalent organization can be used to store the name on a secondary storage device. The address of an array element can be calculated with the starting address of the array and the element's index. For example, if you want to retrieve the third character in the array, you can compute its address as the sum of the first element's address plus the index value, assuming index values start at 0.

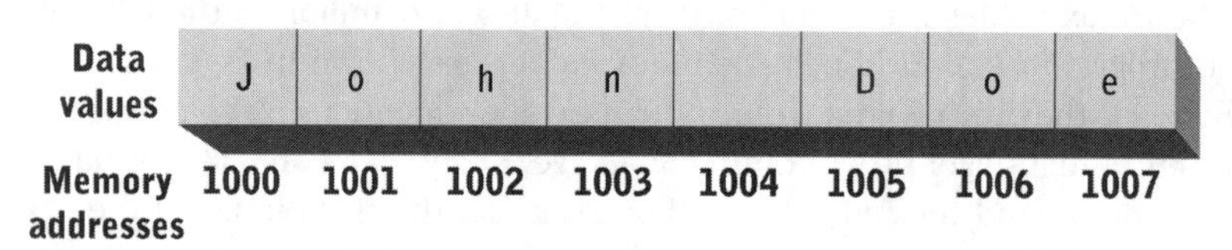

FIGURE 3.8 A character array in contiguous storage locations
Courtesy of Course Technology/Cengage Learning

Using contiguous storage locations, especially in secondary storage devices, complicates the allocation of storage locations. For example, adding new elements to the end of an array might be difficult if these storage locations are already allocated to other data items. For this reason, contiguous storage allocation is generally used only for fixed-length arrays.

For a variety of reasons, you might need to store array elements in widely dispersed storage locations. A **linked list** is a data structure that uses pointers so that list elements can be scattered among nonsequential storage locations. Figure 3.9 shows a generic example of a **singly linked list**. Each list element occupies two storage locations: The first holds the list element's data value, and the second holds the address of the next list element as a pointer.

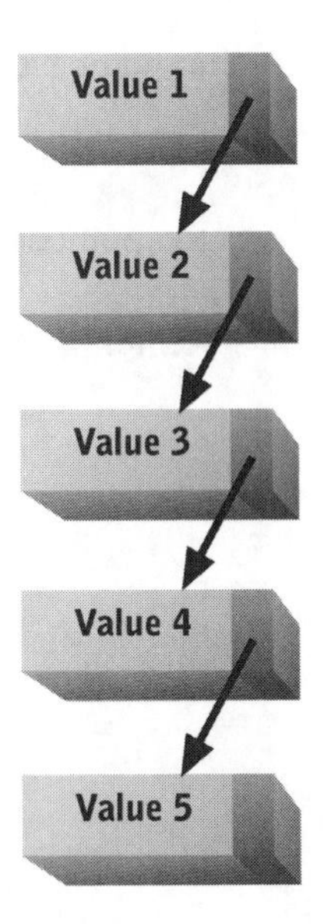

FIGURE 3.9 Value and pointer fields of a singly linked list
Courtesy of Course Technology/Cengage Learning

Figure 3.10 shows a character string stored in a linked list. Note that characters are scattered among nonsequential, or noncontiguous, storage locations.

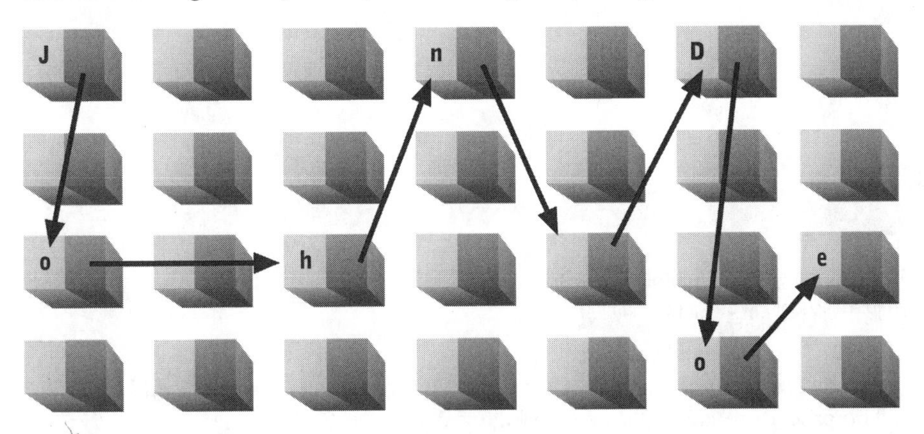

FIGURE 3.10 A character array stored in noncontiguous memory locations, with pointers connecting the array elements
Courtesy of Course Technology/Cengage Learning

More storage locations are required for a linked list than for an array with equivalent content because both data and pointers must be stored. Using pointers also complicates the task of locating array elements. References to specific array elements must be resolved by following the chain of pointers, starting with the first array element. This method can be inefficient if the array contains a large number of elements. For example, accessing the 1000th element means reading and following pointers in the first 999 elements.

Linked lists are easier to expand or shrink than arrays are. The procedure for adding a new element is as follows (see Figure 3.11):

1. Allocate storage for the new element.
2. Copy the pointer from the element preceding the new element to the pointer field of the new element.
3. Write the address of the new element in the preceding element's pointer field.

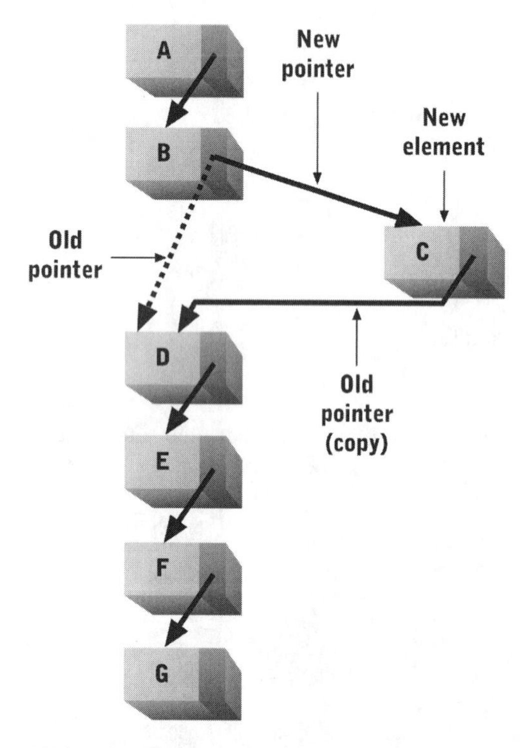

FIGURE 3.11 Inserting a new element in a singly linked list
Courtesy of Course Technology/Cengage Learning

In contrast, inserting an element in a list stored in contiguous memory can be time consuming. The procedure is as follows (see Figure 3.12):

1. Allocate a new storage location at the end of the list.
2. For each element past the insertion point, copy its value to the next storage location, starting with the last element and working backward to the insertion point.
3. Write the new element's value in the storage location at the insertion point.

Inserting an element near the beginning of the array is highly inefficient because of the many required copy operations.

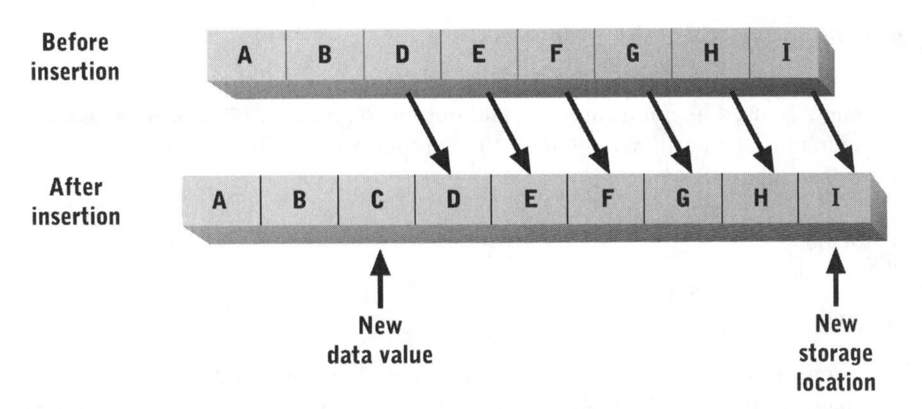

FIGURE 3.12 Inserting a new element in an array stored in contiguous memory locations
Courtesy of Course Technology/Cengage Learning

Figure 3.13 depicts a more complicated linked list called a **doubly linked list**. Each element of a doubly linked list has two pointers: one pointing to the next element in the list and one pointing to the previous element in the list. The main advantage of doubly linked lists is that they can be traversed in either direction with equal efficiency. The disadvantages are that more pointers must be updated each time an element is inserted into or deleted from the list, and more storage locations are required to hold the extra pointers.

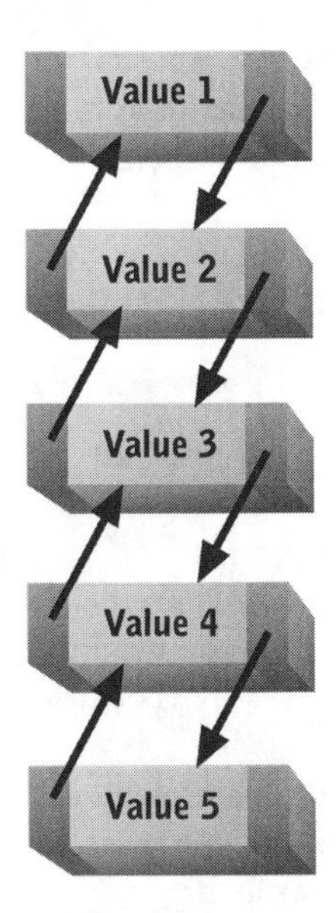

FIGURE 3.13 A doubly linked list
Courtesy of Course Technology/Cengage Learning

Records and Files

A **record** is a data structure composed of other data structures or primitive data elements. Records are commonly used as a unit of input and output to and from files or databases. For example, the following data items might be the contents of the data structure for a customer record (see Figure 3.14):

Account-Number	Street-Address
Last-Name	City
First-Name	State
Middle-Initial	Zip-Code

Each component of the record is a basic data element (for example, Middle-Initial) or another data structure (for example, a character array for Street-Address). To speed input and output, records are usually stored in contiguous storage locations, which restricts the record's array components to a fixed length.

Account-Number	Last-Name	First-Name	Middle-Initial	Street-Address	City	State	Zip-Code

FIGURE 3.14 A record data structure
Courtesy of Course Technology/Cengage Learning

A sequence of records on secondary storage is called a **file**. A sequence of records stored in main memory is normally called a table, although its structure is essentially the same as a file. Files can be organized in many different ways, the most common being sequential and indexed.

In a sequential file, records are stored in contiguous storage locations. As with arrays stored in contiguous storage, accessing a specific record is simple. The address of the nth record in a file can be computed as follows:

$$\text{address-of-first-record} + ((n - 1) \times \text{record-size})$$

If the first byte of the first record is at address 1 and the record size is 200 bytes, the address of the fourth record is 601.

Sequential files suffer the same problems as contiguous arrays when records are being inserted and deleted. A copy procedure similar to the one in Figure 3.12 must be used to add a record to a file. The procedure is even less efficient for files than for arrays because of the large size of the records that must be copied.

One method of solving this problem is to use linked lists. With files, the data elements of a linked list are entire records instead of basic data elements. The methods for searching, inserting records, and deleting records are essentially the same as described previously.

Another method of organizing files uses an **index**, an array of pointers to records. The pointers can be ordered in any sequence you need. For example, a file of customer records can be ordered by ascending account number, as shown in Figure 3.15.

FIGURE 3.15 An indexed file
Courtesy of Course Technology/Cengage Learning

The advantage of using an index lies in the efficiency of record insertion, deletion, and retrieval. When a record is added to a file, storage is allocated for the record, and data is placed in the storage location. The index is then updated by inserting the new record's address. An index update follows the same procedure as an array update. Because the array contains only pointers, it's small and fast to update.

Classes and Objects

Up to this point, data and programs have been discussed as two fundamentally different things. Programs contain instructions, which transform data inputs into data outputs when executed. Data items are held in storage devices, moved to CPU registers when needed, transformed into data outputs by executing instructions, and sent to output or storage devices again. In this view of computer systems and software behavior, data items are mainly static, and programs are the active means for transforming and updating data items.

In the 1980s and 1990s, computer researchers developed a different view of computer and software behavior: combining program instructions and data into a single data structure. A **class** is a data structure containing both traditional (static) data elements and programs that manipulate the data. The programs in a class are called **methods**. A class combines related data items in much the same way a record does, but it extends the record to include methods for manipulating data items.

Think of the customer record in Figure 3.14 as a starting point for defining a Customer class. The record contains some primitive data items and data structures describing features of a customer. (Others, such as account balance, could be added to create a more complete representation of a customer.)

To turn the customer record into a Customer class, methods for manipulating or modifying data elements in the record must be added. For example, you can add a simple method for updating a data element's content called Update-First-Name. You can check for legal values, such as making sure the value supplied for Zip-Code is a valid U.S. zip code in the customer's state. Functions, such as Print-Mailing-Label, or methods for modifying the customer's account balance to reflect transactions, such as Apply-Payment, can be added. Figure 3.16 shows some possible data elements and methods for the Customer class.

<table>
<tr><th>Account-Number</th><th>Last-Name</th><th>First-Name</th><th>Middle-Initial</th><th>Street-Address</th><th>City</th><th>State</th><th>Zip-Code</th><th>Account-Balance</th></tr>
<tr><td colspan="5">Update-Account-Number</td><td colspan="4">Update-Zip-Code</td></tr>
<tr><td colspan="5">Update-Last-Name</td><td colspan="4">Validate-Zip-Code</td></tr>
<tr><td colspan="5">Update-First-Name</td><td colspan="4">Apply-Purchase</td></tr>
<tr><td colspan="5">Update-Middle-Initial</td><td colspan="4">Apply-Payment</td></tr>
<tr><td colspan="5">Update-Street-Address</td><td colspan="4">Apply-Credit</td></tr>
<tr><td colspan="5">Update-City</td><td colspan="4">Print-Mailing-Label</td></tr>
<tr><td colspan="5">Update-State</td><td colspan="4">Print-Account-Statement</td></tr>
</table>

FIGURE 3.16 A Customer class containing traditional data elements (darker boxes) and methods (lighter boxes)

Courtesy of Course Technology/Cengage Learning

An **object** is one instance, or variable, of a class. Each person who's a customer is represented by one variable or object of the Customer class. Each object can be stored in a storage device, and each object's data elements occupy a separate part of the storage device.

Viewed as just another data structure, an object differs little from a record. Much as the data element Street-Address can be represented with a character array, a method can be represented as an array of CPU instructions. Methods can be represented in other ways, including linked lists of instructions or pointers to files containing sequential sets of instructions. In essence, the only new primitive data type needed to represent a method in an object is an instruction. As you see in Chapter 4, instructions, like data, have specific representation and storage formats.

Summary

- Data can be represented in many ways. To be processed by any device, data must be converted from its native format into a form suitable for the processing device. Modern computers represent data as electronic signals and implement processing devices by using electronic circuitry. Electronic processing devices exploit the physical laws of electricity. Because the laws of electricity can be stated as mathematical equations, electronic devices can perform the computational functions embedded in these equations. Other ways of implementing processors, such as mechanics and optics, use similar mathematically stated physical laws.
- All data, including nonnumeric data, are represented in modern computers as strings of binary digits, or bits. Bits are used because 0 and 1 can be encoded in clearly recognizable electrical states, such as high and low voltage. Binary electrical states can be processed and stored by reliable, cheap electrical devices. A handful of primitive data types are represented and processed by a CPU, including integers, real numbers, characters, Boolean values, and memory addresses.
- Numeric values other than 0 and 1 are represented by combining multiple bits into larger groups, called bit strings, much as the decimal digits 2, 5, and 9 can be combined to form the larger value 592. Each bit string has a specific data format and coding method. There are many different data formats and coding methods. CPU designers choose formats and methods that represent the best balance among compactness, range, ease of manipulation, accuracy, and standardization. The optimal balance can vary, depending on the type of data and its intended uses.
- Integers have no fractional component and can use simple data formats. Twos complement is the most common integer format, although excess notation is used sometimes. Real numbers have both whole and fractional components. Their complex structure requires a complex data format called floating-point notation that represents a numeric value as a mantissa multiplied by a positive or negative power of 2. A value can have many digits of precision in floating-point notation for very large or very small magnitudes but not both large and small magnitudes at the same time. Floating-point formats used in modern computers typically follow IEEE standards, which include binary and decimal representations ranging from 32 to 128 bits in length. Floating-point formats are less accurate and more difficult to process than twos complement format.
- Character data isn't inherently numeric. Characters are converted to numbers by means of a coding table, in which each character occupies a sequential position. A character is represented by converting it to its integer table position. Characters are extracted from integers by the reverse process. Many different tables are possible. Most computers use the ASCII or Unicode coding tables. ASCII is an older standard geared toward the English language. Unicode is a more recent standard with a character set large enough to encompass ASCII and all written languages.
- A Boolean data value must be true or false. Memory addresses can be simple or complex numeric values, depending on whether the CPU uses a flat or segmented memory model. Flat memory addresses can be represented as a single integer. Segmented memory

addresses require multiple integers. Many CPUs also provide double-precision numeric data types, which double the number of bits used to store a value.

- Programs often define and manipulate data in larger and more complex units than primitive CPU data types. A data structure is a related group of primitive data elements organized for some type of common processing and is defined in software. To enable a CPU to manipulate a data structure, software decomposes operations on the data structure into equivalent operations on its primitive components. Commonly used data structures include arrays, linked lists, records, tables, files, indexes, and objects. Many data structures use pointers, which are stored memory addresses, to link primitive data components.

In this chapter, you've learned the different ways data is represented in modern computers, focusing particularly on the relationship between data representation and the CPU. This chapter and the preceding chapters have introduced basic concepts and terminology of systems architecture. The understanding you've gained in the first three chapters lays the foundation for the more detailed discussion of the hardware implementations of data processing, storage, and communication in upcoming chapters.

Key Terms

address
American Standard Code for Information Interchange (ASCII)
array
base
binary number
bit
bit string
Boolean data type
Boolean logic
byte
character
class
collating sequence
data structure
decimal point
double-precision
doubly linked list
excess notation
Extended Binary Coded Decimal Interchange Code (EBCDIC)
file
flat memory model
floating-point notation
hexadecimal notation
high-order bit
index
integer
International Alphabet 5 (IA5)
International Organization for Standardization (ISO)
Latin-1
least significant digit
linked list
long integer
low-order bit
machine data type
manipulation
method
most significant digit
multinational character
numeric range
object
octal notation
overflow
pointer
primitive data type
radix

radix point
real number
record
segmented memory model
signed integer
singly linked list
string
truncation
twos complement notation
underflow
Unicode
unsigned integer

Vocabulary Exercises

1. An element in a(n) ________ contains pointers to both the next and previous list elements.
2. ________ notation encodes a real number as a mantissa multiplied by a power (exponent) of 2.
3. A(n) ________ is an integer stored in double the normal number of bit positions.
4. Increasing a numeric representation format's size (number of bits) increases the ________ of values that can be represented.
5. Assembly (machine) language programs for most computers use ________ notation to represent memory address values.
6. A(n) ________ is a data item composed of multiple primitive data items.
7. In older IBM mainframe computers, characters were encoded according to the ________ coding scheme.
8. A(n) ________ is the address of another data item or structure.
9. In a positional numbering system, the ________ separates digits representing whole number quantities from digits representing fractional quantities.
10. A(n) ________ is an array of characters.
11. Most Intel CPUs use the ________, in which each memory address is represented by two integers.
12. A set of data items that can be accessed in a specified order by using pointers is called a(n) ________.
13. A(n) ________ contains 8 ________.
14. A(n) ________ list stores one pointer with each list element.
15. The result of adding, subtracting, or multiplying two integers might result in overflow but never ________ or ________.
16. A(n) ________ is a sequence of primitive data elements stored in sequential storage locations.
17. A(n) ________ is a data structure composed of other data structures or primitive data elements, commonly used as a unit of input and output to and from files or databases.
18. A(n) ________ data item can contain only the values true or false.
19. A(n) ________ is an array of data items, each of which contains a key value and a pointer to another data item.

20. Many computers implement ________ numeric data types to increase accuracy and prevent overflow and underflow.
21. Unlike ASCII and EBCDIC, ________ is a 16-bit or 32-bit character coding table.

22. The ________ is the bit of lowest magnitude in a byte or bit string.
23. ________ occurs when the result of an arithmetic operation exceeds the number of bits available to store it.
24. In a CPU, ________ arithmetic generally is easier to implement than ________ arithmetic because of a simpler data coding scheme and data manipulation circuitry.
25. In the ________, memory addresses consist of a single integer.
26. The ________ has defined a character-coding table called ________, which combines the ASCII-7 coding table with an additional 128 Western European multinational characters.
27. Data represented in ________ is transmitted accurately between computer equipment from different manufacturers if each computer's CPU represents real numbers by using an IEEE standard notation.
28. The ordering of characters in a coding table is called a(n) ________.
29. A(n) ________ is a data structure containing both static data and methods.
30. A(n) ________ is one instance or variable of a class.

Review Questions

1. What are the binary, octal, and hexadecimal representations of the decimal number 10?
2. Why is binary data representation and signaling the preferred method of computer hardware implementation?
3. What is excess notation? What is twos complement notation? Why are they needed? In other words, why can't integer values be represented by ordinary binary numbers?
4. What is the numeric range of a 16-bit twos complement value? A 16-bit excess notation value? A 16-bit unsigned binary value?
5. What is overflow? What is underflow? How can the probability of their occurrence be minimized?
6. Why are real numbers more difficult to represent and process than integers?
7. Why might a programmer choose to represent a data item in IEEE binary128 floating-point format instead of IEEE binary64 floating-point format? What additional costs might be incurred at runtime (when the application program executes) as a result of using the 128-bit instead of the 64-bit format?
8. Why doesn't a CPU evaluate the expression `'A' = 'a'` as true?
9. What are the differences between ASCII and Unicode?
10. What primitive data types can normally be represented and processed by a CPU?
11. What is a data structure? List several types of common data structures.
12. What is an address? What is a pointer? What purpose are they used for?

13. How is an array stored in main memory? How is a linked list stored in main memory? What are their comparative advantages and disadvantages? Give examples of data that would be best stored as an array and as a linked list.
14. How does a class differ from other data structures?

Problems and Exercises

1. Develop an algorithm or program to implement the following function:

   `insert_in_linked_list (element,after_pointer)`

   The `element` parameter is a data item to be added to a linked list, and the `after_pointer` parameter is the address of the element after which the new element will be inserted.
2. Develop an algorithm or program to implement the following function:

   `insert_in_array (element,position)`

   The `element` parameter is a data item to be added to the array, and the `position` parameter is the array index at which the new element will be inserted. Make sure to account for elements that must be moved over to make room for the new element.
3. Consider the following binary value:

   `1000 0000 0010 0110 0000 0110 1101 1001`

   What number (base 10) is represented if the value is assumed to represent a number stored in twos complement notation? Excess notation? IEEE binary32 floating-point notation?
4. How are the values 515_{10} and -515_{10} represented as ordinary binary numbers? How are they represented as octal and hexadecimal numbers? How are they represented in 16-bit excess notation and in 16-bit twos complement notation?
5. How is the binary value $101101 \times 2^{-101101}$ represented in IEEE binary32 floating-point notation?
6. Convert the following values represented in base 12 to their equivalent representations in base 2, base 5, and base 10:

   $+1A78_{12}$

   $-90B2_{12}$

Research Problems

1. Choose a commonly used microprocessor, such as the Intel Core (*www.intel.com*) or IBM POWER6 (*www.ibm.com*). What data types are supported? How many bits are used to store each data type? How is each data type represented internally?
2. Most personal and office productivity applications, such as word-processing applications, are marketed internationally. To minimize development costs, software producers develop a single version of the program but have separate configurations for items such as menus and error messages that vary from language to language. Investigate a commonly used development tool for applications (called an "integrated development environment"), such

as Microsoft Visual Studio (*www.microsoft.com*). What tools and techniques (for example, Unicode data types and string tables) are supported for developing multilingual programs?

3. Object-oriented programming has been adopted widely because of its capability to reuse code. Most application development software provides class libraries and extensive support for complex data structures, including linked lists. Investigate one of these libraries, such as the Microsoft Foundation Class (*www.microsoft.com*) or the Java 2 Platform (*http://java.sun.com*) application programming interface (API). What data structures are supported by the library? What types of data are recommended for use with each data structure object? Which classes contain which data structures, and what methods does the library provide?

CHAPTER 4

PROCESSOR TECHNOLOGY AND ARCHITECTURE

CHAPTER GOALS

- Describe CPU instruction and execution cycles
- Explain how primitive CPU instructions are combined to form complex processing operations
- Describe key CPU design features, including instruction format, word size, and clock rate
- Describe the function of general-purpose and special-purpose registers
- Explain methods of enhancing processor performance
- Describe the principles and limitations of semiconductor-based microprocessors
- Summarize future processing trends

Chapter 2 gave a brief overview of computer processing, including the function of a processor, general-purpose and special-purpose processors, and the components of a central processing unit (CPU). This chapter explores CPU operation, instructions, components, and implementation (see Figure 4.1). It also gives you an overview of future trends in processor technology and architecture.

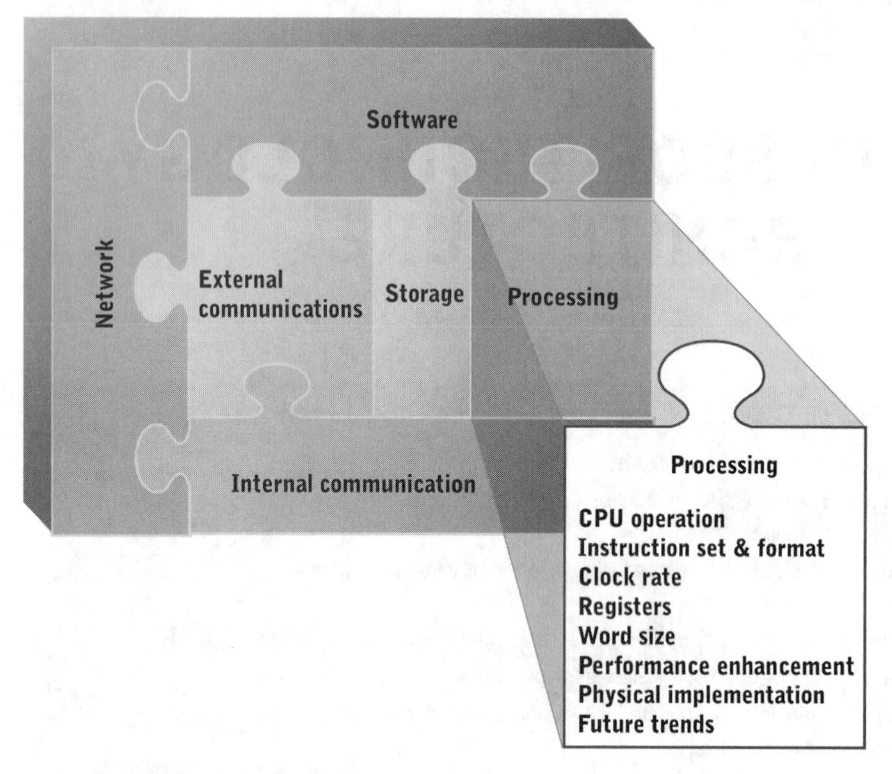

FIGURE 4.1 Topics covered in this chapter
Courtesy of Course Technology/Cengage Learning

CPU OPERATION

Recall from Chapter 2 that a CPU has three primary components: the control unit, the arithmetic logic unit (ALU), and a set of registers (see Figure 4.2). The control unit moves data and instructions between main memory and registers, and the ALU performs all computation and comparison operations. Registers are storage locations that hold inputs and outputs for the ALU.

A complex chain of events occurs when the CPU executes a program. To start, the control unit reads the first instruction from primary storage. It then stores the instruction in a register and, if necessary, reads data inputs from primary storage and stores them in registers. If the instruction is a computation or comparison instruction, the control unit signals the ALU what function to perform, where the input data is located, and where to store the output data. The control unit handles executing instructions to move data to memory, I/O devices, or secondary storage. When the first instruction has been executed, the next instruction is read and executed. This process continues until the program's final instruction has been executed.

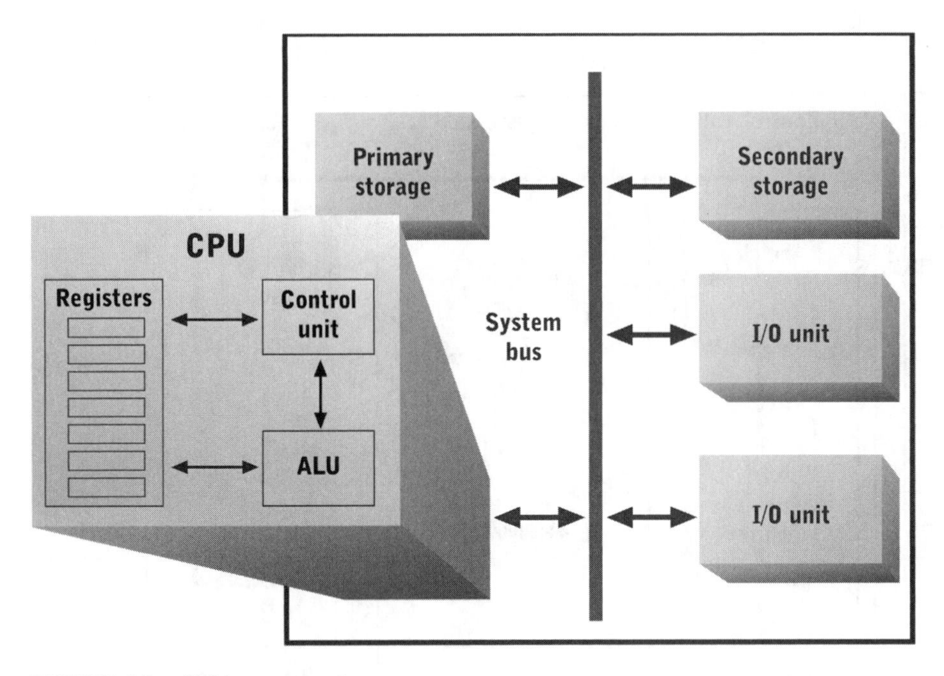

FIGURE 4.2 CPU components
Courtesy of Course Technology/Cengage Learning

The actions the CPU performs can be divided into two groups—the **fetch cycle** (or **instruction cycle**) and the **execution cycle**. During the fetch cycle, data inputs are prepared for transformation into data outputs. During the execution cycle, the transformation takes place and data output is stored. The CPU alternates constantly between fetch and execution cycles. Figure 4.3 shows the flow between fetch and execution cycles (denoted by solid arrows) and data and instruction movement (denoted by dashed arrows).

During the fetch cycle, the control unit does the following:

- Fetches an instruction from primary storage
- Increments a pointer to the location of the next instruction
- Separates the instruction into components—the instruction code (or number) and the data inputs to the instruction
- Stores each component in a separate register

During the execution cycle, the ALU does the following:

- Retrieves the instruction code from a register
- Retrieves data inputs from registers
- Passes data inputs through internal circuits to perform the addition, subtraction, or other data transformation
- Stores the result in a register

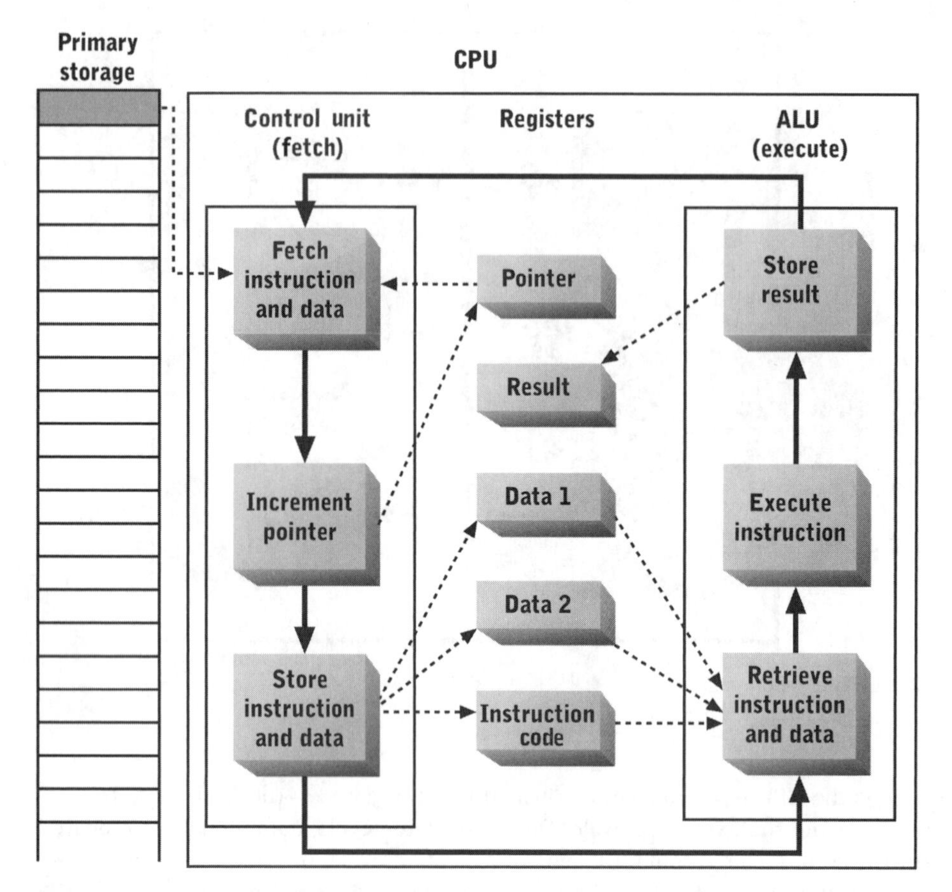

FIGURE 4.3 Control and data flow during the fetch and execution cycles
Courtesy of Course Technology/Cengage Learning

At the conclusion of the execution cycle, a new fetch cycle is started. The control unit keeps track of the next program instruction location by incrementing a pointer after each fetch. The second program instruction is retrieved during the second fetch cycle, the third instruction is retrieved during the third fetch cycle, and so forth.

INSTRUCTIONS AND INSTRUCTION SETS

As you learned in Chapter 2, an instruction is a command to the CPU to perform a primitive processing function on specific data inputs. It's the lowest-level command that software can direct a processor to perform. As stored in memory, an instruction is merely a bit string logically divided into a number of components. The first group of bits represents the instruction's unique binary number, commonly called the **op code**. Subsequent groups of bits hold the instruction's input values, called **operands**. The operand can contain a data item (such as an integer value) or the location of a data item (such as a memory address, a register address, or the address of a secondary storage or I/O device).

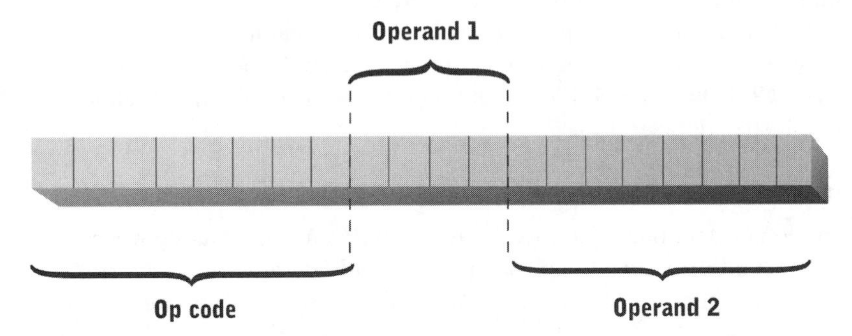

FIGURE 4.4 An instruction containing one op code and two operands
Courtesy of Course Technology/Cengage Learning

An instruction directs the CPU to route electrical signals representing data inputs through predefined processing circuits that implement the appropriate function. Data inputs are accessed from storage or extracted directly from the operands and stored in one or more registers. For computation and logic functions, the ALU accesses the registers and sends the corresponding electrical signals through the appropriate processing circuitry. This circuitry transforms input signals into output signals representing the processing result. This result is stored in a register in preparation for movement to a storage device or an I/O device or for use as input to another instruction.

The control unit executes some instructions without assistance from the ALU, including instructions for moving or copying data, as well as simple tasks, such as halting or restarting the CPU.

The collection of instructions that a CPU can process is called the CPU's **instruction set**. CPU instruction sets vary in CPUs in the following ways:

- Number of instructions
- Size of instructions, op codes, and operands
- Supported data types
- Number and complexity of processing operations performed by each instruction

The instruction sets of different CPUs reflect differences in design philosophy, processor fabrication technology, class of computer system, and type of application software. CPU cost and performance depend on these characteristics.

The full range of processing operations expected of a computer can be implemented with approximately one dozen instructions. This instruction set can perform all the computation, comparison, data movement, and branching functions for integer and Boolean data types. Computation and comparison of real numbers can be accomplished by software with complex sequences of integer instructions operating separately on the whole and fractional parts. Small instruction sets were common in early CPUs and microprocessors, but most modern CPUs have much larger instruction sets. The following sections describe the minimal instruction set.

Data Movement

A **MOVE** instruction copies data bits to storage locations and can copy data between any combination of registers and primary storage locations. A **load** operation is a data transfer from main memory into a register. A **store** operation is a data transfer from a register into primary storage.

MOVE tests the bit values in the source location and places copies of these values in the destination location. The former bit values in the destination are overwritten. At the completion of the MOVE, both sending and receiving locations hold identical bit strings. The name "move" is a misnomer because the data content of the source location is unchanged.

MOVE is also used to access storage and I/O devices. An input or storage device writes to a specific memory location, and the CPU retrieves the input by reading that memory location and copying its value into a register. Similarly, data is output or stored by writing to a predefined memory address or range of addresses. The output or storage device continually monitors the content of its assigned memory addresses and reads newly written data for storage or output.

Data Transformations

The most primitive data transformation instructions are based on Boolean logic:

- NOT
- AND
- OR
- XOR

These four Boolean instructions and the ADD and SHIFT instructions (discussed in the following sections) are the basic building blocks of all numeric comparisons and computations. They're summarized in Table 4.1.

TABLE 4.1 Primitive data transformation instructions

Instruction	Function
NOT	Each result bit is the opposite of the operand bit.
AND	Each result bit is true if both operand bits are true.
OR	Each result bit is true if either or both operand bits are true.
XOR	Each result bit is true if either (but not both) operand bit is true.
ADD	Result is the arithmetic sum of the operands.
SHIFT	Move all bit values left or right, as specified by the operand.

NOT

A **NOT** instruction transforms the Boolean value true (1) into false (0) and the value false into true. The rules that define the output of NOT on single bit inputs are as follows:

NOT 0 = 1
NOT 1 = 0

With bit strings, NOT treats each bit in the bit string as a separate Boolean value. For example, executing NOT on the input 10001011 produces the result 01110100. Note that NOT has only one data input, whereas all other Boolean instructions have two.

AND

An **AND** instruction generates the result true if both of its data inputs are true. The following rules define the result of AND on single bit data inputs:

0 AND 0 = 0
1 AND 0 = 0
0 AND 1 = 0
1 AND 1 = 1

The result of AND with two bit string inputs is shown in the following example:

```
      10001011
AND   11101100
--------------
      10001000
```

OR

There are two types of OR operations in Boolean logic. An **inclusive OR** instruction (the word "inclusive" is usually omitted) generates the value true if either or both data inputs are true. The rules that define the output of inclusive OR on single bit inputs are as follows:

0 OR 0 = 0
1 OR 0 = 1
0 OR 1 = 1
1 OR 1 = 1

The result of inclusive OR with two bit string inputs is shown in the following example:

```
      10001011
OR    11101100
      --------
      11101111
```


An **exclusive OR (XOR)** instruction generates the value true if either (but not both) data input is true. The following rules define the output of XOR on single bit inputs:

0 XOR 0 = 0
1 XOR 0 = 1
0 XOR 1 = 1
1 XOR 1 = 0

Note that if either operand is 1, the result is the complement of the other operand. Specific bits in a bit string can be inverted by XORing the bit string with a string containing 0s in all positions except the positions to be negated. For example, the following XOR inverts only the right four bit values and leaves the left four bit values unchanged:

```
      10001011
XOR   00001111
      --------
      10000100
```

Every bit in a bit string can be inverted by XORing with a string of 1s:

```
      10001011
XOR   11111111
      --------
      01110100
```

Note that XORing any input with a string of 1s produces the same result as executing NOT. Therefore, NOT isn't required in a minimal instruction set.

ADD

An **ADD** instruction accepts two numeric inputs and produces their arithmetic sum. For example, the following ADD shows the binary addition of two bit strings:

```
      10001011
ADD   00001111
      --------
      10011010
```

Note that the mechanics of the addition operation are the same, regardless of what the bit strings represent. In this example, if the bit strings represent unsigned binary numbers, the operation is as follows:

$$139_{10} + 15_{10} = 154_{10}$$

If the bit strings represent signed integers in twos complement format, the operation is the following:

$$-117_{10} + 15_{10} = -102_{10}$$

Binary addition doesn't work for complex data types, such as floating-point and double-precision numbers. If the CPU supports complex numeric data types, a separate ADD instruction must be implemented for each type.

SHIFT

Figure 4.5 shows the effect of a **SHIFT** instruction. In Figure 4.5(a), the value 01101011 occupies an 8-bit storage location. Bit strings can be shifted to the right or left, and the number of positions shifted can be greater than one. Typically, the second operand holds an integer value indicating the number of bit positions to shift the value. Positive or negative operand values indicate shifting to the left or right.

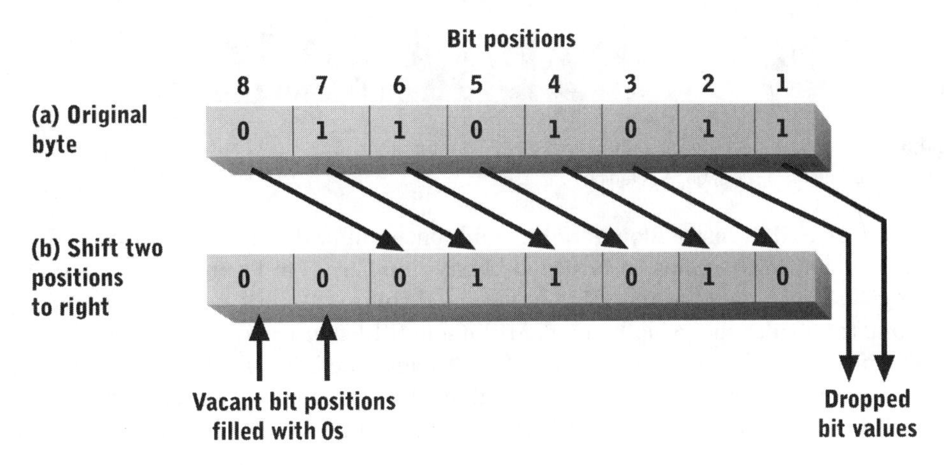

FIGURE 4.5 Original data byte (a) shifted 2 bits to the right (b)
Courtesy of Course Technology/Cengage Learning

Figure 4.5(b) shows the result of shifting the value two positions to the right. The resulting value is 00011010. In this case, the values in the two least significant positions of the original string have been lost, and the vacant bit positions are filled with 0s.

Figure 4.5 is an example of a **logical SHIFT**, which can extract a single bit from a bit string. Figure 4.6 shows how shifting an 8-bit value (a) to the left by 4 bits (b) and then to the right by 7 bits creates a result (c) with the fourth bit of the original string in the rightmost position. Because all other bit positions contain 0s, the entire bit string can be interpreted as true or false. Shifting an 8-bit twos complement value to the right by seven positions is a simple way to extract and test the sign bit.

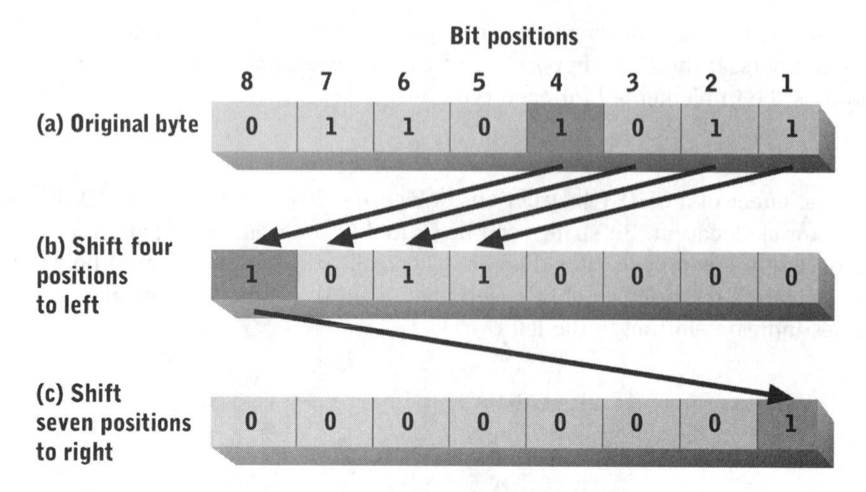

FIGURE 4.6 Extracting a single bit with logical SHIFT instructions
Courtesy of Course Technology/Cengage Learning

An **arithmetic SHIFT** instruction performs multiplication or division, as shown in Figure 4.7. If a bit string contains an unsigned binary number, as in Figure 4.7(a), shifting to the left by 1 bit (b) multiplies the value by two, and shifting the original bit (a) to the right by 2 bits (c) divides the value by four. Arithmetic SHIFT instructions are more complex when applied to twos complement values because the leftmost bit is a sign bit and must be preserved during the SHIFT instruction. Most CPUs provide a separate arithmetic SHIFT instruction that preserves the sign bit of a twos complement value.

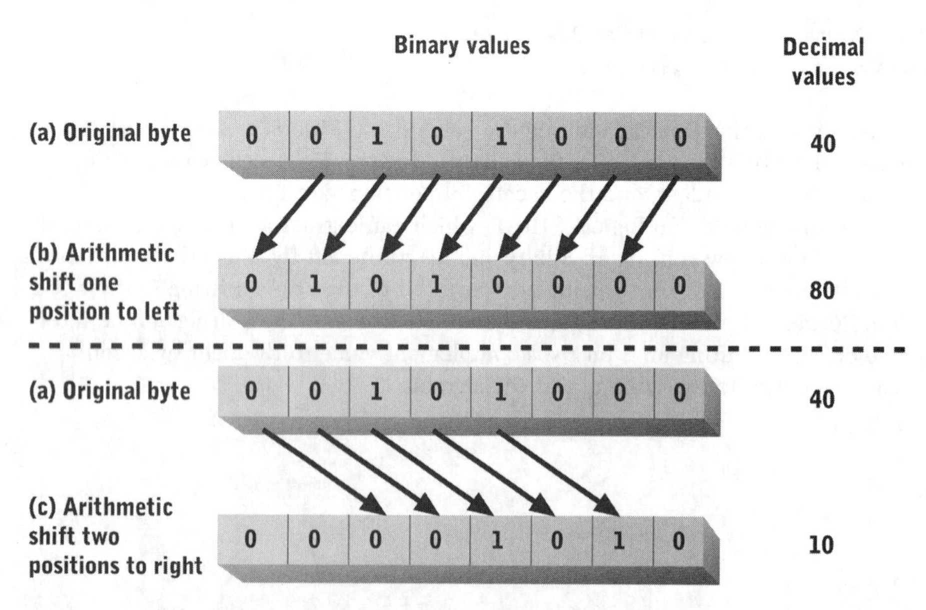

FIGURE 4.7 Multiplying and dividing unsigned binary values with SHIFT instructions
Courtesy of Course Technology/Cengage Learning

Sequence Control

Sequence control operations, which alter the flow of instruction execution in a program, include the following:

- Unconditional BRANCH
- Conditional BRANCH
- HALT

BRANCH

A **BRANCH** (or **JUMP**) instruction causes the processor to depart from sequential instruction order. Recall that the control unit fetches the next instruction from memory at the conclusion of each execution cycle. The control unit consults a register to determine where the instruction is located in primary storage. BRANCH has one operand containing the memory address of the next instruction. It's actually implemented as a MOVE instruction. The BRANCH operand is loaded into the register that the control unit uses to fetch the next instruction.

In an **unconditional BRANCH**, the processor always departs from the normal execution sequence. For example, if the user of a word-processing program clicks the Print menu item, the program always branches to the corresponding instruction sequence. In a **conditional BRANCH**, the BRANCH occurs only if a specified condition is met, such as the equivalence of two numeric variables. The condition must be evaluated, and the Boolean result must be stored in a register. The conditional BRANCH instruction checks this register's contents and branches only if the value in it is true. As discussed in Chapter 2, conditional BRANCH instructions enable a program to vary its behavior depending on the data values being processed, as when computing taxes with progressive tax rates.

HALT

A **HALT** instruction suspends the normal flow of instruction execution in the current program. In some CPUs, it causes the CPU to cease all operations. In others, it causes a BRANCH to a predetermined memory address. Typically, a portion of the operating system (OS) is stored at this address, effectively transferring control to the OS and terminating the previously executing program.

Complex Processing Operations

TIP

Reviewing the twos complement discussion in Chapter 3 before studying the following examples might be helpful.

Complex processing operations can be performed by combining the simpler instructions. For example, subtraction can be implemented as complementary addition. That is, the operation A - B can be implemented as A + (-B). As described in Chapter 3, a negative twos complement value can be derived from its positive counterpart by adding 1 to the complement of the positive value. A bit string's complement can be generated by XORing it with a string of binary 1 digits.

For example, the complement of 0011 (3_{10}), represented as a twos complement value, can be derived as follows:

$$XOR(0011,1111) + 0001 = 1100 + 0001 = 1101 = -3_{10}$$

This result is added to implement a subtraction operation. For example, the result of subtracting 0011 from 0111 can be calculated as follows:

$$\begin{aligned} 7_{10} - 3_{10} &= ADD(ADD(XOR(0011,1111),0001),0111) \\ &= ADD(ADD(1100,0001),0111) \\ &= ADD(1101,0111) \\ &= 10100 \end{aligned}$$

Because 4-bit values are used, the result of 10100 is truncated from the left, resulting in the value 0100.

Comparison operations can be implemented in much the same way as subtraction. A comparison operation generates a Boolean output value. Typically, an integer value of 0 is interpreted as false, and any nonzero value is interpreted as true. The comparison $A \neq B$ can be performed by generating the complement of B and adding it to A. If the two numbers are equal, the result of the addition is a string of 0s (interpreted as false). An equality comparison can be performed by negating the Boolean result of an inequality comparison.

Greater-than and less-than comparisons can also be performed with subtraction followed by extraction of the sign bit. For the condition $A < B$, subtracting B from A generates a negative result if the condition is true. In twos complement notation, a negative value always has a 1 in the leftmost position (that is, the sign bit). SHIFT can be executed to extract the sign bit. For example, the twos complement value 10000111 is a negative number. The sign bit is extracted by shifting the value 7 bits to the right, resulting in the string 00000001. The SHIFT result is interpreted as a Boolean value (1, or true, in this case).

For example, the comparison 0111 < 0011 can be evaluated as follows:

$$\begin{aligned} 0111 < 0011 &= SHIFT(ADD(0111,ADD(XOR(0011,1111),0001)),0011) \\ &= SHIFT(ADD(0111,ADD(1100,0001)),0011) \\ &= SHIFT(ADD(0111,1101),0011) \\ &= SHIFT(0100,0011) \\ &= 0000 \end{aligned}$$

The second operand of the SHIFT instruction is a binary number representing the direction and number of bit positions to shift (+3, or right 3, in this example). The result is 0, interpreted as the Boolean value false.

A Short Programming Example

Take a look at the following high-level programming language statement:

```
IF (BALANCE < 100) THEN
    BALANCE = BALANCE - 5
ENDIF
```

This computation might be part of a program that applies a monthly service charge to checking or savings accounts with a balance below \$100. Figure 4.8 shows a program performing this computation by using only the simpler CPU instructions. Figure 4.9 shows the register contents after each instruction is executed when the account balance is \$64. The coding format for all numeric data is 8-bit twos complement.

```
1   MOVE    M1, R1      'load BALANCE
2   MOVE    M2, R2      'load minimum balance 100₁₀
3   MOVE    M3, R3      'load service charge 5₁₀
4   MOVE    M4, R4      'load constant 1₁₀
5   MOVE    M5, R5      'load constant 7₁₀
6   NOT     R2, R2      'start < comparison
7   ADD     R2, R4, R2  '
8   ADD     R1, R2, R0  '
9   SHIFT   R0, R5      'end < comparison
10  XOR     R0, R4, R0  'invert comparison result
11  BRANCH  R0, 16      'branch if comparison false
12  NOT     R3, R3      'start negation of service charge
13  ADD     R3, R4, R3  'end negation of service charge
14  ADD     R1, R3, R0  'subtract service charge from balance
15  MOVE    R0, M1      'store new balance
16  HALT                'terminate program
```

FIGURE 4.8 A simple program using primitive CPU instructions
Courtesy of Course Technology/Cengage Learning

Instruction	R0	R1	R2	R3	R4	R5
1		01000000				
2			01100100			
3				00000101		
4					00000001	
5						00000111
6			10011011			
7			10011100			
8	11011100					
9	00000001					
10	00000000					
11						
12				11111010		
13				11111011		
14	00111011					
15						
16						

FIGURE 4.9 Register contents after executing each instruction in Figure 4.8 when the account balance in memory location M1 is $64

Courtesy of Course Technology/Cengage Learning

The following list describes the instructions used in the program:

1. Instructions 1 through 5 load the account balance, minimum balance, service charge, and needed binary constants from memory locations M1 through M5.
2. A less-than comparison is performed in instructions 6 through 9:
    - The right side of the comparison is converted to a negative value by executing a NOT instruction (instruction 6) and adding 1 to the result (instruction 7).
    - The result is added to the account balance (instruction 8), and the sum is shifted seven places to the right to extract the sign bit (instruction 9). At this point, register R0 holds the Boolean result of the less-than comparison.

    For this example, all BRANCH instructions are assumed to be conditional on the content of a register. The BRANCH is taken if the register holds a Boolean true value and otherwise ignored. To jump beyond the code implementing the service charge if the account balance is above the minimum, the Boolean result of the condition must be inverted before branching.
3. Instruction 10 inverts the sign bit stored in the rightmost bit of R0 by XORing it against 00000001 (stored in R4).

4. Instruction 11 executes a conditional BRANCH. Because the original sign bit was 1, the inverted value is 0. Therefore, the BRANCH is ignored, and processing proceeds with instruction 12.
5. Instructions 12 through 14 subtract the service charge stored in register R3 from the account balance. Instructions 12 and 13 convert the positive value to a negative value, and instruction 14 adds it to the account balance.
6. Instruction 15 saves the new balance in memory.
7. Instruction 16 halts the program.

Instruction Set Extensions

The instructions described to this point are sufficient for a simple general-purpose processor. All the more complex functions normally associated with computer processing can be implemented by combining these primitive building blocks. Most CPUs provide a much larger instruction set, including advanced computation operations (such as multiplication and division), negation of twos complement values (NOT followed by ADD 1), and testing a sign bit and other single-bit manipulation functions. These instructions are sometimes called **complex instructions** because they represent combinations of primitive processing operations.

Complex instructions represent a tradeoff between processor complexity and programming simplicity. For example, the three-step process of NOT ADD ADD performs subtraction in Figure 4.8, instructions 12 through 14. Because subtraction is a commonly performed operation, most CPU instruction sets include one or more subtraction instructions. Providing a separate NOT instruction complicates the CPU by requiring extra processing circuitry for the additional instruction but simplifies machine-language programs.

Complex instructions also represent a tradeoff between CPU complexity and program execution speed. Multistep instruction sequences execute faster if they're performed in hardware as a single instruction, avoiding the overhead of fetching multiple instructions and accessing intermediate results in registers. Other efficiencies might also be achieved by hard-wiring the steps together. However, these efficiencies have limits, as described later in "RISC and CISC."

Additional instructions are required when new data types are added. For example, most CPUs provide instructions for adding, subtracting, multiplying, and dividing integers. If double-precision integers and floating-point numbers are supported, additional computation instructions must be included for these data types. It's not unusual to see a half-dozen different ADD instructions in an instruction set to support integers and real numbers in single- and double-precision as well as signed and unsigned "short" integers.

Some instruction sets also include instructions that combine data transformations with data movement. The primitive instructions in Figure 4.8 use registers for both data input and data output. In some CPUs, data transformation instructions allow one or more operands to be a memory address, thus combining data movement with data transformation.

INSTRUCTION FORMAT

Recall that an instruction consists of an op code (instruction number) and zero or more operands (representing data values or storage locations). An **instruction format** is a template that specifies the number of operands and the position and length of the op code and operands. Instruction formats vary between CPUs in many ways, including the op code size, the meaning of specific op code values, the data types used as operands, and the length and coding format of each type of operand. The term "instruction format" erroneously implies that each CPU uses a single format. Because instructions vary in the number and type of operands they require, a single CPU must support multiple instruction formats for various combinations of operand types.

Figure 4.10 shows an instruction format consisting of an op code and three operands with a total length of 20 bits. This sample format is typical of instructions that use register inputs and outputs. The op code occupies the first 8 bits. Most CPUs represent the op code as an unsigned binary number. Therefore, an 8-bit op code provides 256 possible instructions numbered 0 through 255. Each operand is a 4-bit unsigned binary number identifying 1 of 16 possible registers. Larger instruction sets or more registers would require more bits to represent the op code and operands and, therefore, a longer instruction format.

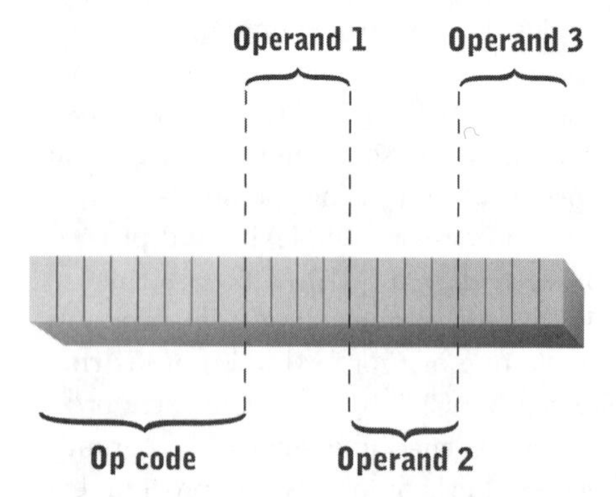

FIGURE 4.10 An instruction format with three register operands
Courtesy of Course Technology/Cengage Learning

Figure 4.11 shows a three-operand instruction format for load and store instructions using segmented memory addresses. The first operand contains the number of the register to or from which data is moved. The second operand contains the number of the register that identifies a memory segment, and the third operand contains the segment offset.

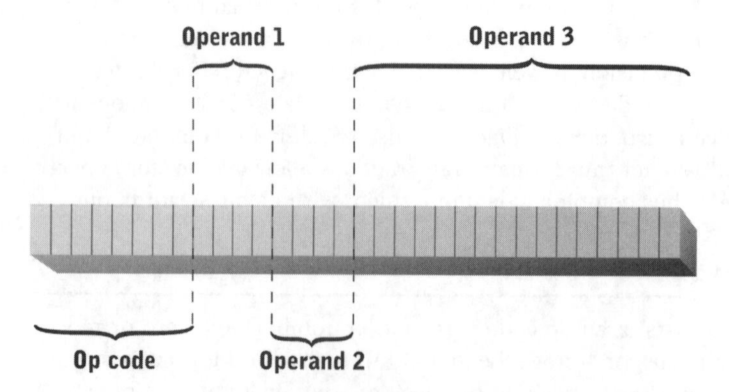

FIGURE 4.11 An instruction format that includes a segmented memory address
Courtesy of Course Technology/Cengage Learning

Recall from Chapter 3 that some CPUs use a segmented memory-addressing scheme, in which a memory address consists of two parts—a memory segment identifier and a segment offset. For segmented memory schemes, the second operand identifies a register that holds the segment identifier, and the third operand is the segment offset. The control unit constructs a complete memory address, using the contents of the register identified in the second operand and the content of the third operand. The instruction format in Figure 4.11 is typical of early Intel CPUs that segmented memory into as many as 16 (2^4) segments, each containing 64 KB (2^{16}).

Instruction Length

Instruction formats in a single CPU can be fixed or variable in length. **Fixed-length instructions** simplify the instruction-fetching process in the control unit. If the instruction format is fixed length, the amount by which the instruction pointer must be incremented after each fetch is a constant. This increment is the length of an instruction.

Instruction format lengths can be equalized by "padding" shorter instruction formats with trailing 0 bits. For example, the format in Figure 4.10 could be padded with 12 trailing 0 bits to increase its length so that it matches the format in Figure 4.11. The CPU ignores the extra bits when processing instructions that use the padded format.

With a **variable-length instruction**, the amount by which the instruction pointer is incremented after a fetch is the length of the most recently fetched instruction. The control unit must check the op code of each fetched instruction to determine the correct increment value. If you think of a table with op code values on the left and format lengths on the right, the control unit looks up the op code in the table and extracts the appropriate value to add to the instruction pointer.

Variable-length instructions also complicate instruction fetching because the number of bytes to be fetched isn't known in advance. One method of addressing this uncertainty is to always fetch the number of bytes in the longest instruction format, but this method can result in many unnecessary memory accesses. Another method is to fetch only the number of bytes in the shortest format. Then the op code is examined to determine the instruction's length and the number of additional bytes to fetch, if any. In either case, extra computer resources are used in the fetch operation that wouldn't have been used if the instruction length were fixed and known before the fetch.

Although fixed-length instructions and fields simplify control unit functions, they do so at the expense of efficient memory use. Some instructions have no operands, and others have one, two, or three operands. If fixed-length instructions are used, the instruction length must be the length of the longest instruction—for example, an instruction with two or three large operands. Smaller instructions stored in memory must be padded with empty bit positions to extend their length to the maximum. The memory used to pad short instructions is wasted, and programs require more memory during execution and more time to load from secondary storage.

RISC and CISC

Reduced instruction set computing (RISC) is a philosophy of processor and computer system design first used in the late 1980s. The primary architectural feature of a RISC processor is the absence of some, but not all, complex instructions from the instruction set. In particular, RISC processors avoid instructions that combine data transformation and data movement operations. For example, a non-RISC processor might provide a data transformation instruction in this form:

Transform(Address1,Address2,Address3)

Address1 and Address2 are the memory addresses of data inputs, and Address3 is the memory address to which the result is stored. In a RISC processor, transformation operations always use register inputs and outputs. The single complex instruction shown previously requires four separate RISC instructions:

```
MOVE(Address1,R1)
MOVE(Address2,R2)
Transform(R1,R2,R3)
MOVE(R3,Address3)
```

Each instruction is loaded and executed independently, consuming at least four execution cycles.

Although the lack of many complex instructions is the main distinguishing feature of a RISC processor, other differences typically include fixed-length instructions, short instruction length, and a large number of general-purpose registers. To describe these differences fully, RISC must be contrasted with its opposite design philosophy—**complex instruction set computing (CISC)**.

Complex instruction sets were developed because early computers had limited memory and processing power. Memory was expensive, and many computers had barely enough to hold an entire application program. To compensate for limited memory, many CPU designers provided complex instructions that would do more work per instruction. As a result, programs required less memory, and complex operations executed more quickly.

For example, assume a floating-point addition operation, implemented as a single complex instruction, can be executed in one processor cycle. In addition, assume an equivalent operation can be performed with a sequence of five integer math instructions, each requiring one processor cycle. In this example, a direct implementation of the complex floating-point instruction saves four processor cycles each time it's executed.

If complex instructions are so beneficial, why would anyone want to eliminate them? The most important reason is that each complex instruction provides benefits if measured in isolation, but as more complex instructions are added, the instruction set becomes large and complex, which creates two problems. First, it complicates the control unit because there are more instructions to interpret and usually more instruction formats and data types to deal with. Also, a large set of complex instructions often goes hand in hand with variable-length instruction formats. This format adds even more complexity to the job of fetching and decoding instructions, which in turn increases fetching and decoding time. Because every simple or complex instruction must be fetched before it's executed, a performance penalty is applied to every instruction, even the simple ones.

The second problem arising from a large, complex instruction set is microprocessor size. As discussed later in "Current Technology Capabilities and Limitations," speed improvements in microprocessors are achieved mainly by miniaturization. The simpler the processor, the easier the task of shrinking it, and the more reliably smaller versions can be fabricated. Because CISC processors are much more complex than RISC processors, they're more difficult to fabricate reliably in smaller sizes. RISC processor design reduces the instruction set's size and complexity to increase raw speed in fetching and executing instructions. RISC design follows a "less is more" strategy and extends the instruction set only when the benefits are very high.

The main disadvantage of RISC CPUs is the extra memory required for program storage and execution. Because many complex instructions aren't present, programs must use multiple simple instructions in their place. The equivalent sequences of simple instructions occupy more storage than complex instructions do and increase program memory requirements. Although increased memory is a disadvantage for RISC, it's not currently a major factor because memory cost has fallen rapidly.

Compared with CISC processors, RISC processors are also inefficient at executing programs that do many of the functions complex instructions are designed for, such as transforming data items stored in memory and immediately storing the results back in memory. Although this inefficiency is a potential disadvantage, detailed studies of typical program behavior have shown that many complex instructions aren't used frequently. Typical programs spend most of their time executing primitive instructions, such as load, store, add, and compare. In many cases, the speed advantage of complex instructions isn't realized often enough to make up for the performance penalty applied to every instruction.

CPU performance depends heavily on the construction of the programs being executed, regardless of whether the CPU is RISC, CISC, or a hybrid. Software should be optimized to take advantage of the presence or absence of specific CPU capabilities. In modern software development, most optimization takes place when a program written in a high-level language, such as C or Java, is compiled to produce CPU instructions.

NOTE

Compilation is discussed in Chapter 10.

Despite the inherent advantages of RISC design, CISC CPUs from Intel, including the Core2, Xeon, and Itanium, dominate the microcomputer, workstation, and midrange computer classes and have made major inroads into mainframes and supercomputers. RISC-based designs, such as the IBM POWER6, are used mainly in large servers and supercomputers. How can Intel CISC CPUs compete against supposedly faster RISC CPUs? Why does Intel stick with CISC technology?

One part of the answer is backward compatibility. Early Intel desktop CPUs were developed when CISC was clearly dominant. Software developed for these CPUs was typically sold to customers in the form of CPU instructions (for example, .exe program files in MS-DOS or Windows). Because customers didn't want to abandon old software when upgrading to new CPUs, newer Intel CPUs needed to execute the same complex instructions supported by older CPUs. Although there are several ways to provide backward compatibility, one of the easiest is to continue expanding on older CISC designs in newer chips.

Another part of the answer is Intel's dominance in CPU fabrication technology. In essence, Intel can build more complex CPUs economically because it has more advanced fabrication technology and a large market share over which to spread the technology costs. By using the latest fabrication technology, Intel can produce CISC CPUs with clock speeds that rival RISC competitors despite their more complex circuitry.

Yet another part of the answer is the advanced RISC techniques used by Intel and most other microprocessor manufacturers, covered later in "Enhancing Processor Performance." For now, realize that just because a CPU supports complex instructions doesn't mean it can't use RISC design principles "under the hood." In a similar vein, few processors that claim to be RISC avoid all CISC characteristics. The terms "RISC" and "CISC" are overused in marketing literature to paint stark contrasts between CPU designs, but the realities are seldom as far apart as a strict interpretation of these terms might imply.

CISC and RISC are different ends of a design continuum representing tradeoffs in several design factors. The economics and technology of program design and behavior, processor fabrication, and memory cost made CISC an optimal solution in the early days of computing. In the 1990s, as memory became cheaper and CPU fabrication technology improved, the optimal tradeoff shifted toward RISC. At the moment, there's no clear "best" approach to processor design, partly because processors of each type borrow heavily from the best features of the other type. Future changes in technology and in how computers are used might favor RISC, CISC, or a completely different approach to CPU design.

CLOCK RATE

The **system clock** is a digital circuit that generates timing pulses, or signals, and transmits the pulses to other devices in the computer. (Note: It's not the clock used to keep track of the current date and time.) It's generally a separate device with a dedicated communication line monitored by all devices in the computer system. All actions, especially the CPU's fetch and execution cycles, are timed according to this clock. Storage and I/O devices are timed by the clock signal, and all devices in a computer coordinate their activities with the system clock.

The frequency at which the system clock generates timing pulses is the system's **clock rate**, and each "tick" of the clock begins a new **clock cycle**. CPU and computer system clock rates are expressed in **hertz (Hz)**. One hertz corresponds to one clock cycle per second. Modern CPUs and computer systems have clocks that generate millions or billions of timing pulses per second. The frequency of these clocks is measured in **megahertz (MHz)**, meaning millions of cycles per second, or **gigahertz (GHz)**, billions of cycles per second.

The inverse of the clock rate is called the CPU cycle time. In most CPUs, the **cycle time** is the time required to fetch and execute the simplest instruction in the instruction set. For example, if the CPU clock rate is 5 GHz and NOT is the simplest instruction, the time required to fetch and execute a NOT instruction can be computed as the inverse of the clock rate:

$$\text{cycle time} = \frac{1}{\text{clock rate}} = \frac{1}{5{,}000{,}000{,}000} = 0.0000000002 \text{ second} = 0.2 \text{ nanosecond}$$

Clock rate and cycle time are important CPU performance measures. However, they tell only part of the performance story for a CPU or computer system. CPU clock rate is frequently misinterpreted by equating it with the following:

- Instruction execution rate
- Overall computer system performance

From the perspective of program execution speed, the most important CPU performance consideration is the rate at which instructions are executed. This rate is generally stated in units called **millions of instructions per second (MIPS)**, although modern CPUs and computers can execute billions or trillions of instructions per second. MIPS are assumed to measure CPU performance when manipulating single-precision integers. When manipulating single-precision floating-point numbers, CPU performance is measured in **millions of floating-point operations per second (megaflops or MFLOPS)**, **billions of floating-point operations per second (gigaflops or GFLOPS)**, or **trillions of floating-point operations per second (teraflops or TFLOPS)**. The latest generation of supercomputers can now execute 10^{15} floating-point operations per second, called **petaflops (PFLOPS)**.

If all instructions, regardless of function or data type, were executed in a single clock cycle, then clock rate, MIPS, and MFLOPS would be equivalent. However, execution time varies with the processing function's complexity. Instructions for simple processing functions, such as Boolean functions and equality comparisons, typically require complete execution in one clock cycle. Instructions for functions such as integer multiplication and division typically require two to four clock cycles to complete execution. Instructions that perform complex functions on complex data types, such as double-precision floating-point numbers, typically require up to a dozen clock cycles to complete execution.

The number of instructions executed in a given time interval depends on the mix of simple and complex instructions in a program. For example, assume a program executes 100 million integer instructions; 50% are simple instructions requiring a single clock cycle, and 50% are complex instructions requiring an average of three clock cycles. The average program instruction requires two clock cycles—in this case, $(1 \times 0.50) + (3 \times 0.50)$—and the CPU MIPS rate is 50% of the clock rate when executing this program. Different mixes of simple and complex instructions and data types result in different ratios of clock rate to MIPS and MFLOPS. For all but the simplest programs, MIPS and MFLOPS are much smaller than the CPU clock rate.

The previous MIPS calculation assumes that nothing hinders the CPU in fetching and executing instructions, but the CPU relies on slower devices to keep it supplied with instructions and data. For example, main memory is typically 2 to 10 times slower than the processor. That is, the time required for a main memory read or write operation is typically 2 to 10 CPU clock cycles. Accessing secondary storage is thousands or millions of times slower than the CPU, so the CPU might be idle while waiting for access to storage and I/O devices. Each clock cycle the CPU spends waiting for a slower device is called a **wait state**.

Unfortunately, a CPU can spend much of its time in wait states, during which no instructions are being executed. Therefore, a computer system's effective MIPS rate is much lower than the MIPS rate of the CPU measured in isolation because of delays imposed by waiting for storage and I/O devices.

NOTE

Chapter 6 discusses methods of minimizing wait states.

MIPS and MFLOPS are poor measurements for comparing computer system performance because processor instruction sets vary so widely and because CPU performance depends so heavily on memory access speed and other aspects of computer design. A better measure of comparative computer system performance is how quickly a specific program or set of programs is executed.

TECHNOLOGY FOCUS

Benchmarks

A **benchmark** is a measure of CPU or computer system performance when carrying out one or more specific tasks. Benchmarks can be used in several ways, including the following:

- Compare the performance of multiple computers from one or many vendors.
- Measure the performance of an existing computer system and determine how its performance might be improved.
- Determine the computer system or configuration that best matches specific application requirements.

A **benchmark program** performs specific tasks that can be counted or measured. For example, a benchmark program might test CPU computational speed by performing the task of multiplying two large matrixes or computing the value of pi to many digits. Because the number of computations making up the task is known in advance, the number of computations executed per second can be computed based on the total time required to execute the benchmark program.

A key issue in choosing and using benchmarks is determining the characteristics of software that will run on a computer system and choosing benchmarks that most closely match these characteristics. Because computer systems can run a variety of application programs, benchmark programs are often combined into **benchmark suites**. For example, a benchmark suite for evaluating a PC might include programs that simulate word processing, photo editing, game playing, Web browsing, and downloading and installing large software updates. A benchmark suite for a server should match the tasks the server will perform (for example, handling database queries and updates or processing online order transactions). A benchmark suite for a supercomputer typically includes programs that test integer and floating-point computational speed, particularly on large and complex problems.

Developing benchmark programs and updating them to match rapidly changing software and hardware are complex tasks. Computer manufacturers and software vendors have banded together to create and support nonprofit organizations that develop benchmarks, perform tests, and publish results. Two large benchmarking organizations are the following:

- Standard Performance Evaluation Corporation (SPEC, *www.spec.org*)
- Transaction Processing Performance Council (TPC, *www.tpc.org*)

SPEC was founded in 1988 to develop benchmarks for high-performance workstations. Its mission has grown to encompass high-performance computing and computers running open-source software in a variety of application environments. Table 4.2 summarizes some commonly used SPEC benchmarks.

(*continued*)

TABLE 4.2 Sample SPEC benchmarks

Benchmark	Description
SPEC CPU	A suite of two benchmark programs, CINT and CFP, that measure the performance of computation-intensive programs with integer (CINT) and floating-point (CFP) data. This benchmark is widely used to compare the performance of supercomputer nodes and high-performance workstations.
SPEC MPI	Measures the computational performance of clustered computers. This suite uses large computational problems distributed across multiple computing nodes with substantial use of message passing, shared memory, and shared file systems.
SPECviewperf	Measures workstation performance on graphics-intensive tasks and applications, such as computer-aided design and animation.
SPECmail	Measures the performance of a computer system acting as an e-mail server using standard Internet mail protocols. This benchmark simulates the processing load of a corporate e-mail server supporting 40,000 users.

TPC was also founded in 1988 with a mission to develop benchmarks useful for measuring online transaction processing and database performance. This application area was growing rapidly in the 1980s as more companies were automating accounting, sales, production control, and other business functions. Its importance increased during the 1990s and 2000s when companies moved commerce to Web-based systems and groups of companies interconnected their business systems to improve operational efficiency.

Benchmarking for online transaction processing is considerably more complex than for scientific applications. For example, the performance of a typical Web-based storefront depends on multiple hardware elements, such as CPU, memory, storage subsystems, and network interfaces, and software elements beyond application software, such as OSs, database management systems, and Web server software. Table 4.3 summarizes some commonly used TPC benchmarks.

TABLE 4.3 Sample TPC benchmarks

Benchmark	Description
TPC-C	Measures the response time and throughput of five typical business transactions (for example, sales order entries and stock level queries) processed against a relational database.
TPC-E	Simulates the online transaction processing load of a brokerage firm, including real-time customer orders and inquiries and back-end interaction with stock trading and reporting systems.
TPC-App	Simulates a 24/7 Web storefront. This suite exercises the Web server front-end customer interface and back-end database interaction and business-to-business functions with a large number of transactions and simultaneous online sessions.
TPC-H	Simulates complex database queries against a large database typical of decision support and data-mining applications.

(*continued*)

SPEC, TPC, and other benchmark programs and suites are updated constantly to reflect changes in computer systems and changes in the tasks these systems perform. For example, SPEC MPI is a recent benchmark developed to reflect the move in supercomputer configuration from large stand-alone supercomputers to massively parallel computation with grids and clusters. SPEC CPU has been updated recently to better measure performance issues for multicore processors. TPC-App updates the older TPC-C to reflect the modern reality of 24/7 e-commerce.

CPU REGISTERS

Registers play two primary roles in CPU operation. First, they provide a temporary storage area for data the currently executing program needs quickly or frequently. Second, they store information about the currently executing program and CPU status—for example, the address of the next program instruction, error messages, and signals from external devices.

General-Purpose Registers

General-purpose registers are used only by the currently executing program. They typically hold intermediate results or frequently used data values, such as loop counters or array indexes. Register accesses are fast because registers are implemented in the CPU. In contrast, storing and retrieving from primary storage is much slower. Using registers to store data needed immediately or frequently increases program execution speed by avoiding wait states.

Adding general-purpose registers increases execution speed but only up to a point. Any process or program has a limited number of intermediate results or frequently used data items, so CPU designers try to find the optimal balance among the number of general-purpose registers, the extent to which a typical process will use these registers, and the cost of implementing these registers. As the cost of producing registers has decreased, their number has increased. Current CPUs typically provide several dozen general-purpose registers.

Special-Purpose Registers

Every processor has **special-purpose registers** used by the CPU for specific purposes. Some of the more important special-purpose registers are as follows:

- Instruction register
- Instruction pointer
- Program status word

When the control unit fetches an instruction from memory, it stores it in the **instruction register**. The control unit then extracts the op code and operands from the instruction and performs any additional data movement operations needed to prepare for execution. The process of extracting the op code and operands, loading data inputs, and signaling the ALU is called instruction **decoding**.

The **instruction pointer (IP)** can also be called the "program counter." Recall that the CPU alternates between the instruction (fetch and decode) and execution (data movement or transformation) cycles. At the end of each execution cycle, the control unit starts the next fetch cycle by retrieving the next instruction from memory. This instruction's address is stored in the instruction pointer, and the control unit increments the instruction pointer during or immediately after each fetch cycle.

The CPU deviates from sequential execution only if a BRANCH instruction is executed. A BRANCH is implemented by overwriting the instruction pointer's value with the address of the instruction to which the BRANCH is directed. An unconditional BRANCH instruction is actually a MOVE from the branch operand, which contains the branch address, to the instruction pointer.

The **program status word (PSW)** contains data describing the CPU status and the currently executing program. Each bit in the PSW is a separate Boolean variable, sometimes called a **flag**, representing one data item. The content and meaning of flags vary widely from one CPU to another. In general, PSW flags have three main uses:

- Store the result of a comparison operation.
- Control conditional BRANCH execution.
- Indicate actual or potential error conditions.

The sample program shown previously in Figure 4.8 performed comparison by using the XOR, ADD, and SHIFT instructions. The result was stored in a general-purpose register and interpreted as a Boolean value. This method was used because the instruction set was limited.

Most CPUs provide one or more COMPARE instructions. COMPARE takes two operands and determines whether the first is less than, equal to, or greater than the second. Because there are three possible conditions, the result can't be stored in a single Boolean variable. Most CPUs use two PSW flags to store a COMPARE's result. One flag is set to true if the operands are equal, and the other flag indicates whether the first operand is greater than or less than the second. If the first flag is true, the second flag is ignored.

To implement program branches based on a COMPARE result, two additional conditional BRANCH instructions are provided: one based on the equality flag and the other based on the less-than or greater-than flag. Using COMPARE with related conditional BRANCH instructions simplifies machine-language programs, which speeds up their execution.

Other PSW flags represent status conditions resulting from the ALU executing instructions. Conditions such as overflow, underflow, or an attempt to perform an undefined operation (dividing by zero, for example) are represented by PSW flags. After each execution cycle, the control unit tests PSW flags to determine whether an error has occurred. They can also be tested by an OS or an application program to determine appropriate error messages and corrective actions.

WORD SIZE

A **word** is a unit of data containing a fixed number of bytes or bits and can be loosely defined as the amount of data a CPU processes at one time. Depending on the CPU, processing can include arithmetic, logic, fetch, store, and copy operations. For example, a statement such as "The Intel Core2 is a 64-bit processor" implies that logic and arithmetic operations use 64-bit operands and load and store operations use 64-bit memory addresses. Word size normally matches the size of general-purpose registers and is a fundamental CPU design decision with implications for most other computer components.

In general, a CPU with a large word size can perform a given amount of work faster than a CPU with a small word size. For example, a processor with a 128-bit word size can add or compare 128-bit integers by executing a single instruction because the registers holding the operands and the ALU circuitry are 128 bits wide.

Now consider manipulating 128-bit data in a CPU with a 64-bit word size. Because the operands are larger than the word size, they must be partitioned and the operations carried out on the pieces. For example, in a comparison operation, the CPU compares the first 64 bits of the operands and then, in a second execution cycle, compares the second 64 bits. This process is inefficient because multiple operands are loaded from or stored to memory, and multiple instructions are executed to accomplish what's logically a single operation.

Because of these inefficiencies, a 128-bit CPU usually is more than twice as fast as a 64-bit processor when processing 128-bit data values. Inefficiencies are compounded as the operation's complexity increases. For example, division and exponentiation of 128-bit data might be four or five times slower on a 64-bit processor than on a 128-bit processor.

CPU word size also has implications for system bus design. Maximum CPU performance is achieved when the bus width is at least as large as the CPU word size. If the bus width is smaller, every load and store operation requires multiple transfers to or from primary storage. For example, moving 16 bytes of character data to contiguous memory locations requires two separate data movement operations on a 64-bit bus, even if the processor's word size is 128 bits. Similarly, fetching a 128-bit instruction requires two separate transfers across the bus.

The physical implementation of memory is likewise affected by word size. Although the storage capacity of memory is always measured in bytes, data movement between memory and the CPU is generally performed in multiples of the word size. For a 64-bit CPU, memory should be organized to read or write at least 64 contiguous bits in a single access operation. Lesser capabilities incur extra wait states.

As with many other CPU design parameters, word size increases yield performance improvements only up to a certain point. Two main issues are the usefulness of the extra bits and their cost. The extra bits are useful if the larger word is less than or equal to the size of data items that application programs normally manipulate.

The added benefit of increased word size drops off sharply past 64 bits. Recall from Chapter 3 that integers are typically stored in 32-bit or 64-bit twos complement format, and real numbers are typically stored in IEEE binary32 or binary64 floating-point format. Of what use are the extra 64 bits of a 128-bit CPU when adding two 64-bit data items? They're of no use at all. The extra 64 bits simply store 0s that are carried through the computational process and are beneficial only when manipulating 128-bit data items. However, many current word-processing programs, financial applications, and other programs never manipulate 128-bit data items. Only a few types of application programs can use the additional bits. Examples include numerical-processing applications that need very high precision (such as navigation systems), numerical simulations of complex phenomena, some database- and text-processing applications, and programs that manipulate continuous streams of audio or video data. For other programs, the increased computational power of the larger word size is wasted.

The waste of extra word size wouldn't be a major concern if cost weren't an issue. However, doubling word size generally increases the number of CPU components by 2.5 to 3 times, thus increasing CPU cost and complicating microprocessor fabrication. Costs tend to rise at nonlinear rates when approaching the limits of current fabrication technology. Until recently, these limits made 64-bit CPUs an expensive luxury—generally used only in larger and more expensive computer systems. Now technology has reached the point where placing billions of transistors on a single chip is cost effective. Recent microprocessor generations, including the Intel Core2, IBM POWER6, and AMD Opteron, feature multiple CPUs on a single chip and a 64-bit word size and include 128-bit data formats, registers, and ALUs optimized for processing continuous streams of multimedia information.

ENHANCING PROCESSOR PERFORMANCE

Modern CPUs use a number of advanced techniques to improve performance, including the following:

- Memory caching
- Pipelining
- Branch prediction and speculative execution
- Multiprocessing

Memory caching is discussed in Chapter 5. The remaining techniques are forms of parallel processing and are discussed in the following sections.

Pipelining

Refer back to Figure 4.3 to review the steps in fetching and executing an instruction:

1. *Fetch* from memory.
2. *Increment* and store instruction pointer (IP).
3. *Decode* instruction and store operands and instruction pointer.
4. *Access* ALU inputs.
5. *Execute* instruction in the ALU.
6. *Store* ALU output.

Note that each function in this list is performed by a separate portion, or stage, of the CPU circuitry. All computation and comparison instructions must pass through each stage of this circuitry in sequence.

Pipelining is a method of organizing CPU circuitry so that multiple instructions can be in different stages of execution at the same time. A pipelining processor operates similarly to a car assembly line. Cars are assembled by moving them sequentially through dozens or hundreds of production stages. Each stage does a small part of the assembly process (for example, install a door, an engine, or a dashboard) and then passes the car onto the next stage. As each car passes to the next stage, another arrives from the previous stage.

Assume a typical computation or comparison instruction requires six CPU cycles to pass through all six stages—one cycle for each stage. In a traditional processor, each instruction must pass through all six stages before the next instruction can enter the first stage. Therefore, a sequence of 10 instructions requires $10 \times 6 = 60$ CPU cycles to complete execution. What if it were possible to overlap the processing stages, similar to a car assembly line (see Figure 4.12)? For example, the first instruction could be in the store stage, the second instruction in the execute stage, the third instruction in the access stage, and so forth. In theory, allowing instructions to overlap in different processor stages would reduce execution time for all 10 instructions to 15 CPU cycles—a 400% improvement!

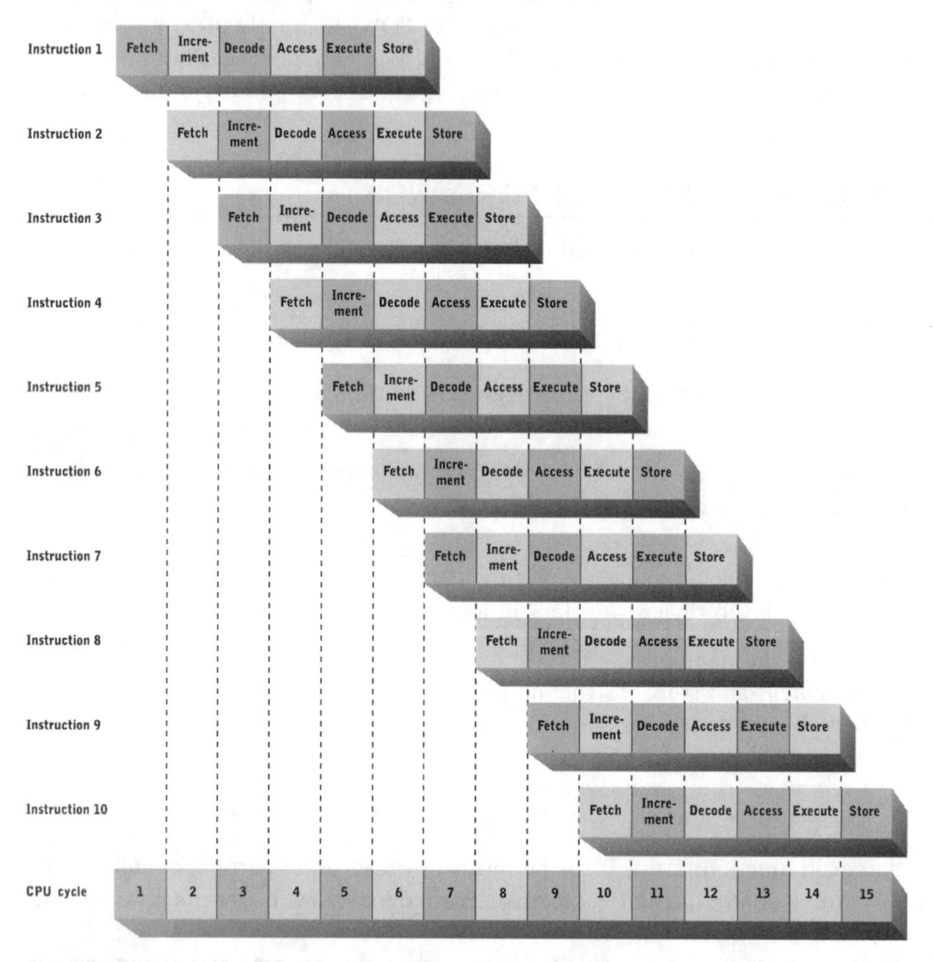

FIGURE 4.12 Overlapped instruction execution via pipelining
Courtesy of Course Technology/Cengage Learning

The theoretical improvement shown in Figure 4.12 is unlikely to be realized for several reasons. One reason is that building a processor in which all six stages take exactly one clock cycle would be nearly impossible. Operations such as fetching the next instruction from main memory and performing a computation in the ALU usually take longer than simpler operations, such as incrementing the instruction pointer. Pipelining's efficiency is reduced if some stages take longer than others.

Another reason that a 400% performance improvement is unlikely is that instructions aren't always executed sequentially. When the instruction sequence is altered by a conditional BRANCH instruction, the CPU empties the pipeline's content and starts over with the first instruction of a new sequence. This problem and methods of addressing it are described next.

Branch Prediction and Speculative Execution

As discussed in Chapter 2, one feature that enables a general-purpose processor to emulate intelligent behavior is its capability to execute algorithms that alter their behavior based on different data inputs. The example you saw was a progressive tax computation, in which the tax rate increases as income increases. The processor applied the progressive tax algorithm by executing a series of comparison operations followed by conditional BRANCHes, corresponding to a program such as the one in Figure 4.13.

```
10 INPUT INCOME
20 IF INCOME > 8350 THEN GOTO 50
30 TAX = INCOME * 0.10
40 GOTO 180
50 IF INCOME > 33950 GOTO 80
60 TAX = 835.00 + (INCOME - 8350) * 0.15
70 GOTO 180
80 IF INCOME > 82250 GOTO 110
90 TAX = 4675.00 + (INCOME - 33950) * 0.25
100 GOTO 180
110 IF INCOME > 171550 GOTO 140
120 TAX = 16750.00 + (INCOME - 82250) * 0.28
130 GOTO 180
140 IF INCOME > 372950 GOTO 170
150 TAX = 41754.00 + (INCOME - 171550) * 0.33
160 GOTO 180
170 TAX = 108216.00 + (INCOME - 372950) * 0.35
180 OUTPUT TAX
190 END
```

FIGURE 4.13 The income tax program from Chapter 2
Courtesy of Course Technology/Cengage Learning

Refer back to the six-stage pipeline shown in Figure 4.12. If instruction 4 is a conditional BRANCH and the condition controlling a conditional BRANCH is true, the branch overwrites the instruction pointer in CPU cycle 9 with a new value. At that point, instructions 5 through 9 have been fetched based on incrementing an IP value that's now invalid. Because the branch condition is true, all work on these instructions must be abandoned, and the processor must start anew to fill the pipeline and produce useful processing results. The opportunity for parallelism in completing instructions after instruction 4 has been lost, and it will be six more CPU cycles before another processing result is stored.

This situation is similar to deciding to switch models being built on a car assembly line because customers no longer want the current model. The assembly line must be stopped, and all cars currently in production must be removed. In addition, no new models will be available until one car proceeds through every stage of the assembly process from start to finish.

There are various approaches to dealing with the problem of conditional BRANCHes. One approach is fetching several instructions ahead of the current instruction and examining them to see whether any are conditional BRANCH instructions. By analyzing the order of pending instructions, the CPU can determine that it can evaluate the branch condition early. This approach is complicated, and not all branches can be evaluated out of sequence. However, if it succeeds often enough, it can improve processor speed by redirecting the CPU's efforts to the "right" set of instructions.

Under another approach, called **branch prediction**, the CPU guesses whether a branch condition will be true or false based on past experience. To provide a performance improvement, the guess must be made during or soon after the conditional BRANCH instruction is loaded, before it's completely executed. Branch prediction is best applied to conditional BRANCH instructions embedded in a program loop. For example, branch prediction could be used to compute withholding tax for all employees of a single filing status—that is, using a loop to compute the tax for all employees whose tax is computed with the same algorithm.

Using the payroll program in Figure 4.13, and assuming that most single employees have incomes greater than $8350 but less than or equal to $33,950, the CPU guesses that the branch condition in line 20 will be true and the branch condition in line 50 will be false. After the CPU loads and examines the conditional BRANCH in line 20, it immediately loads the conditional BRANCH instruction from line 50, bypassing machine instructions for lines 30 and 40. After the CPU loads and examines the conditional BRANCH instruction in line 50, it predicts false for the branch condition and begins work on the instructions that implement line 60 in the next CPU cycle.

As the CPU executes conditional BRANCH instructions in the loop, it "keeps score," tracking how often the condition for each branch instruction has been true or false. Based on the scores, the CPU determines whether to load the next sequential instruction past the conditional BRANCH or the instruction at the branch operand's address. The program path the CPU chooses is speculative—the CPU doesn't know whether its guess is correct until the conditional BRANCH instruction proceeds through all CPU stages. Therefore, the term **speculative execution** describes instructions executed after the guess but before the final result is known with certainty.

The processing needed to keep track of conditional BRANCH instructions and their scores can add considerable overhead to early CPU stages, which might slow the entire processor. If the CPU's guess is later found to be incorrect, instructions in intermediate stages of processing must be flushed from the pipeline.

Another approach that avoids the complexities of branch prediction and speculative execution is simultaneous execution (multiprocessing) of both program paths after a conditional BRANCH. Rather than guess which path will be taken, the CPU assumes that either path is possible and executes instructions from both paths until the conditional BRANCH is finally evaluated. At that point, further work on the incorrect path is abandoned, and all effort is directed to the correct path.

Simultaneous execution of both paths requires redundant CPU stages and registers. The necessary processing redundancy can be achieved with parallel stages in a single CPU or with multiple CPUs sharing memory and other resources.

Multiprocessing

The term **multiprocessing** describes any CPU architecture in which duplicate CPUs or processor stages can execute in parallel. There's a range of possible approaches to implementing multiprocessing, including the following:

- Duplicate circuitry for some or all processing stages in a single CPU
- Duplicate CPUs implemented as separate microprocessors sharing main memory and a single system bus
- Duplicate CPUs on a single microprocessor that also contains main memory caches and a special bus to interconnect the CPUs

These approaches are listed from less difficult to more difficult implementation. The first approach was made possible by microprocessor fabrication technology improvements in the late 1980s and early 1990s. The second approach became cost effective with fabrication technology advances in the late 1990s and early 2000s. The third approach has been implemented in general-purpose microprocessors since the mid-2000s.

Duplicate processing stages of a single CPU enable some, but not all, instructions to execute in parallel. For example, take a look at these formulas:

1. $((a + b)\ c) - d) \div e$
2. $((a + b) \times (c + d)) - ((e + f) \div (g + h))$

Formula 1 provides no opportunities for parallel execution because the computation operations must be implemented from left to right to produce the correct result. In formula 2, however, all four addition operations can be performed in parallel before the multiplication and division operations, which can also be performed in parallel before performing the subtraction operation. Assuming each operation requires a single ALU cycle, a processor with a single ALU stage consumes seven ALU cycles to calculate the result of formula 2. If two ALUs are available, formula 2 can be calculated in four ALU cycles by performing two additions in parallel, then two more additions in parallel, then the multiplication and division in parallel, and finally the subtraction.

The performance improvement assumes enough general-purpose registers to hold all intermediate processing results. Earlier processing stages must also be duplicated to keep the multiple ALUs supplied with instructions and data. As with pipelined processing, conditional BRANCH instructions limit the CPU's capability to execute nearby instructions in parallel, but the performance gains are substantial for complex algorithms, such as computing bank account interest and image processing.

Embedding multiple CPUs in a single computer system and sharing resources, such as main memory, among them offers more possibilities for executing parallel instructions. Because a typical computer system executes multiple programs at the same time, multiple CPUs enable instructions from two separate programs to execute simultaneously. However, the CPUs must cooperate to ensure that they don't interfere with one another. For example, the CPUs shouldn't try to write to the same memory location, send a message over the system bus, or access the same I/O device at the same time. To keep interference from happening, processors must "watch" one another continuously, which adds complexity to each processor.

Current microprocessor fabrication technology enables placing multiple CPUs on the same chip. The coordination complexities are similar to those of CPUs on separate chips, but implementing CPUs on the same chip allows them to communicate at much higher speeds and allows placing shared memory and other common resources on the same chip. This topic is discussed in more detail in Chapter 6.

TECHNOLOGY FOCUS

Intel Core Processor Family

Intel Core processors are direct descendents of the Intel 8086/8088 processors used in the original IBM PC. Current Core processors maintain backward compatibility with these processors and intervening processor generations in all areas, including instruction set, data types, and memory addressing. The first Core processor was introduced in 2006 as the successor to several generations of the Pentium processor. Later Core processor generations include the Core2 Duo, Core2 Extreme, Core2 Quad, and Core-i7.

Data Types

Core processors support a wide variety of data types (see Table 4.4). For purposes of describing data types, a word is considered to be 16 bits long (as it was in the 8086/8088). Data types are of five sizes, including 8 (byte), 16 (word), 32 (doubleword), 64 (quadword), and 128 bits (double quadword). Two variable-length data types are also defined. Note that MMX stands for "multimedia extensions," SSE stands for "streaming SIMD extensions," and SIMD stands for "single-instruction multiple-data."

TABLE 4.4 Data types supported in Core processors

Data type	Length (bits)	Coding format
Unsigned integer	8, 16, 32, or 64	Ordinary binary
Signed integer	8, 16, 32, or 64	Twos complement
Real	32, 64, or 80	IEEE floating-point standards (1985)
Bit field	32	32 binary digits
Bit string	Variable up to 2^{32} bits	Ordinary binary
Byte string	Variable up to 2^{32} bytes	Ordinary binary
Memory address	32, 48, 64, or 80	Ordinary binary
MMX (integer)	8 × 8, 4 × 16, 2 × 32	Packed twos complement
SSE (integer)	16 × 8, 8 × 16, 4 × 32, 2 × 64	Packed twos complement
SSE (real)	4 × 32, 2 × 64	Packed IEEE 32-bit floating-point

No special data types are provided for Boolean or character data, which are assumed to use appropriate integer data types. A bit field data type consists of up to 32 individually accessible bits. Two variable-length data types are also defined: bit string and byte

(continued)

string. A bit string is a sequence of 1 to 2^{32} bits commonly used for large binary objects, such as compressed audio or video files. A byte string consists of 1 to 2^{32} bytes. Elements of a byte string can be bytes, words, or doublewords, thus allowing storage of long ASCII and Unicode character sequences.

Multimedia data can be stored in a "packed" format, in which multiple twos complement or IEEE floating-point values are stored. The earliest packed data formats (called MMX data types) were defined for the Pentium processor, and several generations of extensions (called SSE data types) were defined later. Packed data types enable some CPU instructions to operate on up to 16 embedded data items in parallel, which improves performance substantially for operations such as photo editing and decoding streaming video.

Six different memory address formats are provided. Two are segmented memory addresses that are backward compatible with the 8088, 8086, and 80286 processors. Two flat memory address formats are carried over from the 80386, 80486, and Pentium processors. The remaining formats support 64-bit addressing, which allows memory sizes larger than 4 GB.

Instruction Set

As of this writing, the instruction set of a Core processor with the latest multimedia and virtual memory extensions includes 678 different instructions, with approximately 60% dedicated to packed (multimedia) data types. It's one of the largest instruction sets in modern microprocessors. As expected, such a large instruction set has many instruction formats of varying length.

A detailed discussion of the Core processor instruction set fills more than 1600 pages of documentation. The large number of instructions represents both the rich variety of supported data types and the desire to implement as many processing functions as possible in hardware, clearly placing the Core in the CISC camp of CPU design.

Word Size

There's no clear definition of word size for Core processors. For backward compatibility with 8088/8086 microprocessors, a word is defined as 16 bits. Register sizes range from 32 bits for many special-purpose registers to 128 bits for the multimedia ALU. Core processors have multiple ALUs dedicated to integer, floating-point, and packed data types, each with its own registers and internal processing paths. Internal data paths between CPU and on-chip memory components are 256 bits wide.

Clock Rate

The original Core processor debuted at a clock rate of 1.2 GHz. The fastest Core processor at this writing has a 3.33 GHz clock rate. ALUs execute at twice the processor frequency, with substantial use of parallel processing for packed data types and pipelining, branch prediction, speculative execution, and out-of-order execution for all data types. With the right kinds of data and programming, these technologies can yield multiple executed instructions per clock cycle.

THE PHYSICAL CPU

A CPU is a complex system of interconnected electrical switches. Early CPUs contained several hundred to a few thousand switches, and modern CPUs contain millions of switches. In this section, you look at how these switches perform basic processing functions and then at how switches and circuits have been implemented physically.

Switches and Gates

The basic building blocks of computer processing circuits are electronic switches and gates. Electronic **switches** control electrical current flow in a circuit and are implemented as transistors (described in more detail later in "Transistors and Integrated Circuits"). Switches can be interconnected in various ways to build gates. A **gate** is a circuit that can perform a processing function on a single binary electrical signal, or bit. The 1-bit and 2-bit processing functions performed by gates include the logical functions AND, OR, XOR, and NOT. Figure 4.14 shows the electrical component symbols for NOT, AND, OR, XOR, and NAND (NOT AND). A NOT gate is also called a signal inverter because it transforms a value of 0 or 1 into its inverse (opposite).

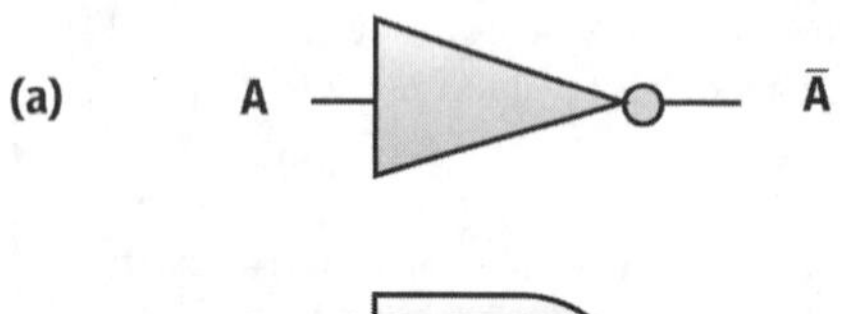

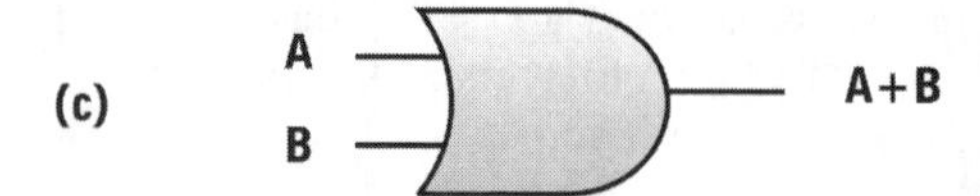

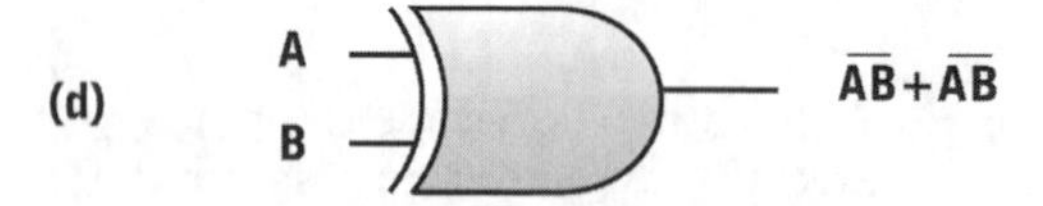

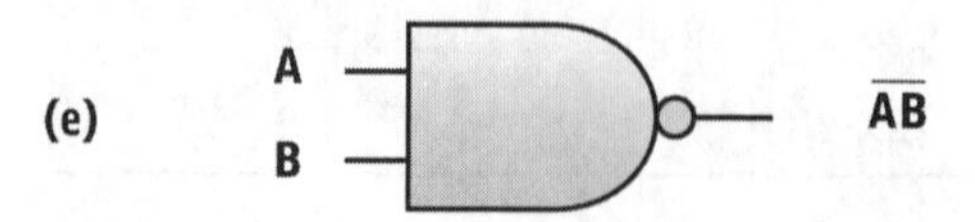

FIGURE 4.14 Electrical component symbols for a signal inverter or NOT gate (a), an AND gate (b), an OR gate (c), an XOR gate (d), and a NAND gate (e)

Courtesy of Course Technology/Cengage Learning

Processing circuits for more complex operations are constructed by combining logic gates. One example is the XOR gate. Although it's represented by a single symbol in Figure 4.14(d), it's actually constructed by combining NOT, AND, and OR gates.

Figure 4.15 shows the components of half adder and full adder circuits for single bit inputs. Addition circuits for multiple bit inputs are constructed by combining a half adder for the least significant bit position with full adder circuits for the remaining bit positions. More complex processing functions require more complicated interconnections of gates. Modern CPUs have millions of gates to perform a wide variety of processing functions on a large number of bits.

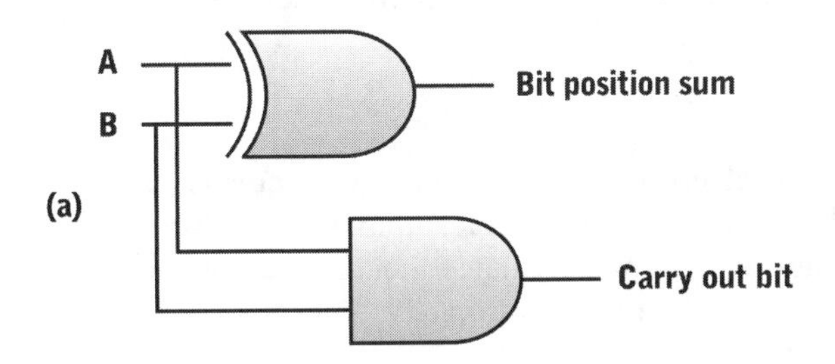

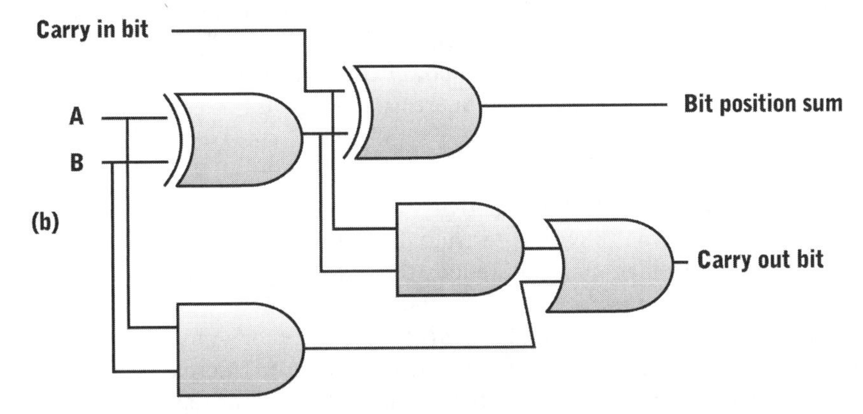

FIGURE 4.15 Circuit diagrams for half adder (a) and full adder (b)
Courtesy of Course Technology/Cengage Learning

Electrical Properties

Gates and their interconnections carry and transform electrical signals representing binary 1s and 0s. The construction of these switches and their connections is important in determining the CPU's speed and reliability. The speed and reliability of a CPU are affected not only by the materials used in its fabrication, but also by the properties of electricity, such as conductivity, resistance, and heat.

Conductivity

Electrical current is the flow of electrons from one place or device to another. An electron requires a sufficient energy input to excite it to move. After electrons are excited, they move from place to place, using molecules as stepping stones. Conductive molecules are typically arranged in straight lines, generically called **wires** or **traces**. Because each switch is usually interconnected with many other switches, a CPU contains many more traces than switches. Traces are as important as switches in determining CPU speed and reliability.

The capability of an element or a substance to enable electron flow is called **conductivity**. Substances that electrons can flow through are called **conductors**. Electrons travel through a perfect conductor with no loss of energy. With less than perfect conduction, energy is lost as electrons pass through. If enough energy is lost, electrons cease to move, and the flow of electrical current is halted. Commonly used conductors include aluminum, copper, and gold.

Resistance

The loss of electrical power that occurs as electrons pass through a conductor is called **resistance**. A perfect conductor would have no resistance. Unfortunately, all substances have some degree of resistance. Conductors with low resistance include some well-known and valuable metals, such as silver, gold, and platinum. Fortunately, some cheaper materials, such as copper, serve nearly as well.

The laws of physics state that energy is never really lost but merely converted from one form to another. Electrical energy isn't really lost because of resistance. Instead, it's converted to heat, light, or both, depending on the conductive material. The amount of generated heat depends on the amount of electrical power transmitted and the conductor's resistance. Higher power and/or higher resistance increase the amount of generated heat.

Heat

Heat has two negative effects on electrical conductivity. The first is physical damage to the conductor. To reduce heat and avoid physical destruction of a switch or trace, manufacturers must use a very low-resistance material, reduce the power input, or increase the size of the switch or trace.

The second negative effect of heat is that it changes the conductor's inherent resistance. The resistance of most materials increases as their temperature increases. To keep operating temperature and resistance low, some method must be used to remove or dissipate heat. Although many ways to do this are possible, the simplest is providing a cushion of moving air around the device. Heat migrates from the surface of the device or its coating to the air and then is transported away by fans or through ventilation openings.

A **heat sink** (see Figure 4.16) is an object specifically designed to absorb heat and rapidly dissipate it via air or water movement. It's placed in direct physical contact with an electrical device to dissipate heat from it, exposing a large surface area to the moving air to allow more rapid dissipation.

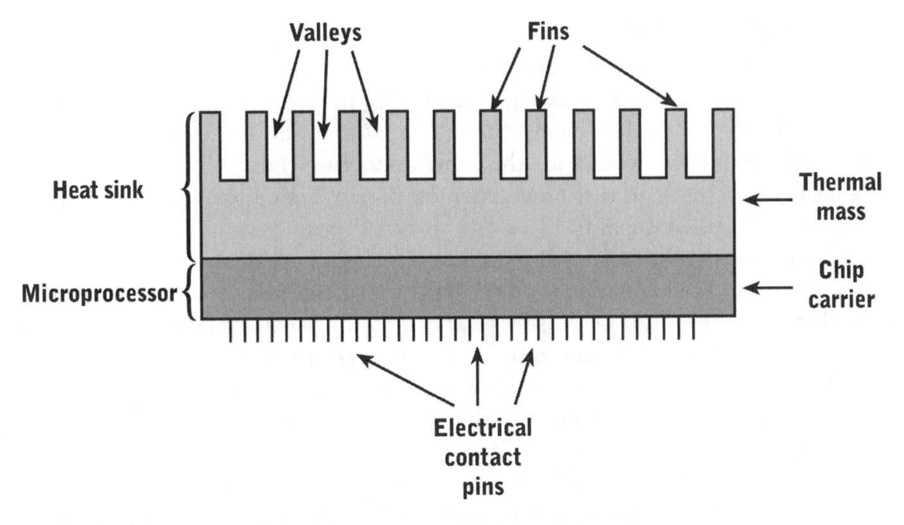

FIGURE 4.16 A heat sink attached to a surface-mounted microprocessor
Courtesy of Course Technology/Cengage Learning

Speed and Circuit Length

To perform processing, electrons must move through the traces and switches of a processing circuit. In a vacuum, electrical energy travels at the (constant) speed of light, which is approximately 180,000 miles per second. Electricity travels through a trace at approximately 70% of the speed of light. There's a fundamental relationship between circuit length and processing speed: The time required to perform a processing operation is a function of circuit length and the speed of light.

Because circuit length is the only variable, the path to faster processing is clear: Reduce circuit length. Shorter circuits require smaller switches and shorter and narrower traces. Miniaturization has been the primary basis for CPU speed and clock rate improvement since the first electrical computer. The first IBM PC used an Intel 8086/8088 microprocessor with a clock rate of 4.77 MHz. As of this writing, commercial microprocessors have clock rates as high as 5 GHz.

Processor Fabrication

Reliable, efficient, and cost-effective electrical circuits must balance power requirements, resistance, heat, size, and cost. The earliest computers were constructed with ordinary copper wire and vacuum tube switches and were unreliable because of the heat the vacuum tubes generated. They were also quite large, typically filling an entire room with processing circuitry less powerful than what's in a cheap calculator today. Improvements in materials and fabrication techniques have vastly increased processor performance and reliability.

Transistors and Integrated Circuits

In 1947, researchers at Bell Laboratories discovered a class of materials called **semiconductors**. The conductivity of these materials varies in response to the electrical inputs applied. **Transistors** are made of semiconductor material that has been treated, or doped, with chemical impurities to enhance the semiconducting effects. Silicon and germanium are basic elements with resistance characteristics that can be controlled or enhanced with chemicals called dopants.

A transistor is an electrical switch with three electrical connections. If power is applied to the second (middle) connection, power entering at the first connection flows out through the third connection. If no power is applied to the second connection, no power flows from the first connection to the third. Transistors are combined to implement the gates shown in Figure 4.14, which are combined to make larger circuits, such as adders and registers. Therefore, transistors and the traces that interconnect them are the fundamental building blocks of all CPUs.

In the early 1960s, photolithography techniques were developed to fabricate miniature electronic circuits from multiple layers of metals, oxides, and semiconductor materials. This new technology made it possible to fabricate several transistors and their interconnections on a single chip to form an **integrated circuit (IC)**. Integrated circuits reduced manufacturing cost per circuit because many chips could be manufactured in a single sheet, or wafer. Combining multiple gates on a single chip also reduced the manufacturing cost per gate and created a compact, modular, and reliable package. As fabrication techniques improved, it became possible to put hundreds, then thousands, and today, billions of electrical devices on a single chip. The term **microchip** was coined to refer to this new class of electronic devices.

Microchips and Microprocessors

A **microprocessor** is a microchip containing all the circuits and connections that implement a CPU. The first microprocessor (see Figure 4.17) was designed by Ted Hoff of Intel and introduced in 1971. Microprocessors ushered in a new era in computer design and manufacture. The most important part of a computer system could now be produced and purchased as a single package. Computer system design was simplified because designers didn't have to construct processors from smaller components. Microprocessors also opened an era of standardization as a small number of microprocessors became widely used. The PC revolution wouldn't have been possible without standardized microprocessors.

FIGURE 4.17 The Intel 4004 microprocessor containing 2300 transistors
Courtesy of Intel Corporation

Current Technology Capabilities and Limitations

Gordon Moore, an Intel founder, made an observation during a speech in 1965 that has come to be known as **Moore's Law**. He observed that the rate of increase in transistor density on microchips had increased steadily, roughly doubling every 18 to 24 months. He further observed that each doubling was achieved with no increase in unit cost. Moore's Law has proved surprisingly durable. Figure 4.18 shows that increases in transistor density of Intel microprocessors have followed Moore's Law. Software developers and users have come to expect CPU performance to double every couple of years with no price increase.

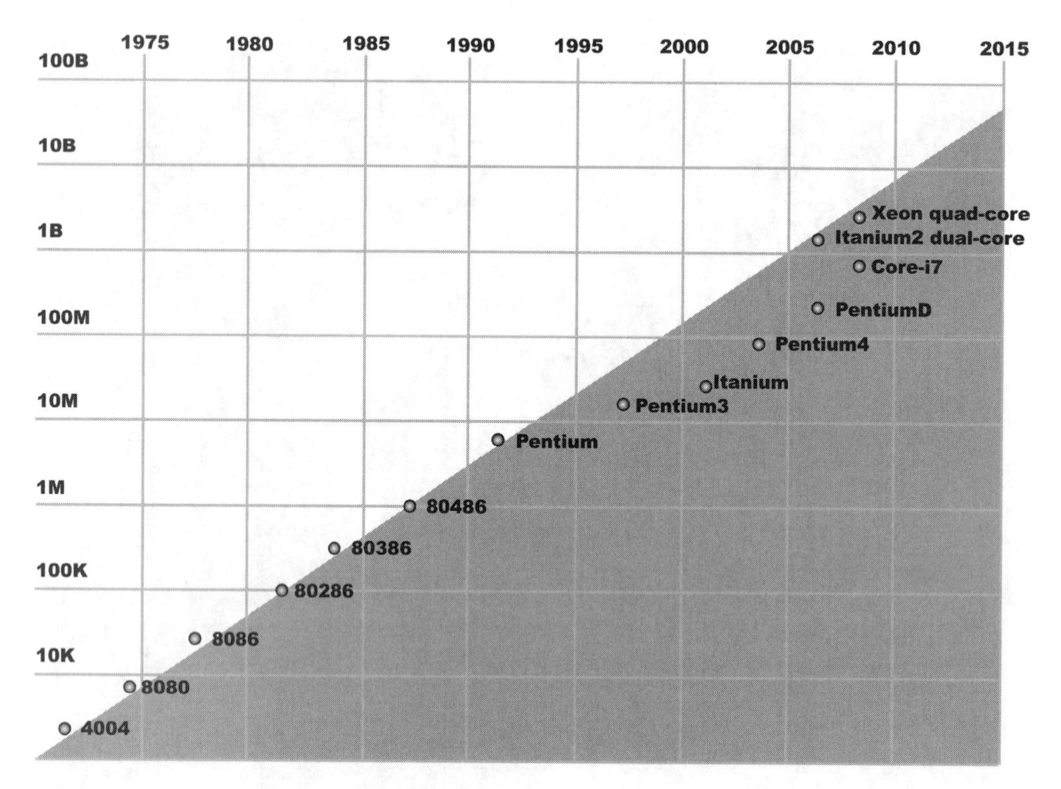

FIGURE 4.18 Increases in transistor count for Intel microprocessors
Courtesy of Course Technology/Cengage Learning

Arthur Rock, a venture capitalist, made a short addendum to Moore's Law. **Rock's Law** states that the cost of fabrication facilities for the latest chip generation doubles every four years. This law has also proved quite durable. A fabrication facility using the latest production processes currently costs at least $10 billion. Improvements in chip fabrication have been achieved by increasingly expensive and exotic techniques. The added expense hasn't led to price increases, however, because new fabrication methods produce larger numbers of chips, and the demand for chips exceeds the production capacity of any single factory.

Current fabrication technology is capable of squeezing more than one billion transistors onto a wafer of silicon approximately 1 square centimeter (see Figure 4.19). The process starts with a wafer, which is a flat disk of pure silicon, and a thin layer of conductive or semiconductive material is spread over its surface. The material is exposed to ultraviolet light focused through a patterned map, similar to an overhead transparency. An etching chemical is then applied, which removes the exposed portion of the layer. The portion that remains has a specific pattern, or map, corresponding to one layer of the microprocessor circuitry. Additional layers are added and then etched, gradually building up complete transistors and wires.

FIGURE 4.19 A wafer of processors with 410 million transistors each
Courtesy of Intel Corporation

The wafer contains hundreds or thousands of microprocessors, which are cut apart, encased in carriers, and tested. It's not unusual for half or more of the microprocessors to fail initial testing. Testing failures can be the result of impurities in materials, contamination of layers as they're laid down on the wafer, errors in etching, and variations in production process parameters. Those that survive initial tests are then subjected to further tests at successively higher clock rates. A microprocessor that fails to perform reliably at a higher clock rate might still be usable and rated for sale at a lower clock rate.

Increases in the number of circuits and packing density on a chip have been achieved by shrinking the size of transistors and the traces connecting them and by increasing the number of layers added to the wafer. Shrinking the size of transistors and traces is done with an etching process that leaves thinner and thinner lines of conductive and semiconductive material on the wafer; increasing the number of layers puts additional components on top of one another on the chip.

Increased packing density causes some problems, however. As the size of devices and wiring decreases, the capability to transmit electrical power is reduced. This decrease in transmission requires reduced operating voltage (currently as low as 0.8 volts, down from 5 volts in the 1980s). Low voltage circuits are more susceptible to damage from voltage surges and static electricity. Also, low voltage signals are more susceptible to contamination and noise from nearby circuits and external interference. On the plus side, lower operating voltage also reduces power requirements—a benefit for all computers but especially for battery-powered devices, such as laptops, netbooks, and cell phones.

Electrical resistance is also a problem because circuits must have uniform resistance properties throughout the chip. A slight increase in resistance in one part of the chip might prevent enough electrical power from reaching another part of the chip. Uniform resistance requires very high and uniform purity of the conductive and semiconductive material in the chip. Resistance can also cause problems due to heat. Because a chip is a sealed package, there's no way to circulate air or any other substance around its circuits. Heat must migrate through the chip carrier to be dissipated by a fan or heat sink. Heat generation per circuit is minimized by the circuits' small size, low operating voltage, and low-resistance materials, but each device and trace does generate heat. Higher clock rates generate more heat as more electrons are forced through the circuits. Inadequate or barely adequate cooling can substantially shorten a microprocessor's operating life.

BUSINESS FOCUS

Intel Server CPUs

Beginning with the Pentium line of CPUs, Intel has developed products along two parallel development paths, one targeted toward ordinary desktop computers and the other targeted toward workstations and servers. Table 4.5 summarizes the history of both development paths. In essence, each release of a desktop processor has been followed quickly by similar, but more powerful, processors targeted to the workstation and server market.

TABLE 4.5 Parallel development of Intel desktop and server processors

Desktop processor	Server processor
Pentium (1993)	Pentium Pro (1995)
Pentium II (1997)	Pentium II Xeon (1998)
Pentium III (1999)	Pentium III Xeon (1999)
Pentium 4 (2000)	Xeon (2001), Xeon MP (2002)
Pentium D (2005)	Xeon 7000 series (2005)
Core (2006)	Xeon 5100 series and 7200 series (2006)
Core2 (2006)	Xeon 5400 series and 7300 series (2007)
Core-i7 (2008)	Xeon 5500 series and 7400 series (2009)

(continued)

Intel Xeon processors share much technology with their Pentium and Core counterparts, including their instruction sets, integer and floating-point execution units, approaches to managing memory caches, and manufacturing processes. However, Xeon processors have extra features not found in related desktop processors, including the following:

- Larger memory caches
- The capability to access more main memory
- The capability to share access to main memory, the system bus, and other resources between multiple CPUs

Early Xeon processors filled an important niche in the small server market but weren't powerful enough to perform processing duties in many large-scale servers. Later Xeon versions have fared better in the large server market, although they still hold a small market share. Generally, large-scale servers use processors with higher computational and memory performance, on-chip error checking, and support for more than four CPUs per server. IBM, Sun Microsystems, and Hewlett-Packard currently dominate this market, although Sun Microsystems's future is uncertain.

In the mid-1990s, Intel teamed with Hewlett-Packard to develop a new CPU architecture called explicitly parallel instruction computing (EPIC). Its characteristics include 64-bit word size, pipelined execution, prefetching of instructions and operands, and parallel execution of multiple instructions.

Intel's first CPU based on the EPIC architecture, called the Itanium, was released in 2001, and the Itanium2 was released in 2002. Although the Itanium is well suited to large-scale servers, computer manufacturers have been slow to build systems based on it. Hewlett-Packard is the only vendor producing Itanium-based servers in significant quantities. IBM POWER CPUs have been more successful, partly because of IBM's entrenched position in the large-scale server market and its long history of developing and supporting server OSs.

To match IBM's success, Intel will have to convince computer manufacturers to develop new mainframe-class servers with the Itanium and convince large-scale server customers to buy them and adopt compatible operating systems. Itanium market penetration is expected to be slow, given the time and cost of mainframe development and the high costs and risks of purchasing and operating large-scale servers.

Questions:

- You're purchasing new servers for a medium-sized business in which the main server functions are file and printer sharing. Which CPU—Xeon, Itanium, or POWER—is the best choice?
- You're purchasing new servers for a server farm. The servers host e-commerce Web sites and supporting software, such as an Oracle DBMS. Which CPU is the best choice?

FUTURE TRENDS

Will Moore's Law continue to hold for the foreseeable future? Will transistor size and cost continue to shrink at the present rate? Many industry experts don't think so. They predict hitting a technological wall between 2010 and 2020, when the limits of current microprocessor design and fabrication will be reached.

Beyond a certain point, further miniaturization will be more difficult because of the nature of the etching process and the limits of semiconducting materials. The size of traces and devices depends on the etching beam wavelength, but there are limits to the ability to generate and focus short wavelength lasers and X-rays. The width of chip traces is also rapidly approaching the molecular width of the materials for constructing traces; a trace can't be less than one molecule wide.

Subsequent improvements will require fundamentally different approaches to microprocessor design and fabrication. These approaches might be based on a number of technologies, including optics and quantum physics. However, none of these technologies has yet been proved on a commercial scale.

Optical Processing

In an electrical computer, computation is performed by forcing electricity to flow through millions of electrical switches and wires. Optical computing could eliminate interconnections and simplify fabrication problems because photon pathways can cross without interfering with one another. Although fiber-optic wires are usually used to connect optical devices, connecting them without wires is possible. If the sending and receiving devices are aligned precisely, light signals sent between them can travel through free space. Eliminating wires would vastly improve fabrication cost and reliability.

Commercially available and affordable optical computers have yet to become a reality. As long as semiconductors continue to get faster and cheaper, there isn't a sufficient economic incentive to develop optical technologies. However, there's an economic incentive to pursue specialized processing applications in the areas of telecommunications and networking. Practical optical processors will probably appear first as dedicated communication controllers and later evolve into full-fledged computer processors.

Electro-Optical Processing

As silicon-based processors become faster and as more are integrated into single chips, the problem of supplying them with enough data becomes more acute. Traditional bus interfaces are simply too slow and power hungry to provide sufficient data transfer capacity between many processors or between processors and primary storage.

As their name implies, electro-optical transistors are a bridge between electrical and optical computer components. Early versions were used primarily in fiber-optic switches and generally constructed of gallium arsenide. Unfortunately, producing gallium arsenide devices and interfacing them with silicon-based processors proved exceptionally difficult. For the past decade, researchers have shifted their efforts to materials such as indium phosphide alone and in combination with silicon. They have constructed electro-optical devices by using fabrication methods more similar to those currently used to manufacture microprocessors. These devices haven't yet found their way into commercial products but are expected to in a few years.

Quantum Processing

Quantum computing, introduced in a Technology Focus in Chapter 2, uses quantum states to simultaneously encode two values per bit, called a qubit, and uses quantum processing devices to perform computations. In theory, quantum computing is well suited to solving problems that require massive amounts of computation, such as factoring large numbers and calculating private encryption keys.

Current computers solve many different types of problems, only a few of which require massive computation. As currently envisioned, quantum computing offers no benefit compared with conventional electrical processors for problem types that don't require massively parallel computation. Quantum processors might never replace conventional processors in most computer systems, or perhaps they will develop in ways that can't be imagined now. Much as Bell Labs researchers couldn't have imagined 5 GHz microprocessors when they developed the first transistor, current researchers probably can't foresee the quantum devices that will be built later in the 21st century or how they'll be applied to current and future computing problems.

Summary

- The CPU alternates continuously between the fetch (instruction) cycle and the execution cycle. During the fetch cycle, the control unit fetches an instruction from memory, separates the op code and operands, stores the operands in registers, and increments a pointer to the next instruction. During the execution cycle, the control unit or the ALU executes the instruction. The ALU executes the instruction for an arithmetic or a logical operation. The control unit executes all other instruction types.
- Primitive CPU instructions can be classified into three types: data movement, data transformation, and sequence control. Data movement instructions copy data between registers, primary storage, secondary storage, and I/O devices. Data transformation instructions implement simple Boolean operations (NOT, AND, OR, and XOR), addition (ADD), and bit manipulation (SHIFT). Sequence control instructions control the next instruction to be fetched or executed. More complex processing operations are implemented by sequences of primitive instructions.
- An instruction format is a template describing the op code's position and length and the position, type, and length of each operand. Most CPUs support multiple instruction formats. CPUs can support fixed-length or variable-length instruction formats. Fixed-length formats are simpler for the control unit to fetch and decode. Variable-length instructions use primary and secondary storage more efficiently. RISC CPUs use fixed-length instructions and generally avoid complex instructions, particularly those combining data movement and data transformation. RISC CPUs are simpler than CISC CPUs, but they're less efficient than CISC CPUs when performing some complex operations.
- The CPU clock rate is the number of instruction and execution cycles potentially available in a fixed time interval. Typical CPUs have clock rates measured in GHz (1 GHz = 1 billion cycles per second). A CPU generally executes fewer instructions than its clock rate implies. The rate of actual or average instruction execution is measured in MIPS, MFLOPS, GFLOPS, or TFLOPS. Many instructions, especially floating-point operations, require multiple execution cycles to complete. The CPU can also incur wait states pending accesses of storage or I/O devices.
- CPU registers are of two types—general-purpose and special-purpose. Programs use general-purpose registers to hold intermediate results and frequently needed data items. General-purpose registers are implemented in the CPU so that their contents can be read or written quickly. Within limits, increasing the number of general-purpose registers decreases program execution time. The CPU uses special-purpose registers to track the status of the CPU and the currently executing program. Important special-purpose registers include the instruction register, instruction pointer, and program status word.
- Word size is the number of bits a CPU can process simultaneously. Within limits, CPU efficiency increases with word size because inefficient piece-by-piece operations on large data items are avoided. For optimal performance, other computer system components, such as the system bus, should match or exceed CPU word size.
- Techniques to enhance processor performance include pipelining, branch prediction, speculative execution, and multiprocessing. Pipelining improves performance by allowing multiple instructions to execute simultaneously in different stages. Branch prediction and

speculative execution ensure that the pipeline is kept full while executing conditional BRANCH instructions. Multiprocessing provides multiple CPUs for simultaneous execution of different processes or programs.

- CPUs are electrical devices implemented as silicon-based microprocessors. Like all electrical devices, they're subject to basic laws and limitations of electricity, including conductivity, resistance, heat generation, and speed as a function of circuit length. Microprocessors use a small circuit size, low-resistance materials, and heat dissipation to ensure fast and reliable operation. They're fabricated by using expensive processes based on ultraviolet or laser etching and chemical deposition. These processes, and semiconductors themselves, are approaching fundamental physical size limits that will stop further improvements. Technologies that might move performance beyond semiconductor limitations include optical processing, hybrid optical-electrical processing, and quantum processing.

Now that you've learned about the inner workings of computer processors, it's time to move on to other computer system components. In Chapter 5, you learn the characteristics and inner functions of primary and secondary storage devices, including magnetic and optical storage technology—the bases of all modern storage devices.

Key Terms

ADD
AND
arithmetic SHIFT
benchmark
benchmark program
benchmark suite
billions of floating-point instructions per second (gigaflops or GFLOPS)
BRANCH
branch prediction
clock cycle
clock rate
complex instruction
complex instruction set computing (CISC)
conditional BRANCH
conductivity
conductor
cycle time
decoding
exclusive OR (XOR)
execution cycle
fetch cycle
fixed-length instruction
flag
gate
general-purpose register
gigahertz (GHz)
HALT
heat sink
hertz (Hz)
inclusive OR
instruction cycle
instruction format
instruction pointer (IP)
instruction register
instruction set
integrated circuit (IC)
JUMP
load
logical SHIFT
megahertz (MHz)
microchip
microprocessor
millions of floating-point operations per second (megaflops or MFLOPS)
millions of instructions per second (MIPS)

Moore's Law
MOVE
multiprocessing
NOT
op code
operand
petaflops (PFLOPS)
pipelining
program status word (PSW)
reduced instruction set computing (RISC)
resistance
Rock's Law
semiconductor
SHIFT
special-purpose register
speculative execution
store
switch
system clock
trace
transistor
trillions of floating-point operations per second (teraflops or TFLOPS)
unconditional BRANCH
variable-length instruction
wait state
wire
word

Vocabulary Exercises

1. The ________ time of a processor is 1 divided by the clock rate (in hertz).
2. A CPU typically uses multiple ________ to account for differences in the number and type of operands in instructions.
3. ________ generates heat in electrical devices.
4. ________ is a semiconducting material with optical properties.
5. A(n) ________ is an electrical switch built of semiconducting materials.
6. A(n) ________ improves heat dissipation by providing a thermal mass and a large thermal transfer surface.
7. One ________ is one cycle per second.
8. Applying a(n) ________ OR transformation to input bit values 1 and 1 generates true. Applying a(n) ________ OR transformation to the same inputs generates false.
9. When an instruction is first fetched from memory, it's placed in the ________ and then ________ to extract its components.
10. Using ________ instructions simplifies the process of instruction fetching and decoding.
11. A(n) ________ is an electrical circuit that implements a Boolean or other primitive processing function on single bit inputs.
12. A microchip containing all the components of a CPU is called a(n) ________.
13. A(n) ________ instruction transforms the bit pairs 1/1, 1/0, and 0/1 into 1.
14. The address of the next instruction to be fetched by the CPU is held in the ________.
15. The contents of a memory location are copied to a register while performing a(n) ________ operation.
16. A(n) ________ or ________ contains multiple transistors or gates in a single sealed package.

17. A(n) _______ instruction always alters the instruction execution sequence. A(n) _______ instruction alters the instruction execution sequence only if a specified condition is true.
18. A(n) _______ processor doesn't directly implement complex instructions.
19. A(n) _______ instruction copies data from one memory location to another.
20. The CPU incurs one or more _______ when it's idle, pending the completion of an operation by another device in the computer system.
21. A(n) _______ is the number of bits the CPU processes simultaneously. It also describes the size of a single register.
22. In many CPUs, a register called the _______ stores bit flags representing CPU and program status, including those representing processing errors and the results of comparison operations.
23. The components of an instruction are its _______ and one or more _______.
24. Two 1-bit values generate a 1 result value when a(n) _______ instruction is executed. All other input pairs generate a 0 result value.
25. A(n) _______ CPU typically uses variable-length instructions and has a large instruction set.
26. A(n) _______ operation transforms a 0 bit value to 1 and a 1 bit value to 0.
27. _______ predicts that transistor density will double every two years or less.
28. A(n) _______ is a measure of CPU or computer system performance when performing specific tasks.
29. _______ is a CPU design technique in which instruction execution is divided into multiple stages and different instructions can execute in different stages simultaneously.

Review Questions

1. Describe the operation of a MOVE instruction. Why is the name "move" a misnomer?
2. Why does program execution speed generally increase as the number of general-purpose registers increases?
3. What are special-purpose registers? Give three examples of special-purpose registers and explain how each is used.
4. What are the advantages and disadvantages of fixed-length instructions compared with variable-length instructions? Which type is generally used in a RISC processor? Which type is generally used in a CISC processor?
5. Define "word size." What are the advantages and disadvantages of increasing word size?
6. What characteristics of the CPU and primary storage should be balanced to achieve maximum system performance?
7. How does a RISC processor differ from a CISC processor? Is one processor type better than the other? Why or why not?
8. What factors account for the dramatic improvements in microprocessor clock rates over the past three decades?

9. What potential advantages do optical processors offer compared with electrical processors?
10. Which is the better measure of computer system performance—a benchmark, such as SPEC CINT, or a processor speed measure, such as GHz, MIPS, or MFLOPS? Why?
11. How does pipelining improve CPU efficiency? What's the potential effect on pipelining's efficiency when executing a conditional BRANCH instruction? What techniques can be used to make pipelining more efficient when executing conditional BRANCH instructions?
12. How does multiprocessing improve a computer's efficiency?

Problems and Exercises

1. Develop a program consisting of primitive CPU instructions to implement the following procedure:

```
integer i,a;
i=0;
while (i < 10) do
  a=i*2;
  i=i+1;
endwhile
```

2. If a microprocessor has a cycle time of 0.5 nanoseconds, what's the processor clock rate? ✓ If the fetch cycle is 40% of the processor cycle time, what memory access speed is required to implement load operations with zero wait states and load operations with two wait states?
3. Processor R is a 64-bit RISC processor with a 2 GHz clock rate. The average instruction requires one cycle to complete, assuming zero wait state memory accesses. Processor C is a CISC processor with a 1.8 GHz clock rate. The average simple instruction requires one cycle to complete, assuming zero wait state memory accesses. The average complex instruction requires two cycles to complete, assuming zero wait state memory accesses. Processor R can't directly implement the complex processing instructions of Processor C. Executing an equivalent set of simple instructions requires an average of three cycles to complete, assuming zero wait state memory accesses.

   Program S contains nothing but simple instructions. Program C executes 70% simple instructions and 30% complex instructions. Which processor will execute program S more quickly? Which processor will execute program C more quickly? At what percentage of complex instructions will the performance of the two processors be equal?
4. You have a CPU with a 4.8 GHz clock rate. Both the fetch and execution cycles are 50% of the clock cycle. The average instruction requires 0.5 nanoseconds to complete execution. Main memory access speed for a single instruction is 2 nanoseconds on average. What is the expected average MIPS rate for this CPU? What modern microprocessor architectural features might be added to the CPU to improve its MIPS rate?

Research Problems

1. Investigate the instruction set and architectural features of a modern RISC processor, such as the IBM POWER6 (*www.ibm.com/technology*). In what ways does it differ from the architecture of the Intel Core and Itanium processor families? Which processors use advanced techniques, such as pipelining, speculative execution, and multiprocessing?

2. AMD (*www.amd.com*) produces microprocessors that execute the same CPU instructions as Intel desktop and laptop processors. Investigate its current product offerings. Do they offer true Intel compatibility? Is their performance comparable with that of Intel processors?
3. The Web site Top500.org publishes a list, updated monthly, of the world's most powerful supercomputers. Go to this Web site and investigate the top 10 or 20 computers. What CPUs are used in these computers? What are their clock rates and word sizes? Why do Intel CPUs seem to be underrepresented in the list?

CHAPTER 5

DATA STORAGE TECHNOLOGY

CHAPTER GOALS

- Describe the distinguishing characteristics of primary and secondary storage
- Describe the devices used to implement primary storage
- Compare secondary storage alternatives
- Describe factors that affect magnetic storage devices
- Explain how to choose appropriate secondary storage technologies and devices

In Chapter 2, you were briefly introduced to the topic of storage, including the role of storage in a computer system and the differences between primary and secondary storage. In this chapter, you explore storage devices and their underlying technologies in depth (see Figure 5.1). This chapter also outlines characteristics common to all storage devices and compares primary and secondary storage technologies.

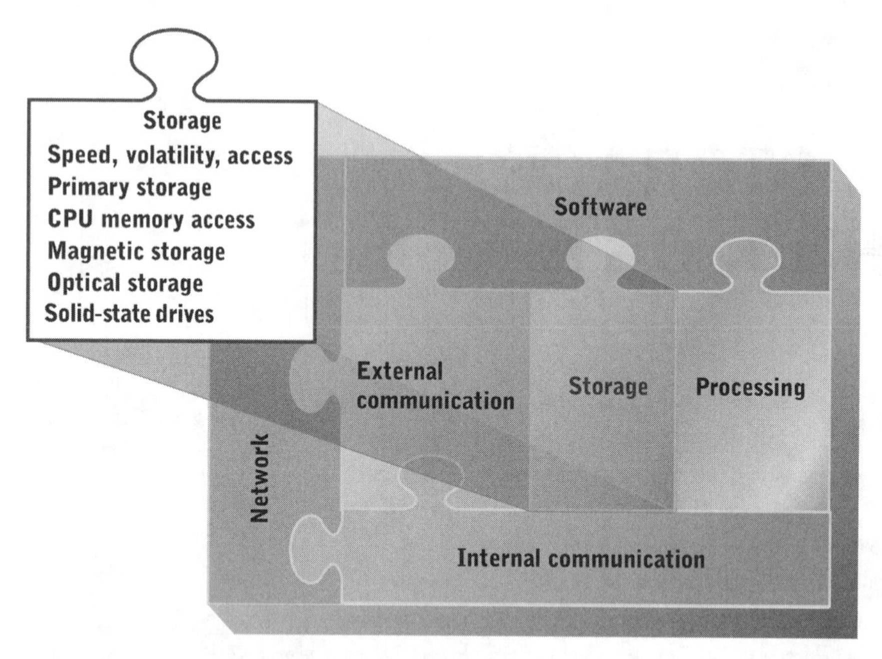

FIGURE 5.1 Topics covered in this chapter
Courtesy of Course Technology/Cengage Learning

STORAGE DEVICE CHARACTERISTICS

A storage device consists of a read/write mechanism and a storage medium. The **storage medium** is the device or substance that actually holds data. The read/write mechanism is the device used to read or write data to and from the storage medium. A device controller provides the interface between the storage device and system bus.

NOTE

Device controllers are discussed in detail in Chapter 6.

In some storage devices, the read/write mechanism and storage medium are a single unit using the same technology. For example, most types of primary storage use electrical circuits implemented with semiconductors for both the read/write mechanism and storage medium. In other storage devices, the read/write mechanism and the storage medium use fundamentally different technologies. For example, tape drives use an electromechanical device for the read/write mechanism and a magnetic storage medium composed of polymers and metal oxides.

A typical computer system has many storage devices (see Figure 5.2). Storage devices and technologies vary in several important characteristics, which include the following:

- Speed
- Volatility

- Access method
- Portability
- Cost and capacity

No single device or technology is optimal in all characteristics, so any storage device optimizes some characteristics at the expense of others. A computer system has a variety of storage devices, each offering a cost-effective solution to a particular storage requirement.

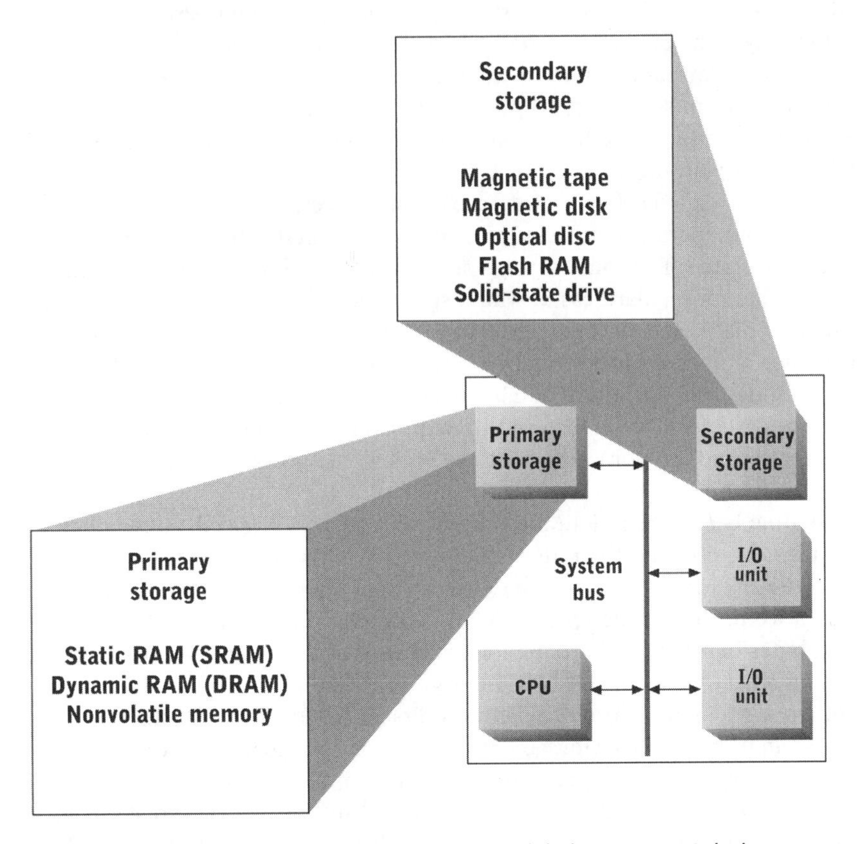

FIGURE 5.2 Primary and secondary storage and their component devices
Courtesy of Course Technology/Cengage Learning

Speed

Speed is the most important characteristic differentiating primary and secondary storage. It's essential because the CPU must be continuously supplied with instructions and data to keep it busy. For example, a CPU with a 1 GHz clock rate needs a new instruction and supporting data every nanosecond. As you learned in Chapter 4, a wait state is a CPU cycle spent waiting for access to an instruction or data. Wait states reduce CPU and computer system performance.

As discussed in Chapter 2, registers in the CPU are storage locations for instructions and data. Their location enables zero wait states for access, but CPUs have a limited number of registers—far fewer than are needed to hold typical programs and their data. Primary storage extends the limited capacity of CPU registers. The CPU moves data and instructions continually between registers and primary storage. To ensure that this movement incurs few or no wait states, all or part of primary storage is implemented with the fastest available storage devices. With current technology, primary storage speed is typically faster than secondary storage speed by a factor of 10^5 or more.

Speed is also an important issue for secondary storage. Many information system applications need access to large databases to support ongoing processing. Program response time in these systems depends on secondary storage access speed, which also affects overall computer performance in other ways. Before a program can be executed, its executable code is copied from secondary to primary storage. The delay between a user request for program execution and the first prompt for user input depends on the speed of both primary and secondary storage.

Storage device speed is called access time. **Access time** is the time required to perform one complete read or write operation. It's assumed to be the same for both reading and writing unless otherwise stated. For some storage devices, such as random access memory (RAM), access time is the same regardless of which storage location is accessed. For other storage devices, such as disks, access time varies with storage location and is typically expressed as an average of access times for all storage locations, called **average access time**. (Access times are described in more detail later in "Magnetic Disk"). Access times of primary storage devices are generally expressed in nanoseconds (ns, billionths of a second). Access times for secondary storage devices are typically expressed in milliseconds (ms, thousandths of a second).

By itself, access time is an incomplete measure of storage device speed. A complete measure consists of access time and the unit of data transfer to or from the storage device. Data transfer unit size varies from one storage device to another. For primary storage devices, it's usually a word. Depending on the CPU, a word can represent 2, 4, 8, or more bytes, with 4 and 8 bytes being the most common word sizes. Data transfer unit size is sometimes used as an adjective, as in "64-bit memory."

Secondary storage devices read or write data in units larger than words. **Block** is a generic term for describing secondary storage data transfer units. Block size is normally stated in bytes and can vary widely between storage devices and even in a single storage device. A 512-byte block is the most common data transfer unit for magnetic disks. The term **sector** describes the data transfer unit for magnetic disk and optical disc drives. As with blocks, sector size is generally stated in bytes and can vary from one device to another.

A storage device's **data transfer rate** is computed by dividing 1 by the access time (expressed in seconds) and multiplying the result by the unit of data transfer (expressed in bytes). For example, the data transfer rate for a primary storage device with 15 ns access time and a 64-bit word data transfer unit can be calculated as follows:

$$\frac{1 \text{ second}}{15 \text{ ns}} \times 64 \text{ bits} = \frac{1}{0.000000015} \times 8 \text{ bytes} = 533,333,333 \text{ bytes/second}$$

Using average access time, this formula can be used to calculate an average data transfer rate for devices with variable access times.

Volatility

A storage device or medium is **nonvolatile** if it holds data without loss over long periods and is **volatile** if it can't hold data reliably for long periods. Primary storage devices are generally volatile, and secondary storage devices are generally nonvolatile.

Volatility is actually a matter of degree and conditions. For example, RAM is nonvolatile as long as external power is supplied continuously. However, it's generally considered a volatile storage device because continuous power can't be guaranteed under normal operating conditions (for example, during a system restart after installing an OS update). Magnetic tape and disk are considered nonvolatile storage media, but data on magnetic media is often lost after a few years because of natural magnetic decay and other factors. Data stored on nonvolatile media might also be lost because of compatible read/write devices becoming obsolete, which has happened with many older diskette and tape formats.

Access Method

The physical structure of a storage device's read/write mechanism and storage medium determines the ways in which data can be accessed. Three broad classes of access are recognized; a single device can use multiple access methods:

- Serial access
- Random access
- Parallel access

Serial Access

A **serial access** storage device stores and retrieves data items in a linear, or sequential, order. Magnetic tape is the only widely used form of serial access storage. Data is written to a linear storage medium in a specific order and can be read back only in that same order. For example, viewing the past few minutes recorded on a video cassette tape requires playing or fast-forwarding past all the minutes preceding it. Similarly, if n digital data items are stored on a magnetic tape, the nth data item can be accessed only after the preceding $n - 1$ data items have been accessed.

Serial access time depends on the current position of the read/write mechanism and the position of the target data item in the storage medium. If both positions are known, access time can be computed as the difference between the current and target positions multiplied by the time required to move from one position to the next.

Because of their inefficient access method, serial access devices aren't used for frequently accessed data. Even when data items are accessed in the same order as they're written, serial access devices are much slower than other forms of storage. Serial access storage is used mainly to hold backup copies of data stored on other storage devices—for example, in a weekly tape backup of user files from magnetic disk.

Random Access

A **random access** device, also called a **direct access** device, isn't restricted to any specific order when accessing data. Rather, it can access any storage location directly. All primary storage devices and disk storage devices are random access devices.

Access time for a random access storage device may or may not be a constant. It's a constant for most primary storage devices but not for disk storage because of the physical, or spatial, relationships of the read/write mechanism, storage media, and storage locations on the media. This issue is discussed in more detail later in "Magnetic Storage."

Parallel Access

A **parallel access** device can access multiple storage locations simultaneously. Although RAM is considered a random access storage device, it's also a parallel access device. This confusion comes from differences in defining the unit of data access. If you consider the unit of data access to be a bit, access is parallel. That is, random access memory circuitry can access all the bits in a byte or word at the same time.

Parallel access can also be achieved by subdividing data items and storing the component pieces on multiple storage devices. For example, some OSs can store file content on several disk drives. Different parts of the same file can be read at the same time by issuing parallel commands to each disk drive.

Portability

Storage device portability is typically implemented in one of two ways:

- The entire storage device—storage medium, read/write mechanism, and possibly controller—can be transported between computer systems (for example, a USB flash drive).
- The storage medium can be removed from the storage device and transported to a compatible storage medium on another computer (for example, a DVD).

Portable secondary storage devices and devices with removable storage media typically have slower access speeds than permanently installed devices and those with nonremovable media. High-speed access requires tight control of environmental factors and high-speed communication channels to connect the device to the system bus. For example, high-speed access for magnetic disks is achieved, in part, by sealed enclosures for the storage media, thus minimizing or eliminating dust and air density variations. Removable storage media come in contact with a wider variety of damaging environmental influences than isolated media do. Also, an internal magnetic disk drive is usually connected to the system bus by a high-speed channel, whereas a portable magnetic disk drive typically uses a slower channel, such as FireWire or USB.

Cost and Capacity

Each storage device attribute described so far is related to device cost per unit of storage (see Table 5.1). For example, improving speed, volatility, or portability increases cost per unit if all other factors are held constant. Cost per unit also increases as an access method moves from serial to random to parallel. Primary storage is generally expensive compared with secondary storage because of its high speed and combination of parallel and random access methods.

TABLE 5.1 Storage device characteristics and their relationship to cost

Characteristic	Description	Cost
Speed	Time required to read or write a bit, byte, or larger unit of data	Cost increases as speed increases.
Volatility	Capability to hold data indefinitely, particularly in the absence of external power	For devices of a similar type, cost decreases as volatility increases.
Access method	Can be serial, random, or parallel; parallel devices are also serial or random access	Serial is the least expensive; random is more expensive than serial; parallel is more expensive than nonparallel.
Portability	Capability to easily remove and reinstall storage media from the device or the device from the computer	For devices of a similar type, portability increases cost, if all other characteristics are held constant.
Capacity	Maximum data quantity the device or storage medium holds	Cost usually increases in direct proportion to capacity.

Secondary storage devices are usually thought of as having much higher capacity than primary storage devices. In fact, the capacity difference between primary and secondary storage in most computers results from a compromise between cost per unit and other device characteristics. For example, if cost weren't a factor, most users would opt for solid-state drives rather than magnetic disk drives. However, most users need hundreds of gigabytes of secondary storage, and solid-state drives of this capacity cost more than most users can afford. So users sacrifice speed and parallel access to gain the capacity they need at an acceptable cost.

Memory-Storage Hierarchy

A typical computer system has a variety of primary and secondary storage devices. The CPU and a small amount of high-speed RAM usually occupy the same chip. Slower RAM on separate chips composes the bulk of primary storage. One or more magnetic disk drives are usually complemented by an optical disc drive and at least one form of removable magnetic storage.

The range of storage devices in a single computer system forms a memory-storage hierarchy, as shown in Figure 5.3. Cost and access speed generally decrease as you move down the hierarchy. Because of lower cost, capacity tends to increase as you move down the hierarchy. A computer designer or purchaser attempts to find an optimal mix of cost and performance for a particular purpose.

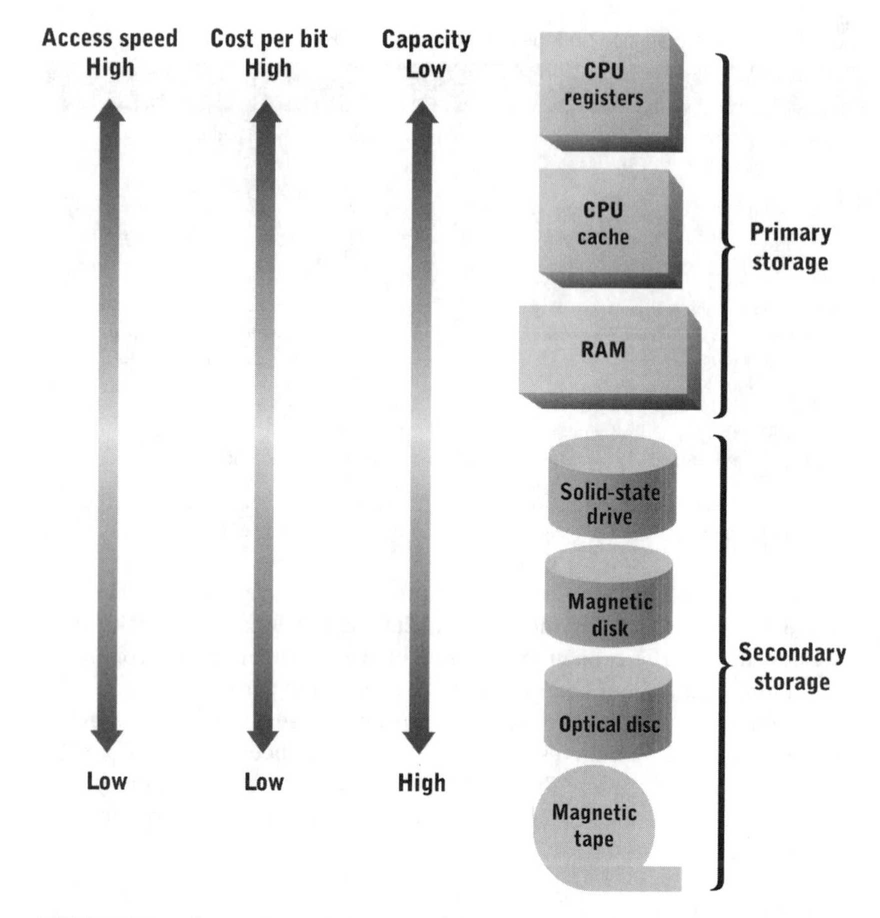

FIGURE 5.3 Comparison of storage devices in terms of cost and access speed
Courtesy of Course Technology/Cengage Learning

PRIMARY STORAGE DEVICES

As discussed, the critical performance characteristics of primary storage devices are access speed and data transfer unit size. Primary storage devices must closely match CPU speed and word size to avoid wait states. CPU and primary storage technologies have evolved in tandem—in other words, CPU technology improvements are applied to the construction of primary storage devices.

Storing Electrical Signals

Data is represented in the CPU as digital electrical signals, which are also the basis of data transmission for all devices attached to the system bus. Any storage device or controller must accept electrical signals as input and generate electrical signals as output.

Electrical power can be stored directly by various devices, including batteries and capacitors. Unfortunately, there's a tradeoff between access speed and volatility. Batteries are quite slow to accept and regenerate electrical current. With repeated use, they also lose their capability to accept a charge. Capacitors can charge and discharge much faster than batteries. However, small capacitors lose their charge fairly rapidly and require a fresh injection of electrical current at regular intervals (hundreds or thousands of times per second).

An electrical signal can be stored indirectly by using its energy to alter the state of a device, such as a mechanical switch, or a substance, such as a metal. An inverse process regenerates an equivalent electrical signal. For example, electrical current can generate a magnetic field. The magnetic field's strength can induce a permanent magnetic charge in a nearby metallic compound, thus "writing" the bit value to the metallic compound. To read the stored value, the stored magnetic charge is used to generate an electrical signal equivalent to the one used to create the original magnetic charge. Magnetic polarity, which is positive or negative, can represent the values 0 and 1.

Early computers implemented primary storage as rings of ferrous material (iron and iron compounds), a technology called **core memory**. These rings, or cores, are embedded in a two-dimensional wire mesh. Electricity sent through two wires induces a magnetic charge in one metallic ring. The charge's polarity depends on the direction of electrical flow through the wires.

Modern computers use memory implemented with semiconductors. Basic types of semiconductor memory include random access memory and nonvolatile memory. There are many variations of each memory type, described in the following sections.

Random Access Memory

Random access memory (RAM) is a generic term describing primary storage devices with the following characteristics:

- Microchip implementation with semiconductors
- Capability to read and write with equal speed
- Random access to stored bytes, words, or larger data units

RAM is fabricated in the same manner as microprocessors. You might assume that microprocessor clock rates are well matched to RAM access speeds. Unfortunately, this isn't the case for many reasons, including the following:

- Reading and writing many bits in parallel requires additional circuitry.
- When RAM and microprocessors are on separate chips, there are delays when moving data from one chip to another.

There are two basic RAM types and several variations of each type. **Static RAM (SRAM)** is implemented entirely with transistors. The basic storage unit is a flip-flop circuit (see Figure 5.4). Each flip-flop circuit uses two transistors to store 1 bit. Additional transistors (typically two or four) perform read and write operations. A flip-flop circuit is an electrical switch that remembers its last position; one position represents 0 and the other represents 1. These circuits require a continuous supply of electrical power to maintain their positions. Therefore, SRAM is volatile unless a continuous supply of power can be guaranteed.

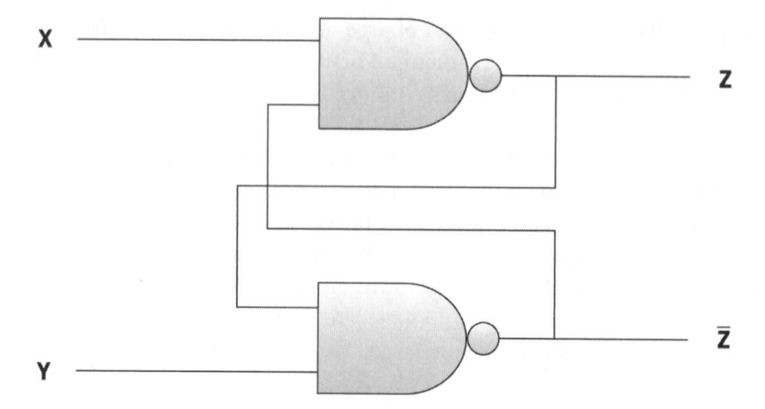

FIGURE 5.4 A flip-flop circuit composed of two NAND gates: the basic component of SRAM and CPU registers
Courtesy of Course Technology/Cengage Learning

Dynamic RAM (DRAM) stores each bit by using a single transistor and capacitor. Capacitors are the dynamic element. However, because they lose their charge quickly, they require a fresh infusion of power thousands of times per second, so DRAM chips include circuitry that performs refresh operations automatically. Each refresh operation is called a **refresh cycle**. Unfortunately, a DRAM chip can't perform a refresh operation at the same time it performs a read or write operation.

Because DRAM circuitry is simpler, more memory cells can be packed into each chip. (In other words, DRAM has higher density.) In contrast, fewer SRAM bits can be implemented in each chip, effectively making SRAM more expensive than DRAM. Despite its simpler circuitry, DRAM is slower than SRAM because of its required refresh cycles and less efficient circuitry for accessing bits. With current fabrication technology, typical access times are 10 to 20 ns for DRAM and 1 to 5 ns for SRAM. Improvements in fabrication technology can decrease both access times but can't change the performance difference.

Note that neither RAM type can match current microprocessor clock rates, which range from 3 to 4 GHz at this writing. For zero wait states in memory accesses, these clock rates require the following memory access speeds:

$$\frac{1}{3\text{ GHz}} = \frac{1}{3{,}000{,}000{,}000} \approx 0.00000000033 \approx 0.33\text{ ns}$$

$$\frac{1}{4\text{ GHz}} = \frac{1}{4{,}000{,}000{,}000} = 0.00000000025 = 0.25\text{ ns}$$

The fastest DRAM is one to two orders of magnitude slower than current microprocessors. SRAM is up to one order of magnitude slower. The following technologies are used to bridge the performance gap between memory and microprocessors:

- Read-ahead memory access
- Synchronous read operations

Computer memory contains more than just SRAM or DRAM circuits. Memory circuits are grouped in memory modules containing tens or hundreds of megabytes. Each module contains additional circuitry to implement read and write operations to random locations—circuitry that must be activated before data can be accessed. Activating the read/write circuitry consumes time, which lengthens the time required to access a memory location.

Programs usually access instructions and data items sequentially. Read-ahead memory access takes advantage of this tendency by activating the read/write circuitry for location $n + 1$ during or immediately after an access to memory location n. If the CPU subsequently issues an access command for location $n + 1$, the memory module has already performed part of the work needed to complete the access.

Synchronous DRAM (SDRAM) is a read-ahead RAM that uses the same clock pulse as the system bus. Read and write operations are broken into a series of simple steps, each of which can be completed in one bus clock cycle (similar to pipelined processing, described in Chapter 4). Several clock cycles are required to complete a single random access. However, if memory is being written or read sequentially, accesses occur once per system bus clock tick after the first access has completed.

SDRAM was improved steadily through the late 1990s and 2000s with a series of technologies that doubled the data transfer rate of the previous technology. These technologies are called **double data rate (DDR)** and have been named DDR, DDR2, DDR3, and a proposed DDR4. DDR3 SDRAM supports clock rates up to 1 GHz and reads or writes eight 64-bit words per clock cycle.

Nonvolatile Memory

Memory manufacturers have worked for decades to develop semiconductor and other forms of RAM with long-term or permanent data retention. The generic term for these memory devices is **nonvolatile memory (NVM)**. Of course, manufacturers and consumers would like NVM to cost the same or less than conventional SRAM and DRAM and have similar or faster read/write access times. So far, these goals have proved elusive, but some current and emerging memory technologies show considerable promise.

Even though NVM currently lacks the access speed of RAM, it has many applications in computing and other industries, including storage of programs and data in portable devices, such as handheld computers, netbook computers, and cell phones; portable secondary storage in devices such as digital cameras; and permanent program storage in motherboards and peripheral devices. One of the oldest uses in computers is storing programs such as boot subroutines—the system BIOS, for example. These instructions can be loaded at high speed into main memory from NVM. Software stored in NVM is called **firmware**.

Early NVM technology has evolved through several generations of devices. **Read-only memory (ROM)** is the earliest type of NVM, with data content written permanently during manufacture. **Erasable programmable ROM (EPROM)** is manufactured blank, written (programmed) with a special EPROM writer, and erased by exposure to ultraviolet light. The latest (and only currently used) form of ROM is **electronically erasable programmable ROM (EEPROM)**. An EEPROM device can be programmed, erased, and reprogrammed by signals sent from a CPU. The main drawbacks of EEPROM technology are low density, high cost, and a write speed that's much too slow to be used in primary storage devices.

The most common NVM in use today is **flash RAM** (also called "flash memory"). It's competitive with DRAM in storage density (capacity) and read performance. Unfortunately, write performance is much slower than in DRAM. Also, each write operation is mildly destructive, resulting in storage cell destruction after 100,000 or more write operations. Because of its slower write speed and limited number of write cycles, flash RAM currently has limited applications. It's used to store firmware programs, such as the system BIOS, that aren't changed frequently and are loaded into memory when a computer powers up. Flash RAM is also used in portable secondary storage devices, such as compact flash cards in digital cameras and USB flash drives. These storage devices typically mimic the behavior of a portable magnetic disk drive when connected to a computer system. Flash RAM is also beginning to challenge magnetic disk drives as the dominant secondary storage technology (see "Technology Focus: Solid-State Drives" later in this chapter).

Other NVM technologies under development could overcome some shortcomings of flash RAM. Two promising candidates are magnetoresistive RAM and phase-change memory. **Magnetoresistive RAM (MRAM)** stores bit values by using two magnetic elements, one with fixed polarity and the other with polarity that changes when a bit is written. The second magnetic element's polarity determines whether a current passing between the elements encounters low (a 0 bit) or high (a 1 bit) resistance. The latest MRAM generations have read and write speeds comparable with SRAM and densities comparable with DRAM, which make MRAM a potential replacement for both these RAM types. In addition, MRAM doesn't degrade with repeated writes, which gives it better longevity than conventional flash RAM.

Phase-change memory (PCM), also known as phase-change RAM (PRAM or PCRAM), uses a glasslike compound of **germanium, antimony, and tellurium (GST)**. When heated to the correct temperatures, GST can switch between amorphous and crystalline states. The amorphous state exhibits low reflectivity (useful in rewritable optical storage media) and high electrical resistance. The crystalline state exhibits high reflectivity and low electrical resistance. PCM has lower storage density and slower read times than conventional flash RAM, but its write time is much faster, and it doesn't wear out as quickly.

Memory Packaging

Memory packaging is similar to microprocessor packaging. Memory circuits are embedded in microchips, and groups of chips are packed on a small circuit board that can be installed or removed easily from a computer system. Early RAM and ROM circuits were packaged in **dual inline packages (DIPs)**. Installing a DIP on a printed circuit board is a tedious and precise operation. Also, single DIPs mounted on the board surface occupy a large portion of the total surface area.

In the late 1980s, memory manufacturers adopted the **single inline memory module (SIMM)** as a standard RAM package. Each SIMM incorporates multiple DIPs on a tiny printed circuit board. The edge of the circuit board has a row of electrical contacts, and the entire package is designed to lock into a SIMM slot on a motherboard. The **double inline memory module (DIMM)**, a newer packaging standard, is essentially a SIMM with independent electrical contacts on both sides of the module, as shown in Figure 5.5.

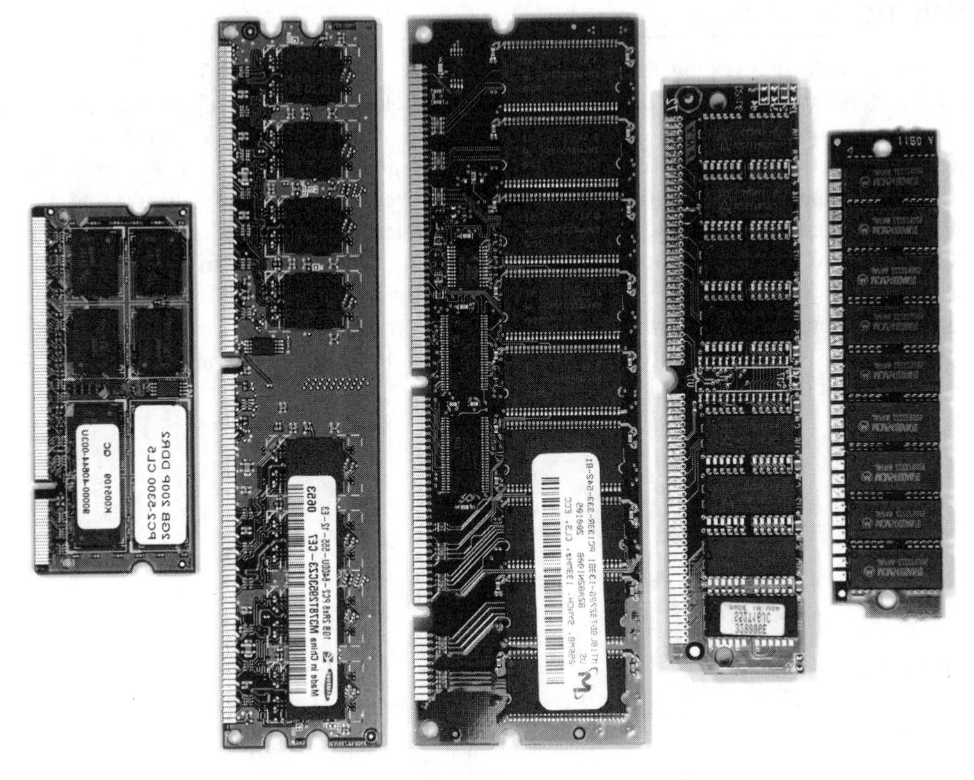

FIGURE 5.5 From left to right, 30-pin SIMM, 72-pin SIMM, DDR DIMM, DDR2 DIMM, and DDR2 DIMM for a laptop computer
Courtesy of Course Technology/Cengage Learning

Current microprocessors include a small amount of on-chip memory (described in Chapter 6). As fabrication techniques improve, the amount of memory that can be packaged with the CPU on a single chip will grow. The logical extension of this trend is placing a CPU and all its primary storage on a single chip, which would minimize or eliminate the current gap between microprocessor clock rates and memory access speeds.

NOTE

Although main memory isn't currently implemented as part of the CPU, the CPU's need to load instructions and data from memory and store processing results requires close coordination between both devices. Specifically, the physical organization of memory, the organization of programs and data in memory, and the methods of referencing specific memory locations are critical design issues for both primary storage devices and processors. These topics are discussed in Chapter 11.

MAGNETIC STORAGE

Magnetic storage devices exploit the duality of magnetism and electricity. That is, electrical current can generate a magnetic field, and a magnetic field can generate electricity. A magnetic storage device converts electrical signals into magnetic charges, captures the magnetic charge on a storage medium, and later regenerates an electrical current from the stored magnetic charge. The magnetic charge's polarity represents the bit values 0 and 1.

Figure 5.6 illustrates a simple magnetic storage device. A wire is coiled around a metallic **read/write head**. To perform a write operation, an electrical current is passed through the wire, which generates a magnetic field across the gap in the read/write head. The direction of current flow through the wire determines the field's polarity—that is, the position of the positive and negative poles of the magnetic field. Reversing the current's direction reverses the polarity.

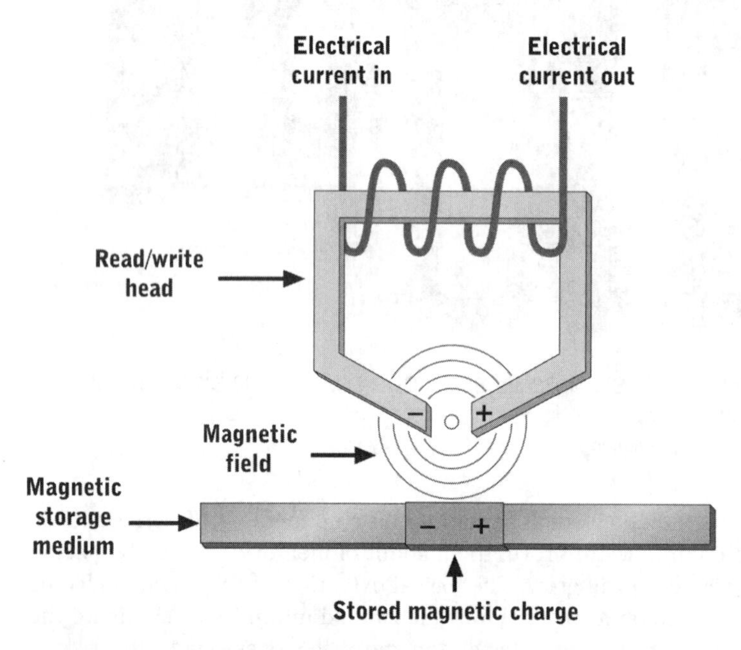

FIGURE 5.6 Principles of magnetic data storage
Courtesy of Course Technology/Cengage Learning

A magnetic storage medium is placed next to the gap, and the portion of the medium's surface closest to the magnetic field is permanently charged. The stored charge's polarity is identical to the polarity of the generated magnetic field. In addition, the strength of the stored charge is directly proportional to the magnetic field's strength, which is determined by several factors, including number of coils in the wire, strength of the electrical current, and mass of the metallic read/write head. The storage medium must be constructed of or coated with a substance capable of accepting and storing a magnetic charge, such as a metallic compound.

A read operation is the inverse of a write operation. The portion of the storage medium with the data being read is placed near the gap of the read/write head, and the stored charge radiates a small magnetic field. When the magnetic field comes in contact with the read/write head, a small electrical current is induced in the read/write head and the wire wrapped around it. The current direction depends on the stored charge's polarity. Electrical switches at either end of the coiled wire detect the direction of current flow and sense a 0 or 1 bit value.

Magnetic storage devices must control or compensate for some undesirable characteristics of magnetism and magnetic storage media, including the following:

- Magnetic decay
- Magnetic leakage
- Minimum threshold current for read operations
- Storage medium coercivity
- Long-term storage medium integrity

A magnetic storage device must balance all these factors carefully to achieve cost-effective and reliable storage. Different storage device requirements, such as speed versus cost, dictate different tradeoffs. The following sections explain factors affecting data loss with magnetic storage media.

Magnetic Decay and Leakage

The tendency of magnetically charged particles to lose their charge over time is called **magnetic decay**. Magnetic decay is constant over time and proportional to the power of the charge. The electrical switches used in a read operation require a minimum, or threshold, current level. Because induced current power is proportional to the magnetic charge's strength, a successful read operation requires a magnetic charge above a certain threshold, called the read threshold.

Over time, the stored charge decays below the read threshold. At that point, the data content of the storage medium is effectively lost. This phenomenon is the main reason that data stored on disks and tapes is usually unreadable after a few years. To minimize potential data loss, data bits must be written with a charge high enough to compensate for decay. Magnetic storage devices write data at a substantially higher charge than the read threshold, thus ensuring long-term readability. However, there are limits to how much charge a read/write head can generate and how much the storage medium can hold.

The strength of a bit charge can also decrease because of **magnetic leakage** from adjacent bits. Any charged area continuously generates a magnetic field that might affect nearby bit areas. If the polarity of adjacent bits is opposite, their magnetic fields tend to cancel out the charge of both areas, and the strength of both charges falls below the read threshold. Magnetic leakage is counteracted by lowering areal density (thus increasing the charge per bit) or by placing uncharged buffer areas between bit areas. Unfortunately, both methods reduce storage capacity.

Areal Density

Coercivity is the capability of a substance or magnetic storage medium to accept and hold a magnetic charge. This property varies widely between elements and compounds. In general, metals and metallic compounds offer the highest coercivity at a reasonable cost. For any material, coercivity is directly proportional to mass. In magnetic storage media, mass is a function of the surface area in which a bit is stored, the thickness of the medium or its coercible coating, and the density of the chargeable material in the medium or coating. Larger, thicker, and denser areas can hold more charge because of their higher coercible mass.

Most users want to store as much data as possible on a storage medium. The simplest way to do this is to reduce the surface area used to store a single bit value, thus increasing the total number of bits that can be stored on the medium. For example, the areal density of a two-dimensional medium can be quadrupled by halving the length and width of each bit area (see Figure 5.7). The surface area allocated to a bit is called the **areal density**, recording density, or bit density. Areal density is typically expressed in bits, bytes, or tracks per inch, abbreviated as bpi, Bpi, and tpi.

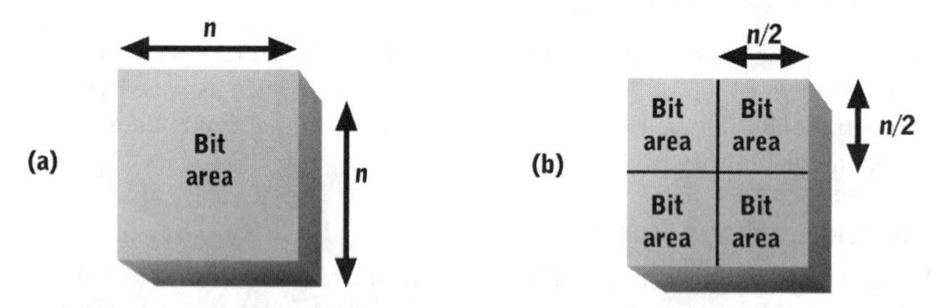

FIGURE 5.7 Areal density is a function of a bit area's length and width (a); density can be quadrupled by halving the length and width of bit areas (b)
Courtesy of Course Technology/Cengage Learning

Some magnetic storage devices, such as hard disks, have fixed areal density, but others, such as tape, have variable areal density. Because reductions in area per bit reduce coercible mass per bit, the problems of magnetic decay and leakage increase as areal density increases. Designers and purchasers of magnetic media and devices must find a suitable balance between high areal density and storage longevity.

Media Integrity

A magnetic storage medium's integrity depends on its construction and the environmental factors it's subjected to. Magnetic media, such as hard disk drives and MRAM, have high media integrity. Other magnetic media, such as magnetic tape, have thin coatings of coercible material layered over a plastic or other substrate. Age and environmental stress can loosen the bond between the coating and substrate, causing the coating to wear away. Physical stress on the medium from fast-forwarding and rewinding tape can accelerate the process, as can temperature and humidity extremes.

Loss of the coercible coating results in a loss of strength in stored magnetic charges. As with magnetic leakage and decay, data becomes unreadable when the remaining charge falls below the read threshold. Magnetic media must be protected from physical abuse and temperature and humidity extremes to extend their life. To ensure long-term media integrity, magnetic tapes are sometimes stored in climate-controlled vaults, for example.

Table 5.2 summarizes the factors leading to data loss in magnetic storage media.

TABLE 5.2 Factors leading to data loss in magnetic storage devices

Factor	Description
Magnetic decay	Natural charge decay over time; data must be written at a higher charge than the read threshold to avoid data loss.
Magnetic leakage	Cancellation of adjacent charges of opposite polarity and migration of charge to nearby areas; data must be written at a higher charge than the read threshold to avoid data loss.
Coercivity	Capability of the storage medium to hold a charge; the medium must have enough mass and coercivity to hold charges strong enough to counteract decay and leakage.
Recording density	The coercible material per bit decreases as the areal density increases; higher areal density makes stored data more susceptible to loss caused by decay and leakage.
Media integrity	Stability of coercible material and its attachment to the substrate; physical stress and extremes of temperature and humidity must be avoided to prevent loss of coercible material.

Magnetic Tape

A **magnetic tape** is a ribbon of plastic with a coercible (usually metallic oxide) coating. Tapes are mounted in a **tape drive**, which is a slow serial access device, for reading and writing. A tape drive contains motors that wind and unwind the tape, one or more read/write heads, and related circuitry. Tapes are used primarily to make backup copies of data stored on faster secondary storage devices and to physically transport large data sets.

Tapes and tape drives come in a variety of shapes and sizes. Tape widths have shrunk over time because technology improvements have increased areal density. Older tape drives used tapes up to 1 inch wide wound on large open reels up to 10 inches in diameter. Modern tape drives use much narrower tapes mounted permanently in plastic cassettes or cartridges (see Figure 5.8). When a cassette is inserted into a tape drive, an access door opens automatically, and the read/write head is positioned around or next to the exposed tape.

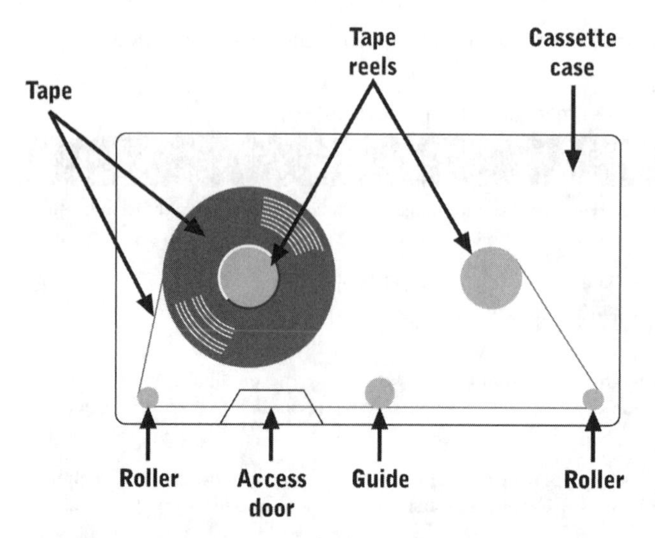

FIGURE 5.8 Components of a typical cassette or cartridge tape
Courtesy of Course Technology/Cengage Learning

Tapes compound magnetic leakage because the tape is wound upon itself. Leakage can occur from adjacent bit positions on the same area of the tape as well as from the layer of tape wound above or below on the reel. Tapes are also susceptible to problems caused by stretching, friction, and temperature variations. As a tape is wound and unwound, it tends to stretch. If enough stretching occurs, the distance between bit positions can be altered, making bits difficult or impossible to read.

There are two geometric approaches to recording data onto a tape surface. **Linear recording** places bits along parallel tracks that run the entire length of the tape, as shown in Figure 5.9(a). Areal density depends on bit spacing in each track and on the number of tracks. **Helical scanning**, as shown in Figure 5.9(b), reads and writes data to or from a tape by rotating the read/write head at an angle to the tape and moving from tape edge to tape edge. This geometry requires more complex read/write mechanisms.

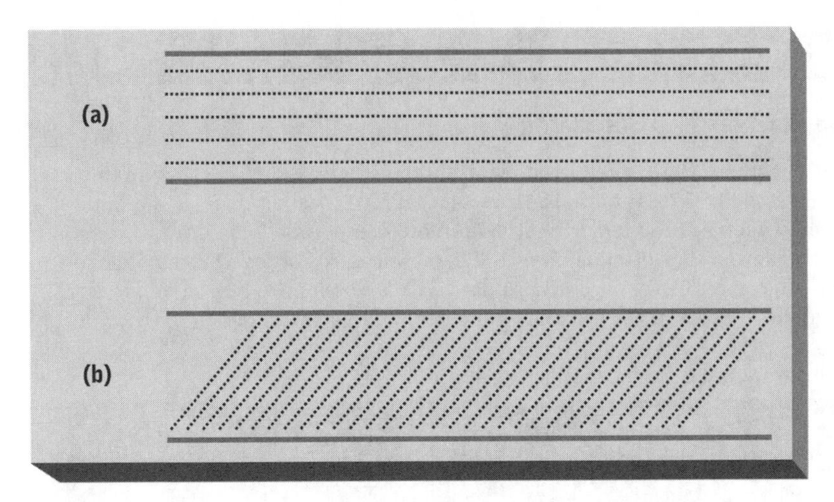

FIGURE 5.9 Data recorded in linear parallel tracks (a) and with helical scanning (b)
Courtesy of Course Technology/Cengage Learning

Helical scanning was originally developed for video recording, such as on VHS tapes. For drives of equal cost, helical scanning can pack more data into the same tape area than linear recording can. The main disadvantages of helical scanning are slower tape motor speeds and direct contact of read/write heads with the tape. The slower tape motor speed makes searching for specific data on a helically scanned tape slower. The direct contact of read/write heads with the tape creates friction that makes helically scanned tapes subject to wear. Linear recording drives usually maintain a small gap between the read/write head and the tape to eliminate friction.

There are at least five competing magnetic tape technology/format combinations and several variations on each (see Table 5.3). The tape storage industry has had little cooperation on standards since the early 1990s. Modern standards are proprietary and subject to licensing fees. For the consumer, the result has been a maze of competing technologies with lots of hype, lots of misinformation, and little compatibility.

TABLE 5.3 Comparison of modern tape formats

Format	Manufacturer	Geometry	Maximum capacity (GB)
Digital Data Storage (DDS)	HP, Sony	Helical	40
Advanced Intelligent Tape (AIT)	Sony	Helical	400
Mammoth	Tandberg	Helical	160
Super Digital Linear Tape (SDLT)	Quantum	Linear	300
Linear Tape Open (LTO)	HP, IBM, Seagate	Linear	800

TECHNOLOGY FOCUS

Magnetic Tape Formats and Standards

Before the mid-1990s, there were several widely used tape storage standards. Mainframe tape drives typically used 0.5- or 1-inch open-reel tape. IBM set standards for mainframe tapes because it manufactured the majority of mainframe computers. Tape drives for smaller computers followed IBM's standards or the open standards of the **Quarter Inch Committee (QIC)**, shown in Table 5.4. In those days, IBM's dominance helped create an "us versus IBM" mentality that perhaps explains the cooperative spirit that led to the QIC.

TABLE 5.4 Sample QIC tape format specifications

Format	Year	Cartridge size (in)	Capacity (GB)	Tracks	Areal density (bpi)
QIC-80	1988	4 × 6	.08	28	14,700
QIC-120	1991	4 × 6	.125	15	10,000
QIC-525	1992	4 × 6	.525	26	20,000
QIC-2100	1993	4 × 6	2.1	30	50,800
QIC-3095	1995	3.25 × 2.5	4.0	72	67,733
QIC-3220	1997	3.25 × 2.5	10.0	108	106,400

The explosion in the market for PCs, small servers, and minicomputers in the 1980s and 1990s led to a corresponding surge in the tape market. Many companies entered the business and guarded their technology closely to gain a competitive advantage. Cooperation became the exception, and when companies did cooperate, they created proprietary standards and charged substantial licensing fees for other companies to produce compatible media and drives. Technology quickly outpaced the QIC standards, and no vendors were willing to contribute proprietary technology to update them.

The **Digital Data Storage (DDS)** standards were developed by Hewlett-Packard and Sony from an earlier technology called **Digital Audio Tape (DAT)**. All DDS tapes are 4 mm wide, and all DDS drives use helical scanning. DDS tapes are cheap, but they aren't designed for precise alignment. DDS drives compensate for cheap tape cassettes with elaborate technology for tape and environmental control, making the drives more expensive. DDS storage was used widely in workstations and small servers.

Super Digital Linear Tape (SDLT) is a standard developed by Quantum Corporation. (Early versions omit the word "super.") SDLT tape is 0.5 inches wide, and the cartridge contains only one reel. The other reel is located in the drive, which uses an elaborate mechanism to grab a plastic tape leader from the cartridge and wind it onto the take-up reel. SDLT records in parallel linear tracks, one at a time, from one end of the tape to another. At the end of the tape, the drive motors reverse direction, and the next track is recorded. The end-to-end track format has advantages when searching for specific data, but it requires many passes to fill the tape completely.

Sony and Exabyte breathed new life into DAT with Sony's **Advanced Intelligent Tape (AIT)** and Exabyte's **Mammoth** standards. (Exabyte merged with Tandberg in the

(continued)

2000s.) Both standards use helical scanning on 8 mm tapes and a more expensive and precisely manufactured tape cartridge. Both standards also improve tape drive technology to pack more data onto a single tape. AIT includes a small RAM cache in each cartridge, which stores directory information to speed searching and data access.

Hewlett-Packard, IBM, and Seagate developed the **Linear Tape Open (LTO)** standard. Despite its name and claims, LTO is a proprietary standard. It uses linear recording and has technology improvements in tape cartridges, coercible materials, read/write heads, and tape control. All this technology comes at a substantial price, so LTO is designed mainly for enterprise servers supporting mission-critical applications. It's currently the market leader and appears to be the only tape format that will thrive in the near future. Drives and media for other formats are still manufactured, although their markets are smaller and based mainly on legacy uses.

Magnetic storage technology has improved rapidly in both price and performance, and many large data centers now rely on redundant disks for storage backup and use tape storage sparingly, if at all. This trend reflects both the shifting cost/performance characteristics of magnetic tape and disk and the trend toward around-the-clock data availability. Many organizations simply can't afford to have data unavailable during a recovery operation from tape. At the same time, tape's role as an archival storage format is under attack from optical discs of ever larger capacity. Tape still leads optical discs in recovery speed and storage capacity, but these advantages continue to shrink.

Magnetic Disk

Magnetic disk media are flat, circular **platters** with metallic coatings that are rotated beneath read/write heads (see Figure 5.10), and data is normally recorded on both sides of a platter. Multiple platters can be mounted and rotated on a common spindle. A **track** is one concentric circle of a platter, or the surface area that passes under a read/write head when its position is fixed. A **cylinder** consists of all tracks at an equivalent distance from the edge or spindle on all platter surfaces.

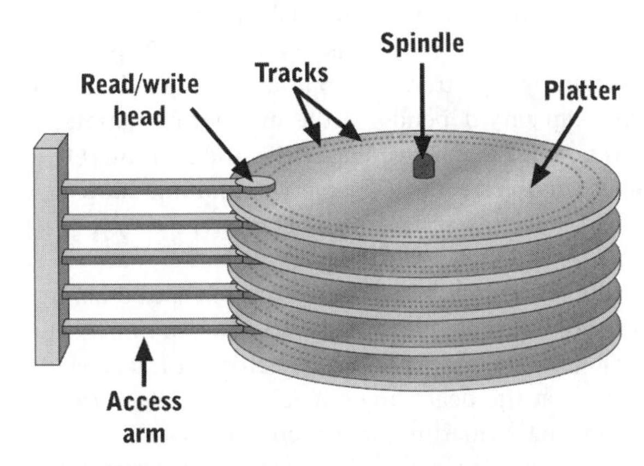

FIGURE 5.10 Primary components of a typical magnetic disk
Courtesy of Course Technology/Cengage Learning

A sector is a fractional portion of a track (see Figure 5.11). A single sector usually holds 512 bytes, which is also the normal data transfer unit to or from the drive.

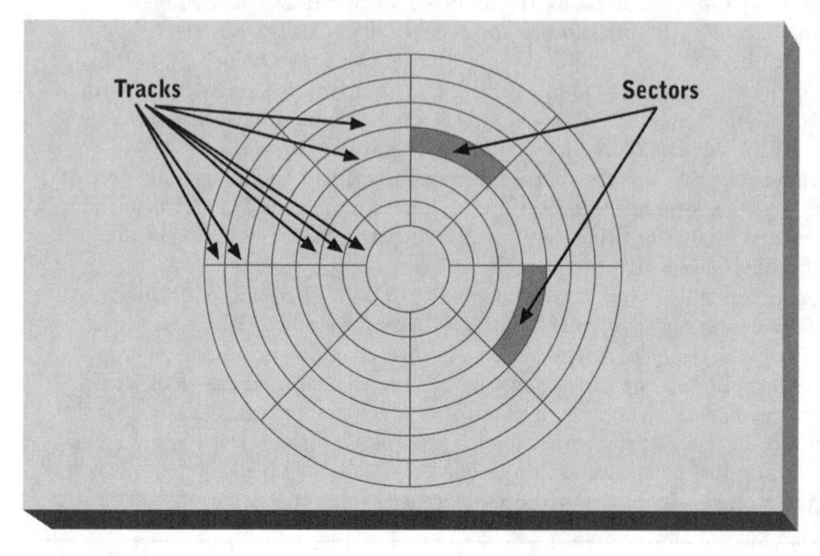

FIGURE 5.11 Organization of tracks and sectors on one surface of a disk platter
Courtesy of Course Technology/Cengage Learning

There's a read/write head mounted on an **access arm** for each side of each platter. Access arms of equal length are attached to a positioning servo. This servo moves the access arms so that read/write heads can be positioned anywhere between the outermost track, closest to the platter's outer edge, and the innermost track, closest to the center of the disk. Read/write heads don't normally make contact with a platter's recording surface. Rather, they float above the platter on a thin cushion of air as the disk spins.

A **hard disk** is a magnetic disk medium with a rigid metal base (substrate). Typically, platter diameter is between 1 and 5 inches. Multiple platters are spun at high speeds, up to 15,000 revolutions per minute (rpm). Drive capacity depends on the number of platters, platter size, and areal density. At this writing, hard disk drive capacities range from 160 GB to 2 TB, but maximum capacity typically doubles every year or two. Multiple hard drives can be enclosed in a single storage cabinet; this arrangement is referred to as a **drive array**.

All magnetic disks are blank when manufactured, although many are formatted before sale. The location of tracks in a platter is fixed, but sector locations in a track are not. One operation performed when formatting a disk is fixing the location of sectors in tracks, which is done by writing synchronization data at the beginning of each sector. Synchronization data can also be written in a sector during formatting or subsequent write operations. Read/write circuitry uses synchronization data to compensate for minor variations in rotation speed and other factors that might disturb the precise timing needed for reliable reading and writing.

A read or write command identifies the platter, track in the platter, and sector in the track. Controller circuitry switches activate the read/write head for the platter, and the positioning servo moves the access arms over the track. The controller circuitry then waits for the correct sector to rotate beneath the read/write head; at this point, the read or write operation is started. The operation is completed as the last portion of the sector passes under the read/write head. The access arms are typically left in their last position until the next read or write command is received.

Disk access time depends on several factors, including the time required to do the following:

- Switch between read/write heads.
- Position read/write heads over a track.
- Wait for the correct sector to rotate beneath read/write heads.

Disk drives share one set of read/write circuits among all read/write heads. The read/write circuitry must be electronically attached, or switched, to the correct read/write head before a sector can be accessed. The time needed to perform this switching, referred to as **head-to-head (HTH) switching time**, is no more than a few nanoseconds. Switching occurs in sequence among read/write heads, and multiple switching operations might be required to activate a head. For example, if there are 10 read/write heads, switching from the first head to the last requires 9 separate switching operations (first to second, second to third, and so forth). HTH switching time is the least important component of access time.

The positioning servo that moves access arms and read/write heads is a bidirectional motor that can be started and stopped quickly to position read/write heads precisely over a track. The time needed to move from one track to another is called **track-to-track (TTT) seek time**, typically measured in milliseconds. It's an average number because of variations in the time required to start, operate, and stop the positioning servo.

After the read/write heads are positioned over the track, the controller waits for the sector to rotate beneath the heads. The controller keeps track of the current sector position by continually scanning the sector synchronization data written on the disk during formatting. When the correct sector is sensed, the read or write operation is started. The time the disk controller must wait for the right sector to rotate beneath the heads is called **rotational delay**, which depends on platter rotation speed. Increasing the spin rate reduces rotational delay.

Both TTT seek time and rotational delay vary from one read or write operation to another. For example, if two consecutive read requests access adjacent sectors on the same track and platter, rotational delay and TTT seek time for the second read operation are zero. If the second read access is to the sector preceding the first, rotational delay is high, almost a full platter rotation. If one read request references a sector on the outermost track and the other request references the innermost track, TTT seek time is high.

The amount of TTT seek time and rotational delay can't be known in advance because the location of the next sector to be accessed can't be known in advance. Access time for a disk drive is not a constant. Instead, it's usually stated by using a variety of performance numbers, including raw times for HTH switching and rotation speed. The most important performance numbers are average access time, sequential access time, and sustained data transfer rate.

Average access time is computed by assuming that two consecutive accesses are sent to random locations. Given a large number of random accesses, the expected value of HTH switching time corresponds to switching more than half the number of recording surfaces. The expected value of TTT seek time corresponds to movement over half the tracks, and the expected value of rotational delay corresponds to one-half of a platter rotation. If HTH switching time is 5 ns, the number of read/write heads is 10, TTT seek time is 5 microseconds (μs, millionths of a second), each recording surface has 1000 tracks, and platters are rotated at 7500 rpm, the average access delay is computed as follows:

$$\begin{aligned}
\text{average access delay} &= \text{HTH switch} + \text{TTT seek} + \text{rotational delay} = \\
&= \left(5 \text{ ns} \times \frac{10}{2}\right) + \left(\frac{1000}{2} \times 5\ \mu\text{s}\right) + \left(\frac{60 \text{ seconds}}{7500 \text{ rpm}} \div 2\right) \\
&= .000000025 + (500 \times .000005) + (.008 \div 2) \text{ seconds} \\
&= .000000025 + .0025 + .004 \text{ seconds} \\
&\approx 6.5 \text{ ms}
\end{aligned}$$

Average access time is the sum of average access delay and the time required to read a single sector, which depends entirely on the disk's rotational speed and the number of sectors per track. Using the previous figures and assuming there are 24 sectors per track, the average access time is computed as follows:

$$\begin{aligned}
\text{average access time} &= \text{average access delay} + \left(\frac{60 \text{ seconds}}{7500 \text{ rpm}} \div 24\right) \\
&= .0065 + (.008 \div 24) \\
&\approx 6.83 \text{ ms}
\end{aligned}$$

Sequential access time is the time required to read the second of two adjacent sectors on the same track and platter. HTH switching time, TTT seek time, and rotational delay are zero for this access. The only component of access time is the time required to read a single sector—that is, the second half of the preceding equation, or 0.33 ms in that example.

Because sequential access time is so much faster than average access time, disk performance is improved dramatically if related data is stored in sequential sectors. For example, loading a program from disk to memory is fast if the entire program is stored in sequential locations on the same track. If the program won't fit in a single track, performance is maximized by storing the program in a single cylinder or group of adjacent cylinders.

If portions of a program or data file are scattered around the disk in random sectors, performance is reduced because of switching read/write heads, positioning the access arm, and waiting for sectors to rotate beneath the heads. A disk with many program and data files scattered on it is said to be **fragmented.**

NOTE

File storage in sequential sectors improves performance only if file contents are accessed sequentially. If file contents are accessed in random order, sequential storage might not improve performance.

Over time, file contents tend to become fragmented in many nonsequential sectors. Fragmentation is a normal byproduct of adding, deleting, modifying, and moving files and directories. Most OSs include a utility program for performing **disk defragmentation**.

A disk defragmentation utility reorganizes disk content so that a file's contents are stored in sequential sectors, tracks, and platters (see Figure 5.12).

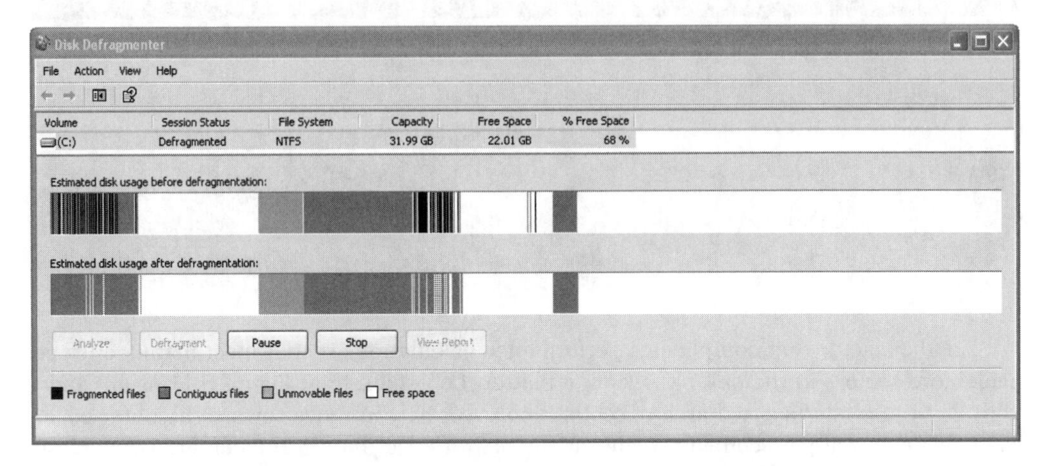

FIGURE 5.12 Display of fragmentation before and after disk defragmentation
Courtesy of Course Technology/Cengage Learning

Effective disk drive performance also depends on the speed at which data can be moved through the disk controller circuitry to and from memory or the CPU. Communication channel capacity is rarely a restriction on a single disk drive's data transfer rate, but it can limit the effective throughput of a drive array.

A disk drive's data transfer rate is a summary performance number combining the physical aspects of data access with the electronic aspects of data transfer to the disk controller or system. As with access times, generally two different numbers are stated: maximum data transfer rate and sustained data transfer rate. The maximum data transfer rate is the fastest rate the drive can support and assumes sequential access to sectors. Because mechanical delays are minimal with sequential accesses, this number is quite high, typically several megabytes per second. The maximum data transfer rate for the previously described performance figures, assuming no controller delays, is as follows:

$$\frac{1}{0.00033 \text{ seconds}} \times 512 \text{ bytes per sector} = 1.5 \text{ MB per second}$$

The **sustained data transfer rate** is calculated with much less optimistic assumptions about data location. The most straightforward method is to use average access time. For the previous example, this is the sustained data transfer rate:

$$\frac{1}{0.00683 \text{ seconds}} \times 512 \text{ bytes per sector} = 75 \text{ KB per second}$$

Because OSs typically allocate disk space to files sequentially, the sustained data transfer rate is generally much higher than the preceding calculation implies. Table 5.5 lists performance statistics for hard disk drive models from Seagate Technology, Inc. and Hitachi, Ltd.

TABLE 5.5 Hard disk drive performance statistics

Manufacturer	Model	Platters	Capacity (GB)	Rotation speed (rpm)	Average access time (ms)
Seagate	ST3500320AS	2	500	7200	8.5
	ST3600002SS	4	600	10,000	3.8
	ST3600057SS	4	600	15,000	3.4
Hitachi	E5K500	3	500	5400	12.0
	C10K300	2	300	10,000	3.9
	15K600	4	600	15,000	3.4

Another factor that complicates performance calculations is that most manufacturers pack more sectors in the outer tracks of a platter. Look closely at Figure 5.11 again, and note that the platter surface area of sectors increases as you move toward the outer edge. Coercible material per sector is greater at the platter edge than in the center. To increase capacity per platter, disk manufacturers divide tracks into two or more zones and vary the sectors per track in each zone (see Figure 5.13).

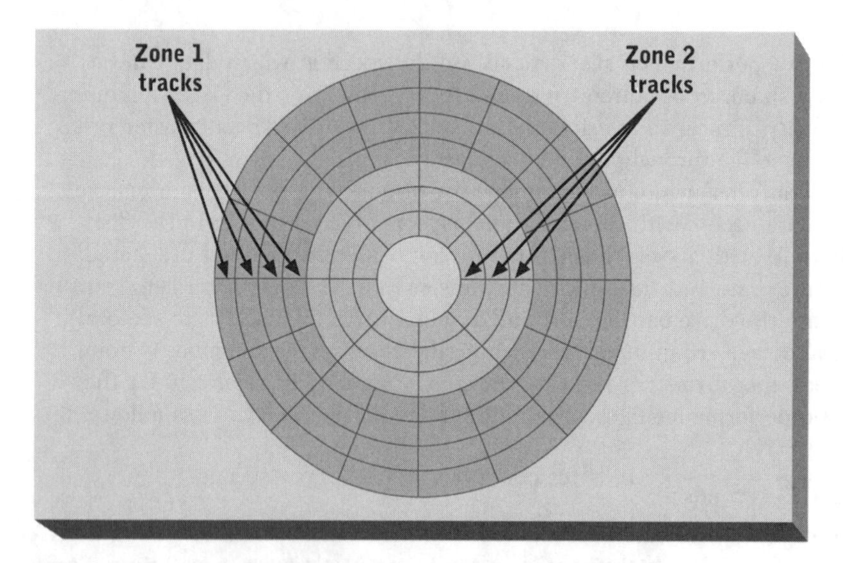

FIGURE 5.13 A platter divided into two zones with more sectors per track in the outer zone
Courtesy of Course Technology/Cengage Learning

Computing the average access time is more complex when sectors are more densely packed on the platter's outer portions because the assumption that an average access requires moving the read/write head over half the tracks is no longer valid. For this reason, most average access time computations assume movement over a quarter to a third of the tracks.

TECHNOLOGY FOCUS

Solid-State Drives

Declining cost and improved performance and capacity of flash RAM has created a new class of storage devices that are expected to supplant magnetic disk as the dominant secondary storage technology. A **solid-state drive (SSD)** is a storage device that mimics the behavior of a magnetic disk drive but uses flash RAM or other NVM devices as the storage medium and read/write mechanism. SSDs are currently implemented with the same physical size and interfaces as magnetic drives, which enables them to compete directly and be used interchangeably.

SSDs compare favorably with magnetic disk drives on some characteristics but not on others (see Table 5.6). However, the comparative advantages and disadvantages are expected to change rapidly under the assumption that magnetic disks are a more mature technology than NVM. The rate of improvement in SSD characteristics is expected to be faster than for magnetic disks, resulting in magnetic disks being gradually replaced with SSDs.

TABLE 5.6 Comparison of solid-state drives and magnetic disks in 2010

Device characteristic	Solid-state drives	Magnetic disks
Read speeds	Typical average access times are less than 5 ms per sector for random or sequential reads.	Typical average access times are less than 5 ms per sector for random reads and less than 1 ms for sequential reads.
Write speeds	Write times depend on the underlying NVM technology, with more expensive technologies having write times comparable with read times and less expensive technologies being two to five times slower.	Write times are typically 5% to 15% slower than read times.
Volatility	Operational life depends on use and NVM technology. Flash RAM wears out after 100,000 or more write operations, which might be less than 5 years, depending on the write frequency.	Typical operational life is 5 to 10 years, with an unlimited number of accesses.
Access method	Access is random by sector, although current inexpensive flash RAM technology requires writes to multiple sectors at once, which slows writing considerably.	Access is random by sector, although faster sequential access occurs if disks are defragmented frequently and when files are accessed sequentially.
Portability	Lack of moving parts provides inherent portability with little or no performance penalty.	Use of moving parts limits the performance of portable drives compared with nonportable drives.
Capacity and cost	Maximum capacity of 64 GB per drive. Cost ranges from $5 to $15 per gigabyte, depending on capacity, interface, and NVM technology.	Maximum capacity is 2 TB per drive. Cost ranges from 2¢ to $2 per gigabyte, depending on capacity, interface, and performance.

(continued)

Current SSDs use flash RAM as the storage medium, although newer NVM types, such as MRAM and PCM, might supplant it in the next decade. At present, the main drawbacks of SSDs are byproducts of the current state of flash RAM technology. Flash RAM's slow write times give magnetic disk drives a major performance advantage in applications that perform many sequential writes, such as digital audio recording and other multimedia applications. New flash RAM technology minimizes the write performance penalty but at a substantially higher cost and with reduced capacity.

Volatility also limits the use of SSDs in write-intensive applications, although modern SSDs use a technique called "wear leveling" to spread write operations around the storage medium, thus evening out the impact of destructive writes and extending the storage device's useful life. SSD cost per gigabyte is also much higher, which currently limits it to applications in which its advantages are worth the higher price.

SSDs are much more tolerant to shock and other negative environmental factors commonly encountered with portable devices, such as multifunction cell phones, netbooks, and laptop computers. Magnetic disk drives used in portable applications typically have much lower performance than in fixed applications to compensate for environmental factors. Therefore, the performance advantage of magnetic disk drives compared with SSDs is much smaller in the portable market, and the cost differential can often be justified by improved reliability.

SSDs also have an advantage over magnetic disk drives in power consumption. It's yet another advantage in the portable market, but it's also an advantage in large data centers that often house tens or hundreds of thousands of disk drives. SSDs are just beginning to be used in large data centers for applications in which their high read performance and low power requirements offset their higher cost, slower write times, and limited write cycles.

OPTICAL MASS STORAGE DEVICES

Optical storage came of age in the 1990s. Its primary advantages, compared with magnetic storage, are higher areal density and longer data life. Typical optical recording densities are at least 10 times higher than for magnetic storage devices. Higher density is achieved with tightly focused lasers that can access a very small storage medium area. Magnetic fields can't be focused on such a small area without overwriting surrounding bit positions.

Most optical storage media can retain data for decades because they aren't subject to magnetic decay and leakage. Optical storage is popular because of its standardized and inexpensive storage media. The CD and DVD formats used for musical and video recordings are also supported by most computer optical storage devices.

Optical storage devices store bit values as variations in light reflection (see Figure 5.14). The storage medium is a surface of highly reflective material, and the read mechanism consists of a low-power laser and a photoelectric cell. The laser is focused on one bit location of the storage medium at a specific angle, and the photoelectric cell is positioned at a complementary angle to intercept reflected laser light. A highly reflective surface spot causes the photoelectric cell to generate a detectable electrical current. A poorly reflective surface spot doesn't reflect enough light to cause the photoelectric cell to fire. The current, or lack of current, the photoelectric cell produces is interpreted as a binary 1 or 0.

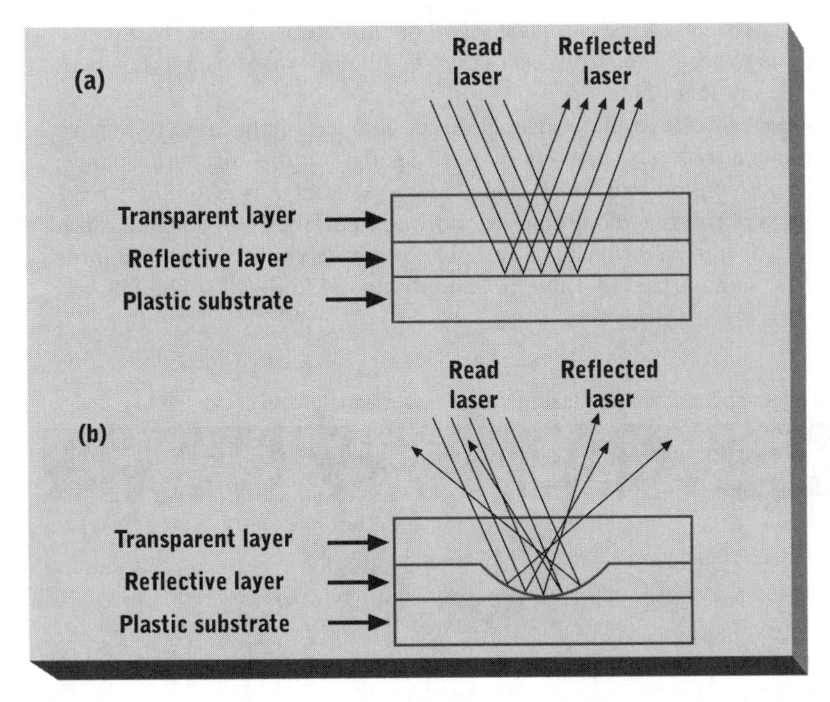

FIGURE 5.14 Optical disc read operations for a 1 bit (a) and a 0 bit (b)
Courtesy of Course Technology/Cengage Learning

There are multiple ways to implement bit areas with low reflectivity. In Figure 5.14, low reflection is a result of a concave dent, or pit, in the reflective layer. Other ways of achieving low reflectivity bit areas include burned areas, dyes, and substances that change from crystalline to amorphous states.

An optimal surface material could be changed rapidly and easily from highly to poorly reflective and then back again an unlimited number of times. The search for this material and a reliable and efficient method of altering its reflectivity has proved difficult. Some materials change quickly in one direction but not another, such as photosensitive sunglass materials. The capability of most materials to change reflectivity degrades over time and with repeated use. Several optical write technologies are currently in use or under development, but there's no clearly dominant technology.

Current optical storage devices use a disc storage medium. (Optical storage media are usually called "discs," and magnetic storage media are called "disks.") Because the recording surface is a disc, there are many similarities to magnetic disk storage, including a read/write mechanism mounted on an access arm, a spinning disc, and performance limitations caused by rotational delay and movement of the access arm.

Although both magnetic and optical devices use disk technology, their performance is quite different. Optical disc storage is slower for a number of reasons, including the use of removable media and the inherent complexity of an optical write operation. The performance difference has narrowed considerably in recent years. Magnetic hard disks typically have access times under 10 ms, and read access time for optical drives is generally more than 100 ms. Write access time is typically two or three times longer than read access time. This gap is expected to shrink as optical technology matures and magnetic technology reaches its physical limits.

Magnetic and optical storage aren't currently direct competitors because they have different cost/performance tradeoffs. Magnetic hard disks are still the cost/performance leader for general-purpose online secondary storage. Optical storage is favored for read-only storage with low performance requirements and when portability in a standardized format is needed. Optical storage devices are available in a wide variety of storage formats and write technologies, summarized in Table 5.7 and discussed in detail in the following sections.

TABLE 5.7 Technologies and storage formats for optical and magneto-optical storage

Technology/ format	Writable?	Description
CD-ROM	No	Adaptation of musical CD technology; 650 or 700 MB capacity.
CD-R	One time only	CD-ROM format with a dye reflective layer that can be written by a low-power laser.
CD-RW	Yes	CD-ROM format with phase-change reflective layer; can be written up to 1000 times.
DVD-ROM	No	Adaptation of DVD video technology; similar to CD-ROM but more advanced; 4.7 GB (single layer) or 8.5 GB (dual layer) capacity.
DVD+/-R	One time only	DVD-ROM single- and dual-layer formats; similar to CD-R with improved performance and capacity. DVD-R and DVD+R are slightly different formats
DVD+/-RW	Yes	DVD-ROM single- and dual-layer formats with phase-change reflective layer. DVD-RW and DVD+RW are slightly different formats.
BD	No	Trade name is Blu-ray disc. Improved DVD technology using higher wavelength lasers; 25 GB (single layer) or 50 GB (dual layer) capacity. Higher capacity discs with more than two layers are under development.
BD-R	One time only	Blu-ray version of DVD+/-R.
BD-RE	Yes	Blu-ray version of DVD+/-RW.
Magneto-optical	Yes	Outdated combination of magnetic and optical technologies; only Sony still manufactures drives; capacity up to 30 GB per disk.
HVD	Yes	Stores bits holographically in the optical recording layer. A 2007 multivendor standard specifies a 500 GB disc, but no drives are currently in production.

CD-ROM, DVD-ROM, and BD

Sony and Philips originally developed **compact disc (CD)** technology for storing and distributing music in the **CD digital audio (CD-DA)** format. **CD read-only memory (CD-ROM)** is compatible with CD-DA but includes additional formatting information to store directory and file information. Both CD-DA and CD-ROM are read-only storage technologies and formats because data is permanently embedded in the disc during manufacture.

A CD is 120 millimeters (approximately 4.75 inches) in diameter and made largely of durable polycarbonate. Bit values are represented as flat areas called "lands" and concave dents called "pits" in the reflective layer. Data is recorded in a single continuous track that spirals outward from the disc's center. Error correction data written along the track at regular intervals helps the CD drive's read mechanism recover from errors introduced by dust, dirt, and scratches.

DVD is an acronym for both "digital video disc" and "digital versatile disc." (The two terms are equivalent.) A consortium of companies developed DVD as a standard format for distributing movies and other audiovisual content. Like CD-ROM, **DVD read-only memory (DVD-ROM)** is an adaptation of audiovisual DVD in which DVD discs can store computer data. Most DVD-ROM drives can also read CD, CD-R, and CD-RW discs.

Blu-ray disc (BD), an important update to DVD-ROM, was originally designed for high-definition video discs. As with CD and DVD, the technology has been adapted to computer data storage. DVDs and BDs are the same size as CDs, and their read and write technologies are similar enough that storage and most drives are backward compatible with all earlier standards.

DVD and BD technologies improve on CD technology in several ways:

- Increased areal density achieved with smaller wavelength lasers and more precise mechanical control
- Higher rotational speeds
- Improved error correction
- Multiple recording sides and layers

CD-ROMs and DVD-ROMs are popular media for distributing software and large data sets, such as maps, census data, phone directories, and other large databases. Standardized formats, high density, and cheap manufacturing cost make them well suited for these purposes. Their primary drawback is that discs can't be rewritten.

Recordable Discs

Recordable versions of the major optical disc formats (CD-R, DVD+R, DVD-R, DVD+R DL, DVD-R DL, and BD-R) use a laser that can be switched between high and low power and a laser-sensitive dye embedded in the disc. The dye is stable when scanned at low power during a read operation but changes its reflective properties when scanned at higher power during a write operation. The write operation is destructive, so recordable disc formats can be written only one time. Because the dyes are sensitive to heat and light, recordable discs should be stored in a dark location at room temperature. Also, because the dyes degrade over time, data might be lost after a few decades.

Phase-Change Optical Discs

Phase-change optical technology enables nondestructive writing to optical storage media. The technology is based on the same GST material used in MRAM, which can change state easily from amorphous to crystalline and then back again. The reflective characteristics of this material are quite different in the amorphous and crystalline states. The difference is less than with manufactured or dye-based discs but enough to be detected by newer optical scanning technologies.

GST changes from an amorphous state to a crystalline state when heated to a precise temperature. Heating the material to its melting point changes it back to an amorphous state. The melting point is low, so high-power lasers aren't required. However, multiple passes are usually required to generate enough heat, so write times are substantially longer than read times. The reflective layer loses its capability to change state with repeated heating and cooling. Current rewritable media wear out after about 1000 write operations.

CD-Rewritable (CD-RW) is a standard for phase-change optical discs that use the CD-ROM format. Rewritable DVDs use one of four different standards: DVD+RW, DVD-RW, DVD+RW DL, and DVD-RW DL. BD-RE is the rewritable version of the Blu-ray standard.

Magneto-Optical Drives

A **magneto-optical (MO) drive** uses a laser and reflected light to sense magnetically recorded bit values. Reading is based on the polarity of the reflected laser light, which is determined by the polarity of the magnetic charge. This laser polarity shift is known as the "Kerr effect."

The disc surface is a material that accepts a magnetic charge when heated to approximately 150 degrees Celsius. To bring the material to a chargeable temperature, a high-power laser is focused on a bit position. A magnetic write head positioned below the media surface then generates the appropriate charge. Because the laser beam focuses on a single bit position, only that bit position is raised to a sufficient temperature. Surrounding bits are unaffected by the magnetic field from the write head because they're cooler.

Magneto-optical technology peaked in the mid-1990s. Few new MO drives have been sold since then because of the introduction of cheaper rewritable CD, DVD, and BD drives. However, there's still a market for MO drives because many organizations created data archives on magneto-optical discs and still need access to these archives. Therefore, the technology probably won't fade away completely for a few more years.

BUSINESS FOCUS

Which Storage Technology?

The Ackerman, Holt, Sanchez, and Trujillo (AHST) law firm has thousands of clients, dozens of partners and associates, and a costly document storage and retention problem. Before 1990, AHST archived legal documents and billing records on paper. As in-office file cabinets filled, the paper was boxed, labeled, and transported to a warehouse for storage. AHST retains all records for at least 20 years and longer in some cases.

In the late 1980s, AHST adopted automated procedures for archiving documents. Most documents were stored as computer files in their original format (WordPerfect, at

(*continued*)

that time). Documents with signatures were scanned and stored as JPEG files. Several possible storage media were considered, including different forms of magnetic tape and removable MO drives. The firm chose Sony MO drives based on the following selection criteria, listed roughly in order of importance:

- Permanence (25-year minimum document recoverability)
- Cost
- High storage capacity
- Ease of use
- Stability and market presence of the technology and selected vendor(s)

AHST has used several different MO media, including 600 MB, 2.3 GB, and 9.1 GB 5.25-inch discs, and several different drives. Document formats have also changed to include Microsoft Word and Adobe Acrobat files. Starting last year, AHST began recording depositions with a digital video camera and archiving them as MPEG files stored on 9.1 GB MO drives. AHST estimates its current annual storage requirements for document and video deposition archives at 1 to 5 TB, approximately 90% of which is video.

Electronic document and secondary storage technologies have evolved considerably since AHST chose magneto-optical storage. Many new tape formats have been developed, and optical formats, such as CD and DVD, have become commonplace. At the same time, magneto-optical technology has evolved, even though its popularity has waned. Drives and media are becoming harder to find, and costs are high compared with more modern alternatives. For these reasons, the firm is reassessing the choice of technology for permanent document storage.

Questions:

- Summarize the pros and cons of the removable secondary storage technologies described in this chapter with respect to the firm's selection criteria.
- What technologies should AHST adopt for future document and video storage? Why?
- How and when might the technologies you recommended become obsolete?

Summary

- A typical computer system has multiple storage devices filling a variety of needs. Primary storage devices support immediate execution of programs, and secondary storage devices provide long-term storage of programs and data. The characteristics of speed, volatility, access method, portability, cost, and capacity vary between storage devices, forming a memory-storage hierarchy. Primary storage tends to be at the top of the hierarchy, with faster access speeds and higher costs per bit of storage, and secondary storage tends to be in the lower portion of the hierarchy, with slower access speeds and lower costs.
- The critical performance characteristics of primary storage devices are their access speed and the number of bits that can be accessed in a single read or write operation. Modern computers use memory implemented with semiconductors. Basic types of memory built from semiconductor microchips include random access memory (RAM) and nonvolatile memory (NVM). The primary RAM types are static and dynamic. Static RAM (SRAM) is implemented entirely with transistors; dynamic RAM (DRAM) is implemented with transistors and capacitors. SRAM's more complex circuitry is more expensive but provides faster access times. Neither type of RAM can match current microprocessor clock rates. NVM is usually relegated to specialized roles and secondary storage because of its slower write speeds and limited number of rewrites.
- Magnetic secondary storage devices store data bits as magnetic charges. Magnetic tapes are ribbons of plastic with a coercible coating. Data is written to or read from tape by passing over a tape drive's read/write head. Magnetic disks are platters coated with coercible material. Platters are rotated in a disk drive, and read/write heads access data at various locations on the platters. Magnetic disk drives are random access devices because the read/write head can be moved to any location on a disk platter. They face increasing competition from solid-state drives based on flash RAM and other forms of nonvolatile memory.
- Optical discs store data bits as variations in light reflection. An optical disc drive reads data bits by shining a laser beam onto a small disc location. High and low reflections of the laser are interpreted as 1s and 0s. The cost per bit of optical storage is less than magnetic storage, at the expense of slower access speed. Types of optical discs (and drives) include different forms of CDs and DVDs. CD-ROM, DVD-ROM, and BD are written during manufacture. CD-R, DVD+/-R, and BD-R are manufactured blank and can be written to once. CD-RW and rewritable DVD and Blu-ray formats use phase-change technology to write and rewrite media.

Now that CPUs, primary storage, and secondary storage have been covered in some detail, it's time to turn your attention to how these devices are integrated in a computer system. Chapter 6 describes how processing, I/O, and storage devices interact and discusses methods of improving overall computer system performance. Chapter 7 describes types of I/O devices.

Key Terms

access arm
access time
Advanced Intelligent Tape (AIT)
areal density
average access time
block
Blu-ray disc (BD)
CD digital audio (CD-DA)

CD read-only memory (CD-ROM)
coercivity
compact disc (CD)
core memory
cylinder
data transfer rate
Digital Audio Tape (DAT)
Digital Data Storage (DDS)
direct access
disk defragmentation
double data rate (DDR)
double inline memory module (DIMM)
drive array
dual inline packages (DIPs)
DVD
DVD read-only memory (DVD-ROM)
dynamic RAM (DRAM)
electronically erasable programmable ROM (EEPROM)
erasable programmable ROM (EPROM)
firmware
flash RAM
fragmented
germanium, antimony, and tellurium (GST)
hard disk
head-to-head (HTH) switching time
helical scanning
linear recording
Linear Tape Open (LTO)
magnetic decay
magnetic leakage
magnetic tape
magnetoresistive RAM (MRAM)
magneto-optical (MO) drive
Mammoth
nonvolatile
nonvolatile memory (NVM)
parallel access
phase-change memory (PCM)
platters
Quarter Inch Committee (QIC)
random access
read-only memory (ROM)
read/write head
refresh cycle
rotational delay
sector
sequential access time
serial access
single inline memory module (SIMM)
solid-state drive (SSD)
static RAM (SRAM)
storage medium
Super Digital Linear Tape (SDLT)
sustained data transfer rate
synchronous DRAM (SDRAM)
tape drive
track
track-to-track (TTT) seek time
volatile

Vocabulary Exercises

1. Dynamic RAM requires frequent ________ to maintain its data content.
2. The ________ rate is the speed at which data can be moved to or from a storage device over a communication channel.
3. Three standard optical storage media that are written only during manufacture are called ________, ________, and ________.
4. ________, ________, ________, and ________ are competing standards for rewritable DVD discs.

5. The _______ of a hard disk drive generate or respond to a magnetic field.
6. Data stored on magnetic media for long periods of time might be lost because of _______ and _______.
7. A(n) _______ stores data in magnetically charged areas on a platter.
8. The contents of most forms of RAM are _______, making them unsuitable for long-term data storage.
9. _______ and _______ are outdated technologies for nonvolatile primary storage. _______ and _______ are promising new technologies for implementing NVM.
10. _______ is typically stated in milliseconds for secondary storage devices and nanoseconds for primary storage devices.
11. The three components that are summed to calculate average access time for a disk drive are _______, _______, and _______.
12. In a magnetic disk drive, a read/write head is mounted on the end of a(n) _______.
13. The access method for RAM is _______ or _______ if words are considered the unit of data access. The access method is _______ if bits are considered the unit of data access.
14. Both DDS and AIT use _______ to record bits and tracks on a magnetic tape.
15. _______ and _______ are two current standards or formats for linear recording on magnetic tapes.
16. A(n) _______ mimics the behavior and physical size of a magnetic disk drive but has no moving parts.
17. A(n) _______ is a series of sectors stored along one concentric circle on a platter.
18. A magnetic disk drive's data transfer rate can be calculated by dividing 1 by the drive's access time and multiplying the result by the _______ size.
19. _______, _______, and _______ are storage formats originally designed for music or video recording that have been applied to computer data storage.
20. Tape drives are _______ devices. _______ are random or direct access devices.
21. Average access time can usually be improved by _______ files stored on a disk.
22. Modern desktop and laptop computers generally use memory packaged on small standardized circuit boards called _______.
23. The _______ of a magnetic or optical storage medium is the ratio of bits stored to a unit of the medium's surface area.
24. For most disk drives, the unit of data access and transfer is a(n) _______ or _______.
25. Software programs stored permanently in ROM are called _______.
26. Many open standards for cartridge tape storage have been defined by the _______.

Review Questions

1. What factors limit the speed of an electrically based processing device?
2. What are the differences between static and dynamic RAM?
3. What improvements are offered by synchronous DRAM compared with ordinary DRAM?

4. Why isn't flash RAM commonly used to implement primary storage?
5. Describe current and emerging nonvolatile RAM technologies. What potential advantages do the emerging technologies offer compared with current flash RAM technology?
6. Describe serial, random, and parallel access. What types of storage devices use each method?
7. How is data stored and retrieved on a magnetic mass storage device?
8. Describe the factors that contribute to a disk drive's average access time. Which of these factors is improved if the spin rate is increased? Which is improved if areal density is increased?
9. What problems contribute to read/write errors on magnetic tapes? Are these problems also present with other magnetic storage media/devices?
10. What are the advantages and disadvantages of helical scanning compared with linear recording?
11. Why is the areal density of optical discs higher than the areal density of magnetic disks? What factors limit this areal density?
12. Describe the processes of reading from and writing to a phase-change optical disc. How does the performance and areal density of these discs compare with magnetic disks?
13. List and briefly describe the standards for recordable and rewritable CDs and DVDs. Are any of the standards clearly superior to their competitors?

Problems and Exercises

1. A magnetic disk drive has the following characteristics:
   - 10,000 rpm spin rate
   - 2 ns head-to-head switching time
   - 3 μs average track-to-track seek time
   - 5 platters, 1024 tracks/platter side recorded on both sides, 50 sectors per track on all tracks

   Questions:

   a. What is this drive's storage capacity?

   b. How long will it take to read the drive's entire contents sequentially?

   c. What is this drive's serial access time?

   d. What is this drive's average (random) access time (assuming movement over half the tracks and half the sectors in a track)?

2. A CPU has a clock rating of 2.4 GHz, and half of each clock cycle is used for fetching and the other half for execution.

   Questions:

   a. What access time is required of primary storage for the CPU to execute with zero wait states?

b. How many wait states per fetch will the CPU incur if all primary storage is implemented with 5 ns SRAM?

c. How many wait states per fetch will the CPU incur if all primary storage is implemented with 10 ns SDRAM?

3. Compute data transfer rates for the following types of devices:

Storage device	Average access time	Data transfer unit size
RAM	4 ns	64 bits
Optical disc	100 ms	512 bytes
Magnetic disk	5 ms	512 bytes

Research Problems

1. Select three to five computers at work or school covering a range of ages and types (for example, PCs, workstations, and servers). Visit a Web site such as *www.kingston.com* or *www.memory4less.com* and locate a 1, 2, or 4 GB memory expansion for each computer. Note the prices and memory type for each upgrade. What factors account for the variation in memory price and type for each computer?
2. IBM has invested heavily in research and development for holographic storage systems. Investigate this research to determine the nature of holographic storage, its applicability to primary and secondary storage, and the likelihood of seeing products based on holographic technology in the near future.
3. You're the chief information officer of a rapidly growing company. You manage several dozen small and medium servers and are about to acquire your first mainframe to set up a data warehouse. You currently have AIT tape drives, but you wonder whether they'll be enough in the future. You're concerned about tape drive capacity, reliability, and cost and are considering switching to another tape format or technology. Which of the current tape formats provides the highest capacity at the lowest cost? Which is most reliable? Which is least likely to become obsolete 5 years from now? Which, if any, should you buy?
4. You're a home computer user and amateur photographer with almost 500 GB of digital photos stored on your hard disk. You're running short of disk space and want to archive photos on removable optical discs. Investigate the capacity and cost of different recordable and rewritable CD and DVD media. Which media type offers the lowest cost? Which offers the best performance? Which offers the longest life? Which would you choose and why? Are there any feasible alternatives to optical discs for this purpose?

CHAPTER 6

SYSTEM INTEGRATION AND PERFORMANCE

CHAPTER GOALS

- Describe the system and subsidiary buses and bus protocols
- Describe how the CPU and bus interact with peripheral devices
- Describe the purpose and function of device controllers
- Describe how interrupt processing coordinates the CPU with secondary storage and I/O devices
- Describe how buffers and caches improve computer system performance
- Compare parallel processing architectures
- Describe compression technology and its performance implications

Earlier chapters discussed the processing and storage components of a computer system. Now it's time to describe how these components communicate with one another. This chapter explores the system bus and device controllers and explains how the CPU uses them to communicate with secondary storage and input/output (I/O) devices. The chapter also covers bus protocols, interrupt processing, buffering, caching, parallel processing architectures, and compression and how these elements affect system performance (see Figure 6.1).

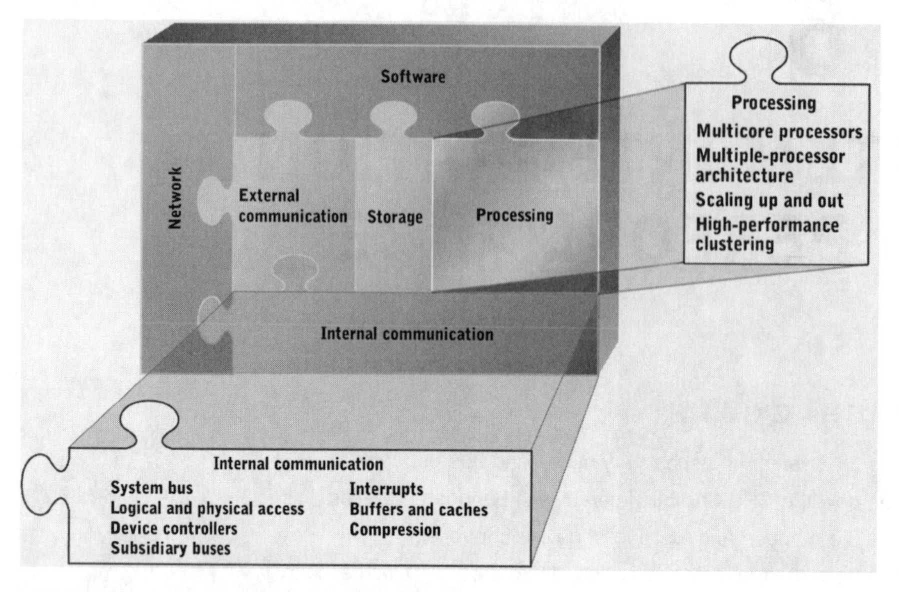

FIGURE 6.1 Topics covered in this chapter
Courtesy of Course Technology/Cengage Learning

SYSTEM BUS

A **bus** is a shared electrical or optical channel that connects two or more devices. As you learned in Chapter 2, a system bus connects computer system components, including the CPU, memory, storage, and I/O devices. There are typically multiple storage and I/O devices, collectively referred to as **peripheral devices**. A system bus can be conceptually or physically divided into specialized subsets, including the data bus, the address bus, and the control bus, as shown in Figure 6.2.

As its name implies, the **data bus** transmits data between computer system components. The **address bus** transmits a memory address when primary storage is the sending or receiving device. The **control bus** carries commands, command responses, status codes, and similar messages. Computer system components coordinate their activities by sending signals over the control bus.

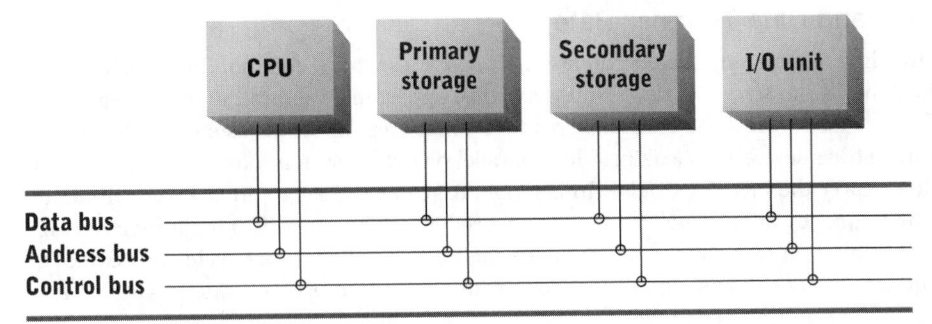

FIGURE 6.2 The system bus and attached devices
Courtesy of Course Technology/Cengage Learning

A bus can be constructed by using parallel communication lines, a smaller number of serial communication lines, or a combination of both types of lines. With parallel communication lines, each line carries only one bit value or signal at a time, and many lines are required to carry data, address, and control bits. With serial communication lines, a single communication line can carry multiple bits, one after another, in rapid succession.

NOTE

Chapter 8 discusses parallel and serial channel construction more fully.

Until the 2000s, system buses were always constructed with parallel electrical lines. Each bit of a data item, memory address, or digital control value was carried on a separate electrical line, and additional lines were used to carry clock signals, electrical power, and return (or ground) wires to complete electrical circuits. The number of bus lines gradually increased over time to account for increases in CPU word and memory address size, increases in control signals to coordinate more complex communication, and more diverse power requirements of peripheral devices and device controllers.

Starting in the 2000s, system buses began to incorporate serial channels to carry data and address bits. Serial channels are more reliable than parallel channels at very high speeds, and they enable physically smaller buses and bus connections, an important advantage as computers and their components continue to shrink. Initial use was confined to mainframes, supercomputers, and video subsystems of workstations and PCs. The conversion from entirely parallel to hybrid serial/parallel system buses in all computer classes is taking place gradually and probably won't be complete until well into the 2010s.

Bus Clock and Data Transfer Rate

Devices attached to a system bus coordinate their activities with a common **bus clock**. Each clock "tick" marks the start of a new communication opportunity across the bus. Having each attached device follow the same clock ensures that a receiving device is "listening" at the same time a transmitting device is "speaking." In a parallel or hybrid system bus, one or more control bus lines carry the bus clock pulse. In a serial bus, clock pulses are intermixed with data signals. The frequency of bus clock pulses is measured in megahertz (MHz) or gigahertz (GHz).

The time interval from one clock pulse to the next is called a **bus cycle**. Bus cycle time is the inverse of the bus clock rate. For example, if the bus clock rate is 400 MHz, each bus cycle's duration is as follows:

$$\begin{aligned}\text{bus cycle time} &= \frac{1}{\text{bus clock rate}} \\ &= \frac{1}{400{,}000{,}000\ \text{Hz}} \\ &= 2.5\ \text{nanoseconds}\end{aligned}$$

By its very nature, the system bus must be long because it connects many different computer system components. Component miniaturization has reduced the length of a typical system bus, but it's still 10 centimeters or longer in most desktop computers. A bus cycle can't be any shorter than the time an electrical signal needs to traverse the bus from end to end. Therefore, bus length imposes a theoretical maximum on bus clock rate and a theoretical minimum on bus cycle time.

In practice, bus clock rate is generally set below its theoretical maximum to compensate for noise, interference, and the time required to operate interface circuitry in attached devices. Slower bus clock rates also enable computer component manufacturers to ensure reliability yet hold costs to acceptable levels. As an analogy, think about the speed limit on a road. Although it's possible to construct a road that supports a maximum speed of 150 miles per hour, few roads have a speed limit this high. The road and vehicles that could travel at this maximum speed would be expensive. Reliable and safe transport would require high standards of road and vehicle maintenance, skilled drivers, and perfect weather and visibility.

The maximum capacity of a bus or any communication channel is the product of clock rate and data transfer unit size. For example, the theoretical capacity of a parallel bus with 64 dedicated data lines and a 400 MHz clock rate is as follows:

$$\begin{aligned}\text{bus capacity} &= \text{data transfer unit} \times \text{clock rate} \\ &= 64\ \text{bits} \times 400\ \text{MHz} \\ &= 8\ \text{bytes} \times 400{,}000{,}000\ \text{Hz} \\ &= 3{,}200{,}000{,}000\ \text{bytes per second}\end{aligned}$$

This measure of communication capacity is called a data transfer rate. Data transfer rates are expressed in bits or bytes per second, such as 1 billion bits per second (Gbps) or 3.2 million bytes per second (MBps). See the appendix for more information on data transfer rate units of measure.

There are only two ways to increase the maximum bus data transfer rate: Increase the clock rate, or increase the data transfer unit size (the number of data lines). The evolution of parallel system buses is marked by steady increases in data bus size and more gradual increases in clock rate. However, these increases work against one another, with each increase in data bus size making clock rate increases more difficult. Increasing clock rate in serial channels is much easier, as explained in Chapter 8.

Bus Protocol

The **bus protocol** governs the format, content, and timing of data, memory addresses, and control messages sent across the bus. Every peripheral device must follow the protocol rules. In the simplest sense, a bus is just a set of communication lines. In a larger sense, it's a combination of communication lines, a bus protocol, and devices that implement the bus protocol.

The bus protocol has two important effects on maximum data transfer rate. First, control signals sent across the bus consume bus cycles, reducing the cycles available to transmit data. For example, a disk drive transfers data to main memory as a result of an explicit CPU command, and sending this command requires a bus cycle. In some bus protocols, the command is followed by an acknowledgment from the disk drive and later by a confirmation that the command was carried out. Each signal (command, acknowledgment, and confirmation) consumes a separate bus cycle.

An efficient bus protocol consumes a minimal number of bus cycles for commands, which maximizes bus availability for data transfers. For example, some bus protocols send command, address, and data signals at the same time (during a single bus cycle) and don't use acknowledgment signals. Unfortunately, efficient bus protocols tend to be complex, increasing the complexity and cost of the bus and all peripheral devices.

The second effect of the bus protocol on maximum data transfer rate is regulation of bus access to prevent devices from interfering with one another. If two peripheral devices attempt to send a message at the same time, the messages collide and produce electrical noise. A collision is a wasted transmission opportunity, and allowing collisions effectively reduces data transfer capacity. A bus protocol avoids collisions by using one of two access control approaches: a master-slave approach or a peer-to-peer approach.

In traditional computer architecture, the CPU is the focus of all computer activity. As part of this role, it's also the **bus master**, and all other devices are **bus slaves**. No device other than the CPU can access the bus except in response to an explicit instruction from the CPU. There are no collisions as long as the CPU waits for a response from one device before issuing a command to another device. In essence, the CPU plays the role of bus traffic cop.

Because there's only one bus master, the bus protocol is simple and efficient. However, overall system performance is reduced because transfers between devices, such as from disk to memory, must pass through the CPU. All transfers consume at least two bus cycles: one to transfer data from the source device to the CPU and another to transfer data from the CPU to the destination device. The CPU can't execute computation and logic instructions while transferring data between other devices.

Performance is improved if storage and I/O devices can transmit data between themselves without explicit CPU involvement. There are two commonly used approaches to implementing these transfers: direct memory access and peer-to-peer buses. Under **direct memory access (DMA)**, a device called a **DMA controller** is attached to the bus and to main memory. The DMA controller assumes the role of bus master for all transfers between memory and other storage or I/O devices. While the DMA controller manages bus traffic, the CPU is free to execute computation and data movement instructions. In essence, the DMA controller and CPU share the bus master role, and all other devices act as slaves.

In a **peer-to-peer bus**, any device can assume control of the bus or act as a bus master for transfers to any other device. When multiple devices want to become a bus master at the same time, a single master must be chosen. A **bus arbitration unit** is a simple processor attached to a peer-to-peer bus that decides which devices must wait when multiple devices want to become a bus master. Peer-to-peer bus protocols are substantially more complex and expensive than master-slave bus protocols, but their complexity is offset by more efficient use of the CPU and the bus.

Subsidiary Buses

Computer system performance would be severely limited if all CPU, memory, and peripheral device communication had to traverse the system bus. For example, when you're playing a game on a PC, data is read continuously from disk files and stored in memory, the CPU accesses the data in memory to manipulate the game environment in response to your commands and control inputs, and video and sound data is sent continuously from memory to the video display and speakers. If the system bus were the only communication pathway, it would become a bottleneck and slow the entire system's performance.

Modern computers use multiple subsidiary buses to address this problem. Subsidiary buses connect a subset of computer components and are specialized for these components' characteristics and communication between them:

- *Memory bus*—Connects only the CPU and memory. The **memory bus** has a much higher data transfer rate than the system bus because of its shorter length, higher clock rate, and (in most computers) large number of parallel communication lines. Because the CPU constantly interacts with memory, moving all CPU memory accesses to the memory bus improves overall computer performance because of the memory bus's higher data transfer rate and reduced contention (competition) for system bus access.
- *Video bus*—Connects only memory and the video display interface. As discussed later in this chapter and in Chapter 7, video display data is buffered through memory. The data transfer rate required for display updates ranges from a few megabytes per second for ordinary desktop use to hundreds or thousands of megabytes per second for continuous full-screen motion video. The **video bus** improves computer system performance by removing this traffic from the system bus and providing a high-capacity one-way

communication channel optimized for video data. Examples of video buses include Accelerated Graphics Port (AGP) and Peripheral Component Interconnect Express (PCIe).

- *Storage bus*—Connects secondary storage devices to the system bus. Desktop computers typically have a handful of permanently installed secondary storage devices. Larger computers might have a dozen to thousands of secondary storage devices. Secondary storage devices are much slower than the system bus, so direct attachment is both a waste of bus capacity and impractical if there are many devices. Using a **storage bus** reduces the length and number of physical connections to the system bus and aggregates the lower data transfer capacity of multiple secondary storage devices to better match the higher capacity of a single system bus connection. Examples of storage buses include Serial Advanced Technology Attachment (SATA), Integrated Drive Electronics (IDE), and Small Computer System Interface (SCSI).
- *External I/O bus*—Connects one or more external devices to the system bus. As with secondary storage, providing a direct connection to the system bus isn't practical for external devices, such as keyboards, mice, iPods, flash drives, and portable hard drives. Instead, a subsidiary bus provides the connection points and aggregates their capacity to better match the capacity of a system bus connection. In many computers, **external I/O buses** also connect internal devices, such as memory card readers. Examples of external I/O buses include Universal Serial Bus (USB) and FireWire (IEEE 1394).

Figure 6.3 shows the devices and bus attachments of a typical PC. The PCI bus is the system bus in this computer, and subsidiary buses, except the memory bus, are layered below it.

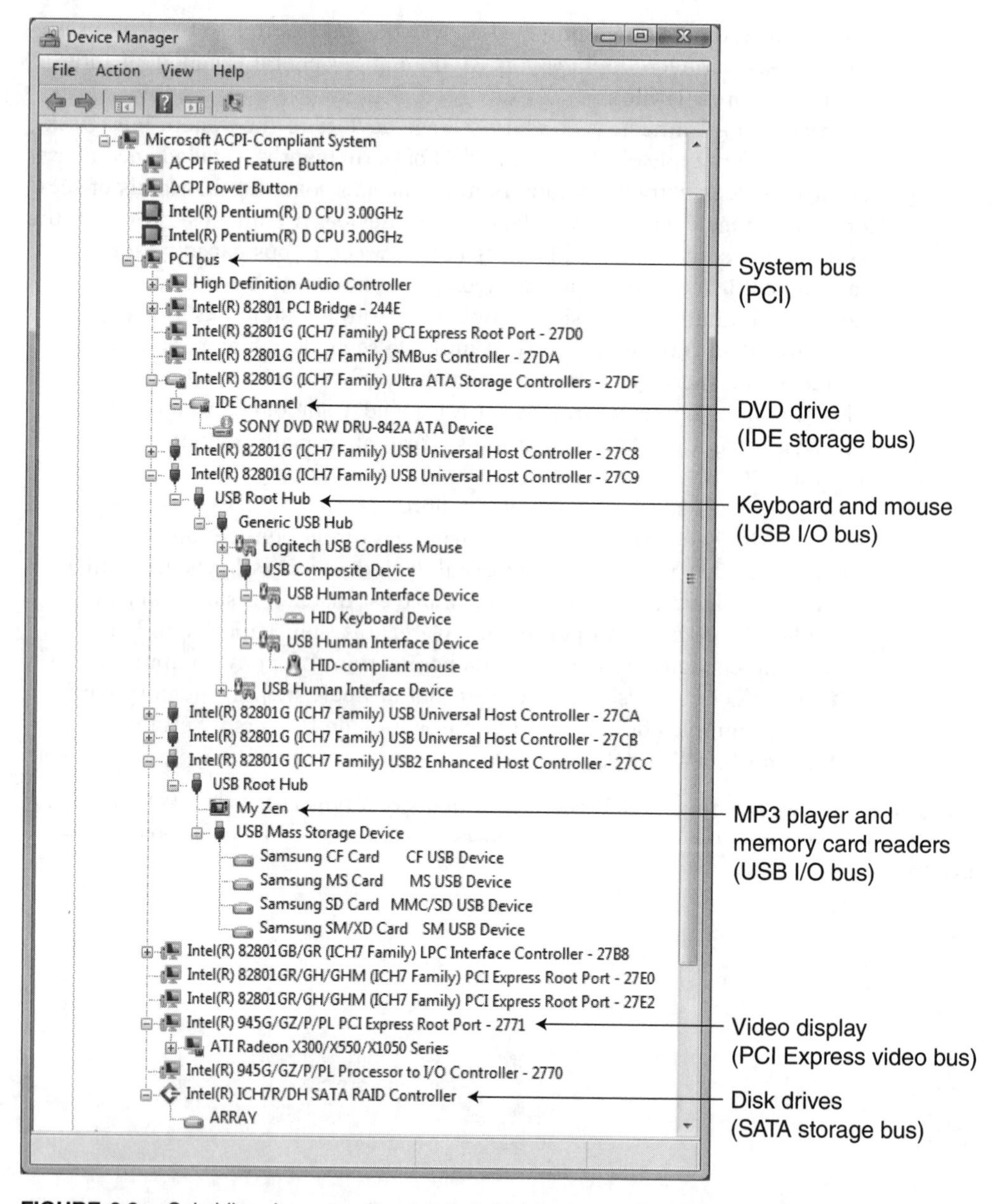

FIGURE 6.3 Subsidiary buses and connected devices in a typical PC
Courtesy of Course Technology/Cengage Learning

TECHNOLOGY FOCUS

PCI

Peripheral Component Interconnect (PCI) is a family of bus standards found in nearly all small and midrange computers and many larger ones. Intel developed the first PCI standard in the early 1990s, and it was widely adopted in the late 1990s. The PCI Special Interest Group (PCI-SIG), formed in 1992, currently includes representatives from most of the world's major computer and computer component manufacturers. The first standard was updated several times through the early 2000s. The PCI-SIG developed two additional standard families, including PCI-X in 1999 and PCI Express (PCIe) in 2002. The PCIe standard is still in development, with version 2.0 published in 2006 and version 3.0 scheduled for release in 2010.

The original PCI 1.0 standard specified a 124-line parallel bus architecture with a 32-bit shared data/address bus, 33 MHz clock rate, and maximum data transfer rate of 133 MBps. Table 6.1 summarizes the lines and their use in the 3.0 standard. Later versions of the PCI standard and the newer PCI-X standard support both 32- and 64-bit data/address buses and have gradually increased clock rate to its current maximum of 533 MHz.

TABLE 6.1 Summary of line functions in the 32-bit PCI 3.0 bus specification

Function	Lines	Comments
Address/data bus	32	Two sequential bus cycles carry address bits in the first cycle and data bits in the second
Control bus	35	Includes lines to carry the clock signal, interrupt codes, source and target device IDs, and error codes
Power	27	+3.3, +5, -12, and +12 volts
Ground	26	Distributed throughout the connector to complete electrical circuits for address, data, control, and power lines
Not defined	4	

By the early 2000s, the need for a fundamentally different bus architecture was becoming apparent. CPUs were clocked at ever higher speeds, and many peripherals, especially video controllers, were chafing under the data transfer rate limitations of parallel bus architectures. PCI and PCI-X bus designs limited overall system performance by forcing every device to operate at the slowest device's speed. The large physical size of PCI and PCI-X bus connections also hampered miniaturization of computer systems and components.

(continued)

The PCIe standard addressed many limitations of the earlier standards by doing the following:

- Adopting serial data transmission for address and data bits
- Providing two-way (full-duplex) communication
- Decreasing the size of bus connections

Data, address, and command bits are transmitted across bus line subsets called "lanes." Each lane contains two wire pairs carrying bits in opposite directions, which enables a device to send and receive bits during the same clock cycle. PCIe 2.0 lanes encode bits in each wire pair at a maximum data transfer rate of 4 Gbps, or 500 MBps. PCIe buses can provide 1, 2, 4, 8, 16, or 32 lanes, yielding a maximum one-way data transmission rate of 16 Gbps for the 2.0 standard. The maximum data transmission rate is somewhat misleading because the bitstream in each lane carries address and command bits in addition to data. Also, cost increases with the number of lanes.

The PCIe standard enables devices to communicate by using all or any subset of available lanes. Devices with low data transfer demand can use a single lane, and devices with higher requirements can increase their available data transfer rate by using additional lanes. The design of the physical bus connector and slot enables devices that support fewer lanes to use bus ports that support more lanes.

PCIe buses are now common as the primary system bus in PCs, with older PCI and PCI-X being phased out gradually. Multilane PCIe connections are widely used for video controllers in PCs and workstations and for high-speed network and secondary storage controllers in servers.

LOGICAL AND PHYSICAL ACCESS

In most computer systems, the system bus is physically implemented on a large printed circuit board with attachment points for various devices (see Figure 6.4). Some devices, such as bus and memory controller circuitry, are permanently mounted to the board. Others, such as the CPU, memory modules, and some peripheral device controllers, are physically inserted into bus ports or dedicated sockets.

An **I/O port** is a communication pathway from the CPU to a peripheral device. In most computers, an I/O port is a memory address, or set of contiguous memory addresses, that can be read or written by the CPU and a single peripheral device. Each peripheral device can have multiple I/O ports and use them for different purposes, such as transmitting data, commands, and error codes. The CPU communicates with a peripheral device by moving data to or from an I/O port's memory addresses. Dedicated bus interface circuitry monitors changes to I/O port memory content and copies updated content to the correct bus port automatically.

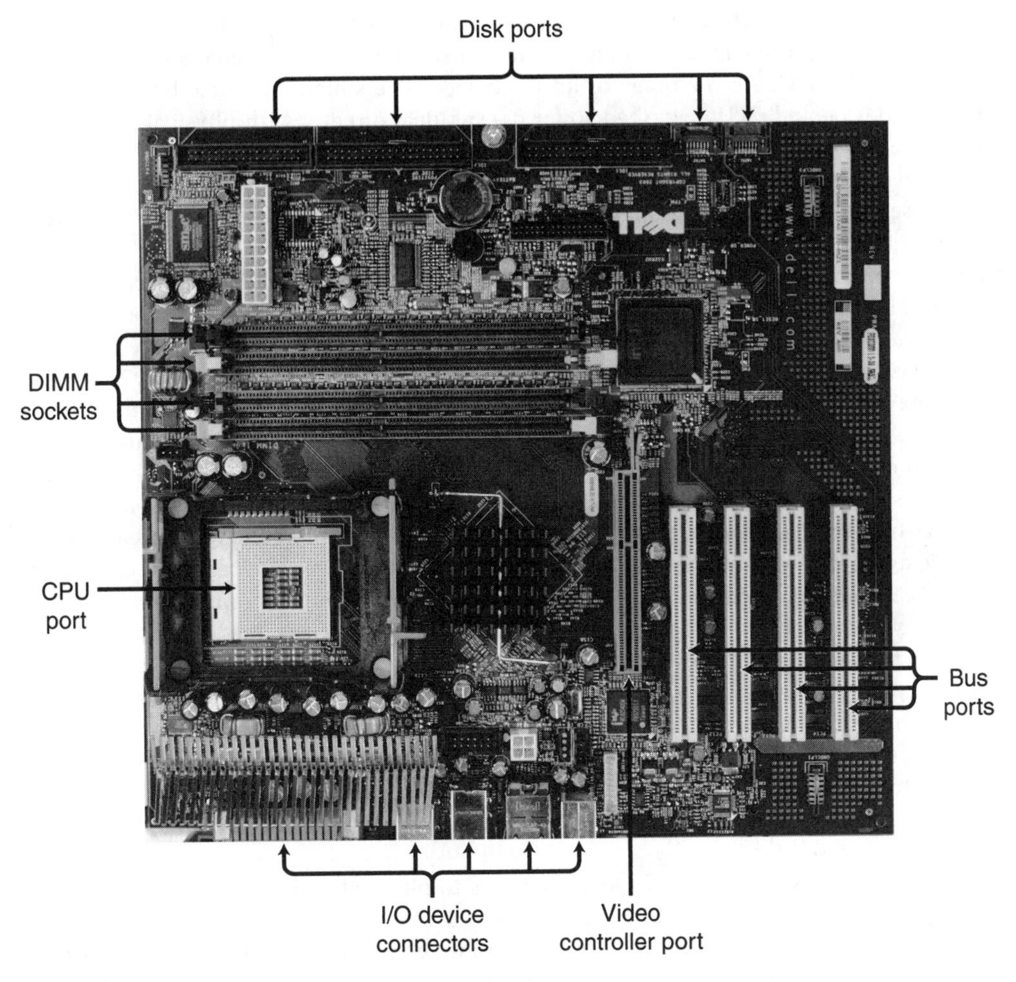

FIGURE 6.4 A typical PC motherboard
Courtesy of Course Technology/Cengage Learning

An I/O port is more than a memory address or data conduit; it's also a logical abstraction used by the CPU and bus to interact with each peripheral device similarly. For example, I/O ports enable the CPU and bus to interact with a keyboard in the same way they interact with a disk drive or video display. I/O ports simplify the CPU instruction set and bus protocol because special instructions and bus circuitry aren't required for each different peripheral device. The computer system is also more flexible because new types of peripheral devices can be incorporated into an older system simply by allocating new I/O ports.

Despite similar interfaces to the CPU and bus, peripheral devices differ in important ways, including storage capacity (if any), data transfer rate, internal data-coding methods, and whether the device is a storage or an I/O device. The simple interface between peripheral device and CPU described so far doesn't deal directly with physical device details, such as how a disk read/write head is positioned, how a certain color is displayed on a video display, or a printer character's font and position.

In essence, the CPU and bus interact with each peripheral device as though it were a storage device containing one or more bytes stored in sequentially numbered addresses. A read or write operation from this hypothetical storage device is called a **logical access**. The set of sequentially numbered storage locations is called a **linear address space**. Logical accesses to a peripheral device are similar to memory accesses. One or more bytes are read or written as each I/O instruction is executed. The address bus lines carry the position in the linear address space being read or written, and the data bus lines carry data. Complex commands and status signals can also be encoded and sent via the data lines.

As an example of a logical access, consider a disk drive organized physically into sectors, tracks, and platters. When the CPU executes an instruction to read a specific sector, it doesn't transmit the platter, track, and sector number as parameters of the read command. Instead, it "thinks" of the disk as a linear sequence of storage locations, each holding one sector of data, and sends a single number to identify the location it wants to read.

To physically access the correct sector, the location in the assumed linear address space must be converted into the corresponding platter, sector, and track, and the disk hardware must be instructed to access this specific location. Linear addresses can be assigned to physical sectors in any number of ways, one of which is shown in Figure 6.5. The disk drive or its device controller (described in the next section) translates the linear sector address into the corresponding physical sector location on a specific track and platter. For example, linear address 43 in Figure 6.5 corresponds to platter 2, track 3, sector 1.

With tape drives, translating logical addresses to physical addresses is more straightforward because blocks on a tape are physically organized in a linear sequence. A logical access to a storage (block) location is translated into commands to position the right part of the tape physically under the read/write head.

Some I/O devices, such as keyboards and sound cards, communicate only in terms of sequential streams of bytes that fill the same memory location. With these devices, the concept of an address or a location is irrelevant. From the CPU's point of view, the I/O device is a storage device with a single location that's read or written repeatedly.

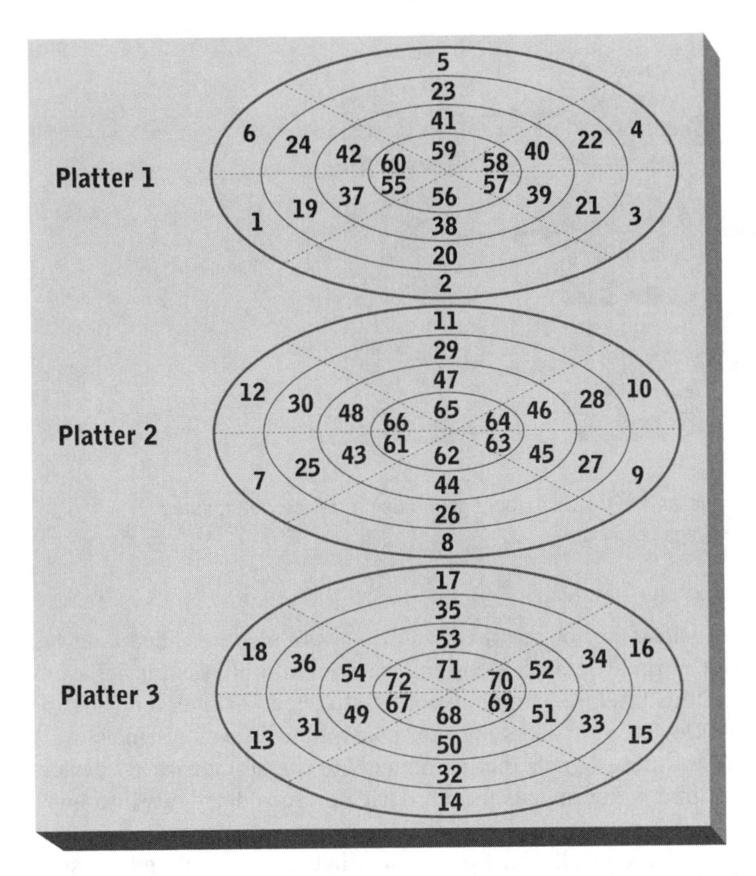

FIGURE 6.5 An example of assigning logical sector numbers to physical sectors on disk platters
Courtesy of Course Technology/Cengage Learning

Other I/O devices do have storage locations in the traditional sense. For example, the character or pixel positions of a printed page or video display are logical storage locations. Each position is assigned an address in a linear address space, and these addresses are used to manipulate pixel or character positions. The device or device controller translates logical write operations into the physical actions necessary to illuminate the corresponding video display pixel or place ink at the corresponding position on the page.

DEVICE CONTROLLERS

Storage and I/O devices are normally connected to the system bus or a subsidiary bus through a **device controller**, as shown in Figure 6.6. Device controllers perform the following functions:

- Implement the bus interface and access protocols.
- Translate logical accesses into physical accesses.
- Enable several devices to share access to a bus connection.

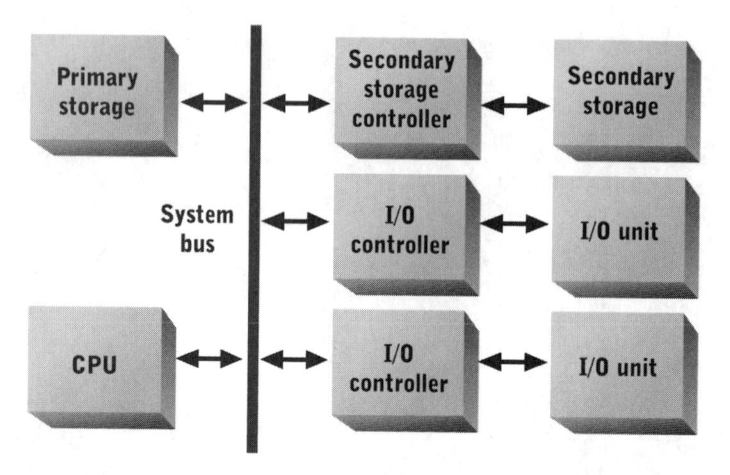

FIGURE 6.6 Secondary storage and I/O device connections using device controllers
Courtesy of Course Technology/Cengage Learning

A device that connects a subsidiary bus to the system bus can also be called a device controller, although it doesn't normally translate logical accesses into physical accesses. A device controller performs all bus interface functions for its attached peripheral devices or between two different buses. Device controllers monitor control bus lines for signals to peripheral devices and translate these signals into commands to the storage or I/O device. Similarly, they translate data and status signals from a device or subsidiary bus into bus control and data signals.

Device controllers perform some or all of the conversion between logical and physical access commands. They "know" the physical details of the attached devices and issue specific instructions to devices based on this knowledge. For example, a disk controller converts a logical access to a specific disk sector in a linear address space into a command to read from a specific head, track, and sector.

Mainframe Channels

In many mainframe computers, a device controller can be a dedicated special-purpose computer called an **I/O channel**, or simply a **channel**. The distinction between an I/O channel and a device controller isn't clear cut. It's a function of power and capability in several key areas, including the following:

- Number of devices that can be controlled
- Variability in type and capability of attached devices
- Maximum communication capacity

NOTE

IBM coined the term "channel" to describe a specific component of its 7000 series mainframe computers, and the term has since gained the generic meaning described in this section. Vendors of other mainframe computers often use different terms—for example, "peripheral processing unit" and "front-end processor"—to describe functionally similar components.

Typical I/O and storage device controllers can control only a few devices of a similar type. For example, the disk controller in most desktop computers can control one or two disk drives. In contrast, a secondary storage channel in a mainframe computer can control several dozen secondary storage devices of different types, such as magnetic disks, optical discs, and magnetic tape drives. Another channel in the same mainframe might control up to 100 video display terminals or point-of-sale I/O devices.

INTERRUPT PROCESSING

Secondary storage and I/O devices have much slower data transfer rates than a CPU does. Table 6.2 lists access times and data transfer rates for typical memory and storage devices. Table 6.3 shows data transfer rates of some I/O devices. Slow access times and data transfer rates are caused primarily by mechanical limitations, such as disk rotation speed. The interval between a CPU's request for input and the moment the input is received can span thousands, millions, or billions of CPU cycles.

TABLE 6.2 Performance characteristics of typical memory and storage devices

Memory or storage device	Maximum data transfer rate	Access time
SDRAM	12.8 GBps	5 nanoseconds (ns)
Magnetic disk	20–100 MBps	3 milliseconds (ms)
Flash drive	20–30 MBps	20–100 ms
Blu-ray disc	50 MBps	100–200 ms
DAT tape	1 MBps	N/A

TABLE 6.3 Typical data transfer rates for I/O devices

I/O device	Maximum data transfer rate
Video display	100–500 MBps
Inkjet printer	30–120 lines/minute
Laser printer	4–20 pages/minute
Network interface	100–10,000 Mbps

If the CPU waits for a device to complete an access request, the CPU cycles that could have been (but weren't) devoted to instruction execution are called **I/O wait states**. To prevent this inefficient use of the CPU, peripheral devices communicate with the CPU by using interrupt signals. In a logical sense, an **interrupt** is a signal to the CPU that some event has occurred that requires the CPU to execute a specific program or process. In a physical sense, an interrupt is an electrical signal that a peripheral device sends over the control bus.

A portion of the CPU, separate from the components that fetch and execute instructions, monitors the bus continuously for interrupt signals and copies them to an **interrupt register**. The interrupt signal is a numeric value called an **interrupt code**, usually equivalent to the bus port number of the peripheral device sending the interrupt. At the conclusion of each execution cycle, the control unit checks the interrupt register for a nonzero value. If one is present, the CPU suspends execution of the current process, resets the interrupt register to zero, and proceeds to process the interrupt. When the interrupt has been processed, the CPU resumes executing the suspended process.

Coordinating peripheral device communication with interrupts allows the CPU to do something useful while it's waiting for an interrupt. If the CPU is executing only a single process or program, there's no performance gain. If the CPU is sharing its processing cycles among many processes, the performance improvement is substantial. When one process requests data from a peripheral device, the CPU suspends it and starts executing another process's instructions. When an interrupt is received, indicating that the access is complete, the CPU retrieves the data, suspends the process it's currently executing, and returns to executing the process that requested the data.

Interrupt Handlers

Interrupt handling is more than just a hardware feature; it's also a method of calling system software programs and processes. The OS provides a set of processing routines, or service calls, to perform low-level processing operations, such as reading data from a keyboard or writing data to a file stored on disk. An interrupt is a peripheral device's request for OS assistance in transferring data to or from a program or a notification that the transfer has already been completed. For example, an interrupt signal indicating that requested input is ready is actually a request to the OS to retrieve the data and place it where the program that requested it can access it, such as in a register or in memory.

There's one OS service routine, called an **interrupt handler**, to process each possible interrupt. Each interrupt handler is a separate program stored in a separate part of primary storage. An interrupt table stores the memory address of the first instruction of each interrupt handler. When the CPU detects an interrupt, it executes a master interrupt handler program called the **supervisor**. The supervisor examines the interrupt code stored in the interrupt register and uses it as an index to the interrupt table. It then extracts the corresponding memory address and transfers control to the interrupt handler at that address.

Multiple Interrupts

The interrupt-handling mechanism just described seems adequate until you consider the possibility of an interrupt arriving while the CPU is busy processing a previous interrupt. Which interrupt has priority? What's done with the interrupt that doesn't have priority?

Interrupts can be classified into the following general categories:

- I/O event
- Error condition
- Service request

The interrupt examples discussed so far have been I/O events used to notify the OS that an access request has been processed and data is ready for transfer. Error condition interrupts are used to indicate errors that occur during normal processing. These interrupts can be generated by software (for example, when attempting to open a nonexistent file) or by hardware (for example, when attempting to divide by zero or when battery power in a portable computer is nearly exhausted).

Application programs generate interrupts to request OS services. An interrupt code is assigned to each service program, and an application program requests a service by placing the corresponding interrupt number in the interrupt register. The interrupt code is detected at the conclusion of the execution cycle, the requesting process is suspended, and the service program is executed.

An OS groups interrupts by their importance or priority. For example, error conditions are normally given higher priority than other interrupts. Critical hardware errors, such as a power failure, are given the highest priority. Interrupt priorities determine whether an interrupt that arrives while another interrupt is being processed is handled immediately or delayed until current processing is finished. For example, if a hardware error interrupt code is detected while an I/O interrupt is being processed, the I/O processing is suspended, and the hardware error is processed immediately.

Stack Processing

While instructions in an application program are being executed, say that an interrupt is received from a pending I/O request. The interrupt is detected, and the corresponding interrupt handler is called. As it executes, the interrupt handler overwrites several general-purpose registers. When the interrupt handler terminates, processing of the application program resumes, but an error occurs because the interrupt handler altered a value in a general-purpose register that the application program needed.

How could this error have been prevented? How did the CPU know which instruction from the application program to load after the interrupt handler terminated? The OS needs to be able to restart a suspended program at exactly the position it was interrupted. When the program resumes execution, it needs all register values to be in the same state they were in when it was interrupted. The mechanism that enables a program to resume execution in exactly the same state as before an interruption is called a stack. A **stack** is a reserved area of main memory accessed on a last-in, first-out (LIFO) basis. LIFO access is similar to a stack of plates in a cafeteria. Items can be added to or removed from only the top of the stack. Accessing the item at the bottom of the stack requires removing all items above it. In a computer system, the stack is a primary storage area that holds register values of interrupted processes or programs; these saved register values are sometimes called the **machine state**. When a process is interrupted, values in CPU registers are added to the stack in an operation called a **push**. When an interrupt handler finishes executing, the CPU removes values on top of the stack and loads them back into the correct registers. This operation is called a **pop**.

Pushing the values of all registers onto the stack isn't always necessary. At a minimum, the current value of the instruction pointer must be pushed because the instruction at this address is needed to restart the interrupted process where it left off. Think of the instruction pointer as a bookmark in a running program. If indirect addressing is in use, the offset register must also be pushed. General-purpose registers are pushed because they might contain intermediate results needed for further processing or values awaiting output to a storage or an I/O device.

Multiple machine states can be pushed onto the stack when interrupts of high precedence occur while processing interrupts of lower precedence. It's possible for the stack to fill to capacity, in which case further attempts to push values onto the stack result in a **stack overflow** error. The stack size limits the number of processes that can be interrupted or suspended. A special-purpose register called the **stack pointer** always points to the next empty address in the stack and is incremented or decremented automatically each time the stack is pushed or popped. Most CPUs provide one or more instructions to push and pop the stack.

Performance Effects

Figure 6.7 summarizes the sequence of events in processing an interrupt. The CPU suspends the application program by pushing its register values onto the stack, and the supervisor suspends its own operation by pushing its register values onto the stack before branching to the interrupt handler. Two pop operations are required to restore the original application program. This sequence is complex and consumes many CPU cycles. The most complex parts are the push and pop operations, which copy many values between registers and main memory. Wait states might occur if the stack isn't implemented in a cache. (Caching is described in the next section.)

Supervisor execution also consumes CPU cycles. The table lookup procedure is fast but still consumes at least 10 CPU cycles. Other steps in the process, discussed in Chapter 11, consume additional CPU cycles. In total, processing an interrupt typically consumes at least 100 CPU cycles in addition to the cycles the interrupt handler consumes.

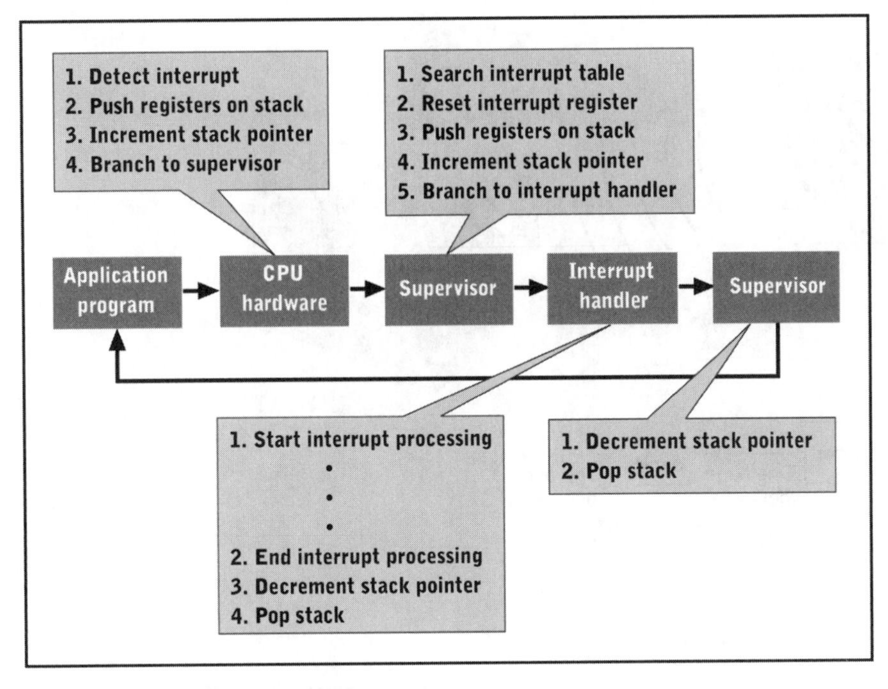

FIGURE 6.7 Interrupt processing
Courtesy of Course Technology/Cengage Learning

BUFFERS AND CACHES

Access to I/O and storage devices is inherently slow. Mismatches in data transfer rate and data transfer unit size are addressed in part by interrupt processing, which consumes substantial CPU resources. RAM can be used to overcome the mismatches in two distinct ways: buffering and caching. The main goal of buffering and caching is to improve overall system performance.

Buffers

A **buffer** is a small reserved area of main memory (usually DRAM or SRAM) that holds data in transit from one device to another and is required to resolve differences in data transfer unit size. It's not required when data transfer rates differ, but using one generally improves performance.

Figure 6.8 shows wireless communication between a PC and a laser printer. A PC usually transmits data 1 bit at a time over a wireless connection, and a laser printer prints an entire page at once. The input data transfer unit from the PC is a single bit, and the output data transfer unit from the laser printer is a full page that can contain up to several million bytes of data. A buffer is required to resolve this difference.

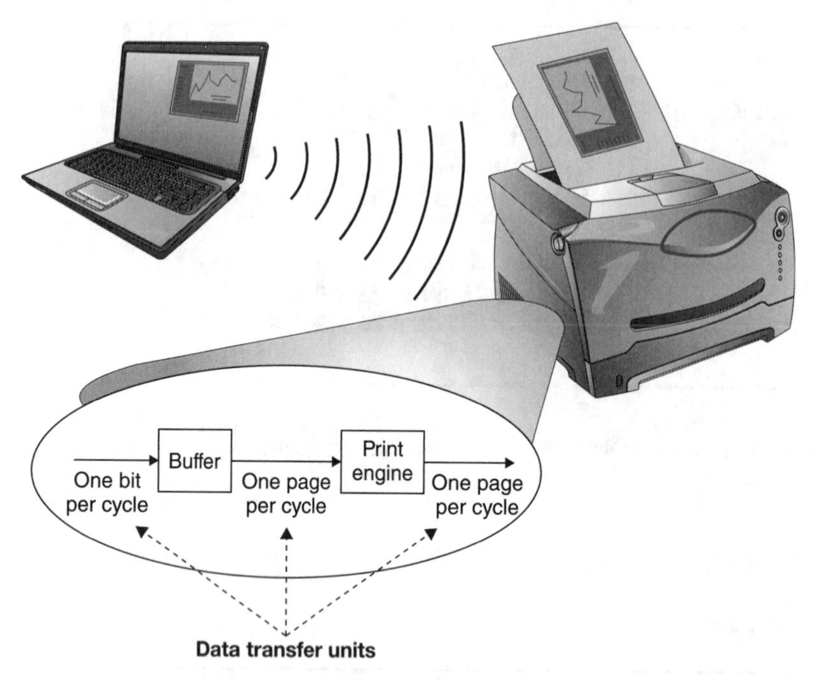

FIGURE 6.8 A buffer resolves differences in data transfer unit size between a PC and a laser printer
Courtesy of Course Technology/Cengage Learning

As single bits are transferred over the wireless connection, they're added to the buffer in sequential order. When all bits of a page have been received, they're removed from the buffer and transferred to the print engine as a single unit. Buffer size must be at least as large as the data output unit. If the buffer isn't large enough to hold a full page, an error called a **buffer overflow** occurs. A buffer for an I/O device is typically implemented in the device, as it is in a laser printer.

A buffer can also improve system performance when two devices have different data transfer rates, as when copying music files from a PC to an iPod via a USB 2.0 connection. The USB controller receives data from primary storage at the system bus's data transfer rate—typically 500 MBps or more. The USB connection transmits data at 480 Mbps, although an iPod typically uses only a fraction of this capacity—a few hundred to a few million megabits per second. Therefore, the USB controller is a bridge between two data transmission channels with a speed difference of at least 80 to 1.

Assume the USB controller has a 1-byte buffer, and after the USB controller receives 1 byte from primary storage and fills the buffer, it sends an interrupt to the CPU to pause the transfer operation. The interrupt prevents more data from arriving and causing a buffer overflow before the USB controller can transmit the current buffer content to the iPod. After the buffer content has been sent to the iPod, the USB controller transmits another interrupt to indicate that it's ready to receive data again.

How much system bus data transfer capacity is used to transmit 1 MB (1,048,576 bytes) via the USB controller? The answer is 3,145,728 bus cycles (1,048,576 × 3). For each data byte transmitted across the system bus, two interrupts are also transmitted: one to stop further transmission and another to restart it. The two interrupts aren't data, but they do consume bus cycles. Further, the CPU must process these interrupts. If you assume 100 CPU cycles are consumed per interrupt, then 209,715,200 CPU cycles are consumed to process interrupts while transferring 1 MB to the iPod.

Now consider the same example with a 1024-byte buffer in the USB controller. A single transfer from main memory to the USB controller would send as many bytes per bus cycle as could be carried on the data bus—for example, 64 bits or 8 bytes. After 131,072 bus transfers (1 MB at 8 bytes per bus cycle), the USB controller would send an interrupt to stop transmission. After the buffer content was transmitted to the iPod, the USB controller would transmit another interrupt to restart data transfer from memory.

Computer system performance improves dramatically with the larger buffer (see Table 6.4). Two interrupts are generated for each 1024 bytes transferred. The total number of bus cycles consumed is as follows:

$$131{,}072 \text{ data transfers} + \left(\frac{1{,}048{,}576 \text{ bytes}}{1024 \text{ bytes per transfer}} \times 2 \text{ interrupts}\right) = 133{,}120 \text{ bus cycles}$$

The number of CPU cycles falls to 204,800 (1,048,576 ÷ 1024 × 2 × 100). Bus and CPU cycles are, therefore, freed for other purposes.

TABLE 6.4 Bus and CPU cycles consumed for a 1 MB data transfer with various buffer sizes

Buffer size	Bus data transfers	Interrupts	Total bus transfers	Improvement	CPU cycles	Improvement
1	1,048,576	2,097,152	3,145,728	N/A	209,715,200	N/A
2	524,288	1,048,576	1,572,864	50%	104,857,600	50%
4	262,144	524,288	786,432	50%	52,428,800	50%
8	131,072	262,144	393,216	50%	26,214,400	50%
16	131,072	131,072	262,144	33.33%	13,107,200	50%
32	131,072	65,536	196,608	25%	6,553,600	50%
64	131,072	32,768	163,840	16.67%	3,276,800	50%
128	131,072	16,384	147,456	10%	1,638,400	50%
256	131,072	8192	139,264	5.56%	819,200	50%
512	131,072	4096	135,168	2.94%	409,600	50%
1024	131,072	2048	133,120	1.52%	204,800	50%

Diminishing Returns

Note the rates of change in performance improvement for total bus transfers and CPU cycles shown in Table 6.4. The first few increases in buffer size improve bus efficiency at a constant rate until the data transfer unit matches the bus width. As buffer size increases above 8 bytes, CPU cycle consumption decreases at a linear rate, but total bus cycles decrease at a diminishing rate. In other words, each doubling of buffer size yields fewer benefits in improved bus efficiency.

If you assume there are no excess CPU cycles, all buffer size increases shown in Table 6.4 provide a substantial benefit. In other words, CPU cycles not used for I/O interrupts are always applied to other useful tasks. However, if you assume extra CPU cycles are available—in other words, the CPU isn't being used at full capacity—the only major benefit of a larger buffer size is reduced bus cycles. Because bus cycle improvement drops off rapidly, there's a point at which further buffer size increases have no real benefit. If you also assume that buffer cost increases with buffer size (a reasonable assumption), there's also a point at which the cost of additional buffer RAM is higher than the monetary value of more efficient bus utilization.

NOTE

If the discussion in the preceding paragraphs sounds vaguely familiar, you've probably studied economics. The underlying economic principle is called the **law of diminishing returns**. Simply stated, this law says that when multiple resources are required to produce something useful, adding more of a single resource produces fewer benefits. It's just as applicable to buffer size in a computer as it is to labor or raw material inputs in a factory. It has many other applications in computer systems, some of which are described elsewhere in this book.

Table 6.5 shows bus and CPU performance improvements for a 16 MB transfer through the USB controller. Because the example starts with a much larger buffer size than in Table 6.4, improvements in bus efficiency are barely measurable. CPU efficiency improves at a constant rate until buffer size exceeds the amount of data being transferred, after which there's no further improvement in bus or CPU efficiency. Why don't increases in buffer size beyond 16 MB result in any improvement? Because in this example, there's nothing to store in the extra buffer space.

TABLE 6.5 Bus and CPU cycles consumed for a 16 MB data transfer with large buffer sizes

Buffer size (bytes)	Bus data transfers	Interrupts	Total bus transfers	Improvement	CPU cycles	Improvement
1,048,576	2,097,152	32	2,097,184	N/A	3200	N/A
2,097,152	2,097,152	16	2,097,168	0.000763%	1600	50%
4,194,304	2,097,152	8	2,097,160	0.000381%	800	50%
8,388,608	2,097,152	4	2,097,156	0.000191%	400	50%
16,777,216	2,097,152	2	2,097,154	0.000095%	200	50%
33,554,432	2,097,152	2	2,097,154	0%	200	0%
67,108,864	2,097,152	2	2,097,154	0%	200	0%

The results in Tables 6.4 and 6.5 do more than show the law of diminishing returns in action. They also show the importance of clearly defined assumptions and an understanding of the work a computer system will be asked to do. Accurate computation of improvements requires understanding all affected components—in this case, the CPU, bus, USB controller, and buffer; clearly stated operational assumptions—for example, whether the CPU is being used at full capacity; and well-defined workload characteristics—size and frequency of data transfers to and from the USB controller, for instance.

Caches

Like a buffer, a **cache** is a reserved area of high-speed memory (usually RAM) that improves system performance. However, a cache differs from a buffer in several important ways, including the following:

- Data content isn't automatically removed as it's used.
- A cache is used for bidirectional data transfer.
- A cache is used only for storage device accesses.
- Caches are usually much larger than buffers.
- Cache content must be managed intelligently.

The basic idea behind caching is simple. Access to data in a high-speed cache can occur much more quickly than access to slower storage devices, such as magnetic disks. The speed difference is entirely a function of the faster access speed of the RAM used to implement the cache. Performance improvements require a large enough cache and the "intelligence" to use it effectively.

Performance improvements are achieved differently for read and write accesses. During a write operation, a cache acts similarly to a buffer. Data is first stored in the cache and then transferred to the storage device. Performance improvements are the same as those of a buffer: reduced bus and CPU overhead as a function of larger continuous data transfers. Because caches are usually much larger than buffers, the performance improvement is increased. However, the law of diminishing returns usually results in only slightly better performance than with a typical buffer.

Write caching can result in more substantial performance improvement when one write access must be confirmed before another can begin. Some programs, such as transaction updates for banks and retailers, require confirmed writes. In Figure 6.9, when data is written to a cache (1), the confirmation signal is sent immediately (2), before the data is written to the storage device (3).

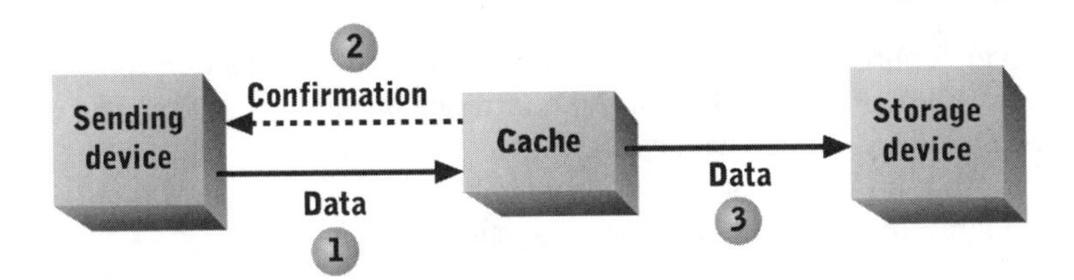

FIGURE 6.9 A storage write operation with a cache
Courtesy of Course Technology/Cengage Learning

Sending a confirmation before data is written to the storage device can improve program performance because the program can proceed immediately with other processing tasks. If the program is performing a series of write operations, performance improvement ceases as soon as the cache is full. As with a buffer, interrupts are used to coordinate the data transfer activity.

Immediate write confirmations are risky because an error might occur while copying data from the cache to the storage device. The biggest danger is total cache failure caused by a power failure or other hardware error. In this case, all data, possibly representing many cached write operations, is lost permanently, and there's no way to inform the programs that wrote the data of the loss.

Data written to a cache during a write operation isn't automatically removed from the cache after it's written to the underlying storage device. Data remaining in the cache can improve performance when the data is reread shortly after it's written. A subsequent read of the same data item is much faster, unless the data item has been purged from the cache for another reason—for example, to make room for other data items.

Most performance benefits of a cache occur during read operations (see Figure 6.10). The key performance-enhancing strategy is to guess what data will be requested and copy it from the storage device to the cache before a read request is received (1). When the read request is received (2), data can be supplied more quickly because access to the cache is much faster than access to the storage device (3). If the requested data isn't in the cache, it must be read from the storage device after the read request is received, resulting in a waiting period for the requester equal to the storage device's access and data transfer times.

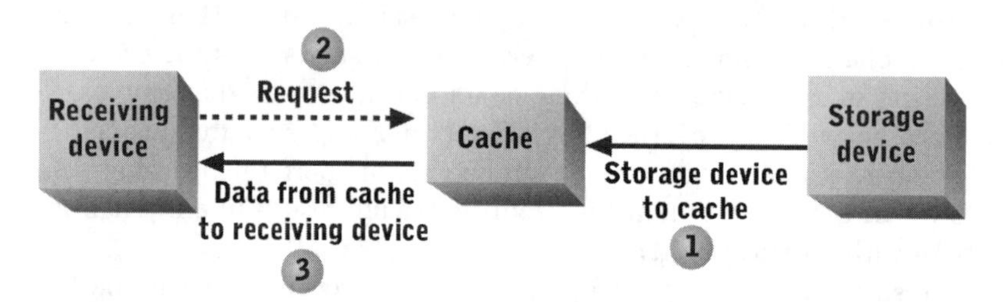

FIGURE 6.10 A read operation when data is already stored in the cache
Courtesy of Course Technology/Cengage Learning

A **cache controller** is a processor that implements the procedures shown in Figures 6.9 and 6.10. It can be implemented in the following:

- A storage device controller or communication channel, as a special-purpose processor controlling RAM installed in the controller or channel; more common with primary storage caches
- The OS, as a program that uses part of primary storage to implement the cache; more common with secondary storage caches

A cache controller guesses what data will be requested in the near future and loads this data from the storage device into the cache before it's actually requested. Guessing methods can be simple or complex. One simple method is to assume linear access to storage locations. For example, while servicing a read request for location n, the cache controller prefetches location $n + 1$ and possibly several subsequent storage locations. More complex guessing methods require more complex processing circuitry or software. The extra complexity and expense are justified only if the guessing accuracy improves and if these improvements lead to substantial overall performance improvement.

When a read operation accesses data already contained in the cache, the access is called a **cache hit**. When the data needed isn't in the cache, the access is called a **cache miss**. The ratio of cache hits to read accesses is called the cache's **hit ratio**. A cache miss requires performing a **cache swap** to or from the storage device. The cache controller guesses which data items are least likely to be needed in the near future, writes them to the storage device, and purges them from the cache. The requested data is then read from the storage device and placed in the cache.

Caches can be implemented in main memory if an OS program serves as the cache controller. However, this approach can reduce overall system performance by reducing memory and CPU cycles available to application programs. Modern device controllers can implement a cache with RAM and the cache controller installed on the device controller. In this case, the CPU and system software are unaware that a cache is in use.

A surprisingly small cache can improve performance dramatically. Typical ratios of cache size to storage device capacity range from 10,000:1 to 1,000,000:1. Primary storage caches with a 10,000:1 size ratio typically achieve cache hits more than 90% of the time. Actual performance gain depends on cache size and the nature of storage device accesses. Frequent sequential read accesses tend to improve performance. Frequent random or scattered read accesses tend to reduce the hit ratio.

Primary Storage Cache

Current processor speeds exceed the capability of dynamic RAM (DRAM) to keep the processor supplied with data and instructions. Although static RAM (SRAM) more closely matches processor speed, it's generally too expensive to use for all primary storage. When the CPU accesses DRAM, it incurs wait states. For example, a 4 GHz processor incurs 10 wait states each time it reads sequentially from 400 MHz SDRAM and many more wait states if the reads are from scattered locations.

One way to limit wait states is to use an SRAM cache between the CPU and SDRAM primary storage. The SRAM cache can be integrated into the CPU, located on the same chip as the CPU, or located between the CPU and SDRAM. Multiple cache levels can be used in the same chip or computer. When three cache levels are in use, the cache closest to the CPU is called a **level one (L1) cache**, the next closest cache is called a **level two (L2) cache**, and the cache farthest from the CPU is called a **level three (L3) cache**. Multicore processors (discussed later in "Processing Parallelism") typically implement all three levels on the same chip, including a small (64 KB) L1 cache in each CPU, a larger (0.5 to 2 MB) L2 cache next to each CPU, and an even larger (2 to 16 MB) L3 cache shared by all CPUs.

Primary storage cache control can be quite sophisticated. The default sequential flow of instructions in most software enables simple guesses based on linear storage access to work some of the time. However, this simple strategy doesn't account for accesses to operands and conditional BRANCH instructions. Current CPUs use sophisticated methods to improve the hit ratio. For example, instead of guessing whether a conditional BRANCH will be taken, the cache controller might prefetch both subsequent instructions. This method consumes more of the cache but guarantees that the next instruction will be in the cache. Another strategy is to examine operands as instructions are loaded into the cache. If an operand is a memory address, the cache controller prefetches the data at that location into the cache.

Secondary Storage Caches

Disk caching is common in modern computer systems, particularly in file and database servers. In a file server, disk-caching performance can be improved if information about file access is tracked and used to guide the cache controller. Specific strategies include the following:

- Give frequently accessed files higher priority for cache retention.
- Use read-ahead caching for files that are read sequentially.
- Give files opened for random access lower priority for cache retention.

The OS is the best source of file access information because it updates information dynamically as it services file access requests. Because the OS executes on the CPU, implementing access-based cache control is difficult if the cache controller is a special-purpose processor in a disk controller.

Many computer system designers now rely on the OS to implement disk caching instead of using specialized disk controller hardware. They believe the extra cost of hardware-based disk-caching solutions is better spent on more primary storage and faster or additional CPUs. The OS uses the extra memory and CPU cycles to implement larger caches and more intelligent access-based cache control. This approach makes sense only if the computer system has enough CPU capacity to devote to cache management.

PROCESSING PARALLELISM

Many applications, such as the following examples, are simply too big for a single CPU or computer system to execute:

- *Large-scale transaction processing applications*—Computing monthly social security checks and electronic fund transfers, for example
- *Data mining*—Examining a year of sales transactions at a large grocery store chain to identify purchasing trends, for example
- *Scientific applications*—Producing hourly weather forecasts for a 24-hour period covering all of North America, for example

Solving problems of this magnitude would require days, months, or years for even the fastest single CPU or computer system. The only way to solve them in a timely fashion is to break the problems into pieces and solve each piece in parallel with separate CPUs. Even for applications that could be performed by a single CPU, parallel processing can improve performance. This section examines parallel-processing hardware architectures and their performance implications.

Multicore Processors

As described in Chapter 5, current semiconductor fabrication techniques are capable of placing billions of transistors and their interconnections in a single microchip. A full-featured 64-bit CPU, even one with multiple ALUs and pipelined processing, typically requires fewer than 100 million transistors. This raises an obvious question—what else can be placed in a single microchip to improve processor and computer system performance?

Until the mid-2000s, the only answer was cache memory, which yields substantial performance enhancements by helping overcome the speed differences between CPU circuitry and off-chip memory. However, as single-chip transistor counts increased into the hundreds of millions, devoting the "extra" transistors entirely to cache memory began to yield fewer performance improvements because the performance benefit of larger caches is subject to rapidly diminishing returns.

The latest trend in high-performance CPU design embeds multiple CPUs and cache memory on a single chip—an approach called **multicore architecture**, in which the term **core** describes the logic, computation, and control circuitry of a single CPU. As of this writing, AMD offers a six-core processor (see Figure 6.11), Intel offers a six-core version of its Xeon processor, and IBM has demonstrated a prototype of an eight-core POWER7 processor. With future advances in semiconductor fabrication, more cores in a single microprocessor will be possible.

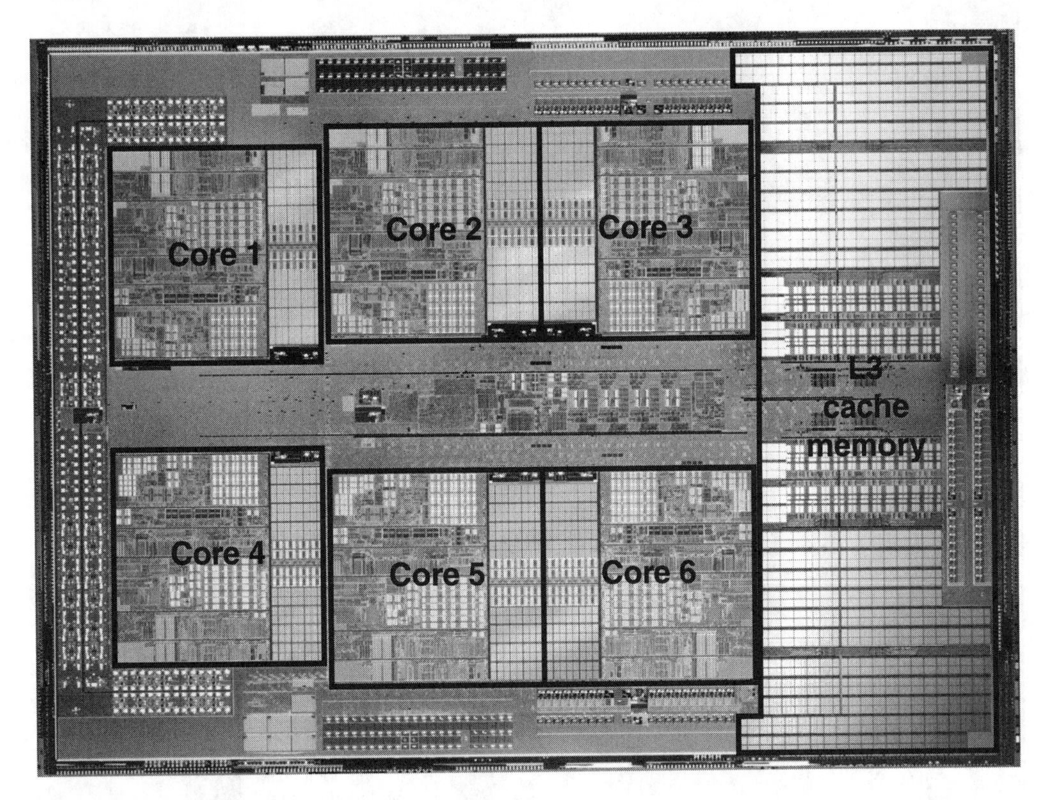

FIGURE 6.11 The six-core AMD Opteron processor
Courtesy of Advanced Micro Devices, Inc.

Multicore architectures typically share memory cache, memory interface, and off-chip I/O circuitry between cores. This sharing reduces the total transistor count and cost and offers some synergistic benefits compared with separate CPUs on a single motherboard. For large-scale computational problems, a common data set can be stored in cache memory, and each core can work on a different part. This arrangement yields substantial performance improvements in many numerical-modeling and image-processing applications because communication and data exchange between CPUs don't have to traverse the slower system bus.

TECHNOLOGY FOCUS

Core-i7 Memory Cache

The Intel Core-i7 processor uses three levels of primary storage caching, as shown in Figure 6.12. All processing cores, caches, and the external memory interface share the same chip, which enables much faster performance than with earlier Intel processors. Four-core processors share the same cache architecture.

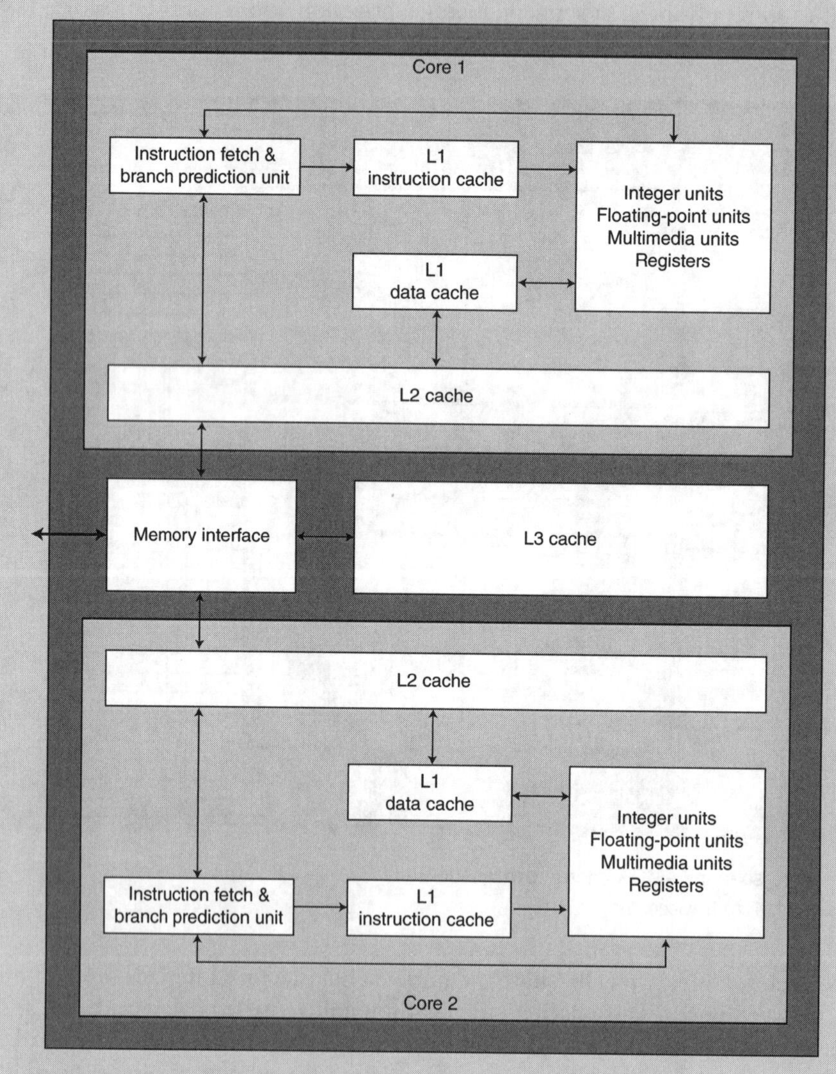

FIGURE 6.12 Memory caches in a dual-core Intel Core-i7 processor
Courtesy of Course Technology/Cengage Learning

(continued)

The L1 cache is divided into two 16 KB segments: one for data and the other for instructions. The L2 cache is 256 KB and holds both instructions and data. L1 data cache contents are swapped automatically with the L2 cache as needed. The instruction fetch and branch prediction unit examines incoming instructions for conditional BRANCHes and attempts to prefetch instructions from both branch paths into the L1 instruction cache. The branch prediction unit also interacts with CPU execution units to determine the likelihood of BRANCH conditions being true or false.

Using separate L1 caches for instructions and data allows some streamlining of processor functions. The L1 instruction cache is read-only. The L1 data cache can be read or written as operands are moved to and from registers. Cache controller functions are optimized for each type of access. However, it's possible for a program to modify its own instructions. In this case, the code section being modified is stored in both L1 caches at the same time. The processor detects this condition automatically and maintains consistency between the two caches.

The memory interface controller acts as a cache controller for the L2 and L3 caches. It controls movement between external memory and the L3 and L2 caches and ensures consistency of their contents. L1 cache management is handled internally in each processing core.

The Core-i7 can assign different caching modes to 4 KB, 2 MB, or 4 MB regions of main memory. Write-back caching enables full caching of both read and write operations. Main memory is updated only when necessary, such as when a modified cache line must be removed to make room for new content. Write-through caching causes all write operations to update caches and main memory at the same time. This mode is useful for regions of memory that can be read directly without knowledge of a processing core by another core or by other devices on the system bus, such as a disk or video controller. All caching functions can be disabled for any memory region.

Multicore processors add complexity to cache management because multiple cores can cache the same memory regions simultaneously. Program execution errors are possible if a program running on one processor updates a cached memory location, and another program running on another processor subsequently reads this same memory location from an outdated cache copy. Core-i7 processors use a technique called "memory snooping" to maintain consistency between caches and main memory. The memory interface controller monitors access to main memory by all cores. An access to memory that's currently cached by another core causes the corresponding cache contents to be marked invalid. The modified cache contents are reloaded from main memory automatically the next time they're accessed. If a memory region cached by multiple cores is written, all cores exchange updated cache contents to ensure consistency.

Multiple-Processor Architecture

Multiple-processor architecture is a more traditional approach to multiprocessing that uses two or more processors on a single motherboard or set of interconnected motherboards. When multiple processors occupy a single motherboard, they share primary storage and a single system bus. When necessary, they can exchange messages over the system bus and transfer data by reading from and writing to primary storage. However, these exchanges are much slower than the exchanges between cores in multicore architecture.

Multiple-processor architecture is common in midrange computers, mainframe computers, and supercomputers; the number of processors and the number of cores per processor tend to increase with computer class. Multiple programs can run in parallel, and the OS can move them between processors to meet changing needs. Multiple-processor architecture is a cost-effective approach to computer system design when a single computer runs many different application programs or services. In this case, there's little need for CPUs to share data or exchange messages, so the performance penalty associated with these exchanges is seldom incurred.

Multiple-processor architecture is also common in workstations. The similarity in processing and I/O power of midrange and workstation computers enables manufacturers to build both types from a common pool of subcomponents, such as motherboards and storage subsystems.

Scaling Up and Scaling Out

The phrase **scaling up** describes approaches to increasing processing and other computer system power by using larger and more powerful computers. Both multicore and multiple-processor architectures are examples of scaling up because they increase the power of a single computer system. The alternative approach is **scaling out**—partitioning processing and other tasks among multiple computer systems. Two of the approaches to multicomputer architecture described in Chapter 2, clusters and grids, are examples of scaling out. Blade servers are a combination of scaling up and scaling out, although some people consider them an example only of scaling up because all blades are normally housed in a single computer system or rack.

Until the 1990s, scaling up was almost always a more cost-effective strategy to increase available computer power because communication between computers was extremely slow compared with communication between a single computer's components. However, beginning in the early 1990s, the speed of communication networks increased rapidly. Today, 10 Gbps speed across networks is common, and much higher speeds are close at hand. As a result, the performance penalty of communication between computer systems has diminished. Also, because networking technology is widely deployed, economies of scale have decreased high-speed network costs. Other changes that have increased the benefits of scaling out include the following:

- Distributed organizational structures that emphasize flexibility
- Improved software for managing multicomputer configurations

Organizations change more rapidly than they did a few decades ago. They need the flexibility to deploy and redeploy all types of resources, including computing resources. In general, scaling out enables them to distribute computing resources across many locations and combine disparate resources to solve large problems.

Until recently, the complexity of multicomputer configurations made administering widely distributed resources difficult and expensive. However, improvements in management software have made it feasible to manage large, diverse collections of computing resources with a minimum of labor. Nonetheless, administering a few large computer systems is still less expensive than administering many smaller ones. So scaling up is still a cost-effective solution when maximal computer power is required and flexibility isn't as important. Examples of environments in which scaling up is cost effective include large data-processing centers that service the transaction-processing needs of multiple organizations, such as banks and insurance companies.

High-Performance Clustering

The largest computational problems, such as those encountered in modeling three-dimensional physical phenomena, can't be solved by a single computer. These problems are typically handled by groups of powerful computers organized into a cluster. Each computer system, or node, in the cluster usually contains multiple CPUs, and each CPU can be dual-core.

Figure 6.13 shows a dual-cluster architecture used by the European Centre for Medium-Range Weather Forecasts. (To see the full-color version, go to *www.ecmwf.int/services/computing/overview/ibm_cluster.html*.) Each cluster contains more than 200 IBM pSeries 575 32-CPU computers operating as compute nodes. Groups of 16 nodes are interconnected by a dedicated network through a network I/O node. The network I/O nodes connect to a shared high-speed network, which enables connections between groups of compute nodes and between compute nodes and secondary storage. Secondary storage is controlled by separate storage nodes.

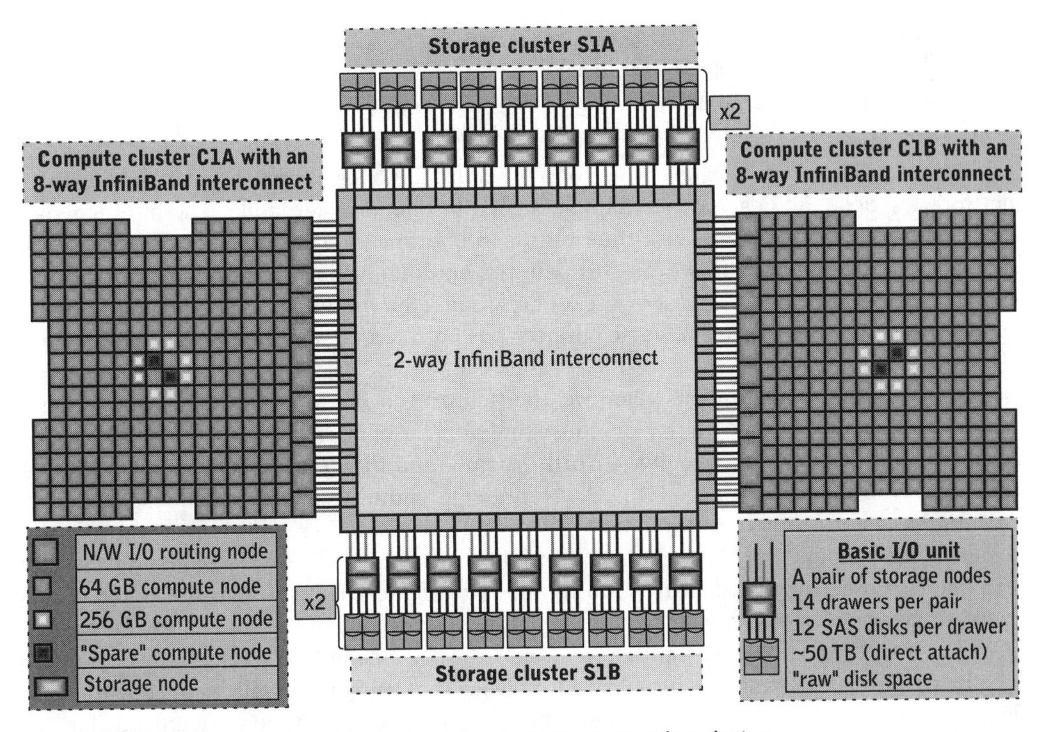

FIGURE 6.13 Organization of two interconnected supercomputing clusters
Courtesy of European Centre for Medium-Range Weather Forecasts

The cluster organization in Figure 6.13 addresses a common problem in supercomputing—data movement between processing nodes. For example, a simplified approach to global weather forecasting breaks up the globe into 10 nonoverlapping regions and further divides each region into four zones (see Figure 6.14).

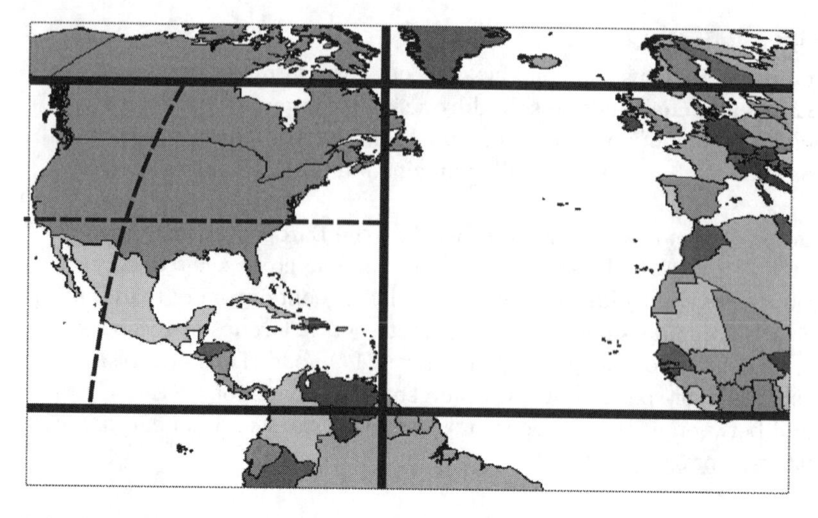

FIGURE 6.14 Sample weather forecast regions (separated by solid lines) and division of a region into zones (separated by dashed lines)

Courtesy of Course Technology/Cengage Learning

Each region can be assigned to a group of compute nodes connected to a single network I/O node, and each zone can be assigned to one compute node. The interdependence of forecast calculations in a zone results in continuous data transfer between CPUs in a single compute node. However, this data traverses high-speed bus and memory connections in a single motherboard or set of interconnected motherboards. Of equal importance is that the data doesn't traverse connections between compute nodes or the network that connects the groups.

Now consider what happens when weather features in the forecast move from zone to zone. For example, look at the region consisting of most of North America, Central America, and the western edge of the North Atlantic and the division of this region into four zones (on the left in Figure 6.14). A low-pressure center or cold front moving from the western to eastern United States moves from one zone to another. This movement creates a shared data dependency between the compute nodes assigned to the region, and these computers must exchange data to perform forecast calculations. A similar data exchange between compute nodes assigned to different zones occurs when modeling a hurricane as it moves westward from the Atlantic Ocean into the Caribbean Sea.

By dividing the forecasting problem into pieces that mirror the cluster organization in Figure 6.13, the problem is divided hierarchically into pieces with increasing needs for data exchange. The largest amount of data is exchanged between CPUs in a single computer dedicated to a single zone. The next largest amount of data is exchanged between compute nodes connected to a single network I/O node, which traverses data connections specific to this group. The least amount of data is exchanged between compute nodes assigned to a region, and this data traverses the network that connects all nodes. Partitioning the problem to match the cluster architecture ensures that most data exchange traverses high-speed paths. Therefore, all CPUs are kept as busy as possible by avoiding long waits for data sent from "distant" processors.

COMPRESSION

We live in a world that's rich in processing power but overwhelmed with data. People routinely download megabytes or gigabytes of data via the Internet and store gigabytes of data on handheld devices, terabytes on desktop computers, and petabytes to exabytes in corporate and government data centers. The demand for data storage and data transfer capacity often exceeds budgets and sometimes exceeds technical feasibility. Further, although processors are faster and cheaper, the ability to feed them with data hasn't kept pace.

Compression is a technique that reduces the number of bits used to encode data, such as a file or a stream of video images transmitted across the Internet. Reducing the size of stored or transmitted data can improve performance whenever there's plenty of inexpensive processing power but a dearth of data storage or data transfer capacity. As this situation is common, compression is used in many devices and applications, including iPods, DVD players, YouTube, video conferencing, and storage of large data files, such as medical images and motion video.

A **compression algorithm** is a mathematical compression technique implemented as a program. Most compression algorithms have a corresponding **decompression algorithm** that restores compressed data to its original or nearly original state. In common use, the term "compression algorithm" refers to both the compression and decompression algorithms. There are many compression algorithms, which vary in the following ways:

- Types of data for which they're best suited
- Whether information is lost during compression
- Amount by which data is compressed
- Computational complexity (CPU cycles required to execute the compression program)

A compression algorithm can be lossless or lossy. With **lossless compression**, any data input that's compressed and then decompressed is exactly the same as the original input. Lossless compression is required in many applications, such as accounting records, executable programs, and most stored documents. Zip files and archives are examples of lossless compression.

With **lossy compression**, data inputs that are compressed and then decompressed are different from, but still similar to, the original input. Lossy compression is usually applied only to audio and video data because the human brain tolerates missing audio and video data and can usually "fill in the blanks." It's commonly used to send audio or video streams via low-capacity transmission lines or networks, such as video conferencing. MP3 audio encoding and video encoding on DVDs are two examples of lossy compression.

The term **compression ratio** describes the ratio of data size in bits or bytes before and after compression. For example, if a 100 MB file is compressed to 25 MB, the compression ratio is 4:1. Some types of data are easier to compress than others. For example, lossless compression ratios up to 10:1 are achieved easily with word-processing documents and ASCII or Unicode text files. Lossless compression ratios higher than 3:1 are difficult or impossible to achieve with audio and video data. Lossy compression of audio and video can achieve compression ratios up to 50:1, although at high ratios, listeners and viewers can easily discern that information has been lost (see Figure 6.15).

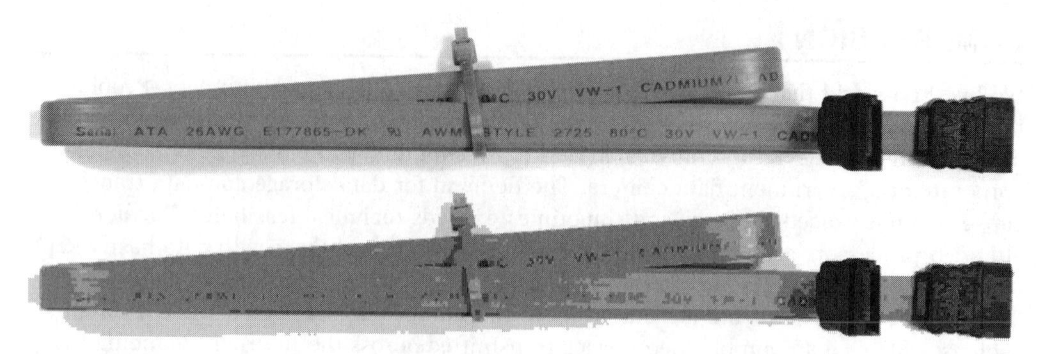

FIGURE 6.15 A digital image before (top) and after (bottom) 20:1 JPEG compression
Courtesy of Course Technology/Cengage Learning

Compression is commonly used to increase the amount of data stored on backup tapes and is sometimes used to reduce disk storage requirements, as shown in Figure 6.16(a). Data sent to the storage device is compressed before it's written. Data read from storage is decompressed before it's sent to the requester. Compression can also increase communication channel capacity, as shown in Figure 6.16(b). Data is compressed as it enters the channel and then decompressed as it leaves the channel. Hardware-based compression is included in all current standards for videoconferencing and is often used for long-distance telephone transmissions.

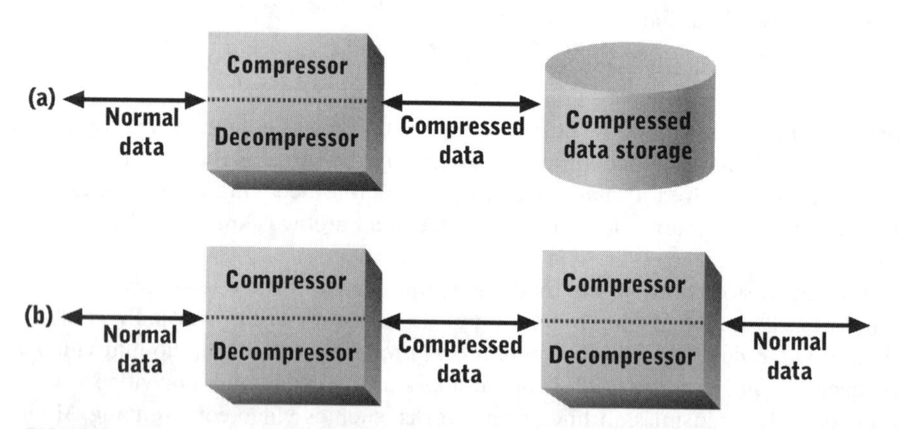

FIGURE 6.16 Data compression with a secondary storage device (a) and a communication channel (b)
Courtesy of Course Technology/Cengage Learning

Using data compression alters the balance of processor resources and communication or storage resources in a computer system. Implementing a compression algorithm consumes processor cycles, and algorithms with high compression ratios often consume more processing resources than those with low compression ratios. The tradeoff between these resources depends on their comparative cost, availability, and degree of use. Using data compression with CPU processing resources might not be cost effective, for example.

Using data compression with special-purpose processors is often preferable. These processors are now widely available and are usually embedded in tape backup devices, video display controllers, and high-speed modems.

TECHNOLOGY FOCUS

MPEG and MP3

The **Moving Picture Experts Group (MPEG)** creates and evaluates standards for motion picture recording and encoding technology, including MPEG-1, MPEG-2, and MPEG-4. Motion pictures usually contain images and sound, so MPEG standards address recording and encoding formats for both data types. Each standard is divided into layers numbered 1 (systems), 2 (video), and 3 (audio). The audio-encoding standard commonly known as **MP3** is actually layer 3 of the MPEG-1 standard. It's a useful encoding method for many types of audio, not just audio tracks in motion pictures.

Analog audio data is converted to digital form by sampling the audio waveform thousands of times per second and then constructing a numerical picture of each sample. On an audio CD, there are 44,100 samples per second per stereo channel. Each sample is represented by a single digital number. Sample quality is improved if many bits are used to encode each sample because of better precision and accuracy. Each sample on an audio CD is encoded in 16 bits, so each second of audio data on a stereo CD requires $44{,}100 \times 16 \times 2 = 1{,}411{,}200$ bits.

A typical motion picture is two hours long and includes far more video than audio data. To encode both data sets in a single storage medium, such as on a DVD-ROM, all data must be compressed. Normal CD-quality music can be compressed by a ratio of 6:1, and most listeners can't distinguish the compressed audio data from the original. MP3, a lossy compression method, discards much of the original audio data and doesn't attempt to reconstruct the data for playback. Therefore, it has a simple decompression algorithm.

MP3 takes advantage of a few well-known characteristics of human audio perception, including the following:

- Sensitivity that varies with audio frequency (pitch)
- Inability to recognize faint tones of one frequency simultaneously with much louder tones in nearby frequencies
- Inability to recognize soft sounds that occur shortly after louder sounds

These three characteristics interact in complex ways. For example, loud sounds at the extremes of the human hearing range, such as 20 Hz or 20 KHz, mask nearby soft frequencies more effectively than do loud sounds in the middle of the human hearing range. MP3 analyzes digital audio data to determine which sound components are masked by others. For example, a bass drum might be masked for a moment by a bass guitar note. MP3 then compresses the audio data stream by discarding information about masked sounds or representing them with fewer bits.

Figure 6.17 shows the conceptual components of an MP3 encoder. A frequency separator divides raw audio data into 32 narrow-frequency audio streams. In parallel, a perceptual modeler examines the entire audio signal to identify loudness peaks. It then determines by how much and for how long these loudness peaks can mask nearby audio data. The masking information and the 32 data streams are sent to an encoder that

(*continued*)

represents data in each frequency. The encoder uses the masking information from the perceptual modeler to determine what data to discard or to represent with less than 16-bit precision.

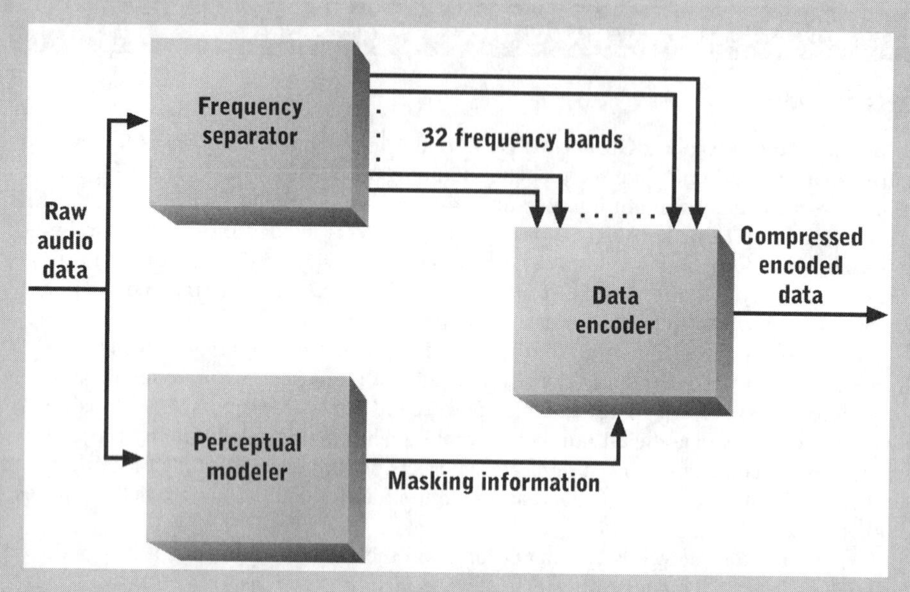

FIGURE 6.17 MP3 encoding components

The MP3 standard is widely implemented in MP3 player software and hardware, Internet broadcasts, and DVD playback devices. There has been much controversy over its widespread use and its effects on music distribution and piracy. Attempts might be made to modify the standard or legally mandate a variant that makes copyright infringement more difficult, although there are doubts about whether that's even possible. The standard will definitely need enhancement to address more advanced forms of audio data, such as theater-quality surround sound, but the coding method and data format have been proved effective by researchers and the marketplace. This success means MP3 will be around in some form for many years to come.

Summary

- The system bus is the communication pathway that connects the CPU with memory and other devices. A bus is a set of data lines, control lines, and status lines. The number and use of these lines as well as the procedures for controlling access to the bus are called the bus protocol. A master-slave bus protocol is simple to implement, but a peer-to-peer bus protocol improves computer system performance.
- The CPU communicates with peripheral devices through I/O ports. For simplicity, the CPU and bus interact with all peripheral devices by using simple data movement instructions. The CPU treats each peripheral device as though it were a storage device with a linear address space. The device, or its controller, translates access commands to the linear address space into whatever physical actions are necessary to perform the access. A peripheral device controller performs several important functions, including logical-to-physical translation, physical bus interface, sharing a single bus connection between multiple devices, and (optionally) caching. A channel is an advanced type of device controller used in mainframe computers. It has higher data transfer capacity, a larger maximum number of attached peripheral devices, and more variability in the types of devices that can be controlled compared with device controllers.
- Application programs use interrupt processing to coordinate data transfers to or from peripheral devices, notify the CPU of errors, and call OS service programs. An interrupt is a signal to the CPU that some condition requires its attention. When an interrupt is detected, the currently executing process is suspended by pushing current register values onto the stack and transferring control to the corresponding interrupt handler. When the interrupt handler finishes executing, the stack is popped, and the suspended process resumes execution from the point of interruption.
- A buffer is a region of memory that holds a single unit of data for transfer to or from a device. Buffers enable devices with different data transfer rates and unit sizes to coordinate data transfer efficiently. A cache is a large buffer implemented in a device controller or primary storage. When used for input, a cache enables more rapid access if the data being requested is already in the cache. The cache controller guesses what data the CPU will request next and loads this data into the cache before it's actually requested.
- Computer system computational capacity can be increased with parallel processing techniques, including multicore processors, multiple-processor architecture, and clustering. A multicore processor includes multiple CPUs and shared memory cache in a single microchip. Multiple-processor architecture uses multiple single-core or multicore processors sharing main memory and the system bus in a single motherboard or computer. Clusters connect computer systems with high-speed links. Each computer in a cluster works on part of a large problem and exchanges data as needed.
- Compression reduces the number of bits required to encode a data set or stream, effectively improving performance by increasing the capacity of a communication channel or storage device. A compression algorithm can be lossless or lossy. With lossless compression, data is exactly the same before and after compression and decompression. With lossy compression, some information is lost during compression and decompression. Data compression requires increased processing resources to implement compression and

decompression algorithms, but it reduces the resources needed for data storage and communication.

Chapter 3 described the ways in which data is represented, and Chapters 4 and 5 described CPU, primary storage, and secondary storage hardware. In this chapter, you've learned how data is transmitted between computer hardware devices. This chapter has also described hardware and software techniques for improving data transfer efficiency and, therefore, overall computer system performance. In Chapter 7, you look at I/O device technology.

Key Terms

address bus
buffer
buffer overflow
bus
bus arbitration unit
bus clock
bus cycle
bus master
bus protocol
bus slaves
cache
cache controller
cache hit
cache miss
cache swap
channel
compression
compression algorithm
compression ratio
control bus
core
data bus
decompression algorithm
device controller
direct memory access (DMA)
DMA controller
external I/O buses
hit ratio
interrupt
interrupt code
interrupt handler
interrupt register
I/O channel
I/O port
I/O wait states
law of diminishing returns
level one (L1) cache
level three (L3) cache
level two (L2) cache
linear address space
logical access
lossless compression
lossy compression
machine state
memory bus
Moving Picture Experts Group (MPEG)
MP3
multicore architecture
multiple-processor architecture
peer-to-peer bus
Peripheral Component Interconnect (PCI)
peripheral devices
pop
push
scaling out
scaling up
stack
stack overflow
stack pointer
storage bus
supervisor
video bus

Vocabulary Exercises

1. A(n) ________________ is a shared electrical or optical communication channel that connects two or more devices.
2. A(n) ________________ cache is generally implemented on the same chip as the CPU.
3. The CPU is always capable of being a(n) ________________, thus controlling access to the bus by all other devices in the computer system.
4. A(n) ________________ is a reserved area of main memory used to resolve differences in data transfer unit size and sometimes data transfer rate.
5. A(n) ________________ is an area of fast memory where data held in a storage device is prefetched in anticipation of future requests for the data.
6. A cache controller is a hardware device that initiates a(n) ________________ when it detects a cache miss.
7. The ________________ transmits command, timing, and status signals between devices in a computer system.
8. If possible, the system bus ________________ rate should equal the CPU's speed.
9. The ________________ is a special-purpose register that always points to the next empty address in the stack.
10. The ________________ transfers control to the interrupt handler at the memory address corresponding to the interrupt code.
11. The set of register values stored in the stack while processing an interrupt is also called the ________________.
12. A(n) ________________ is a program stored in a separate part of primary storage to process a specific interrupt.
13. During interrupt processing, register values of a suspended process are held on the ________________.
14. A(n) ________________ is a signal to the CPU or OS that some device or program requires processing services.
15. A(n) ________________ is a simple processor that intervenes when two devices want control of the bus at the same time.
16. The ________________ has a much higher data transfer rate than the system bus because of its shorter length, higher clock rate, and large number of parallel communication lines.
17. The CPU incurs one or more ________________ if it's idle pending the completion of an I/O operation.
18. The system bus can be divided logically into three sets of transmission lines: the ________________ bus, the ________________ bus, and the ________________ bus.
19. During a(n) ________________ operation, one or more register values are copied to the top of the stack. During a(n) ________________ operation, one or more values are copied from the top of the stack to registers.
20. The comparative size of a data set before and after data compression is described by the compression ________________.

21. If data isn't exactly the same as the original after compressing and decompressing, the compression algorithm is said to be ________________. If data is the same as the original after compressing and decompressing, the compression algorithm is said to be ________________.
22. A(n) ________________ is a special-purpose processor dedicated to managing cache content.
23. A(n) ________________ is a communication pathway from the CPU to a peripheral device.

24. The ________________ carries interrupts, command responses, status codes, and similar messages.
25. The ________________ transmits a memory address when primary storage is the sending or receiving device.
26. The CPU and bus normally view any storage device as a(n) ________________, ignoring the device's physical storage organization.
27. Part of a device controller's function is to translate ________________ into physical accesses.
28. A(n) ________________ controller assumes the role of bus master for all transfers between memory and other storage or I/O devices, leaving the CPU free to execute computation and data movement instructions.
29. A(n) ________________ is a high-capacity device controller used in mainframe computers.
30. When a read operation accesses data already contained in the cache, it's called a(n) ________________.
31. The ________________ defines the format, content, and timing of data, memory addresses, and control messages sent across the bus.
32. In ________________ architecture, multiple CPUs and cache memory are embedded on a single chip.
33. The term ________________ describes methods of increasing processing and other computer system power by using larger and more powerful computers.
34. ________________ architecture is a cost-effective approach to computer design when a single computer runs many different applications or services at once.
35. Examples of a(n) ________________ bus include SATA and SCSI.

Review Questions

1. What is the system bus? What are its main components?
2. What is a bus master? What is the advantage of having devices other than the CPU be bus masters?
3. What characteristics of the CPU and the system bus should be balanced to achieve maximum system performance?
4. What is an interrupt? How is an interrupt generated? How is it processed?

5. What is a stack? Why is it needed?
6. Describe the execution of push and pop operations.
7. What's the difference between a physical access and a logical access?
8. What functions does a device controller perform?
9. What is a buffer? Why might one be used?
10. How can a cache be used to improve performance when reading data from and writing data to a storage device?
11. What's the difference between lossy and lossless compression? For what types of data is lossy compression normally used?
12. Describe how scaling up differs from scaling out. Given the speed difference between a typical system bus and a typical high-speed network, is it reasonable to assume that both approaches can yield similar increases in total computational power?
13. What is a multicore processor? What are its advantages compared with multiple-processor architecture? Why have multicore processors become available only recently?

Problems and Exercises

1. You have a PC with a 2 GHz processor, a system bus clocked at 400 MHz, and a 3 Mbps internal cable modem attached to the system bus. No parity or other error-checking mechanisms are used. The modem has a 64-byte buffer. After it receives 64 bytes, it stops accepting data from the network and sends a "data ready" interrupt to the CPU. When this interrupt is received, the CPU and OS perform the following actions:

   a. The supervisor is called.

   b. The supervisor calls the modem's "data ready" interrupt handler.

   c. The interrupt handler sends a command to the modem, instructing it to copy its buffer content to main memory.

   d. The modem interrupt handler immediately returns control to the supervisor, without waiting for the copy operation to be completed.

   e. The supervisor returns control to the process that was originally interrupted.

   When the modem finishes the data transfer, it sends a "transfer completed" interrupt to the CPU and resumes accepting data from the network. In response to the interrupt, the CPU and OS perform the following actions:

   a. The supervisor is called.

   b. The supervisor calls the "transfer completed" interrupt handler.

   c. The interrupt handler determines whether a complete packet is present in memory. If so, it copies the packet to a memory region of the corresponding application program.

   d. The modem interrupt handler returns control to the supervisor.

   e. The supervisor returns control to the process that was originally interrupted.

   Sending an interrupt requires one bus cycle. A push or pop operation consumes 30 CPU cycles. Incrementing the stack pointer and executing an unconditional BRANCH instruction require one CPU cycle each. The supervisor consumes eight CPU cycles searching the interrupt table before calling an interrupt handler. The "data ready" interrupt handler consumes 50 CPU cycles before returning to the supervisor.

Incoming packets range in size from 64 bytes to 4096 bytes. The "transfer complete" interrupt handler consumes 30 CPU cycles before returning to the supervisor if it doesn't detect a complete packet in memory. If it does, it consumes 30 CPU cycles plus one cycle for each 8 bytes of the packet.

- **Question 1:** How long does it take to move a 64-byte packet from its arrival at the modem until its receipt in the memory area of the target application program or service? State your answer in elapsed time (seconds or fractions of seconds).
- **Question 2:** The computer is running a program that's downloading a large file by using the modem, and all packets are 1024 bytes. What percentage of the computer's CPU capacity is used to manage data transfer from the modem to the program? What percentage of available bus capacity is used to move incoming data from the modem to the program? Assume the bus uses a simple request/response protocol without command acknowledgment.
- **Question 3:** Recalculate your answers to Questions 1 and 2, assuming a modem buffer size of 1024 bytes and all incoming packets being 1024 bytes.

2. A video frame displayed onscreen consists of many pixels, with each pixel, or cell, representing one unit of video output. A video display's resolution is typically specified in horizontal and vertical pixels (such as 800×600), and the number of pixels onscreen is simply the product of these numbers ($800 \times 600 = 480{,}000$ pixels). A pixel's data content is one or more unsigned integers. For a black-and-white display, each pixel is a single number (usually between 0 and 255) representing the intensity of the color white. Color pixel data is typically represented as one or three unsigned integers. When three numbers are used, the numbers are usually between 0 and 255, and each number represents the intensity of a primary color (red, green, or blue). When a single number is used, it represents a predefined color selected from a table (palette) of colors.

   Motion video is displayed onscreen by copying frames rapidly to the video display controller. Because video images or frames require many bytes of storage, they're usually copied to the display controller directly from secondary storage. Each video frame is an entire picture, and its data content, measured in bytes, depends on the resolution at which the image is displayed and the maximum number of simultaneous colors that can be contained in the sequence of frames. For example, a single frame at 800×600 resolution with 256 (2^8) simultaneous colors contains $800 \times 600 \times 1$ byte $= 480{,}000$ bytes of data. Realistic motion video requires copying and displaying a minimum of 20 frames per second; 24 or 30 frames per second are common professional standards. Using fewer frames per second results in a "jerky" motion because the frames aren't being displayed quickly enough to fool the eye and brain into thinking that they're one continuously changing image.

   Assume the computer system being studied contains a bus mastering disk controller and a video controller that copies data to the video display at least as fast as it can be delivered over the bus. Further, the system bus can transfer data at a sustained rate of 100 Mbps, as can both the controllers' bus interfaces. This system will be used to display motion video on a monitor capable of resolutions as low as 640×480 and as high as 1024×768.

   In addition, a single disk drive is attached to the disk controller and has a sustained data transfer rate of 20 MB per second when reading sequentially stored data. The channel connecting the disk drive to the disk controller has a data transfer rate of 80 Mbps. Finally, the files containing the video frames are stored sequentially on the disk, and copying these files' contents from disk to the display controller is the only activity the system will perform (no external interrupts, no multitasking, and so forth).

The video display controller contains 2 MB of 50 ns, 8-bit buffer RAM, and the video image arriving from the bus can be written to the buffer at a rate of 8 bits per 50 ns. The video display's RAM buffer can be written from the bus while it's being read by the display device (sometimes called "dual-porting"). Finally, data can be received and displayed by the display device as fast as the video controller can read it from the RAM buffer.

- **Question 1:** What is the maximum number of frames per second (round down to a whole number) that this system can display in 256 simultaneous colors at a resolution of 640 × 480?
- **Question 2:** What is the maximum number of frames per second (round down to a whole number) that this system can display in 65,536 simultaneous colors at a resolution of 800 × 600?
- **Question 3:** What is the maximum number of frames per second (round down to a whole number) that this system can display in 16,777,216 simultaneous colors at a resolution of 1024 × 768?

3. Employees at your company are watching YouTube videos constantly while using their office computers. The YouTube Web site sends compressed video through the Internet and company network to employees' computers, and the video stream consists of the following:

   - 480 × 360 pixel frames
   - 24-bit color (each pixel is represented by 24 bits)
   - 30 frames per second
   - 10:1 compression ratio

   Employees' computers have a 100 Mbps Ethernet interface and a 400 MHz 32-bit PCI bus. The company has a 100 Mbps internal network and two T1 connections (1.54 Mbps each) to the Internet.

   - **Question 1:** How much of the company's network capacity are employees consuming when watching YouTube videos?
   - **Question 2:** What percentage of available bus cycles on employees' computers are consumed when watching the videos?
   - **Question 3:** Does your answer to Question 1 or Question 2 support forbidding employees to watch YouTube videos? What if there are 500 employees on the company network? What if the company Internet connection is a T3 line (45 Mbps)?

Research Problems

1. Choose two computer systems from a single manufacturer, such as Dell or HP. The first system should be from the "standard" desktop family (for example, Dell Optiplex or HP/Compaq Elite series), and the second system should be from the workstation family (for example, Dell Precision or HP Z-series). For both systems, select a configuration to achieve optimal performance, and examine the computers' technical specifications to

determine the differences underlying the "high-performance" designation for the workstation. Concentrate on the system bus, video bus, storage bus, available processors, memory type, and architecture for interconnecting processors and memory. Is the price difference between these computers commensurate with the performance difference?

2. The trend toward multiple-processor and multicore architectures has placed increasing strain on interconnections between processors, memory, and I/O controllers. Intel's QuickPath Interconnect (QPI) and the HyperTransport Consortium's HyperTransport (HT) are two widely used approaches to addressing the performance bottleneck. Investigate both interconnection standards, and compare similar multicore processors from Intel and AMD that implement each standard. Do the standards address the same basic performance issues? Which offers the best performance? Which do you expect to dominate the marketplace, and why? (See this book's Web site, *www.cengage.com/mis/burd*, for relevant Web links.)

3. Music purchased from iTunes is encoded in the M4P format. Investigate this format and compare it with the MP3 encoding method discussed in this chapter. Which encoding method offers the best tradeoff between compression ratio and audio quality? Which requires the most processing power for audio encoding? Which addresses issues of copy protection, and how are these issues addressed?

CHAPTER 7

INPUT/OUTPUT TECHNOLOGY

CHAPTER GOALS

- Describe basic concepts of text and image representation and display, including digital representation of grayscale and color, bitmaps, and image description languages
- Describe the characteristics and implementation technology of video display devices
- Understand printer characteristics and technology
- Describe the main manual input technologies
- Describe types of optical input devices
- Identify the characteristics of audio I/O devices and explain how they operate

People communicate in many different ways, and these differences are reflected in the variety of methods for interacting with computer systems. This chapter describes the concepts, technology, and hardware used in communication between people and computers (see Figure 7.1). Understanding I/O technology is important because it expands or limits computers' capability to assist people in problem solving.

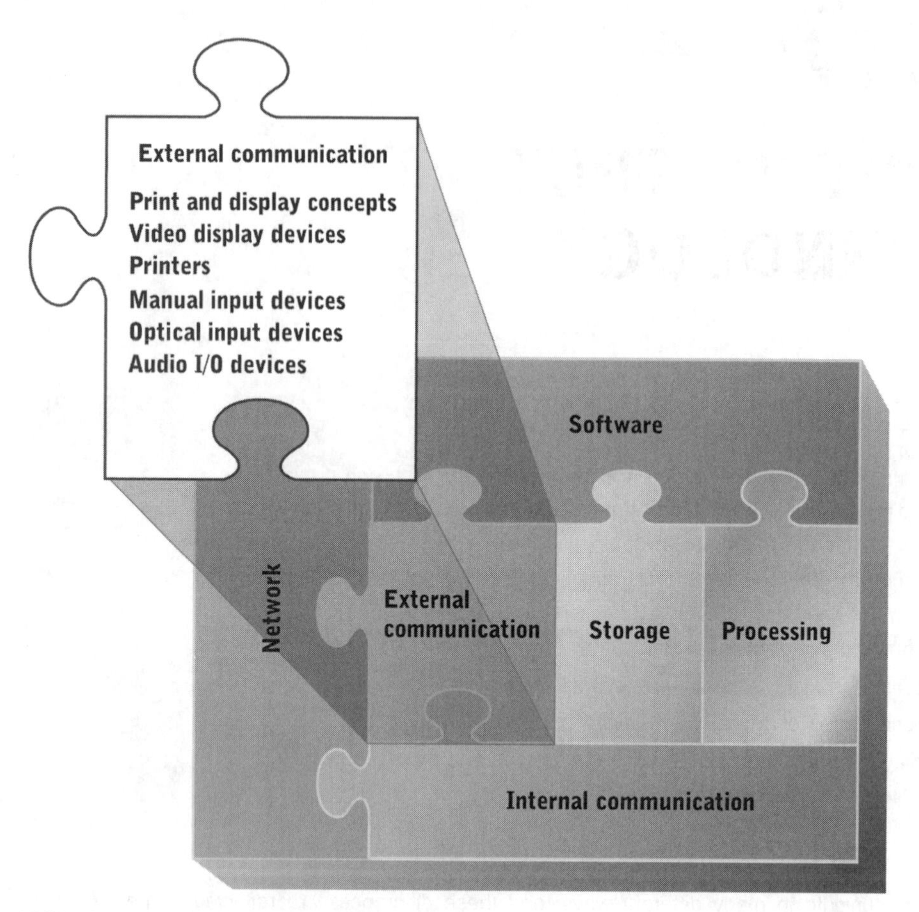

FIGURE 7.1 Topics covered in this chapter
Courtesy of Course Technology/Cengage Learning

BASIC PRINT AND DISPLAY CONCEPTS

Although printing technology dates from the 15th century, communication via video display devices is less than a century old. However, both technologies share many features, including character representation methods, measurement systems, and methods of generating color. These topics are covered first because they're common to many I/O technologies and devices.

Matrix-Oriented Image Composition

Display surfaces vary widely in size and composition. The most common are paper and flat panel displays. Display surfaces have a default background color, usually white for paper and black for video display devices, and are divided into rows and columns, similar to a large table or matrix. Each cell in the matrix represents one part of an image, called a **pixel** (a shortened form of "picture element"). For printed output, a pixel is either empty or contains one or more inks or dyes. For video display, a pixel displays no light or light of a specific color and intensity.

The number of pixels in a display surface depends on these factors:

- Display surface size (height and width)
- Pixel size

For example, a typical 19-inch flat panel display is 28 centimeters (11 inches) high and 37 centimeters (14.5 inches) wide. If display pixels are .25 millimeters square, the total number of pixels in the display is as follows:

$$280 \text{ mm} \times 370 \text{ mm} \div 0.25 \text{ mm} = 414,400$$

The **resolution** of a display is the number of pixels displayed per linear measurement unit. For example, the resolution of the 19-inch flat panel display described previously is 40 pixels per centimeter, or approximately 100 pixels per inch. In the United States, resolution is generally stated in **dots per inch (dpi)**, with a dot equivalent to a pixel. Higher resolutions correspond to smaller pixel sizes. To an observer, the quality of a printed or displayed image is related to the pixel size. As Figure 7.2 shows, smaller pixel size (higher dpi) yields higher print quality because fine details, such as smooth curves, can be incorporated into the image.

Systems Architecture

Systems Architecture

FIGURE 7.2 Text displayed in two resolutions: 50 dpi (top) and 200 dpi (bottom)
Courtesy of Course Technology/Cengage Learning

On paper, pixel size corresponds to the smallest drop of ink that can be placed accurately on the page. Decades ago, printers adopted 1/72 of an inch as a standard pixel size, called a **point**. This term and its use as a printer's measuring system continues to be used, even though modern printing techniques are capable of much higher resolution.

Fonts

Written Western languages are based on systems of symbols called characters. Each character can be represented as a matrix of pixels, as shown in Figure 7.3. A printed character need not exactly match a specific pixel map to be recognizable. For example, people can easily interpret the symbols E, E, **E**, ℰ, and *E* as the letter E, even though their pixel composition varies.

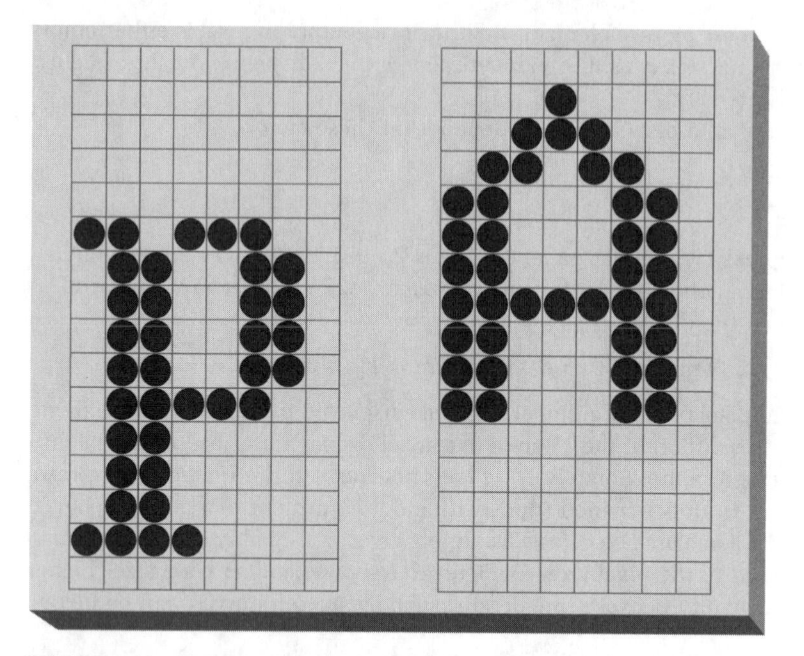

FIGURE 7.3 The letters p and A represented in an 8 × 16 pixel matrix
Courtesy of Course Technology/Cengage Learning

A collection of characters of similar style and appearance is called a **font**. Figure 7.4 shows sample characters from different fonts. Points are the most common unit for measuring font size; point size refers to characters' height, not width, although width is normally scaled to match height. Figure 7.5 shows a single font printed in several point sizes. Notice that characters vary in height and placement. For example, T is taller than a, and letters such as p and q extend below the text baseline. A font's point size is the distance between the top of the highest character and the bottom of the lowest character in the font.

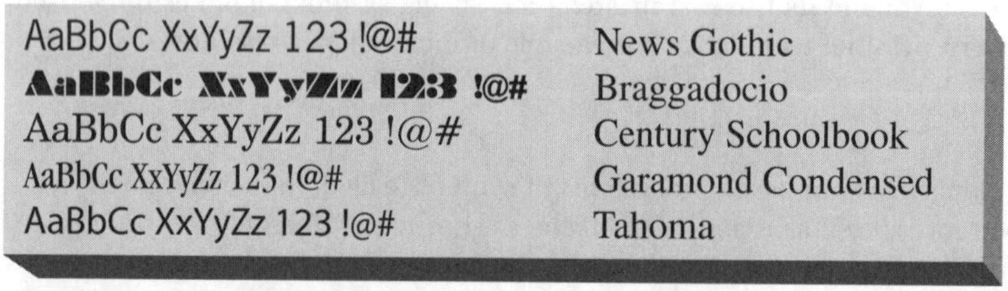

AaBbCc XxYyZz 123 !@# News Gothic
AaBbCc XxYyZz 123 !@# Braggadocio
AaBbCc XxYyZz 123 !@# Century Schoolbook
AaBbCc XxYyZz 123 !@# Garamond Condensed
AaBbCc XxYyZz 123 !@# Tahoma

FIGURE 7.4 Sample characters printed in a variety of fonts
Courtesy of Course Technology/Cengage Learning

8 point Times Roman
10 point Times Roman
12 point Times Roman
16 point Times Roman
24 point Times Roman

FIGURE 7.5 A font (Times Roman) printed in different point sizes
Courtesy of Course Technology/Cengage Learning

Color

The human eye and brain interpret different light frequencies as different colors. Video display devices generate light of a specific color in each pixel. On the printed page, color is the light frequency reflected from the page. Any color in the visible spectrum can be represented as a mixture of varying intensities of primary colors. For example, mixing equal intensities of red and blue produces magenta (a shade of purple). The absence of all three colors is black, and full intensity of all three is white.

Different sets of primary colors are used for different purposes. In paint, red, blue, and yellow are used as primary pigments. Video display devices have used red, green, and blue as primary colors since the earliest color TVs. These primary colors for video display are sometimes referred to by their first letters: **RGB**. A video display that generates color by using mixtures of red, green, and blue is sometimes called an "RGB display."

The printing industry generates color by using the inverse of the primary video display colors. In this context, video display colors are called **additive colors**, and their inverse colors are called **subtractive colors**. The subtractive colors are cyan (absence of red), magenta (absence of green), and yellow (absence of blue), and they're often referred to by the abbreviation **CMY**. Black can be generated by combining all three, but a separate black dye, identified by the letter K, is generally used. This four-dye scheme is called **CMYK** color.

Numeric Pixel Content

Because computers are digital devices, pixel content must be described numerically. A stored set of numbers describing the content of all pixels in an image is called a **bitmap**. The number of bits required to represent a pixel's content depends on the number of different colors the pixel can display.

A **monochrome** display can display one of two colors, so it requires only 1 bit per pixel. A **grayscale** display can display black, white, and many shades of gray in between. The number of gray shades that can be displayed increases with the number of bits used to represent a pixel. If 8 bits per pixel are used, 254 shades of gray are available in addition to pure black (0) and pure white (255). The number of distinct colors or gray shades that can be displayed is sometimes called the **chromatic depth** or **chromatic resolution**.

Most color display schemes represent each pixel's color with three different numbers, each representing the intensity of an additive or subtractive color. Chromatic depth depends on the number of bits used to represent each color's intensity. For example, if each color is represented by an 8-bit number, the chromatic depth is $2^{(3\times8)}$, or around 16 million; this scheme is sometimes called **24-bit color**. Each group of 8 bits represents the intensity (0 = none, 255 = maximum) of one color. For example, with RGB color, the numbers 255:0:0 represent bright red, and the numbers 255:255:0 represent bright magenta.

Another approach to representing pixel color with numbers is to define a color palette. A **palette** is simply a table of colors. The number of bits used to represent each pixel determines the table size. If 8 bits are used, the table has 2^8 (256) entries. Each table entry can contain any RGB color value, such as 200:176:40, but the number of different colors that can be represented is limited by the table's size. Table 7.1 shows the 4-bit color palette for the Color Graphics Adapter (CGA) video driver of the original IBM PC/XT.

TABLE 7.1 4-bit (16-color) palette for the IBM PC/XT Color Graphics Adapter

Table entry number	RGB value	Color name
0	0:0:0	Black
1	0:0:191	Blue
2	0:191:0	Green
3	0:191:191	Cyan
4	191:0:0	Red
5	191:0:191	Magenta
6	127:63:0	Brown
7	223:223:223	White
8	191:191:191	Gray
9	127:127:255	Light blue
10	127:255:127	Light green
11	127:255:255	Light cyan
12	255:127:127	Light red
13	255:127:255	Light magenta
14	255:255:127	Yellow
15	255:255:255	Bright white

Dithering is a process that generates color approximations by placing small dots of different colors in an interlocking pattern. If the dots are small enough, the human eye interprets them as being a uniform color representing a mixture of the dot colors. For example, a small pattern of alternating red and blue dots appears to be magenta. A lower percentage of red and higher percentage of blue is interpreted as violet. An alternating pattern of black and white dots looks gray, which can be confirmed by examining a gray area of one of the figures in this book with a magnifying glass. Grayscale dithering is usually called **half-toning**.

Image Storage and Transmission Requirements

The amount of storage or transmission capacity required for an image depends on the number of bits representing each pixel and the image height and width in pixels. For example, an 800 × 600 image has 480,000 pixels. For a monochrome image, 1 bit represents each pixel, and the image requires 60,000 bytes (480,000 pixels × 1 bit/pixel ÷ 8 bits/byte). Storing or transmitting an 800 × 600 24-bit color image requires 1.44 MB (480,000 pixels × 3 bytes/pixel).

Image storage and transmission requirements are the same regardless of whether the image is stored in primary storage, secondary storage, or an I/O device buffer or is transmitted through a communication channel. Storage and transmission requirements can be reduced with bitmap compression techniques. Common bitmap compression formats include **Graphics Interchange Format (GIF)** and **Joint Photographic Experts Group (JPEG)** for still images and Moving Picture Experts Group (MPEG, introduced in Chapter 6) for moving images. However, all these compression methods are lossy, meaning they result in some loss of image quality.

Image Description Languages

There are two major drawbacks to using bitmaps to represent high-quality images. The first is size. An 8½ × 11-inch, 24-bit color image with 1200 dpi resolution requires approximately 400 MB of uncompressed storage. A good compression algorithm might achieve compression ratios of 10:1 or 20:1, reducing the storage requirements to 20 to 40 MB. Files of this size transmit slowly over communication links, such as a DSL Internet connection.

The second drawback is having no standard storage format for raw bitmaps. Bitmap formats vary across software packages and output device types, manufacturers, and models. Image files stored by software must be converted into a different format before they're printed or displayed. This conversion process consumes substantial processing and memory resources, and multiple conversion utilities or device drivers might be needed.

An **image description language (IDL)** addresses both drawbacks by storing images compactly. Many, but not all, IDLs are device independent. IDLs use compact bit strings or ordinary ASCII or Unicode text to describe primitive image components, such as straight lines and simple shapes. They reduce storage space requirements because a description of a simple image component is usually much smaller than a bitmap of that component. Examples of current IDLs include Adobe's PostScript and Portable Document Format. The internal file formats of many drawing and presentation graphics programs (for example, CorelDRAW and Microsoft PowerPoint) are based on proprietary IDLs.

An IDL can represent image components in several ways:

- Embedded fonts
- Vectors, curves, and shapes
- Embedded bitmaps

Modern I/O devices contain embedded font tables that store symbol bitmaps in ROM or flash RAM, and current OSs store font tables in files. An IDL can represent text in a specific font by storing the text in ASCII or Unicode along with a reference to the font style and size. The CPU, or an embedded processor in an I/O device, can shrink or enlarge a character bitmap to any size according to simple image-processing algorithms. Although fonts can consume considerable storage, each symbol is stored only once and reused every time it's displayed or printed.

In graphics, a **vector** is a line segment with a specific angle and length in relation to a point of origin, as shown in the top half of Figure 7.6. Vectors can also be described in terms of their weight (line width), color, and fill pattern. Line drawings can be described as a **vector list**, a series of concatenated or linked vectors that can be used to construct complex shapes, such as boxes, triangles, and the inner and outer lines of a table or spreadsheet. Images constructed from a vector list resemble connect-the-dots drawings, as shown in the bottom half of Figure 7.6, and can be scaled larger or smaller easily.

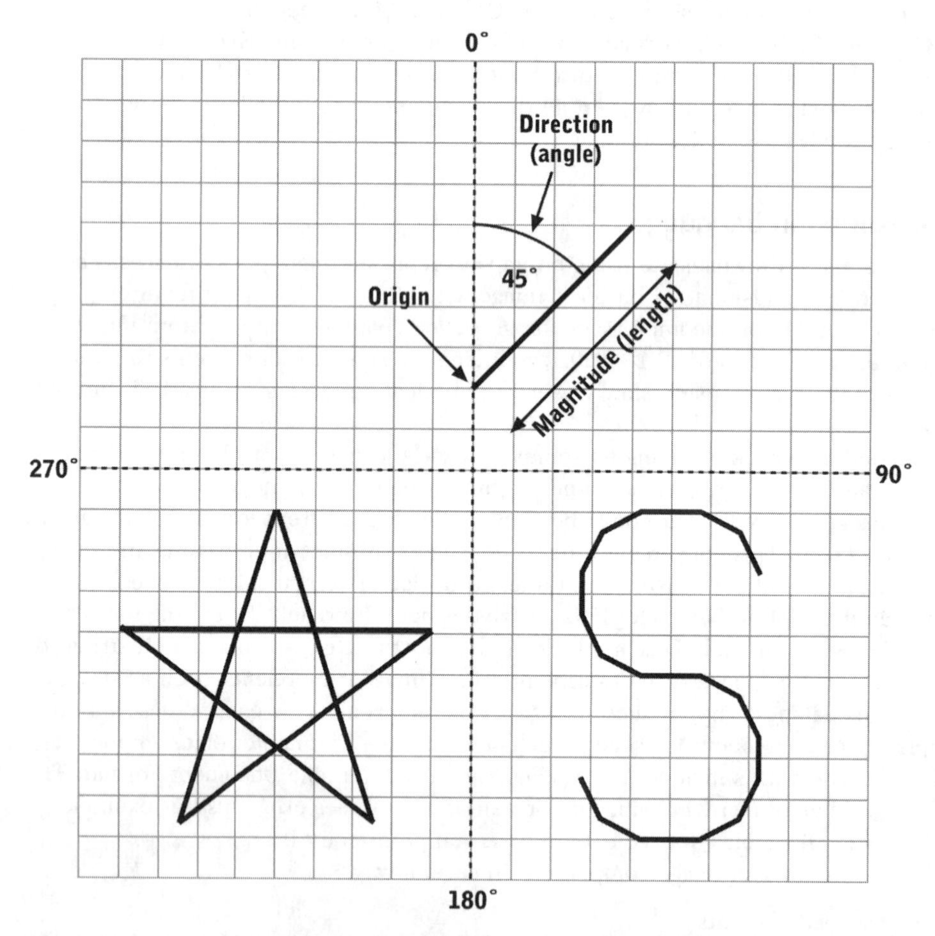

FIGURE 7.6 Elements of a vector (top) and two complex shapes built from vectors (bottom)
Courtesy of Course Technology/Cengage Learning

More complex lines and shapes can be constructed with curves. Like a vector, a curve has a point of origin, length, weight, and color. Unlike a vector, a curve has an angle. Curves can be closed and filled to represent objects such as circles and ovals.

Some image components are simply too complex to represent with vectors or curves. An IDL represents complex image components as embedded bitmaps, which can be compressed with GIF, JPEG, or another compression algorithm.

IDLs are a complex form of compression. As with any compression algorithm, storing images with IDLs substitutes processing resources for storage and communication resources. To display or print an image, a program or device driver must process the IDL image description. For video display, the processing can occur in the CPU, a special-purpose processor embedded in the video controller, or a combination of both. For printed output, processing usually takes place in an embedded processor, although part of the work can be handled by the CPU's execution of device driver instructions.

TECHNOLOGY FOCUS

Adobe PostScript and Portable Document Format

PostScript is an IDL designed mainly for printed documents, although it can also be used to generate video display outputs. It's also a programming language, so a PostScript file is a program that produces a printed or displayed image when an interpreter or a compiler runs it. Images can include display objects, such as characters, lines, and shapes, that can be manipulated in a variety of ways, including rotation, skewing, filling, and coloring.

PostScript commands are composed of normal ASCII or Unicode characters and can include numeric or text data, primitive graphics operations, and procedure definitions. Data constants are pushed onto a stack. Graphic operators and procedures pop their input data from the stack and push results, if any, back onto the stack.

To see an example of PostScript in action, take a look at this short program:

```
newpath
400 400 36 0 360
arc
stroke
showpage
```

The first line declares the start of a new path—straight line, curve, or complex line. The second line lists numeric data items (control parameters) to be pushed onto the stack. As the control parameters are used, they're removed from the stack by the predefined procedure named `arc` (the third line). In the second line, the first two parameters specify the origin (center) of the arc as row and column coordinates. The third parameter, 36, specifies the arc's radius in points. The remaining parameters specify the arc's starting and ending points in degrees. In this example, the starting and ending points represent a circle.

Primitive graphics operations are specified in an imaginary drawing space containing a grid (rows and columns) of points. The first three commands in this program specify a tracing in this space. The fourth command states that this tracing should be stroked (drawn). Because no width is specified, the circle is drawn by using the default line width. The final command, `showpage`, instructs the output processor to display the page's contents as defined so far. A PostScript printer or viewing program would run this program to generate a single page containing one circle.

(*continued*)

Figure 7.7 shows a PostScript program combining text and graphics. The first two sections define the menu box outline and title separator line with primitive vector and shape drawing commands. The third section defines the menu's text content and the display font and font size. The last section builds a complex object (the arrow pointer) by defining its outlines, closing the shape, and filling it.

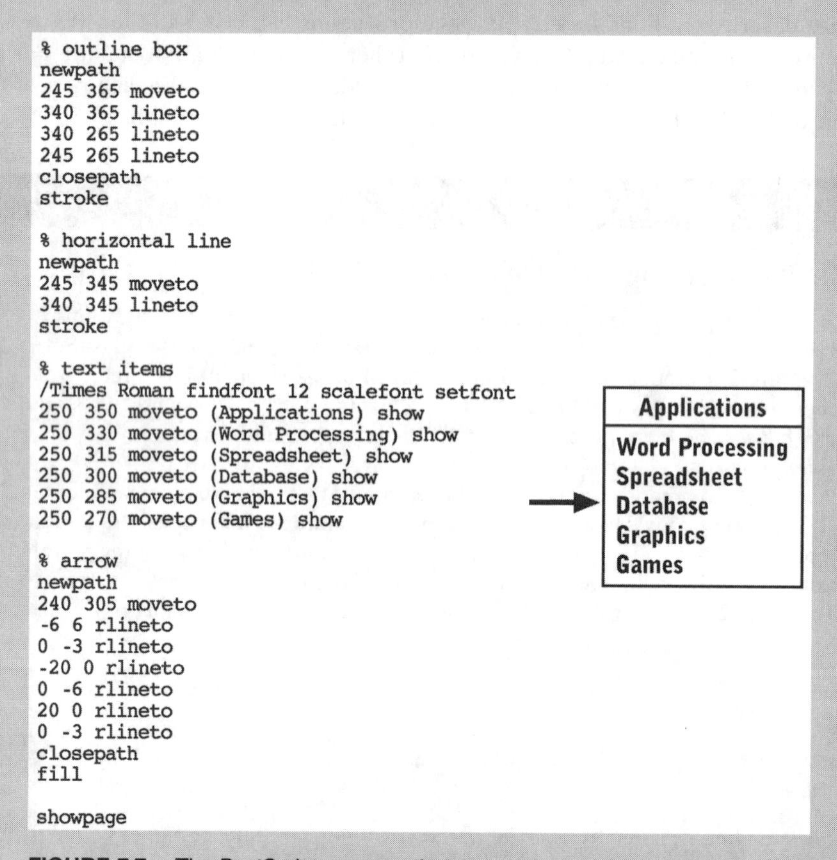

```
% outline box
newpath
245 365 moveto
340 365 lineto
340 265 lineto
245 265 lineto
closepath
stroke

% horizontal line
newpath
245 345 moveto
340 345 lineto
stroke

% text items
/Times Roman findfont 12 scalefont setfont
250 350 moveto (Applications) show
250 330 moveto (Word Processing) show
250 315 moveto (Spreadsheet) show
250 300 moveto (Database) show
250 285 moveto (Graphics) show
250 270 moveto (Games) show

% arrow
newpath
240 305 moveto
-6 6 rlineto
0 -3 rlineto
-20 0 rlineto
0 -6 rlineto
20 0 rlineto
0 -3 rlineto
closepath
fill

showpage
```


FIGURE 7.7 The PostScript program that generates the pop-up menu on the right
Courtesy of Course Technology/Cengage Learning

By 1990, PostScript was widely used in the printing and publishing industries and also used as a graphics file interchange format and an embedded printer technology. PostScript printers have special-purpose computers to run PostScript programs and generate printed pages. Some OSs used PostScript to support GUIs, although no current OSs do so.

Although PostScript was successful, it lacked many features, such as the following, needed to generate and manage documents as an integrated whole rather than a collection of independent images and pages:

- Table of contents with page links
- Hyperlinks to other documents, Web pages, or active objects, such as scripts and executable programs

(*continued*)

- Annotations and bookmarks
- Document properties, such as authorship, copyright, and summary
- Digital signatures

In the early 1990s, Adobe Systems, Inc., developed **Portable Document Format (PDF)**, a superset of PostScript. To ensure that PDF would be widely adopted, Adobe developed the free Adobe Reader program, which can be installed with a variety of OSs and Web browsers, as well as many tools to create PDF documents. For example, Adobe Acrobat Capture enables organizations to convert paper documents to PDF format and store them in searchable archives. With Adobe PDF Writer, users can create PDF documents from desktop word-processing and graphics programs, such as Microsoft Office and CorelDRAW. PDF Writer is installed as a printer driver, and application programs create PDF document files by printing to this driver.

PDF has been widely adopted, especially for documents distributed via the Web. It offers a capability that wasn't available with standard Web protocols—distributing compressed documents with the author's complete control over the format of the printed and displayed document, regardless of the end user's computer, OS, or printer.

VIDEO DISPLAY

The first computer video display devices, introduced in the 1960s, consisted of an integrated keyboard and TV screen called a **video display terminal (VDT)**, or simply a "terminal." VDTs were connected to computer systems through dedicated low-speed communication links and could display only text and primitive graphics. They were the most common form of video display throughout the 1970s and much of the 1980s, until PCs came into common use. Today, they're used mostly in systems such as retail checkout counters and factory floor environments. Modern video displays combine a sophisticated video controller with one or more flat panel displays. These technologies are described in detail in the following sections.

Video Controllers

Video display panels (sometimes called **monitors**) are connected to a **video controller** that's connected to a port on the system bus or a dedicated video bus (see Figure 7.8). The video controller accepts commands and data from the CPU and memory and generates analog or digital video signals, which are transmitted to the monitor.

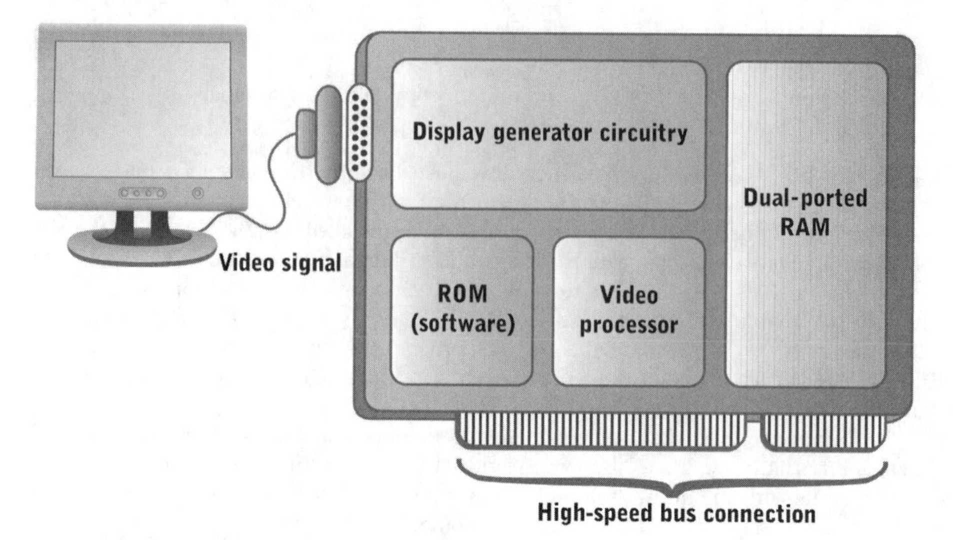

FIGURE 7.8 Video controller with monitor and bus connections
Courtesy of Course Technology/Cengage Learning

In the simplest video controllers, the video controller RAM is a cache for video images in transit from primary storage to the monitor. Software modifies the display by updating the corresponding portions of primary storage and initiating a bus transfer to the video controller. The controller's bus interface circuitry manages the flow of data from primary storage into the cache.

The controller's display generator circuitry reads the cache contents continuously and generates analog or digital video signals, which are transmitted to the monitor. Each transfer of a full screen of data from the display generator to the monitor is called a **refresh cycle**. Monitors require many refresh cycles per second to generate high-quality video images, especially those with rapidly changing content. The number of refresh cycles per second is normally stated in hertz and called the **refresh rate** (for example, a 60 Hz refresh rate).

More complex video controllers are essentially special-purpose graphics computers. Like any computer, they contain a microprocessor and primary storage, and their actions are controlled by dedicated (embedded) software. Unlike a general-purpose computer, they have no secondary storage devices, and their only "I/O devices" are a bus connection to the host computer and one or more connections to display devices.

NOTE

Because protein-folding simulations are similar in many ways to video image processing, client software for Folding@Home (described in Chapter 2) has versions for general-purpose CPUs and for video processors embedded in powerful video controllers. In many computers, a video processor can execute the simulation more quickly than a general-purpose CPU can.

Video RAM (VRAM) is different from ordinary RAM because it can be written by the bus interface circuitry or video processor while being read by display generator circuitry. Simultaneous read/write capability is sometimes called **dual-porting**. Video controllers generally have at least 256 MB of RAM, which can be used as a cache or to support more complex graphics-processing operations.

A video processor and its embedded software can serve multiple functions. At minimum, it serves as a cache controller, deciding when to save and when to overwrite portions of the cache. When displaying complex two-dimensional (2D) and three-dimensional (3D) graphics, the video processor converts incoming IDL commands into cache contents. The IDLs used for video display are generally more complex than those for documents and static images. Video controller IDLs can respond to commands to draw simple and complex shapes, fill them with patterns or colors, and move them from one part of the display to another. The two most widely used video controller IDLs are **Direct3D** and **OpenGL**. Direct3D is part of the Microsoft DirectX suite embedded in Windows OSs. OpenGL was developed by Silicon Graphics, but Khronos Group now maintains it as an open standard. Video processors are optimized to run programs written in these languages.

Video Monitors

Video monitors use a wide variety of technologies to generate displays, with complex tradeoffs between cost and performance characteristics. Until the late 1990s, most computer displays were based on a **cathode ray tube (CRT)**. A CRT is an enclosed glass vacuum tube with an electron gun in the rear that generates a stream of electrons focused in a narrow beam toward the tube's front surface. The interior of the display surface is coated with colored phosphors that emit light when struck by a sufficient number of electrons. Pulsing the electron beam (turning it on and off rapidly) as it travels across the inside of the screen controls the pixel brightness.

CRTs are bulky, heavy, and power-hungry and generate lots of heat. Newer display technologies, collectively called **flat panel displays**, have replaced CRTs because they're thinner, generate higher quality images, and consume less power than CRTs for similar-sized displays. The next sections describe the most common flat panel display technologies.

LCDs

A **liquid crystal display (LCD)** contains a matrix of liquid crystals sandwiched between two polarizing filter panels that block all light except light approaching from a specific angle. The front polarizing panel is rotated 90 degrees from the back panel, as shown in the different orientations in Figure 7.9. Light passing through an uncharged twisted crystal is rotated 90 degrees, which enables light to pass through both panels and the crystal. When an electrical charge is applied to a crystal, it untwists, so light entering the rear panel can't pass through the front panel. An LCD's backlight is usually a white fluorescent bulb. Color display is achieved with a matrix of RGB color filters layered over or under the front panel.

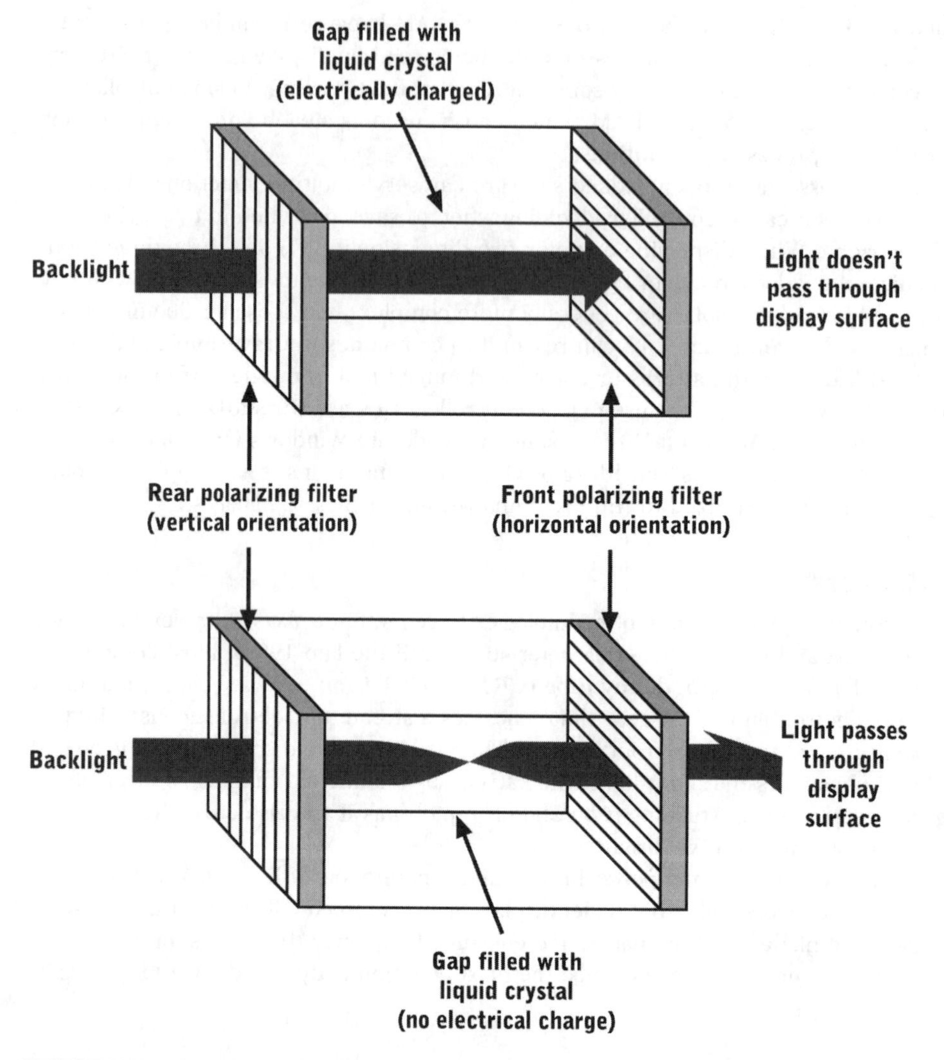

FIGURE 7.9 Light entering the rear filter can't pass through the front filter if the liquid crystal is electrically charged (top); removing the charge returns the liquid crystal to its twisted state, allowing light to pass (bottom)
Courtesy of Course Technology/Cengage Learning

Applying an electrical charge to each display matrix cell requires transistors and wiring. A **passive matrix display** shares transistors among rows and columns of pixels, but an **active matrix display** uses one or more transistors for every pixel. Additional transistors enable pixels to be switched on and off more quickly. Also, with transistors dedicated to each pixel, a continuous charge can be generated, resulting in a brighter display. However, additional transistors increase the display panel's complexity and cost.

Since the early 1990s, active matrix displays have been manufactured with **thin film transistor (TFT)** technology. The wiring and transistors of a TFT display are added in thin layers to a glass substrate. The manufacturing process is similar to that for semiconductors, with successive layering of wiring and electrical devices added with photolithography. With TFT technology, large displays can be manufactured reliably.

LCD displays have less contrast than other flat panel displays because color filters reduce the total amount of light passing through the front of the panel. They also have a more limited viewing angle than CRTs. Acceptable contrast and brightness are achieved with a 30- to 60-degree angle. Outside that angle, contrast drops off sharply and colors shift. In contrast, viewing angles for other flat panel display technologies usually approach 180 degrees, with little loss in contrast or brightness.

Early LCD displays accepted the same analog video signals as CRTs. However, because LCD pixels don't require continuous refresh cycles, analog signals are converted to digital signals. The conversion process adds unnecessary complexity that can result in slower display updating and reduced image quality. Later LCD generations accept both analog and digital video signals, and analog connections are gradually being phased out.

Plasma Displays

A **plasma display** combines elements of CRT and LCD technology. Like LCDs, they're flat panel active matrix display devices. Unlike LCDs, they have no backlight and no color filters. Instead, each pixel contains a gas that emits ultraviolet light when electricity is applied (see Figure 7.10). Each pixel's inner display surface is coated with a color phosphor that emits visible light when struck by ultraviolet light.

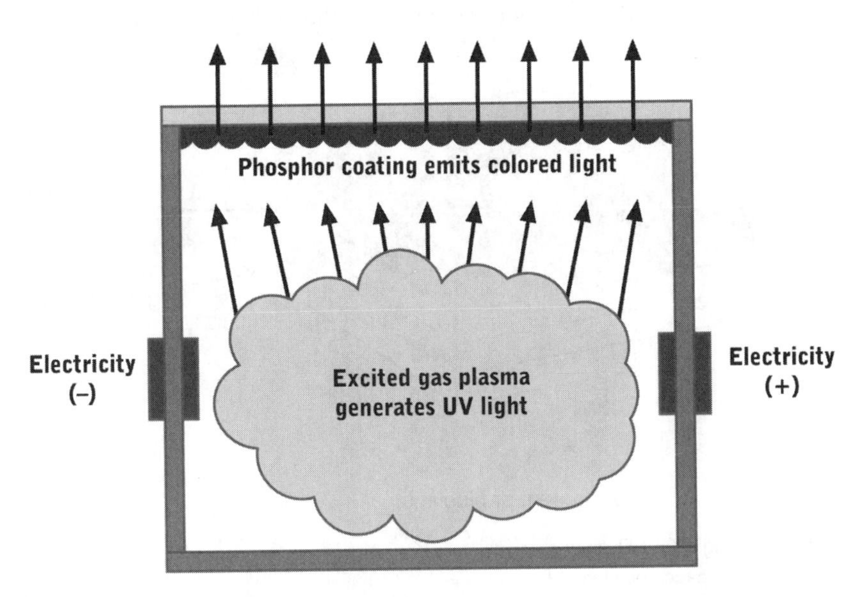

FIGURE 7.10 A plasma display pixel
Courtesy of Course Technology/Cengage Learning

Because plasma displays actively generate colored light near the display surface, they're brighter and have a wider viewing angle than LCDs, but they require more power than LCDs. Plasma displays have some shortcomings that limit their use as general-purpose video monitors. The gases and phosphors that generate light have a limited operational lifetime—currently, up to 100,000 hours. (A year has 8760 hours.) Limited lifetime is a problem with many desktop computer systems and other displays used continuously. Also, pixel size is slightly larger in plasma displays than in LCDs, which reduces image quality when viewed from short distances.

Many companies are investing heavily in plasma display R&D because of the market for large-format flat panel televisions. Some of these R&D results have been applied to computer displays. However, LCDs still dominate the flat panel display market.

LED Displays

Displays based on **light-emitting diodes (LEDs)** are a new entrant in the flat panel display market. Early use of LED technology was limited because of high cost, a complex fabrication process, and difficulties in generating accurate color. Improvements in technology and fabrication have made LED displays more affordable, although they're still more expensive than similar-sized LCD or plasma displays. Modern LED displays, often called **organic LED (OLED)** displays, achieve high-quality color display with organic compounds.

Figure 7.11 shows the layers of an OLED pixel. Electrical current is routed through the cathode layer to the emissive layer, which is composed of organic compounds that emit red, green, or blue light, depending on their chemical composition. The conducting layer transmits electrical current to the anode layer, which is connected to the negative side of the power supply to complete the electrical circuit. In a flexible display, the transparent layer is plastic, and conductive polymers are used for the cathode and anode.

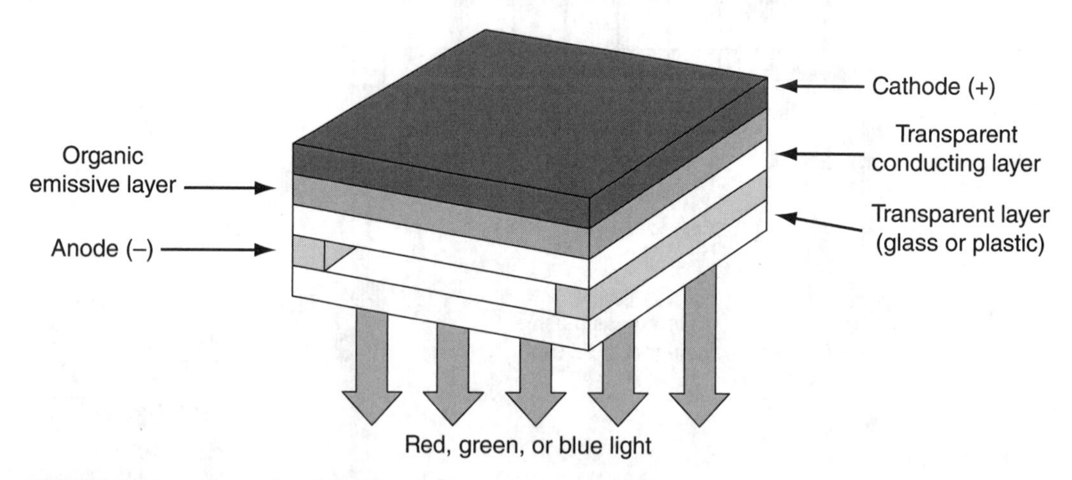

FIGURE 7.11 Layers of a single OLED pixel
Courtesy of Course Technology/Cengage Learning

OLED displays combine many of the best features of LCD and plasma displays. Like LCD and plasma displays, they're manufactured with TFT technology, which makes them thin and lightweight. Like plasma displays, they generate light near the display surface, which provides brightness and a wide viewing angle. OLED displays need less power than

LCD and plasma displays because they require no backlight and don't have to excite gas into a plasma state. The lack of a backlight makes OLED displays thinner than LCD displays, and their emissive layers are thinner than a plasma display's gas cells.

Currently, the most important drawbacks of OLED technology are cost and expected lifetime, but both are expected to improve as the technology matures. The organic emissive compounds have limited lifetimes, so they lose their brightness gradually over periods ranging up to a few hundred thousand hours. More important, the emissive capabilities of different organic compounds degrade at different rates, which causes a gradual color shift over the display lifetime and limits this lifetime to that of the most rapidly degrading emissive compound. Manufacturers have struggled to find combinations of red, green, and blue emissive compounds with similar brightness and degradation rates.

PRINTERS

Printer technology can be classified into three common types:

- Impact
- Inkjet
- Laser

Impact technology began with medieval printing presses: A raised image of a printed character or symbol is coated with ink and pressed against paper. The process was updated in the 1880s with the invention of the typewriter, which enabled precise control over paper position and inserted an ink-soaked cloth ribbon between the raised character image and paper. Early printers were little more than electric typewriters that could accept input from a keyboard or a digital communication port.

Later variations on impact printer technology include the line printer and the dot matrix printer. Line printers had multiple disks or rails containing an entire set of raised character images. Each disk or rail occupied one column of a page, so an entire row or line could be printed at once. A **dot matrix printer** moves a print head containing a matrix of pins over the paper. A pattern of pins matching the character or symbol to be printed is forced out of the print head, generating printed images and characters similar to those shown earlier in Figure 7.3. Some dot matrix printers, customized for high-speed printing of multicopy forms, are still produced.

Inkjet Printers

An **inkjet printer** places liquid ink directly onto paper. It has one or more disposable inkjet cartridges containing large ink reservoirs, a matrix of ink nozzles, and electrical wiring and contact points that enable the printer to control the flow of ink through each nozzle. Each nozzle has a small ink chamber behind it that draws fresh ink from an ink reservoir.

A small drop of ink is forced out of the nozzle in one of two ways:

- *Mechanical movement*—The back of the ink chamber is a piezoelectric membrane, meaning that when electrical current is applied to this membrane, it bends inward, forcing a drop of ink out of the nozzle. When the current is removed, the membrane returns to its original shape, creating a vacuum in the chamber that draws in fresh ink.

- *Heat*—The back of the ink chamber is a resistor that heats rapidly when electrical current is applied (see Figure 7.12a). The ink forms a vapor bubble (see Figure 7.12b), forcing a drop of ink out of the nozzle. When the current is removed, the vapor bubble collapses (see Figure 7.12c), creating a vacuum in the chamber that draws in fresh ink.

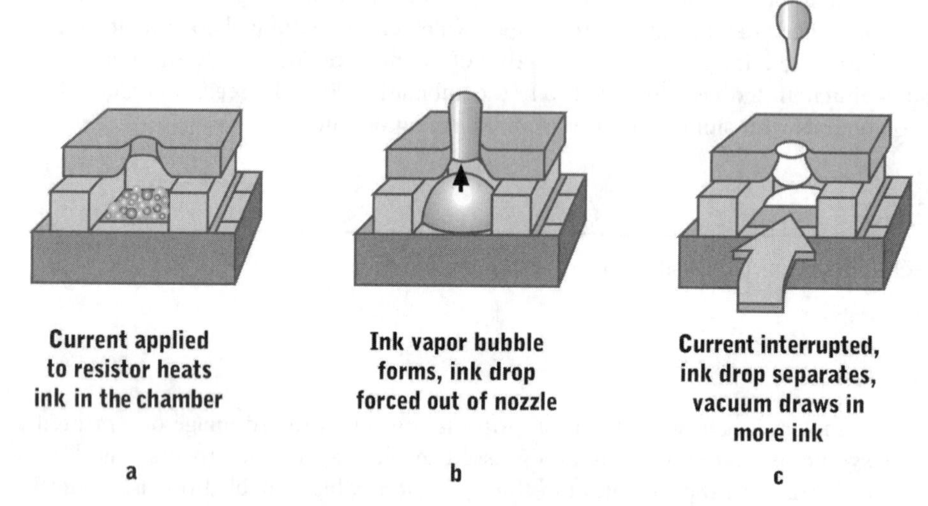
Current applied
to resistor heats
ink in the chamber
a
Ink vapor bubble
forms, ink drop
forced out of nozzle
b
Current interrupted,
ink drop separates,
vacuum draws in
more ink
c

FIGURE 7.12 Ink drop formation and ejection in a thermal inkjet chamber
Courtesy of Hewlett-Packard Company; reproduced with permission

Paper is fed from an input hopper and positioned so that the print nozzles are along its top edge. As the print head is drawn across the paper width by a pulley, the print generator rapidly modulates the flow of current to the ink chamber membranes or resistors, generating a pattern of pixels across the page. The paper is then advanced, and the process is repeated until the entire sheet has been drawn past the moving print head.

Inkjet printer nozzles can produce ink dots as small as 1/600th of an inch in diameter, meaning the resolution can be up to 600 dpi. Color output is generated with CMY inks and alternating rows of nozzles for each ink color. As each row passes a horizontal point on the page, a different color of ink is applied. The inks mix to form printed pixels of various colors. Most color inkjet printers use a separate ink cartridge for black.

Inkjet printer technology was developed in the 1980s and has advanced rapidly to become the most common printing technology. Typical output speeds for inexpensive models are 4 to 6 pages per minute (ppm) for black and 2 to 4 ppm for full color. Black print quality is good to excellent, and color print quality is often better than that of color laser printers. Inkjet printers are the only type of printer capable of producing high-quality color output that's within the budget of a typical home or small-office user.

Printer Communication

Communication between a computer system and an impact printer usually consists of ASCII or Unicode characters because characters and symbols are the fundamental output unit. An impact printer can print each character as it's received, although small buffers are commonly used to improve performance.

Inkjet and laser printers use pixels as the fundamental output unit, so more complex communication methods are required. These printers have large buffers to hold a single line, multiple lines, or an entire page of printed output (refer back to Figure 6.8). An embedded processor and software decide how to store incoming data in the buffer. With bitmaps, data is simply stored in the correct buffer location.

IDLs are commonly used to improve printer performance. The processor interprets the IDL commands for drawing shapes and lines, generates the corresponding bitmaps, and stores them in the correct part of the buffer. With character-oriented data, the processor generates a bitmap for each printed character according to current settings for font style and size.

Laser Printers

A **laser printer** operates with an electrical charge and the attraction of ink to this charge (see Figure 7.13):

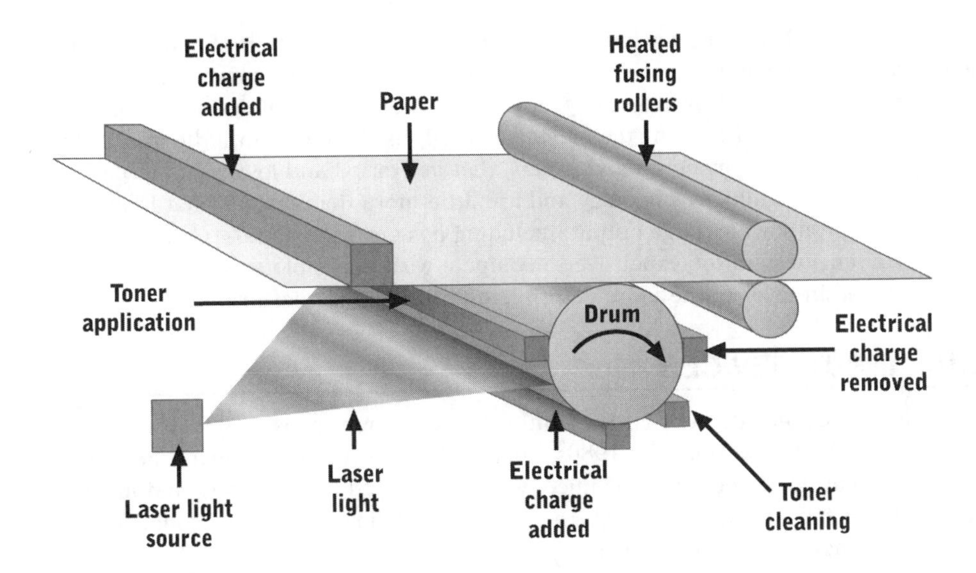

FIGURE 7.13 Components of a laser print engine
Courtesy of Course Technology/Cengage Learning

1. A rotating metal drum is lightly charged over the width of its surface.
2. The print driver reads rows of pixel values from the buffer and modulates a tightly focused laser over the width of the drum.
3. The drum is advanced, and the process is repeated with the next line of pixels.
4. The laser removes the charge wherever it shines on the drum. The drum contains an image of the page, with charged areas representing black pixels and uncharged areas representing white pixels.

5. After charging, the drum then passes a station where fine particles of toner (a dry powder ink) are attracted to the charged areas.
6. In synchronized motion with the drum, paper is fed through a series of rollers and given a high electrical charge. As the paper passes over the drum surface, toner on the drum is attracted to the paper because of its higher charge.
7. The paper and attached toner are fed through heated rollers that fuse the toner to the paper surface.
8. A light source removes all charges from the drum, and excess toner is removed from the drum by a fine blade and/or vacuum.

Color laser output uses three separate print generators (laser, drum, and laser modulators), one for each color. Paper must be aligned precisely as it passes over the three drums. Color laser printers are more complex and expensive than their monochrome counterparts. Color can also be generated with three passes over the same print generator, with different color toner applied during each pass. This method reduces the printer's complexity, cost, and (unfortunately) maximum speed.

Plotters

A **plotter** is a printer that generates line drawings on wide sheets or rolls of paper. Plotters can handle paper widths up to 64 inches, which makes them ideal for producing posters, blueprints, and engineering drawings. They vary in drawing technology, printing speed, and paper width. Pen-based printing technology, which was the norm until the mid-1990s, draws images by using one or more moving pens that are raised and lowered onto paper.

Modern plotters use inkjet technology and are little more than large-format inkjet printers that can produce any large output, including posters and banners. Comparably sized laser print engines are too expensive to compete with inkjet plotters. The term "plotter" has faded in favor of the more descriptive term **large-format printer**.

MANUAL INPUT DEVICES

Manual input devices include keyboards, pointing devices, input pads, and many less familiar but related devices. Until the 1980s, keyboards were the predominant form of manual input. Pointing devices, such as mice, were introduced in the 1980s, and many other technologies have been introduced since then. The following sections concentrate on the most common manual input devices.

Keyboards

Early computer systems accepted input via punched cards or tape. Keypunch machines converted manual keystrokes into punched holes in a cardboard card or paper tape. Another device, called a card reader, converted punched cards into electrical inputs the CPU could recognize. Punched cards were passed over a light source in the card reader, and light shining or not shining through the punched holes was detected and interpreted as input characters.

Modern keyboard devices translate keystrokes directly into electrical signals, eliminating the need for intermediate storage media, such as punched cards or paper tape. They use an integrated microprocessor, called a **keyboard controller**, to generate bitstream outputs. Pressing a key sends a coded signal to the controller, and the controller generates a bitstream output according to an internal program or lookup table.

A keyboard has many special-purpose keys, including function keys, such as F1, Print Screen, and Esc; display control keys, such as ↑, Page Down, and Scroll Lock; and modifier keys, such as Shift, Caps Lock, Ctrl, and Alt. In addition, there are many valid key combinations, such as Ctrl+Alt+Delete. When keys are pressed, a keyboard controller generates output called a **scan code**, a 1- or 2-byte data element representing a specific keyboard event. For most keys, the event is simply a key press. For some keys, pressing and releasing the key represent two different events and generate two different scan codes. The widespread use of IBM-compatible PCs has created a few scan code standards, including those based on keyboard controllers in early IBM PCs. Most keyboards now follow the IBM PS/2 standard.

A keyboard can be connected to a computer in various ways. Older PCs connected to keyboards via a wire and connector based on the IBM PS/2 standard. Most PCs now use USB connections or wireless connection standards, such as Bluetooth.

Over the past few decades, many people have predicted the demise of keyboards, assuming that a combination of speech recognition, optical scanning, and character recognition technologies would soon fill the keyboard's role. (These technologies are discussed later in "Audio I/O Devices.") Although all these technologies have advanced considerably, they aren't developed enough to match the accuracy and speed of a well-trained typist. Pending substantial improvements in competing technologies, keyboards will be around for many years to come.

Pointing Devices

Pointing devices include the mouse, trackball, and joystick. All these devices perform a similar function: translating the spatial position of a pointer or other selection device into numeric values in a system of 2D coordinates. They can be used to enter drawings into a computer system or control the position of a **cursor** (pointer) on a display device.

A mouse is a pointing device that's moved on a flat surface, such as a table, desk, or rubber pad. Its position on the surface corresponds to the pointer's position on a video display. As the mouse is moved left or right, the pointer onscreen moves left or right. As the mouse is moved toward or away from the user, the pointer moves toward the bottom or top of the display.

Mouse position is translated into electrical signals by mechanical or optical components. Most mechanical mice have a roller ball in contact with two wheels mounted on pins. As the mouse moves, the roller ball moves one or both wheels. Wheel movement generates electrical signals that are sent to a device controller and then converted to numeric data describing the direction and distance the mouse has been moved.

An optical mouse uses an optical scanner that constantly scans the surface under the mouse at a high resolution. A microprocessor in the mouse compares many scans per second to determine the direction and speed of movement. This approach works on most surfaces, including those of apparently uniform color and texture. When scanned at high resolutions (such as 600 dpi), most surfaces have discernible texture and color or intensity variation. Optical mice have no moving or exposed parts, as mechanical mice do, that can become contaminated by dust or dirt. Some optical mice have two scanners for more accurate tracking.

A mouse has one to three buttons on top that are clicked to set the cursor's position in the display, select text or menu items, open programs, and display pop-up or context menus. Some also have a scroll wheel for scrolling through a menu or window. Mouse buttons are simple electromechanical switches. Users select text, menu items, and commands by clicking or double-clicking a button or by pressing two buttons at once.

The latest generation of mice and related pointing devices uses embedded gyroscopes and communicates without wires. A gyroscope can detect motion within 3D space, which frees the pointing device from the confines of a 2D surface, such as a desktop. Gyroscopic pointing devices are ideal for lectures and presentations because they give the user more freedom of movement.

A trackball is essentially an upside-down mechanical mouse. The roller ball is typically much larger than in a mechanical mouse, and the selection buttons are mounted nearby. The roller is moved by the fingertips, thumb, or palm. Trackballs require less desktop space than a mouse and are easy to incorporate into other devices, such as keyboards and video games. Because they require minimal finger, hand, and arm dexterity, trackballs are ideal for young children and people with certain physical disabilities, although ease of use generally requires a large ball.

Input Pads

A **digitizer** consists of a digitizing tablet and a pen, stylus, or both. The tablet is sensitive to the stylus or pen's placement at any point on its surface. Drafters and artists trace blueprints and drawings on a digitizer tablet or use it for freehand sketching. Specialized digitizer tablets are used in industries such as mapping, engineering, and medical imaging. With a stylus, drawing, tracing, and command functions can be combined in a single device. Tablet PCs use a similar input and display surface but omit the stylus and add handwriting-recognition software for text entry.

Digitizing tablets and tablet PCs are examples of **input pads**, a general class of input devices. These pads are the basic technology behind other devices, such as signature pads, mouse pads, and touch screens. There are four common input pad technologies:

- Infrared detectors
- Photosensors
- Pressure-sensitive pads
- Magnetic fields

Infrared sensing, the oldest of these technologies, was commonly used in devices such as touch screens, where the pressure-sensing mechanisms must be invisible. An array of infrared beam generators is placed along two adjacent edges of the screen or pad. Arrays of infrared sensors are placed along the other two edges, aligned precisely with the infrared beam generators. When a finger or an object touches the pad, it interrupts one or more of the beams. Each sensor generates a binary input to a microprocessor that compares inputs from all sensors to determine where the touch occurs. Infrared technology is fading in favor of alternative TFT-based technologies.

An input pad can also be composed of a 2D array of photosensors. A **photosensor** converts incoming light energy into outgoing electrical energy. A light pen that generates laser light causes photosensors in the pad to generate a signal when the pen is placed against a point on the pad. Again, a microprocessor compares inputs from all photosensors to determine the pen position. Photosensing provides high precision and resolution but is subject to interference from dirt and other contaminants.

Some large input pads contain four pressure-sensitive ribbons at each edge of the pad. Each ribbon transmits a separate signal, indicating the amount of pressure at one edge of the pad. A microprocessor compares pressure levels for all four ribbons to determine where pressure from a finger, stylus, or pen is being applied on the pad. For example, applying pressure to the center of a square pad results in equal pressure on all four ribbons. Applying pressure at one corner of the pad results in equally high pressure on the two ribbons at that corner and equally low pressure on the two ribbons at the opposite corner.

Smaller pads, such as those used for laptop mouse pads and signature pads, are manufactured with TFTs. One approach uses an array of sensors that detect interruptions in a weak magnetic field. A finger or a metal stylus interrupts the magnetic field at a specific point, generating a pattern of electrical output in the sensors. Another approach uses an array of pressure-sensitive resistors that vary current flow according to the physical pressure an object applies. With either approach, a microprocessor compares input from the entire array of sensors or resistors to determine the precise location of the finger or stylus.

OPTICAL INPUT DEVICES

Optical input devices have come of age in the past decade with the introduction or refinement of devices such as advanced bar-code readers, low-cost optical scanners, and digital still and motion cameras. Optical input devices can be classified into two broad categories based on their application and underlying technology:

- Mark and pattern sensors
- Image capture devices

Mark and pattern sensors capture input from special-purpose symbols placed on paper or the flat surfaces of 3D objects (for example, the shipping label on a box). Specific devices include magnetic mark sensors and bar-code readers. Image capture devices, which are newer, include digital still and motion cameras, which are based on more complex technology than mark sensors and bar-code readers are.

Photosensors are common to both device types. Light reflected from a mark, a symbol, or an object is reflected into a photosensor. A photosensor's current outflow is an analog signal representing the intensity of this reflected light. High light intensities induce large current outflows from the photosensor, and low light intensities produce little or no current. The analog signal is then converted to a digital number. Simple devices, such as bar-code readers, generate a small range of numbers, such as 0 to 7. More complex devices, such as digital cameras, classify the analog input in a larger range, such as 0 to 255, producing a more accurate digital representation of the light input.

Mark Sensors and Bar-Code Scanners

A **mark sensor** scans for light or dark marks at specific locations on a page. Input locations are drawn on a page with circles or boxes and are selected by filling them in with a dark pencil. The mark sensor uses preprinted bars on the edge of the page to establish reference points—for example, the row of boxes corresponding to the possible answers for

Question 5. It then searches for dark marks at specific distances from these reference points. Marks must be of a sufficient size and intensity to be recognized correctly. Older mark sensor devices used magnetic sensing, which required magnetic ink or pencil marks for accurate recognition. Mark sensors now use fixed arrays of photosensors, which don't require magnetic ink or pencil marks.

A **bar-code scanner** detects specific patterns of bars or boxes. The most common type of **bar code** contains a series of vertical bars of equal length but varied thickness and spacing. The order and thickness of bars are a standardized representation of numeric data. To detect these bars, bar-code readers use **scanning lasers** that sweep a narrow laser beam back and forth across the bar code. Bars must have precise width and spacing as well as high contrast for accurate decoding.

When a single scanning laser and photosensor are used, a bar code must be placed at a specific angle in relation to the detector for accurate recognition. Most bar-code scanners have multiple scanning lasers and photosensors mounted at oblique angles, which enables accurate readings at a variety of physical orientations.

Bar-code readers are typically used to track large numbers of inventory items, as in grocery store inventory and checkout, package tracking, warehouse inventory control, and zip code routing for postal mail. The U.S. Postal Service (USPS) uses a modified form of bar coding with evenly spaced bars of equal thickness but varying height. They're placed along the lower edge of an envelope and encode five- and nine-digit zip codes (see Figure 7.14).

The White House
1600 Pennsylvania Avenue NW
Washington, DC 20500

FIGURE 7.14 An address with a standard USPS bar code
Courtesy of Course Technology/Cengage Learning

Modern bar codes encode data in two dimensions. Figure 7.15 shows a sample 2D bar code in the PDF417 format. Data is encoded in patterns of small black and white blocks surrounded by vertical bars marking the left and right boundaries. PDF417 bar codes can hold around 1 KB of data, so descriptive or shipping information can be stored on the shipping container or inventory item.

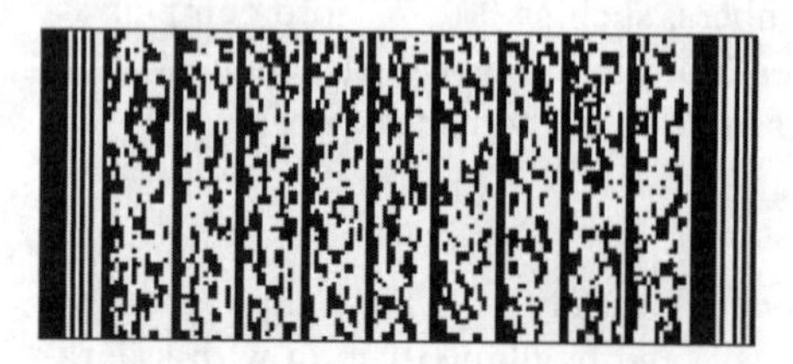

FIGURE 7.15 A PDF417 bar code
Courtesy of Course Technology/Cengage Learning

Optical Scanners

An **optical scanner** generates bitmap representations of printed images. A bright white light shines on the page, and reflected light is detected by an array of photosensors. Chromatic resolution is determined by the photosensors' sensitivity to different light frequencies. High resolution requires small, tightly packed photosensors. Typical desktop scanners have spatial resolution up to 1200 dpi and chromatic resolution up to 24 bits. With manual or hand scanners, a user must move the scanning device over a printed page at a steady rate. Automatic scanners use motorized rollers to move paper past a scanning surface or move the light source and photosensor array under a fixed glass surface. Optical scanners typically communicate with a computer system by using a general-purpose I/O interface, such as USB, Bluetooth, or IEEE FireWire. These interfaces provide more than adequate data transfer rates for most scanners.

Optical character recognition (OCR) devices combine optical-scanning technology with a special-purpose processor or software to interpret bitmap content. After an image has been scanned, the bitmap is searched for patterns corresponding to printed characters. In some devices, input is restricted to characters in prepositioned blocks, such as the "Amount Enclosed" blocks in Figure 7.16. As with mark sensors, preprinted, prepositioned input blocks are a reference point for locating characters, which simplifies the character recognition process.

PIT-ES New Mexico Individual Estimated Tax Payment Voucher   Tax Year________

Indicate quarter the payment is to be applied to: ☐ 1ST (APR 15) ☐ 2ND (JUN 15) ☐ 3RD (SEPT 15) ☐ 4TH (JAN 15)

YOUR FIRST NAME AND INITIAL	LAST NAME	YOUR SOCIAL SECURITY NUMBER
SPOUSE'S FIRST NAME AND INITIAL	LAST NAME	SPOUSE'S SOCIAL SECURITY NUMBER
ADDRESS (Number & Street)	CITY, STATE AND ZIP CODE	☐ Check this box if address is new or changed

Mail payment and voucher to:
New Mexico Taxation and Revenue Department
P.O. Box 8390, Santa Fe, NM 87504-8390

AMOUNT ENCLOSED ☐☐,☐☐☐,☐☐☐.00

FIGURE 7.16 A sample form with OCR input fields
Courtesy of Course Technology/Cengage Learning

More sophisticated OCR software and hardware place no restriction on the position and orientation of symbols, which creates problems in locating and recognizing printed symbols. Recognition is most accurate when text is printed in a single font and style, with all text oriented in the same direction on the page. If the text is handwritten, contains mixed fonts or styles, is combined with images, or has varying orientations on the page, accurate recognition is more difficult. Many inputs are beyond the current capabilities of OCR devices.

NOTE

Error rates of 10% or higher are still common with mixed-font text and even higher with handwritten text. The accuracy and flexibility of input have improved rapidly, but more progress must be made before error rates are acceptable for many types of applications.

Digital Cameras

Digital still cameras, video cameras, and Webcams use a 2D photosensor array placed behind lenses to capture reflected and focused ambient light. Separate sensors for each primary color capture data for a single pixel. Sensors for each pixel are combined on a single sensor array in a row and column geometry and typically produced on a single chip. This sensor array has a native resolution expressed in megapixels, derived by multiplying the number of pixels in the array's rows and columns (for example, 2000 × 3000, or 6 megapixels).

The main difference between digital still and motion video cameras is apparent after an image is captured. A digital still camera captures one image at a time. When the shutter button is pressed, the camera briefly samples the photosensor array's output and generates a bitmap, which is then stored in a memory card (internal or removable). To conserve storage space, the user can configure the camera to store images in a compressed format, such as JPEG.

Both film and digital video cameras capture moving images by rapidly capturing a series of still images called "frames." Moving image quality improves as the number of frames per second (fps) increases. Typically, digital cameras capture 24 to 30 fps. Each frame is stored in a bitmap buffer as it's captured. A separate processor reads buffer content continuously and applies a compression algorithm, such as MPEG, to the data. The processor can generate better quality compressed images if the buffer is large enough to hold several frames. The compression algorithm can also be configured for tradeoffs between compressed image quality and storage requirements.

Portable Data Capture Devices

Since the 1990s, there's been an explosion of portable data capture devices for tasks such as warehouse inventory control and package routing, tracking, and delivery and in retail settings, such as grocery stores and Walmart (see Figure 7.17). Most portable data capture devices combine a keyboard, mark or bar-code scanner, and wireless connection to a wired base station, cash register, or computer system.

FIGURE 7.17 A wireless portable data capture device
Courtesy of Motorola, Inc.

Portable data capture devices differ in the degree of embedded computer power. Simpler devices function only as I/O devices, sending captured data through a wired or wireless communication channel to a computer system. More complex devices are actually portable computers with embedded I/O devices, such as bar-code scanners, radio frequency ID tag readers, and digital cameras. Some devices also include cell phone capabilities.

AUDIO I/O DEVICES

Sound is an analog signal that must be converted to digital form for computer processing or storage. The process of converting analog sound waves to digital representation is called **sampling**, which analyzes the content of the audio sound spectrum many times per second and converts it to a numeric representation. For sound reproduction that sounds natural to people, frequencies between 20 Hz and 20 KHz must be sampled at least 40,000 times per second. Sampling frequencies for digital recordings range from 44,100 to 96,000 samples per second.

Sound varies by frequency (pitch) and intensity (loudness). By using mathematical transformations, a complex sound, such as a human voice, consisting of many pitches and intensities can be converted to a single numeric representation. For natural-sounding reproduction, at least 16 bits (2 bytes) must be used to represent each sample. At present, 24-bit (3-byte) samples are commonly used for better accuracy. For example, one second of sound sampled 96,000 times per second at 24 bits per sample requires 288,000 bytes for digital representation. A full minute of stereo sound at the same frequency and resolution requires about 35 MB of data storage or transmission capacity.

Sampling and playback rely on simple converting devices. An **analog-to-digital converter (ADC)** accepts a continuous electrical signal representing sound (such as microphone input), samples it at regular intervals, and outputs a stream of bits representing the samples. A **digital-to-analog converter (DAC)** performs the reverse transformation, accepting a stream of bits representing sound samples and generating a continuous electrical signal that can be amplified and routed to a speaker. The conversion hardware, processing power, and communication capacity needed to support sampling and playback have existed in consumer audio devices since the 1980s and in ordinary desktop computers since the early 1990s.

In a computer system, sound generation and recognition are used for many purposes, such as the following:

- General-purpose sound output, such as warnings, status indicators, and music
- General-purpose sound input, such as digital recording of voice e-mail messages
- Voice command input
- Speech recognition
- Speech generation

Limited general-purpose sound output has been available in computer systems for many years. VDTs and early PCS could generate a single audible frequency at a specified volume for a specified duration. This type of sound output is called **monophonic** output because only one frequency (note) can be generated at a time. Newer computers use **polyphonic**, or multifrequency, sound generation hardware with built-in amplification and high-quality speakers.

Speech Recognition

Speech recognition is the process of recognizing and responding to the meaning embedded in spoken words, phrases, or sentences. Human speech consists of a series of sounds called **phonemes**, roughly corresponding to the sounds of each letter of the alphabet. A spoken word is a series of interconnected phonemes, and continuous speech is a series of phonemes interspersed with periods of silence. Recognizing single voiced (spoken) phonemes isn't a difficult computational problem. Sound can be captured in analog form by a microphone and converted to digital form by digital sampling, and the resulting digital pattern is compared with a library of patterns corresponding to known phonemes, as shown in Figure 7.18.

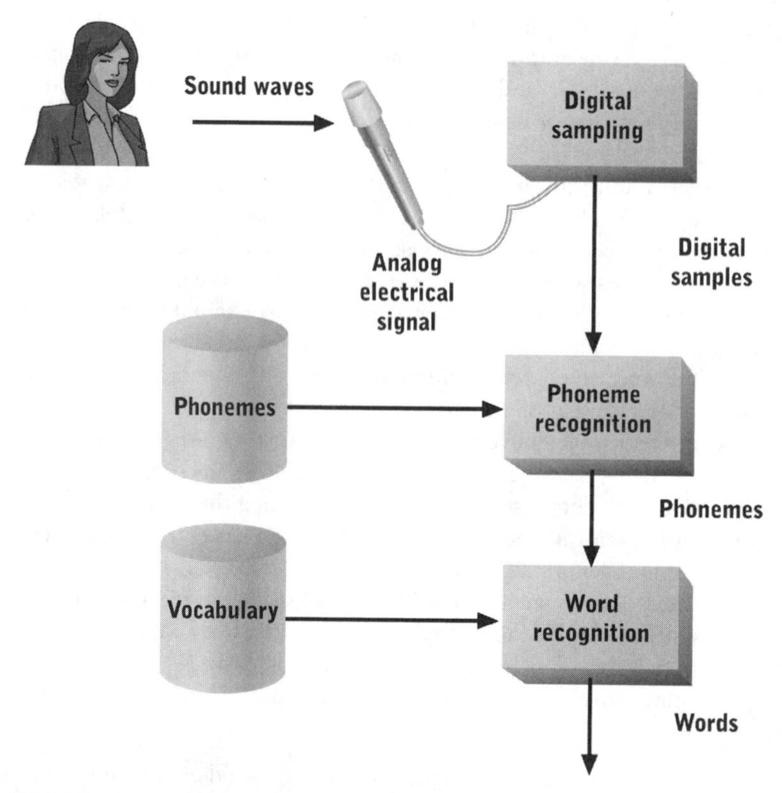

FIGURE 7.18 The process of speech recognition
Courtesy of Course Technology/Cengage Learning

Although the process of speech recognition is simple conceptually, a number of factors complicate it, including the following:

- Speech variability
- Phoneme transitions and combinations
- Real-time processing

The characteristics of voiced phonemes vary widely because of physical, geographic, and situational differences in people. For example, phonemes might differ depending on whether the speaker is male or female, is from Texas or New England, or is making polite conversation or giving a command. Because of these variations, comparison with a library of known phonemes can't be based on an exact match; the closest match must be determined and decisions between multiple possible interpretations must be made.

Another difficult problem in speech recognition results from the continuous nature of speech. Phonemes sound similar when voiced repetitively by the same person. However, when combined with other phonemes in different words, their voicing varies considerably. For example, the letter e is voiced differently in the words "bed" and "neighbor." In addition, a computer must determine where one phoneme ends and another begins as well as where one word ends and another begins. Performing these tasks in real time requires complex software and powerful CPUs.

Most current speech-recognition systems are **speaker dependent**, which means they must be "trained" to recognize the sounds of human speakers. They're also restricted to vocabularies of perhaps only a few thousand words. The most limited are command-recognition systems, designed to recognize up to a few hundred words from a single speaker. These systems are useful when a single user is using the device and manual input is impractical. They have been applied in airplane cockpits, manufacturing control systems, and input systems for people with physical disabilities.

A **digital signal processor (DSP)** is a microprocessor specialized for processing continuous streams of audio or graphical data. DSPs are commonly embedded in audio and video hardware, such as PC sound cards and dedicated audio/video workstations. In some cases, a DSP processes audio input before data is transferred to the CPU. In other cases, the CPU and DSP share processing duties, with each performing the tasks to which it's best suited.

Speech Generation

A device that generates spoken messages based on text input is called an **audio response unit**. Typical applications include delivering limited amounts of information over conventional phones, such as automated phone bank tellers and automated call routing. Audio response units enable a phone to act as an information system's output device.

Simple audio response units store and play back words or word sequences. Words or messages are digitally recorded and stored. To output a message, the stored message is retrieved and sent to a device that converts the digitized voice into analog audio signals. All possible outputs must be recorded in advance. Phrases and sentences can be generated from stored words, although pauses and transitions between words tend to sound unnatural.

A more general and complex approach to speech generation is **speech synthesis**, in which vocal sounds (phonemes) are stored in the system. Character outputs are sent to a processor in the output unit, which assembles corresponding groups of phonemes to generate synthetic speech. The quality of speech output from these units varies considerably. Creating natural-sounding transitions between phonemes within and between words requires sophisticated processing. However, the difficulties of speech generation are far less formidable than those of speech recognition.

General-purpose audio hardware, discussed in the next section, can also be used for speech generation. Digital representations of phoneme waveforms can be combined mathematically to produce a continuous stream of digitized speech, which is then sent to the audio hardware's digital-to-analog converter. After speech is converted to an analog waveform, it's amplified and routed to a speaker or phone.

General-Purpose Audio Hardware

General-purpose audio hardware can be integrated on a PC motherboard or packaged as an expansion card that connects to the system bus (commonly called a **sound card**). At minimum, sound cards include an ADC, a DAC, a low-power amplifier, and connectors (jacks) for a microphone and a speaker or headphones (see Figure 7.19). More elaborate cards might include the following:

- Multichannel surround sound, such as Dolby 5.1
- A general-purpose Musical Instrument Digital Interface (MIDI) synthesizer
- MIDI input and output jacks
- A more powerful amplifier to accommodate larger speakers and generate more volume

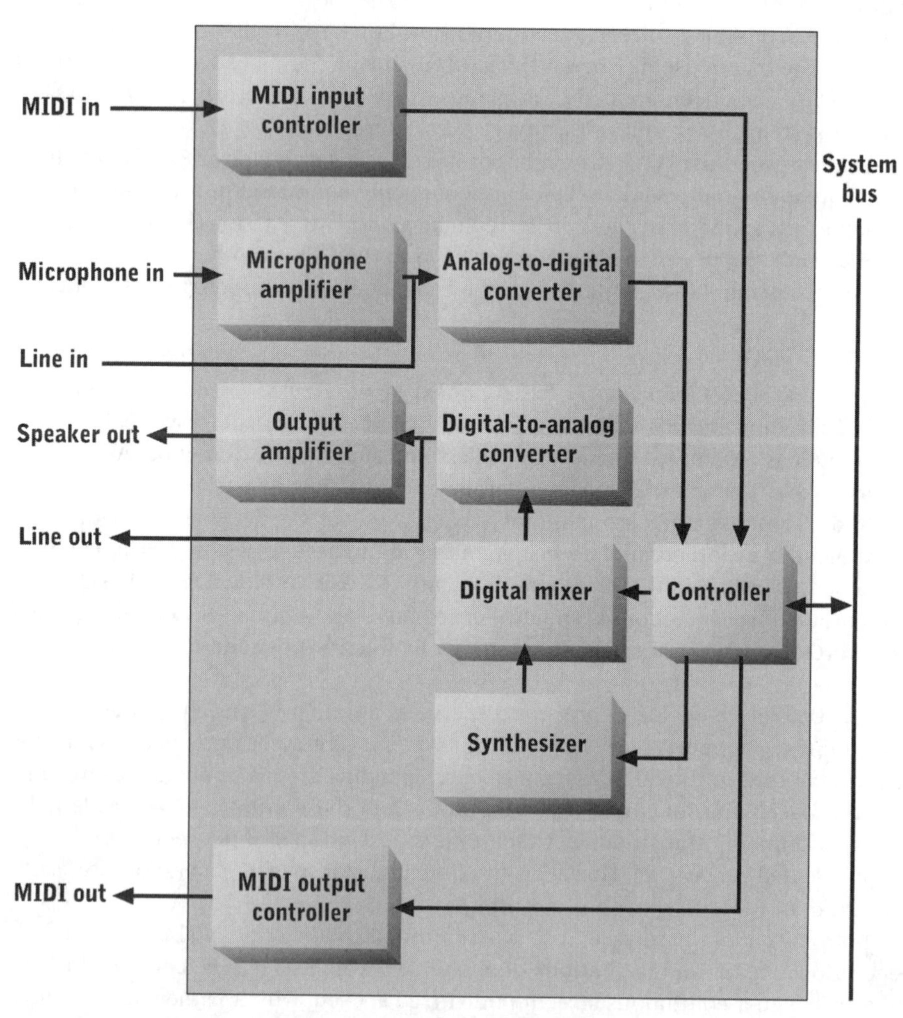

FIGURE 7.19 Components and connections of a typical sound card
Courtesy of Course Technology/Cengage Learning

Other differences in sound cards include degree of polyphony (the number of simultaneous sounds that can be generated), supported sampling rates, and power and speed of embedded DSPs.

Musical Instrument Digital Interface (MIDI) is a standard for storing and transporting control information between computers and electronic musical instruments. Its original purpose was to enable electronic keyboards to control remotely connected synthesizers. The MIDI standard defines a serial communication channel and a standard set of 1- to 3-byte commands for performing these tasks:

- Turn notes on and off.
- Modulate pitch.
- Mimic musical instrument controls, such as a piano's sustain pedal.
- Select specific instruments or artificial sounds.

Up to 16 channels of MIDI data can be sent over the same serial transmission line, so 16 separate instruments or instrument groups can be played at once. Most sound cards include some MIDI capability. At minimum, they provide a simple synthesizer capable of simulating the sound of a few dozen instruments. Files containing MIDI control data are transferred to the sound card over the system bus, and then the synthesizer processes the commands and generates digital waveforms for each instrument. The digital waveforms are combined (mixed) and sent through the DAC, amplifier, and speakers.

More complex sound cards provide more synthesized instruments, better quality waveforms, and the capability to accept or transmit MIDI data through external MIDI connections. The quality of synthesized sound varies widely from one synthesizer and/or sound card to another. The poorest quality sounds are generated by inexpensive DSPs executing simple instrument simulation programs and operating at low sampling rates. The most realistic sounds are achieved with sample-based synthesis, which uses stored digitized recordings of musical instruments.

The main advantage of MIDI is its compact storage format. A command to sound a single note on a piano for 1 second requires 6 bytes: 3 bytes each for the note-on and note-off control messages. A 24-bit digital recording of this sound sampled at 96,000 Hz requires 288,000 bytes of storage. The disadvantage of MIDI is a lack of control over the nature of the generated sound. A MIDI command selects instruments and notes but doesn't control how instrument sounds are generated. Sound quality can vary widely across sound cards and synthesizers, depending on sampling rate, synthesis technique, and hardware capability. The sound for an instrument might not even exist on the synthesizer receiving the MIDI commands.

Summary

- Display surfaces can be divided into rows and columns, similar to a large table or matrix. Each cell, or pixel, in this table represents one simple component of an image. The number of pixels per inch is the display's resolution. A stored set of numeric pixel descriptions is called a bitmap. Monochrome display uses a single bit to indicate whether each pixel is on or off. Multiple bits can represent varying levels of intensity, such as a grayscale.
- Any color in the visible spectrum can be represented as a mixture of varying intensities of the additive (RGB) colors red, green, and blue or the subtractive (CMY) colors cyan, magenta, and yellow. Three numbers represent a pixel in full color. The major drawbacks of bitmaps include their large file size and device dependence. Compression can be used to reduce file size. Image description languages can describe images compactly by using vectors, display objects, or both.
- Video display technologies include liquid crystal displays (LCDs), plasma displays, and light-emitting diode (LED) displays, all of which are packaged in a thin flat panel. Video signals are sent to flat panel displays from a video controller containing a specialized video processor, dual-ported video RAM, embedded software, and display generator circuitry. Video processors accept IDL commands from standards such as Direct3D and OpenGL and generate corresponding bitmaps for output to the display.
- Printer types include impact (dot matrix), inkjet, and laser printers. Dot matrix printers are inexpensive but are slow and noisy and produce poor-quality output. Inkjet printers produce output of excellent quality but are slow. Laser printers are fast and produce excellent-quality output. Plotters, more commonly called "large-format printers," are similar to inkjet printers, except they print on paper up to 64 inches wide.
- Manual input devices include keyboards, pointing devices, and input pads. Keyboards are used to enter text and commands. Mice and other pointing devices are used for pointing to and selecting buttons and menu items, for drawing, and for moving the cursor's position. Input pads can perform many of the same functions as a mouse. Artists and drafters use digitizers, which are large-format input pads. Input pads are also used for devices such as signature pads and touch screens.
- Optical input devices include mark sensors, bar-code readers, optical scanners, digital cameras, and portable data capture devices. All detect light reflected off a printed surface or an object into a photosensor. A mark sensor scans for light or dark marks at specific locations on a page. A bar code is a series of vertical bars of varied thickness and spacing. Optical scanners generate bitmap representations of printed images.
- Optical character recognition (OCR) devices combine optical-scanning technology with intelligent interpretation of bitmap content. Digital still cameras capture single images and store them as raw or compressed bitmaps. Video cameras capture moving images by rapidly capturing a series of still images called "frames" and storing them in a bitmap buffer. Simple portable data capture devices function only as an I/O device, sending captured data through a wired or wireless communication channel to a computer system. More complex devices are actually portable computers with embedded I/O devices, and some have cell phone capabilities, too.

- Sound is an analog waveform that can be sampled and stored as digital data. Computer sound recognition and sound generation hardware and software can be used to play or record speech and music, synthesize speech from text input, and recognize spoken commands. Sound cards include an analog-to-digital converter (ADC), a digital-to-analog converter (DAC), a low-power amplifier, and connectors for a microphone, a speaker, and headphones. MIDI inputs, MIDI outputs, and a music synthesizer can also be provided. Digital signal processors (DSPs) are commonly used for complex sound processing, such as music synthesis and speech recognition.

This chapter discussed how I/O technology supports human-computer interaction. The next two chapters focus on other important classes of I/O technology: data communication and computer networks. These technologies enable computers to interact and allow users to interact simultaneously with multiple computer systems. Computer systems rely heavily on these technologies to deliver powerful information retrieval and processing capabilities to end users.

Key Terms

24-bit color
active matrix display
additive colors
analog-to-digital converter (ADC)
audio response unit
bar code
bar-code scanner
bitmap
cathode ray tube (CRT)
chromatic depth
chromatic resolution
CMY
CMYK
cursor
digital signal processor (DSP)
digital-to-analog converter (DAC)
digitizer
Direct3D
dithering
dot matrix printer
dots per inch (dpi)
dual-porting
flat panel displays
font
Graphics Interchange Format (GIF)
grayscale
half-toning
image description language (IDL)
inkjet printer
input pads
Joint Photographic Experts Group (JPEG)
keyboard controller
large-format printer
laser printer
light-emitting diodes (LEDs)
liquid crystal display (LCD)
mark sensor
monitors
monochrome
monophonic
Musical Instrument Digital Interface (MIDI)
OpenGL
optical character recognition (OCR)
optical scanner
organic LED (OLED)
palette
passive matrix display
phonemes
photosensor
pixel

plasma display
plotter
point
polyphonic
Portable Document Format (PDF)
PostScript
refresh cycle
refresh rate
resolution
RGB
sampling
scan code
scanning lasers
sound card
speaker dependent
speech recognition
speech synthesis
subtractive colors
thin film transistor (TFT)
vector
vector list
video controller
video display terminal (VDT)
video RAM (VRAM)

Vocabulary Exercises

1. Color video display can be achieved by using elements colored _______, _______, and _______.
2. _______ displays have replaced CRTs for video monitors.
3. A(n) _______ display achieves high-quality color display with organic compounds.
4. The printing industry generally uses inks based on the _______ colors, which are _______, _______, and _______. A(n) _______ ink can also be used as a fourth color.
5. A(n) _______ is another name for a large-format printer.
6. _______ and _______ are commonly used image description languages for documents.
7. A(n) _______ printer forces ink onto a page by heating it or forcing it out with a piezoelectric membrane.
8. The _______ or _______ format is commonly used to compress still images.
9. A(n) _______ sound card can generate multiple notes simultaneously. A(n) _______ sound card can generate only one note at a time.
10. A display device's _______ is the number of colors that can be displayed simultaneously.
11. In a computer, an image is stored as a(n) _______, with each pixel represented by one or more numbers.
12. A(n) _______ is a basic component of human speech.
13. A(n) _______ is a small device in optical scanners and digital cameras that converts light into an electrical signal.
14. When a user presses a key, the keyboard controller sends a(n) _______ to the computer.
15. A(n) _______ recognizes input in the form of vertical lines, usually of varied width and spacing.
16. A(n) _______ converts analog sound waves into a digital representation.
17. Each cell in a video display surface's matrix represents a(n) _______.

18. A(n) _______ display illuminates a pixel by using excited gas and a colored phosphor.
19. _______ speech-recognition programs must be trained to recognize one person's voice.
20. A video display's resolution is described by the units _______.
21. Each transfer of a full screen of data from the display generator to the monitor is called a(n) _______.
22. A(n) _______ is 1/72 of an inch and is considered a standard pixel size.
23. A(n) _______ accepts data or commands over a bus and generates output for a video monitor.
24. A collection of characters of similar style and appearance is called a(n) _______.
25. A(n) _______ compresses an image by replacing image components, such as lines and shapes, with equivalent drawing commands.
26. _______ is dual-ported, which enables cache contents to be written by a video processor while they're being read by display generator circuitry.
27. A(n) _______ is an LCD with transistors to control each display pixel.
28. _______ creates an interlocking pattern of colored pixels that fools the eye into thinking a uniform color is being displayed.
29. _______ is a standardized method of encoding notes and instruments for communication with synthesizers and sound cards.

Review Questions

1. Describe the process by which software recognizes keystrokes.
2. What is a font? What is point size?
3. What are the additive colors? What are the subtractive colors? What types of I/O devices use each kind of color?
4. What is a bitmap? How does a bitmap's chromatic resolution affect its size?
5. What is an image description language? What are the advantages of representing images with image description languages?
6. What is JPEG encoding? What is MPEG encoding?
7. Why does a video controller have its own processor and memory?
8. Describe the technologies used to implement flat panel displays. What are their comparative advantages and disadvantages?
9. How does a laser printer's operation differ from that of a dot matrix printer?
10. Describe the types of optical input devices. For what type of input is each device intended?
11. Describe the process of automated speech recognition. What types of interpretation errors are inherent to this process?
12. Describe the components and functions of a typical sound card. How is sound input captured? How is speech output generated? How is musical output generated?

Research Problems

1. Many organizations store all documents, such as medical records, engineering data, and patents, as graphics files. Identify commercial software products that support this form of data storage. Examine the differences between these products and more traditional data storage approaches. Concentrate on efficient use of storage space, internal data representation, and methods of data search and retrieval. What unique capabilities are possible with image-based storage and retrieval? What additional costs and hardware capabilities are required? Examine sites such as *www.databankimx.com* and *www.knowledgetree.com*.

2. Flat panel display technology changes so rapidly that many of the comparative statements made in this chapter might be out of date before the book is published. Investigate current LCD, plasma, and LED display products and answer the following questions:
   - Do LCDs still have a price and performance edge over LED and plasma displays?
   - Have LED and plasma display lifetimes advanced enough to make them viable monitors for most desktop computers?
   - Are any new flat panel display technologies coming into the marketplace?
3. Investigate the Direct3D and OpenGL video standards and answer the following questions:
   - For what types of applications was each originally developed?
   - What OSs support each standard?
   - Does either standard have a clear advantage in any market segments or for specific types of applications?
   - What skills are needed for a programmer to use either standard?
4. Investigate audio interfaces that support recording multichannel digital sound, such as those offered by MOTU (*www.motu.com*), M-Audio (*www.m-audio.com*), and Echo Digital Audio (*www.echoaudio.com*), and answer the following questions:
   - How are the interfaces connected to the computer?
   - What types of audio connections are supported?
   - What are the minimum software and hardware requirements for the host computer?
   - Which are best suited to use at home for simple composing and recording projects?
   - Which are best suited to use in recording studios or for recording a live concert?

CHAPTER **8**

DATA AND NETWORK COMMUNICATION TECHNOLOGY

CHAPTER GOALS

- Explain communication protocols
- Compare methods of encoding and transmitting data with analog or digital signals
- Describe signals and the media used to transmit digital signals
- Describe wireless transmission technology and compare wireless LAN standards
- Describe methods for using communication channels efficiently
- Explain methods of coordinating communication, including clock synchronization and error detection and correction

Successful and efficient communication is a complex endeavor. People have long labored with the intricacies of communicating concepts and data. The method of expression (such as words, pictures, and gestures), language syntax and semantics, and communication rules and conventions vary widely across cultures and contexts. All these areas make successful communication between people a long learning process.

Understanding computer and network communication can be daunting as well because these topics cover a broad range of interdependent concepts and technologies. Chapter 7 addressed some of these topics, and this chapter extends that coverage, as shown in Figure 8.1, and lays the foundation for a detailed discussion of computer networks in Chapter 9.

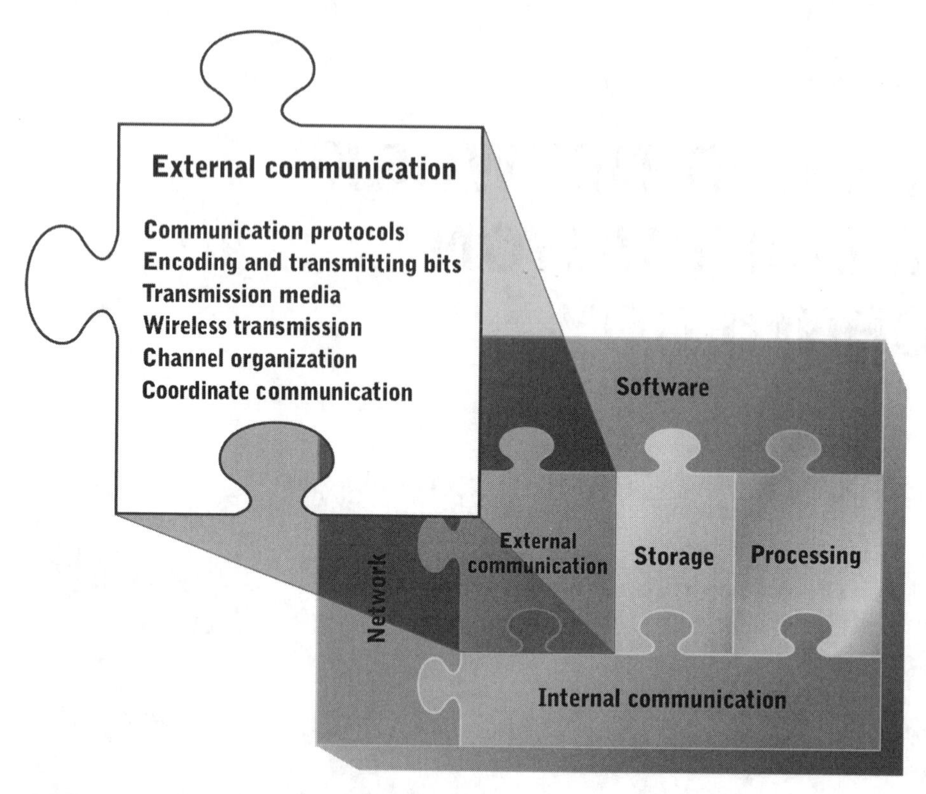

FIGURE 8.1 Topics covered in this chapter
Courtesy of Course Technology/Cengage Learning

COMMUNICATION PROTOCOLS

A **message** is a unit of data or information transmitted from a sender to a recipient. Messages can be loosely categorized into two types—data messages and command messages. Data messages vary in format and content and can include any primitive CPU data types or more complex data types (discussed in Chapter 3). For the purposes of computer communication, a data message's content isn't important; the message is simply moved from one place to another with no attempt to understand its content.

A command message contains instructions that control some aspect of the communication process. Examples include many ASCII device control characters, addressing and routing instructions, and error detection and correction information. Command messages can also be used to transmit information about data messages, such as format, content, length, and other information the receiver needs to interpret the data message correctly.

At the most fundamental level, a message is simply a sequence of bits. Sender and receiver must agree on a common method of encoding, transmitting, and interpreting these bits. This common method is called a **communication protocol**, a set of rules and conventions for communication. Although this definition sounds simple, it encompasses many complex details.

For example, think about the communication protocol that applies during a classroom presentation. First, all participants must agree on a common language, and its grammar and syntax rules are part of the protocol. For spoken languages, the air in the room serves as the transmission medium. Large classrooms might require auxiliary means to improve transmission, such as sound insulation and voice amplification equipment.

Command-and-response sequences ensure efficient and accurate information flow. In this example, the instructor coordinates access to the shared transmission medium. Students use gestures such as raising their hands to ask for permission to speak (transmit messages). In addition, message content is restricted, so the instructor ignores or cuts off messages that are off topic or too long.

Some rules and conventions ensure accurate transmission and reception in this setting. For example, participants must speak loudly enough to be heard by all. The instructor might pause and look at students' faces to determine whether the message is being received and interpreted correctly. In addition, the act of entering the classroom signals participants' acceptance of the communication protocol.

Communication between computers also relies on complex communication protocols. Figure 8.2 organizes the components of a computer communication protocol in a hierarchy. Each node has many possible technical implementations. A complete communication protocol is a complex combination of subsidiary protocols and the technologies to implement them. You might find it helpful to return to this hierarchy often while reading the remainder of this chapter so that you can place topics in their context.

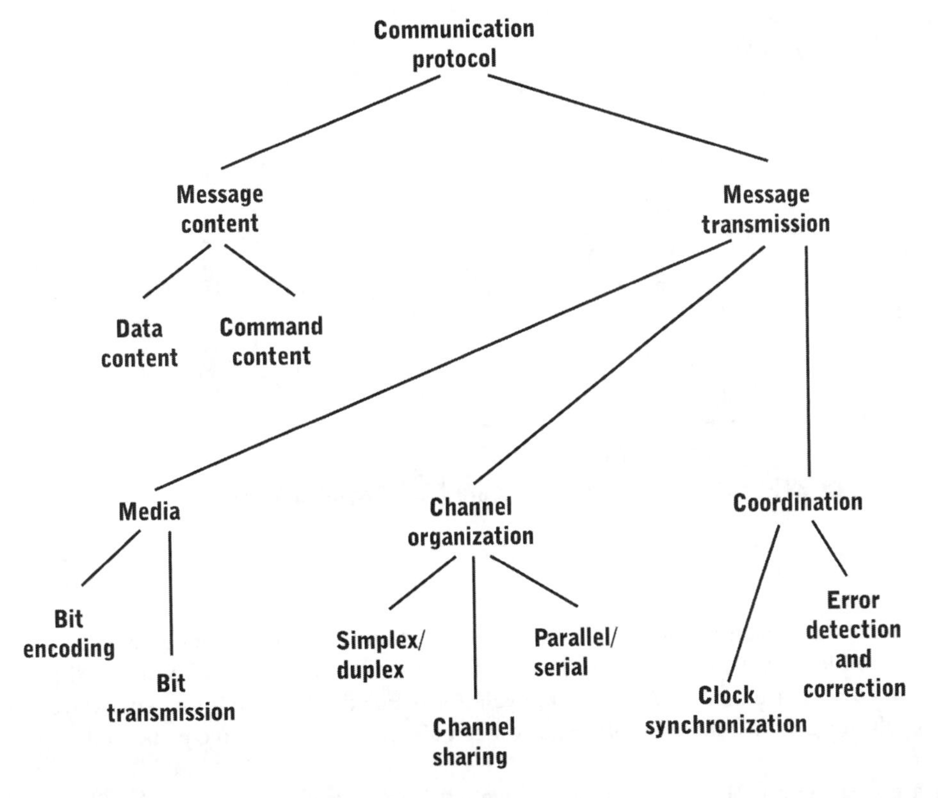

FIGURE 8.2 Components of a communication protocol
Courtesy of Course Technology/Cengage Learning

ENCODING AND TRANSMITTING BITS

Chapter 3 described how data items, such as integers and characters, are encoded in a bit string. This section describes how bits are represented and transported between computer systems and hardware components.

Carrier Waves

Light, radio frequencies, and electricity travel through space or cables as a **sine wave** (see Figure 8.3). A sine wave's energy content varies continuously between positive and negative states. Three characteristics of a sine wave can be manipulated to represent data:

- Amplitude
- Phase
- Frequency

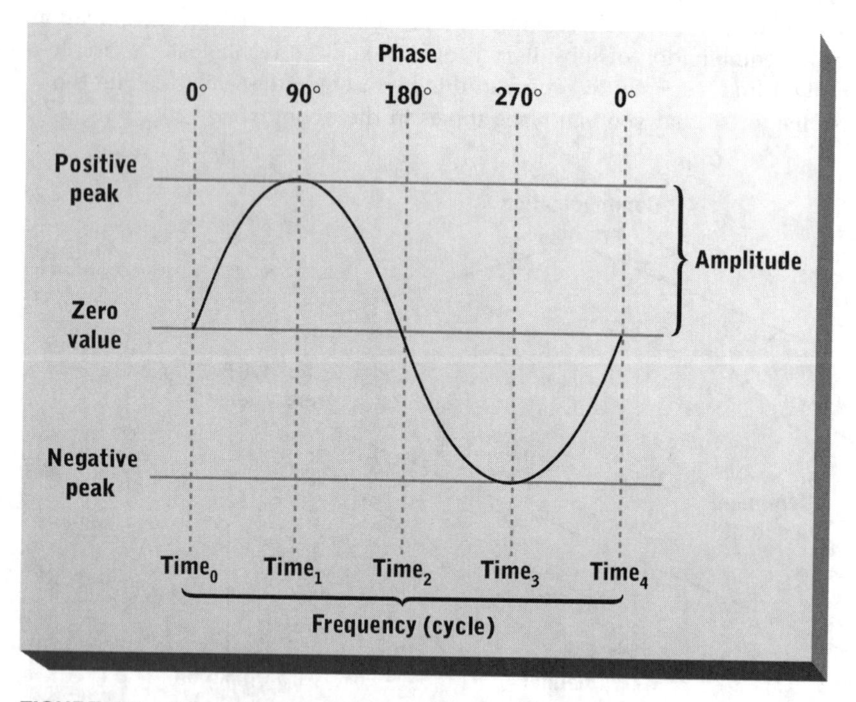

FIGURE 8.3 Characteristics of a sine wave
Courtesy of Course Technology/Cengage Learning

Amplitude is a measure of wave height or power—the maximum distance between a wave's peak and its zero value. It's the same whether it's measured from zero to a positive peak or zero to a negative peak. A complete **cycle** (also called a "period") of a sine wave follows its full range from zero to positive peak, back to zero, to its negative peak, and then back to zero again.

Phase is a specific time point in a wave's cycle. It's measured in degrees, with 0° representing the beginning of the cycle and 360° representing the end. The point of a positive peak is 90°, the zero point between a positive and negative peak is 180°, and the point of a negative peak is 270°.

Frequency is the number of cycles occurring in 1 second and is measured in hertz (Hz). Figure 8.4 shows two sine waves with frequencies of 2 Hz and 4 Hz.

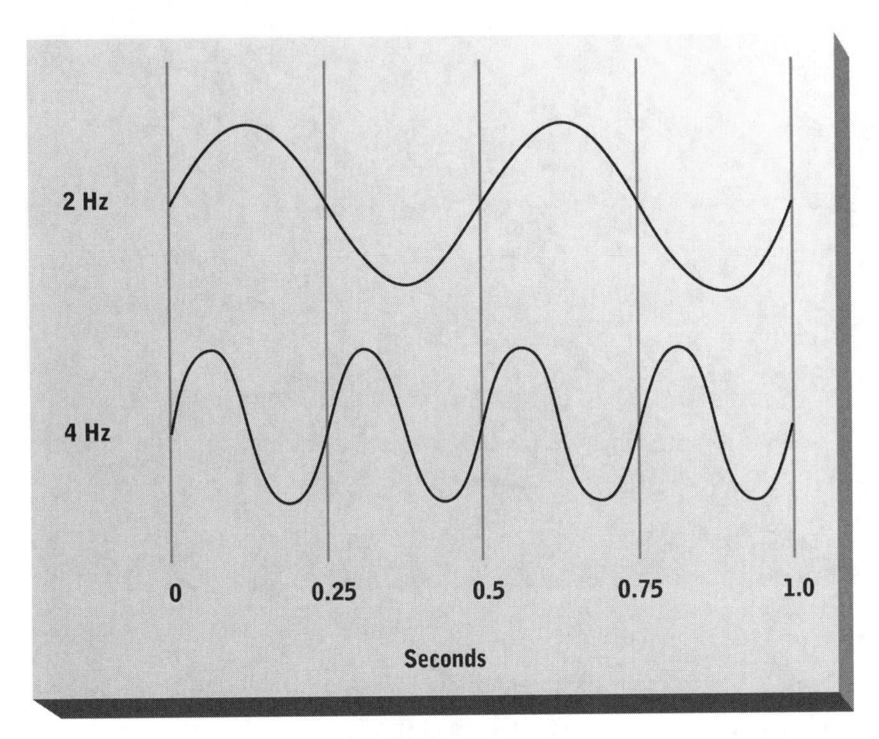

FIGURE 8.4 Sine waves with frequencies of 2 Hz and 4 Hz
Courtesy of Course Technology/Cengage Learning

Waves are important in communication for these reasons:

- Waves travel, or propagate, through space, wires, and fibers.
- Patterns can be encoded in waves.

Bits are encoded in a wave by precisely manipulating, or modulating, amplitude, frequency, or phase. A wave with encoded bits is called a **carrier wave** because it transports (carries) bits from one place to another. The receiver observes a modulated carrier wave characteristic and interprets the modulations as bit values. A **signal** is a data transmission event or group of events representing a bit or group of bits.

You can think of ripples in a pond as a simple example of carrier waves. Using a stick, the sender can generate waves of varying height that travel across the pond to a receiver on the other side. Strong strikes make large waves, and weak strikes make smaller waves. A large wave can be interpreted as a 1 bit and a small wave as a 0 bit, or vice versa. It doesn't matter which interpretation is used, as long as the sender and receiver use the same one.

Data can be encoded as bits (large and small waves) by any shared coding method. For example, text messages can be encoded with Morse code (see Table 8.1), with 1 bits representing dashes and 0 bits representing dots (see Figure 8.5). Data can also be encoded with ASCII, Unicode, or the other coding methods described in Chapter 3.

TABLE 8.1 Morse code

A	•–	N	–•	0	–––––
B	–•••	O	–––	1	•––––
C	–•–•	P	•—•	2	••–––
D	–••	Q	––•–	3	•••––
E	•	R	•–•	4	••••–
F	••–•	S	•••	5	•••••
G	––•	T	–	6	–••••
H	••••	U	••–	7	––•••
I	••	V	•••–	8	–––••
J	•–––	W	•––	9	––––•
K	–•–	X	–••–	?	••––••
L	•–••	Y	–•––	.	•–•–•–
M	––	Z	––••	,	––••––

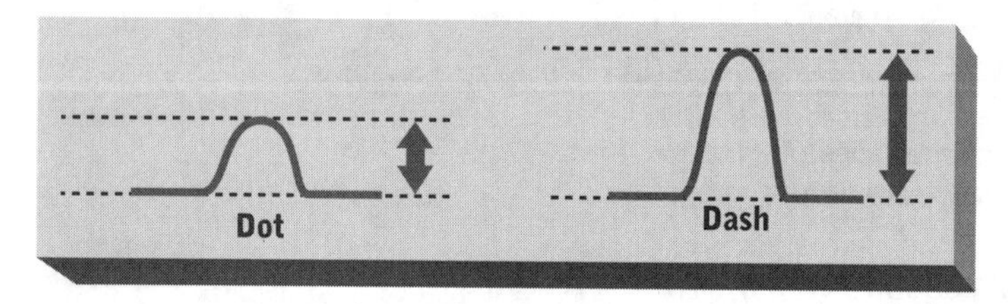

FIGURE 8.5 Wave amplitude representing Morse code dots and dashes
Courtesy of Course Technology/Cengage Learning

Modulation Methods

Amplitude modulation (AM) represents bit values as specific wave amplitudes and is sometimes called **amplitude-shift keying (ASK)**. Figure 8.6 shows an AM scheme using electricity, with 1 volt representing a 0-bit value and 10 volts representing a 1-bit value. The AM method holds frequency constant while varying amplitude to represent data. The amplitude must be maintained for at least one full wave cycle to be interpreted correctly by the receiver.

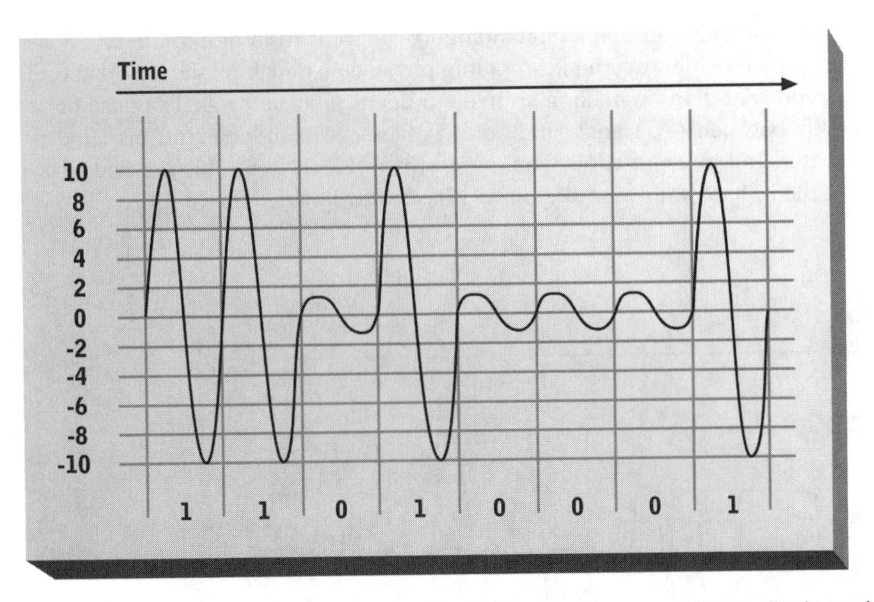

FIGURE 8.6 The bit string 11010001 encoded in a carrier wave with amplitude modulation
Courtesy of Course Technology/Cengage Learning

Frequency modulation (FM) represents bit values by varying carrier wave frequency while holding amplitude constant. It's also called **frequency-shift keying (FSK)**. Figure 8.7 shows an FM scheme, with 2 Hz representing a 0 bit and 4 Hz representing a 1 bit.

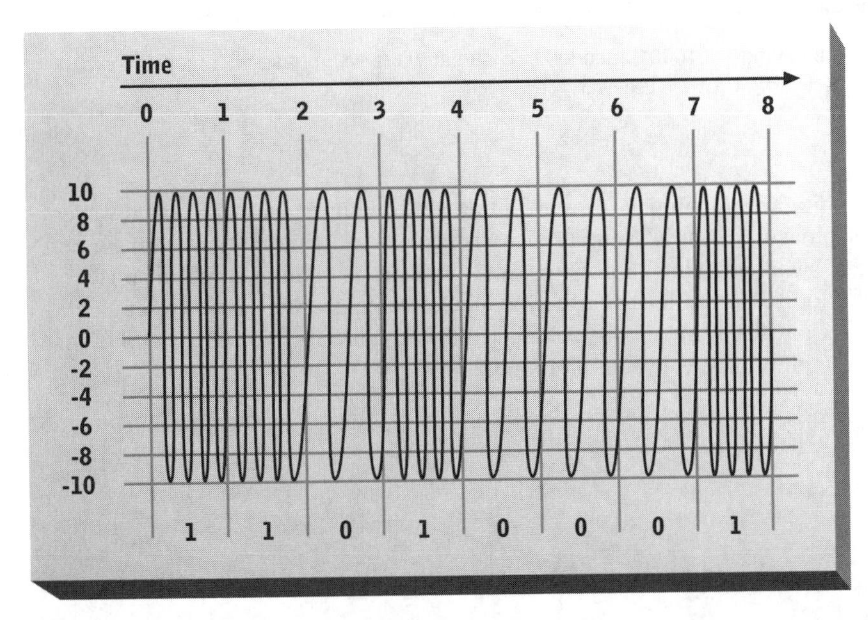

FIGURE 8.7 The bit string 11010001 encoded in a carrier wave with frequency modulation
Courtesy of Course Technology/Cengage Learning

Phase is a wave characteristic that's fundamentally different from amplitude or frequency. The nature of a sine wave makes holding phase constant impossible. However, it can be used to represent data by making an instantaneous shift in a signal's phase or by switching quickly between two signals of different phases. The sudden shift in signal phase can be detected and interpreted as data, as shown in Figure 8.8. This method of data encoding is called **phase-shift modulation** or **phase-shift keying (PSK)**.

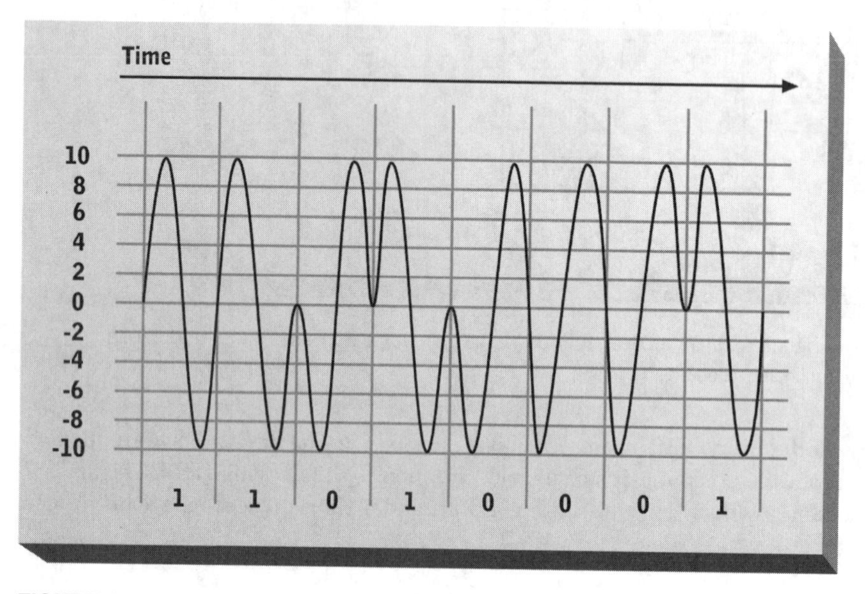

FIGURE 8.8 The bit string 11010001 encoded in a carrier wave with phase-shift modulation
Courtesy of Course Technology/Cengage Learning

Multilevel coding is a technique for embedding multiple bit values in a single wave characteristic, such as frequency or amplitude. Groups of bits are treated as a single unit for the purpose of signal encoding. For example, 2 bits can be combined into a single amplitude level if four different levels of the modulated wave characteristic are defined. Figure 8.9 shows a four-level AM coding scheme. Bit pair values of 11, 10, 01, and 00 are represented by the amplitude values 8, 6, 4, and 2.

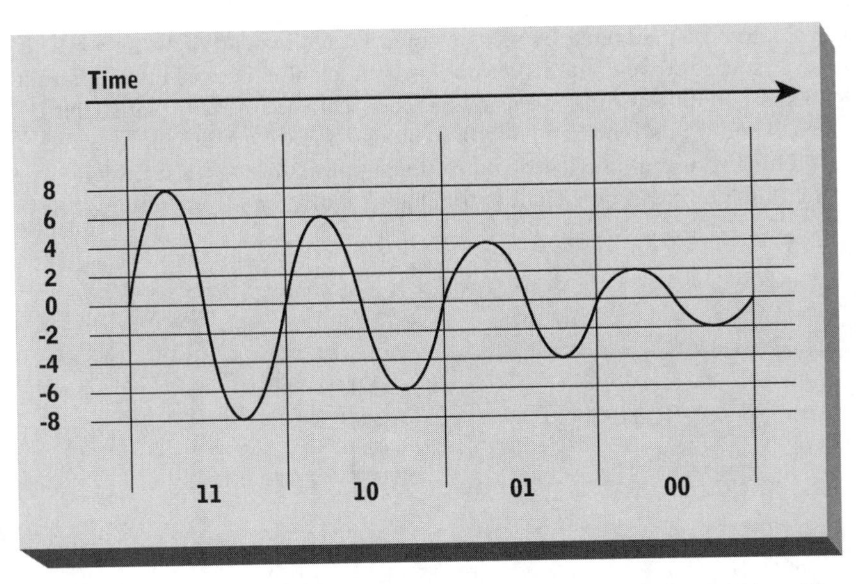

FIGURE 8.9 The bit string 11100100 encoded in a carrier wave with 2-bit multilevel coding
Courtesy of Course Technology/Cengage Learning

Analog Signals

An **analog signal** uses the full range of a carrier wave characteristic to encode continuous data values. A wave characteristic's measured value is equivalent to a data value or can be converted to a data value by a simple mathematical function. For example, say that data is to be encoded as an analog signal by using sound frequency. The numeric value 100 could be encoded by setting the wave frequency to 100 Hz, and the numeric value 9999.9 could be encoded by setting the wave frequency to 9999.9 Hz. Data values anywhere within the range of frequencies that can be generated and detected can be encoded in the signal.

Analog signals are continuous in nature, so they can represent any data value within a continuum (range) of values. In theory, the number of values that can be encoded is infinite. For example, even if the voltage range that can be sent through a wire is limited to between 0 and 10 volts, there are an infinite number of possible voltages in that range—0 volts, 10 volts, and values between 0 and 10, such as 4.1 volts, 4.001 volts, 4.00001 volts, and so forth. The number of different voltages in any range is limited only by the sender's ability to generate them, the transport mechanism's capability to carry them accurately, and the receiver's ability to distinguish between them.

Digital Signals

A **digital signal** can contain one of a finite number of possible values. A more precise term is **discrete signal**, with "discrete" meaning a countable number of possible values. The modulation schemes in Figures 8.6 and 8.7 can encode one of two different values in each signal event and are called **binary signals**. Other possible digital signals include trinary (three values), quadrary (four values), and so forth. The modulation scheme in Figure 8.9 generates a quadrary signal.

Digital signals can also be generated by using a square wave instead of a sine wave as the carrier wave. A square wave contains abrupt amplitude shifts between two different values. Figure 8.10 shows data transmission using an electrical square wave, with 1 bits represented by 5 volts and 0 bits represented by no voltage. Square waves can be generated by rapidly switching (pulsing) an electrical or optical power source—a technique called **on-off keying (OOK)**. In essence, OOK is the digital equivalent of amplitude modulation.

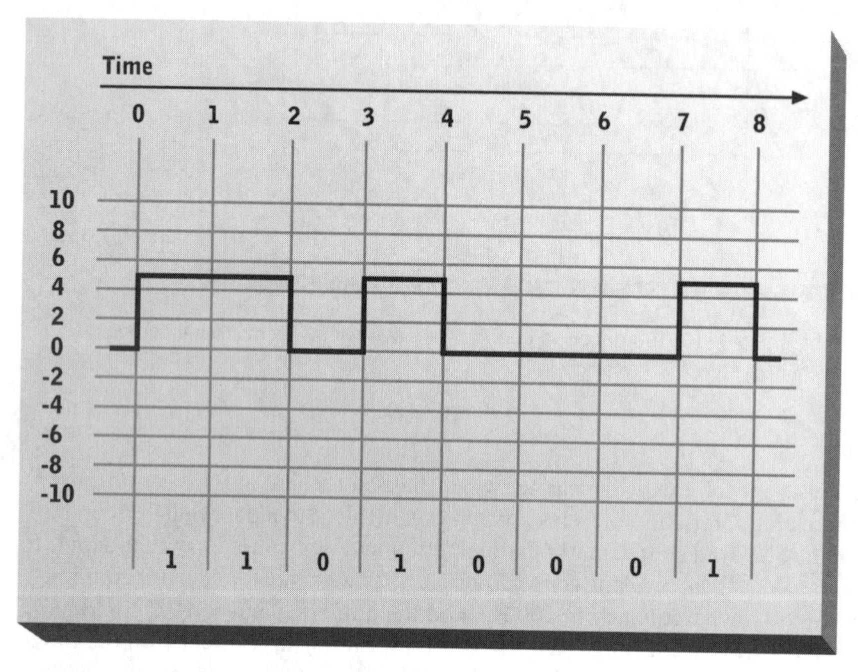

FIGURE 8.10 The bit string 11010001 encoded in square waves (digital signals)
Courtesy of Course Technology/Cengage Learning

Square waves are the preferred method of transmitting digital data over short distances, such as on a system bus. However, electrical square waves can't be transmitted reliably over long distances (more than a kilometer, for example). Power loss, electromagnetic interference, and noise generated in the wiring combine to round off the sharp edges of a square wave (see Figure 8.11). Pulses also tend to spread out because of interactions with the transmission medium. Without abrupt voltage level changes, receivers can't decode data values reliably. Optical square waves are subject to similar problems, although only over much longer distances. They can be transmitted reliably over distances up to a few dozen kilometers through high-quality optical fibers.

NOTE

In most communication and network standards, distance is stated in metric units. One meter is about 39 inches, and one kilometer is approximately 0.625, or 5/8, of a mile.

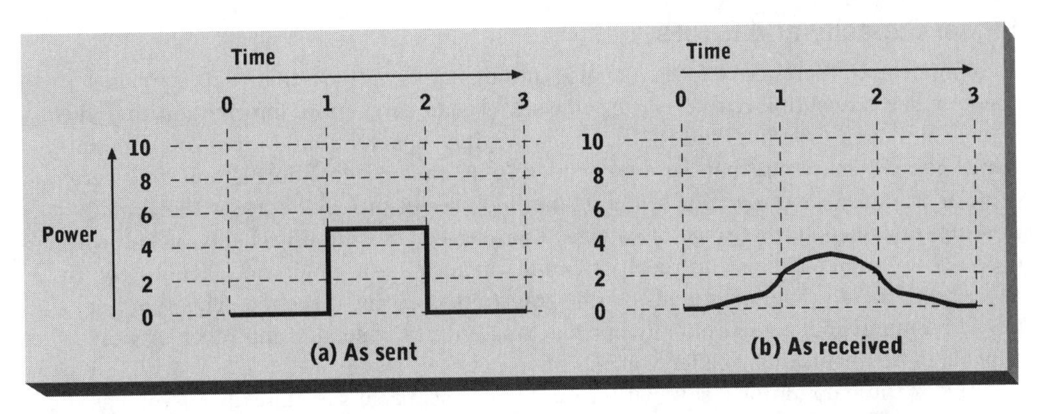

FIGURE 8.11 A square wave before (a) and after (b) transmission over a long distance
Courtesy of Course Technology/Cengage Learning

Wave characteristics, such as frequency and amplitude, aren't inherently discrete. For example, electrical voltage can be 0, 0.1, 0.001, 5, 10, 100, and many other values. To send a digital signal with electrical voltage, sender and receiver choose two values, such as 0 and 5, to represent two different bit values. However, the voltage that reaches the receiver might not be exactly 0 or 5 for a number of reasons, discussed in the next section. How should the receiver interpret voltage that doesn't precisely match the assigned values for 0 and 1 bits? For example, should 3.2 volts be interpreted as 0, 1, or nothing?

A digital signaling scheme defines a range of wave characteristic values to represent each bit value. For example, Figure 8.12 shows two ranges of voltage. Any value in the lower range is interpreted as a 0 bit, and any value in the upper range is interpreted as a 1 bit. The dividing line between the two ranges is called a "threshold." The sender encodes a bit value by sending a specific voltage, such as 0 volts for a 0 bit or 5 volts for a 1 bit. The receiver interprets the voltage by comparing it with the threshold. If the received voltage is above the threshold, the signal is assumed to contain a 1 bit. If it's below the threshold, the signal is assumed to contain a 0 bit. In this figure, 0, 0.1, 1.8, and 2.4999 volts, for example, are interpreted as 0 bits.

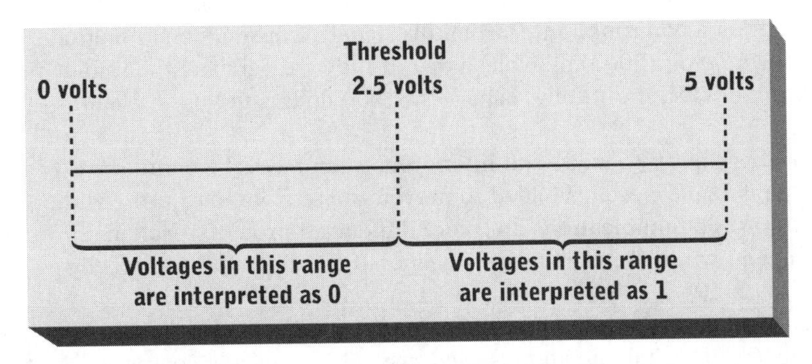

FIGURE 8.12 A binary signaling method using voltage ranges
Courtesy of Course Technology/Cengage Learning

Signal Capacity and Errors

Two important differences between analog and digital signals are their data-carrying capacity and susceptibility to error. Analog signals can carry more information than digital signals in a fixed time interval. Higher data-carrying capacity results from the large number of possible messages that can be encoded in an analog signal during one time period.

For example, say an analog signaling method uses electrical voltage with a signal duration of 1 second, the range of voltages that can be transmitted reliably on the wire is 0 to 127 volts, and sender and receiver can distinguish 1-volt differences accurately. In this situation, 128 different signals can be transmitted during 1 second. This number would be much larger, theoretically approaching infinity, if sender and receiver were capable of distinguishing smaller voltage differences.

Under these conditions, say you have a binary electrical signal, using 64 volts as the threshold. During a transmission event, the number of possible data values sent is only two (a 0 or a 1). In this example, the analog signal's data-carrying capacity is 64 times greater than the binary signal: 128 possible values with analog divided by 2 possible values with binary.

To transmit values larger than 1 with binary signals, sender and receiver combine adjacent signals to form a single data value. For example, the sender could transmit seven different binary signals in succession, and the receiver could combine them to form one numeric value. With this method, it's possible to transmit 128 different messages ($2^7 =$ 128 possible combinations of seven binary signals). However, an analog signal could have communicated any of 128 different values in *each* signal.

Although analog signals have higher data-carrying capacity than digital signals, they're more susceptible to transmission error. If the mechanisms for encoding, transmitting, and decoding electrical analog signals were perfect, transmission error wouldn't be an issue, but errors are always possible. Electrical signals and devices are subject to noise and disruption because of electrical and magnetic disturbances. The noise you hear on a radio during a thunderstorm is an example of these disturbances. Voltage, amperage, and other electrical wave characteristics can be altered easily by this interference.

For example, your bank's computer communicates with its ATMs via analog electrical signals. You're in the process of making a $100 withdrawal. The value $100 is sent from the ATM to the bank's computer by a signal of 100 millivolts. The computer checks your balance and decides you have enough money, so it sends an analog signal of 100 millivolts back to the ATM as a confirmation. During this signal's transmission, a bolt of lightning strikes, inducing a 2-volt (2000 millivolt) surge in the wire carrying the signal. When the signal reaches the ATM, it dutifully dispenses $2000 dollars to you in clean, crisp $10 and $20 bills.

Electrical signals' susceptibility to noise and interference can never be eliminated completely. Computer equipment is well shielded to prevent noise from interfering with internal signals, but external communications are more difficult to protect. Also, errors can result from resistance in wires or magnetic fields generated in a device, such as by a transformer or fan.

A digital electrical signal isn't nearly as susceptible to noise and interference. In the ATM example, what happens if digital encoding is used instead of analog encoding? With a threshold of 2.5 volts (and using 0 volts to represent 0 and 5 volts to represent 1), a voltage surge less than 2.5 volts can be tolerated without misinterpreting a signal (see Figure 8.13). If a 0 is sent at the instant the lightning strikes, the ATM still interprets the signal as a 0 because the 2 volts induced by the lightning are below the threshold value 2.5. If a 1 is sent, the lightning raises the voltage from 5 to 7 volts, a difference that's still above the threshold. Resistance in the wire or other factors degrading voltage aren't a problem as long as the voltage doesn't drop more than 2.5 volts.

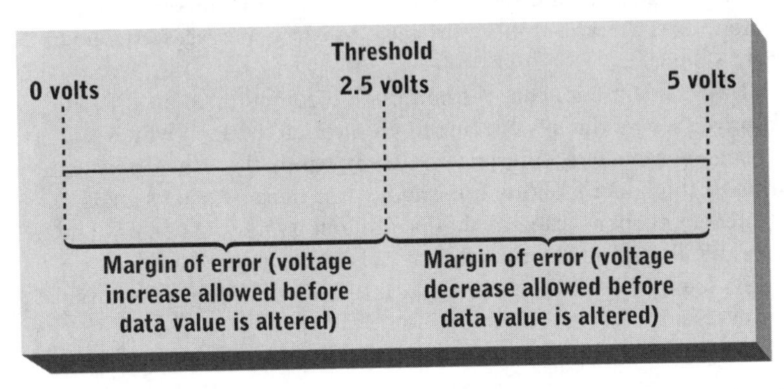

FIGURE 8.13 Margin of transmission error (voltage drop or surge) before the data value encoded in a digital binary signal is altered
Courtesy of Course Technology/Cengage Learning

TRANSMISSION MEDIA

A **communication channel** consists of a sending device, a receiving device, and the transmission medium connecting them. In a less physical sense, it includes the communication protocols used in the channel. Figure 8.14 shows a simple channel design, but most channels have a complex construction, with multiple media segments using different signal types and communication protocols. As a message moves from one media segment to another, it must be altered to conform to the next segment's signal characteristics and protocol.

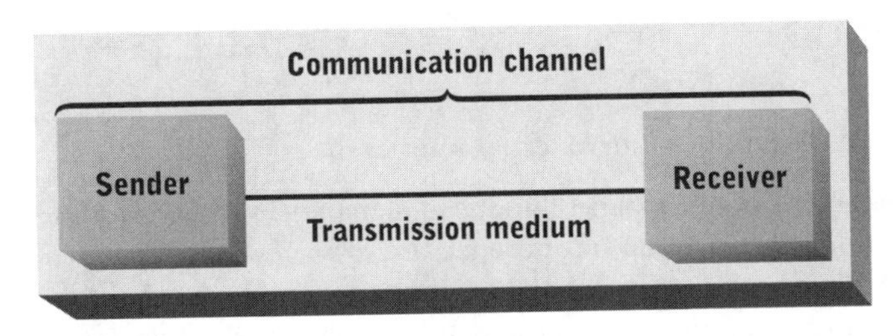

FIGURE 8.14 Elements of a communication channel
Courtesy of Course Technology/Cengage Learning

The communication path that transports signals is called a **transmission medium**. Transmission can be guided or unguided. **Guided transmission** routes signals between two locations through a physical connection, such as copper wire or optical fiber. It's also referred to as **wired transmission**, even when optical fiber (called **fiber-optic cable**) is the transmission medium. Messages must be encoded in signals that can be conducted or propagated through the transmission medium. In copper wires, signals are streams of electrons; in fiber-optic cable, signals are pulses of light. **Unguided transmission**, also called **wireless transmission**, uses the atmosphere or space to carry messages encoded in radio frequency or light signals.

Figure 8.15 shows a DSL modem connection from a home computer to an ISP. Digital electrical signals travel across the system bus to an internal modem, which converts these signals to analog signals in different frequency bands that travel through twisted-pair copper wiring to the house's phone interface. From there, analog signals travel via lower-gauge (thicker) copper wiring to the local phone-switching center. The switching center converts the analog signals to digital signals and routes them via fiber-optic cable to the ISP. Messages sent from the ISP to the home computer travel the same path in reverse.

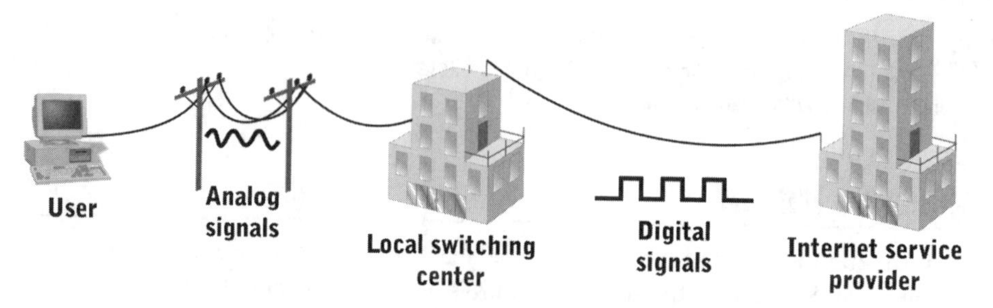

FIGURE 8.15 A DSL modem connection to an ISP
Courtesy of Course Technology/Cengage Learning

Characteristics of transmission media that affect their capability to transmit messages successfully and efficiently include the following:

- Speed and capacity
- Frequency
- Bandwidth
- Noise, distortion, and susceptibility to external interference

In general, transmission media are better when their speed, capacity, and bandwidth are high, and their noise, distortion, and external interference are low. However, media with all these desirable characteristics are costly. Reliability, efficiency, and cost also depend on the signaling method and communication protocol. Varying combinations of transmission medium, signaling method, and protocol are cost effective for different types of communication links. These differences account for the variety of media, signaling methods, and protocols that form the communication channel between a home computer and an ISP or between any sender and receiver.

Speed and Capacity

The fundamental transmission speed limit for any medium is the rate at which a carrier wave propagates through the medium. Electrical signals travel through wires at close to the speed of light. Optical signals transmitted through fibers and radio frequencies transmitted through the atmosphere or space also travel at close to the speed of light. Raw speed varies little in commonly used transmission media. What does vary is length of the media, the ways in which media segments are interconnected, and the rate at which bits are encoded in signals and recognized by the receiver. These factors account for transmission speed variations in different media.

Speed and capacity are interdependent: A faster communication channel has a higher transmission capacity per time unit than a slower channel does. However, capacity isn't solely a function of transmission speed. It's also affected by the efficiency with which the channel is used. Different communication protocols and methods of dividing a channel to carry multiple signals affect transmission capacity.

A channel's speed and data transmission capacity are jointly described as a data transfer rate. A **raw data transfer rate** is the maximum number of bits or bytes per second that the channel can carry. This rate ignores the communication protocol and assumes error-free transmission.

A net or **effective data transfer rate** describes the transmission capacity actually achieved with a communication protocol. It's always less than the raw data transfer rate because no medium is error free and because most communication protocols use part of the channel capacity for tasks other than transmitting raw data. Examples include sending command messages and retransmitting data when errors are detected.

Frequency

Carrier wave frequency is a basic measure of data-carrying capacity. It limits data-carrying capacity because a change in amplitude, frequency, or phase must be held constant for at least one wave cycle to be detected reliably by a receiver. If a single bit is encoded in each wave cycle, the raw data transfer rate is equivalent to the carrier wave frequency. If multiple bits are encoded in each wave cycle via multilevel coding, the raw data transfer rate is an integer multiple of the carrier wave frequency.

Figure 8.16 shows the range of electromagnetic frequencies and the subsets in this range. The term **radio frequency (RF)** describes transmissions using frequencies between 50 Hz and 1 terahertz (THz). Wireless networks and cell phones use the upper end of the RF spectrum, called shortwave radio. Frequencies above 1 THz and below 10,000 THz include infrared, visible, and ultraviolet light. Because these frequencies don't penetrate most materials, they're used less often for data communication. Ships sometimes use light to transmit Morse code messages over short distances, and high-speed digital transmission can use precisely aligned laser light sources and receivers.

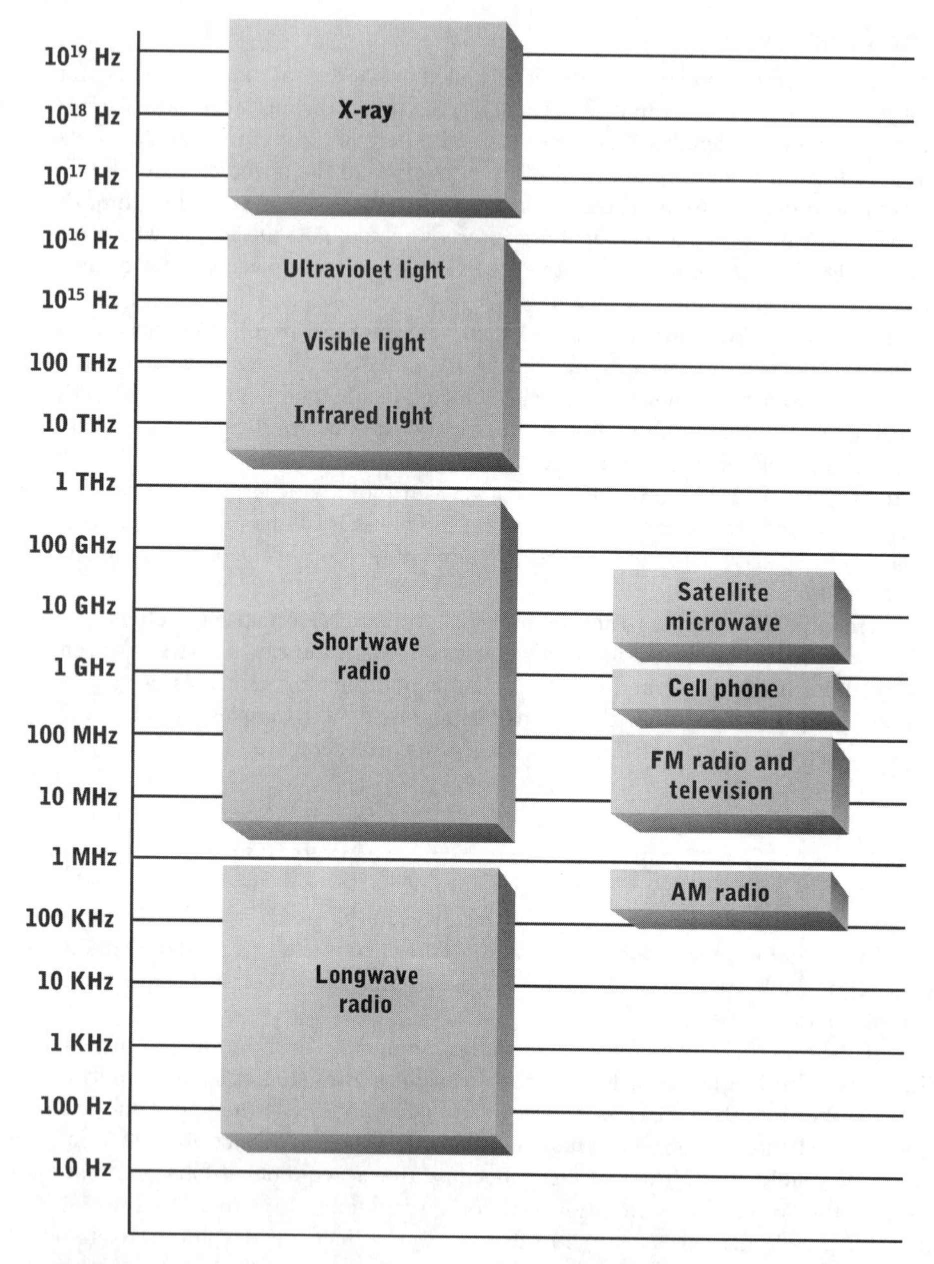

FIGURE 8.16 The spectrum of electromagnetic frequency between 10^1 and 10^{19} Hz
Courtesy of Course Technology/Cengage Learning

Signals can also be encoded in electromagnetic frequencies transmitted through metal, plastic, or glass wires. RF energy is typically transmitted through copper wires, which can propagate frequencies up to around 1 GHz. Light energy is transmitted through plastic or glass fibers; these optical fibers can propagate frequencies up to 10,000 THz.

NOTE

Fiber optics have a higher potential data transmission capacity because light is a higher-frequency carrier wave. However, note the word "potential." Achieving the bit transmission rates implied by carrier wave frequency requires encoding and decoding equipment that can operate at the carrier wave frequency. If the transmission's source or destination is an electrical device, such as a conventional phone or a computer's CPU, signals must be converted between electrical and optical formats, which limits the effective data transmission rate.

Bandwidth

The difference between a signal's maximum and minimum frequencies is called the signal **bandwidth**. The difference between the maximum and minimum frequencies that can be propagated through a transmission medium is called the medium bandwidth. For example, copper wiring in phone circuits can typically carry electromagnetic frequencies between 0 and 100 MHz. The medium bandwidth is computed by subtracting the minimum frequency from the maximum: 100 MHz - 0 Hz = 100 MHz. A similar calculation for an optical fiber carrying visible light is approximately 790 THz - 400 THz = 390 THz.

Like maximum signal frequency, channel (or medium) bandwidth determines the raw data transfer rate. To understand how, you must first understand the concept of a **composite signal** (or complex signal), a signal created by combining multiple simple signals. As an example of the difference between simple and composite acoustic signals, think of a battery-operated smoke alarm, which has an annoying but necessary tendency to emit short high-frequency beeps when its batteries need to be replaced. The amplified signal you hear is a single sine wave, typically at a frequency of 5000 Hz. No one would mistake the beep for music, but it does serve the intended purpose of motivating you to replace the battery.

Now consider the acoustic signal you hear when listening to an orchestra. Instruments, such as violins, produce composite acoustic signals, even when sounding a single note. Each note is more than a single pure frequency; it's a mix of fundamental and harmonic frequencies produced by parts of the instrument vibrating at different frequencies and amplitudes. Each instrument produces notes in a specific range of frequencies and has a unique sonic character defined by the frequency and amplitude of the acoustic waves produced for each note. Each instrument produces a complex signal that merges with those of other instruments to produce an even more complex signal.

Both the smoke alarm's beep and the orchestra's composite signal have information content. The beep carries a single message: "Change the battery." Although the interpretation of music is subjective, clearly it carries a more complex message than a single-frequency beep. You can hear separate instruments and voices in the composite signal, and they can send messages that stir complex emotions, as in a movie soundtrack. Similarly, the bass and drums of popular music can motivate you to dance with their rhythmic low-frequency signals, and vocals deliver other messages at higher frequencies.

Returning to the concept of bandwidth, the smoke alarm signal's bandwidth is very narrow: a specific single frequency—say, 1 Hz. In contrast, an orchestra's bandwidth spans the entire range of human hearing: 20 to 20,000 Hz, or a bandwidth of 19,800 Hz. This larger bandwidth enables transmitting more information per time interval.

Data transmission with analog or digital signals has the same basic relationship between bandwidth and data-carrying capacity. For analog signals, wider bandwidth enables the transmission medium to carry multiple signals of different frequencies at the same time, which merge to form a composite signal that occupies more bandwidth than any of the component signals. Because each frequency carries a different data stream, data transfer rate increases as bandwidth increases.

Higher bandwidth also implies higher data transfer rates for digital signals, although for subtly different reasons. Recall that digital signals are sent by using square waves, which are essentially short, powerful bursts of electrical or optical energy. Also, electrical and optical energy propagate through space, wires, or fibers as sine waves. So how can energy that propagates as a sine wave be used to generate a square wave?

The answer was first described mathematically by Joseph Fourier, a mathematician and physicist in the late 18th and early 19th centuries. He proved that all complex signals could be decomposed into a large or an infinite set of single-frequency signals of varying amplitude and phase by a mathematical technique now known as Fourier analysis. Conversely, a signal such as a complex square wave can be created by combining sine waves of varying frequencies, amplitudes, and phases.

According to Fourier, there are an infinite number of sine waves across an infinite number of frequencies composing a perfectly shaped square wave, so Fourier analysis implies that a square wave requires infinite bandwidth to carry all the component sine waves. Fortunately, narrower bandwidths can produce a usable square wave approximation, but the quality of this approximation does increase with higher bandwidth and maximum signal frequency. Higher-quality square wave approximations can be transmitted at higher rates and travel longer distances without degradation. Therefore, as with analog signals, raw data transmission rate increases as bandwidth increases.

A **modulator-demodulator (modem)** sends digital signals by modulating a carrier wave to embed bits in one or more analog wave characteristics. A common and particularly complex example is a dial-up modem, which carries digital signals over voice-grade phone channels. These channels have a minimum bandwidth of 3100 Hz (3400 Hz - 300 Hz), which provides an acceptable approximation of human speech when using analog signals. Early dial-up modems used four separate frequencies to carry bit values (see Table 8.2). The lowest frequency (1070 Hz) in effect created an upper limit on transmission speed. Assuming a single bit is transmitted per signal cycle, the maximum transmission rate is 1070 bps, although lower bit-encoding rates were used to increase reliability.

TABLE 8.2 Frequency assignments for data transmission over 300 bps analog phone lines

Mode	Signal frequency (Hz)	Binary value
Transmit	1070	0
Transmit	1270	1
Receive	2025	0
Receive	2225	1

Subsequent modem standards have exceeded this theoretical capacity, but they have had to use more complex schemes, such as the following, to achieve raw data transmission rates as high as 56,000 bps:

- Hardware-based data compression
- Multilevel coding of up to 32 bits in a single carrier wave cycle
- An elaborate combination of amplitude and phase modulation

Digital signals transmitted over ordinary phone wires can achieve megabits per second (Mbps) transmission rates, depending on distance, which seems to contradict earlier statements about the capacity of analog and digital signals. This contradiction arises because analog phone-signaling standards don't use all the bandwidth available in a modern phone circuit. Today's phone wiring can transmit frequencies much higher than 3400 Hz, but older analog phone-switching equipment was designed to use only the frequencies between 300 and 3400 Hz. Most of this equipment has been replaced with digital switching equipment that converts electrical voice signals to digital form. Bandwidth above 3400 Hz can be used to transmit digital signals as long as analog voice signals below 3400 Hz are filtered out. Digital subscriber lines use bandwidth above 3400 Hz to encode digital data signals at raw transmission rates up to 20 Mbps.

Signal-to-Noise Ratio

In a communication channel, **noise** refers to unwanted signal components added to the data signal that might be interpreted incorrectly as data. Noise can be heard in many common household devices. Turn on a radio and tune it to an AM frequency (channel) on which no local station is transmitting. The hissing sound is amplified background RF noise. Internally generated noise can also be heard on an iPod or a home stereo system. On an iPod, turn up the volume all the way without playing a song. On a home stereo system, set the amplifier or receiver input to a device that isn't turned on and turn the volume up high. (In either case, be sure to turn the volume back down when you're finished!) The hissing sound you hear is amplified noise in the signal transmission and amplification circuitry. Any sound other than hissing is externally generated noise picked up by the amplifier, speakers, or cabling between them.

Noise can be introduced into copper, aluminum, and other wire types by **electromagnetic interference (EMI)**. EMI can be produced by a variety of sources, including electric motors, radio equipment, and nearby power or communication lines. In an area dense with wires and cables, EMI problems are compounded because each wire both generates and receives EMI. Shielding transmission wires with an extra layer of metallic material, such as a braided fabric or metal conduit (pipe), reduces but doesn't entirely eliminate EMI problems.

Attenuation is the reduction in signal strength (amplitude) caused by interactions between the signal's energy and the transmission medium. It occurs with all signal and transmission medium combinations, although different combinations exhibit different amounts of attenuation. Electrical resistance is the main cause of attenuation with electrical signals and transmission media. Both optical and radio signals exhibit high attenuation when transmitted through the atmosphere; optical signals exhibit much lower attenuation when transmitted through fiber-optic cable. For all signal types, attenuation is proportional to the medium's length. Doubling the length doubles the total attenuation.

Another source of errors in communication is **distortion**—changes to the data signal caused by interaction with the communication channel. The transmission medium is a main source of distortion, although auxiliary equipment, such as amplifiers, repeaters, and switches, can also distort the signal. There are many types of distortion, including resonance, echoes, and selective attenuation. Resonances in a transmission medium amplify certain portions of a complex signal in much the same way that certain bass notes can make a car or room vibrate. Echoes occur when a signal bounces off a hard surface and intermixes with later portions of the same signal. Attenuation in any medium varies with signal frequency. In general, high-frequency analog signals and the high-frequency components of digital signals attenuate more rapidly than lower frequency signals do.

For a receiving device to interpret encoded data correctly, it must be able to distinguish encoded bits from noise and distortion. Distinguishing valid signals from extraneous noise becomes more difficult as transmission speed increases. For example, when listening to a speech, the difficulty or ease of understanding it is related to the speed at which it's delivered and its volume (amplitude) in relation to background noise. Accurate speech reception is impaired by sources of noise, such as other people talking, an air-conditioning fan, or a nearby construction project. The speaker can compensate for noise by increasing speech volume, thus increasing the signal-to-noise ratio. Accurate reception is also impaired if the speaker talks too quickly because the listener has too little time to interpret one word or speech fragment before receiving the next.

A channel's effective speed limit is determined by the power of the data-carrying signal in relation to the power of noise in the channel. This relationship is called the channel's **signal-to-noise (S/N) ratio**. S/N ratio is measured at the channel's receiving end or at a specific distance from the sender, usually in units of signal power called decibels (dB), and is computed as follows:

$$\text{S/N ratio} = \text{signal power} - \text{noise power}$$

The term "ratio" usually implies dividing one value by another. However, the decibel scale used to measure signal and noise power is logarithmic. When logarithms are used, division is reduced to subtraction as follows:

$$\frac{x}{y} = \log x - \log y$$

S/N ratio incorporates the effects of signal attenuation and added noise. For example, say a channel has the following characteristics:

- The sender generates a signal at 80 dB of power.
- The signal attenuates at a rate of 10 dB per kilometer.
- The channel has a noise level of 20 dB at the sending end.
- The channel adds 5 dB of noise per kilometer.

In this example, S/N ratio is 60 dB (80 dB - 20 dB) at the sending end. S/N ratio gradually decreases as the signal moves through the channel because signal power drops and noise power rises. Figure 8.17 plots signal power with the solid line and noise power with the dashed line over various transmission distances. At 2 kilometers, S/N ratio is 30 dB, computed as follows:

$$
\begin{aligned}
\text{S/N ratio} &= \text{signal power - noise power} \\
&= (\text{initial signal power - attenuation}) - (\text{initial noise power} + \text{added noise}) \\
&= (80 - (10 \times 2)) - (20 + (5 \times 2)) \\
&= 60 - 30 \\
&= 30 \text{ dB}
\end{aligned}
$$

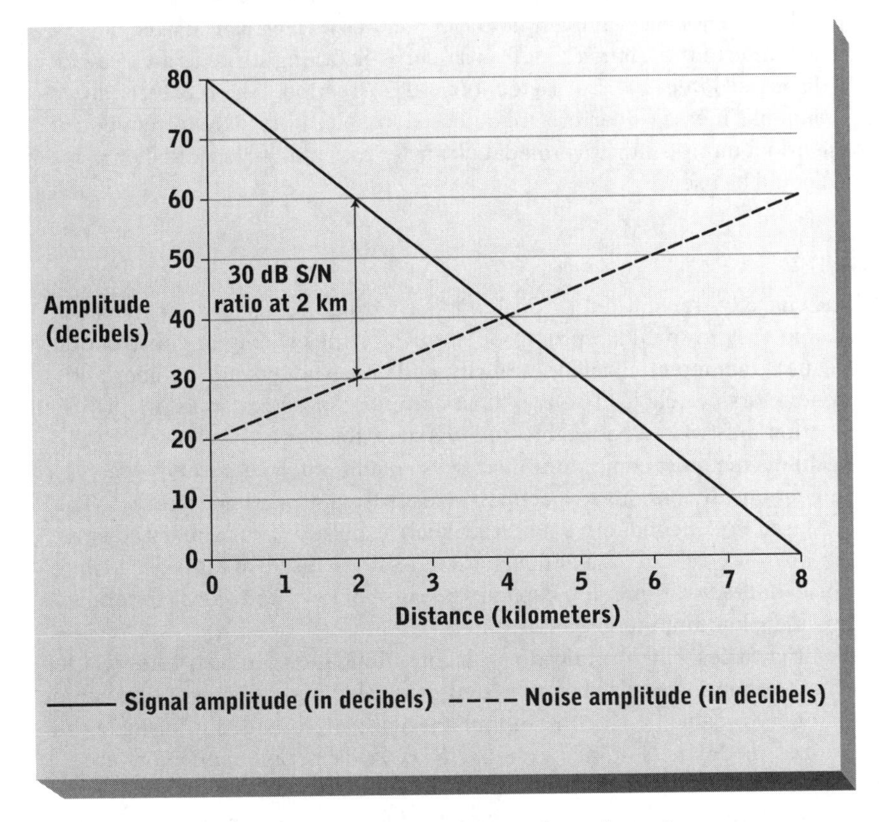

FIGURE 8.17 S/N ratio as a function of distance for a channel
Courtesy of Course Technology/Cengage Learning

Successful transmission is theoretically impossible beyond 4 kilometers because the S/N ratio is negative. In theory, the maximum possible (usable) channel length is 4 kilometers. However, receiving devices usually need some amount of positive S/N ratio for reliable operation. If the receiving device in the channel described previously requires at least 15 dB for error-free operation, the maximum usable channel length is 3 kilometers.

Each transmitted bit is a period of time during which a signal representing a 0 or 1 is on the channel. As transmission speed is increased, the duration of each bit in the signal, known as the **bit time**, decreases. If signal-generating equipment could generate a full-amplitude signal instantaneously, raw data transfer rates could always be increased by shortening bit time. However, no device, including lasers, electrical circuits, and the human voice, can go from zero amplitude to full power instantaneously. Short bit times don't give the signal-generating device enough time to ramp up to full power before the next bit must be transmitted. S/N ratio decreases because the signal amplitude during each bit time slice is decreased. Eventually, a limit is reached at which the bit time is so short that the signal is no louder than background noise. At this point, the S/N ratio becomes zero.

As higher speeds are attempted, the error rate increases. A transmission error represents a wasted opportunity to send a message, thus reducing the effective data transfer rate. Another difficulty is that the noise level might not be constant. In electrical signals, noise usually occurs in short intermittent bursts, such as a nearby lightning strike or an electric motor starting. As discussed later in "Error Detection and Correction," the receiver can request retransmission of a message if errors are detected. If noise bursts are infrequent, retransmissions might not diminish the overall data transfer rate much, and a higher raw transmission speed could be used.

Electrical Cabling

Electrical signals are usually transmitted through copper wiring because it's inexpensive, highly conductive, and easy to manufacture. Copper wiring varies widely in construction details, including gauge (diameter), insulation, purity and consistency, and number and configuration of conductors per cable. The two most common forms of copper wiring used for data communication are twisted-pair cable and coaxial cable.

Twisted-pair cable is the most common transmission medium for phone and local area network (LAN) connections. It contains two copper wires twisted around one another. The wires are usually encased in nonconductive material, such as plastic. The primary advantages of twisted pair are low cost and ease of installation. Its main disadvantages are high susceptibility to noise, limited transmission capacity because of low bandwidth (generally less than 250 MHz), and a low amplitude (voltage) limit.

The Electronics Industries Alliance and the Telecommunications Industry Association have defined the most common standards for twisted-pair cable and connectors, including **Category 5** and **Category 6** cable. Both contain four twisted pairs bundled in a single thin cable, and standardized modular RJ-45 jacks similar to modular phone connectors are used at both ends (see Figure 8.18). Category 6 is the latest widely used standard, although Category 5 is more common because it was used for many years, and pulling new cable through walls and ceilings is an expensive task. Category 6 cable can transmit at speeds up to 1 Gbps (250 Mbps on each wire pair) over long distances and 10 Gbps over distances up to a few meters. Under optimal conditions, Category 5 cable can achieve similar speeds, although not as reliably as Category 6 cable.

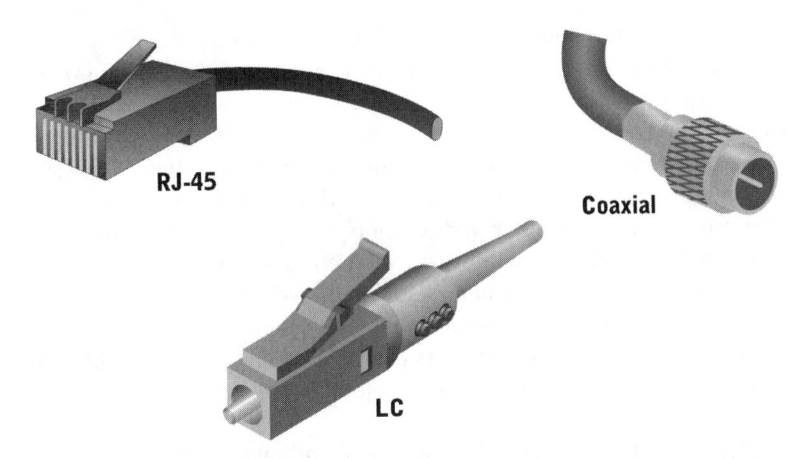

FIGURE 8.18 Common cable connectors
Courtesy of Course Technology/Cengage Learning

Coaxial cable contains a single copper conductor surrounded by a thick plastic insulator, a metallic shield, and a tough plastic outer wrapping. Because of its heavy shielding, coaxial cable is very resistant to EMI. It has a high bandwidth of up to 500 MHz, lower attenuation than twisted pair, and high data transmission capacity, especially over long distances. These characteristics make coaxial cable well suited for cable TV wiring.

Its main disadvantages are that it's more expensive and harder to install than Category 6 twisted-pair cable. Coaxial modular connectors (see Figure 8.18) aren't as easy to install as RJ-45 connectors. Coaxial cable is also stiff, making it more difficult to route through conduits and around corners than Category 6 cable. Coaxial cable isn't used often in networks now because twisted-pair quality has improved, and costs of twisted-pair and fiber-optic cable and devices have declined.

Twin-axial cable is similar to coaxial cable, except that it contains two internal conductors and is thinner. It supports higher data transmission rates and is less susceptible to EMI because the signal and return wire are heavily shielded. Bundles of twin-axial cables are sometimes used for high-speed network connections over short distances.

Optical Cabling

Fiber-optic cable contains strands of light-conducting filaments made of plastic or glass. Electrical shielding isn't used because light waves neither generate nor respond to EMI. A tough plastic coating protects the inner fibers from physical damage. In cables with many glass fiber strands, special construction techniques are required to avoid breaking the glass. Thick, inflexible plastic coatings are used to prevent stretching and breaking the glass fibers, and additional plastic or gels can be embedded in the cable for more protection.

Fiber-optic cables come in three basic types: multimode step-index, multimode graded-index, and single-mode. Multimode cables have an optical fiber with a large diameter, compared with the wavelength of the light signals they carry. The density of optical fibers is different from the surrounding material (cladding). This density difference causes light that might otherwise "escape" the optical fibers to reflect back into the fiber. Light can travel straight down the center of a multimode optical fiber, but more often it follows a zigzag path, reflecting many times per centimeter. Each reflection attenuates the signal because the cladding absorbs energy.

Multimode optical fibers can be constructed with glass or plastic. Plastic fibers are less expensive, although they generally have higher attenuation and dispersion and are suitable only for short-distance communication. However, plastic fibers have some performance advantages over glass fibers, including more capability to bend and more tolerance for scratches and other physical damage. Multimode cables support lower transmission rates and distances than single-mode cables because they disperse and attenuate signals more rapidly. Dispersion occurs because different components of the optical signal travel different lengths as they move through the cable, resulting in a signal shape similar to the one in Figure 8.11.

In **multimode step-index cable**, both the optical fiber and cladding have different but uniform densities throughout the cable, which results in many reflections. **Multimode graded-index cable** uses more expensive optical fibers with slightly lower density in the center than at the edge. This variation in density from center to edge tends to bend light waves away from the cladding before they're reflected, thus reducing, but not eliminating, the number of reflections per centimeter.

Single-mode cable uses optical fibers with a much smaller diameter that vary continuously in density from center to edge. Although these fibers are more expensive than multimode fibers, they eliminate reflections and reduce the variation in distance that different components of the optical signal must travel. The result is much lower attenuation and dispersion, which increases raw data transmission rates and/or transmission distances.

All optical cables are more expensive than copper cables for equivalent distances and are more difficult to install and connect. Therefore, copper cabling is preferred for most short-haul connections, such as within buildings. Fiber-optic cables are generally used where their superior data transmission rate or transmission distance is most valuable. Examples include network wiring between high-capacity servers in the same rack or room, high-capacity connections in or between buildings, and communication over distances more than a few hundred meters.

NOTE

Despite their higher cost, optical cables are often used in new or refurbished buildings because cable cost is only one part of wiring's overall cost. For new networks, designing the network and pulling cable through ceilings, walls, and conduits are a larger part of the total cost than cable cost is. Further, optical cable cost continues to decline, so optical cabling is expected to gradually supplant copper wire for all data communication applications over the next decade.

Amplifiers and Repeaters

No cable type can carry signals more than a few dozen kilometers at high transmission rates with low error rates. Transmitting data over longer distances requires devices such as amplifiers and repeaters to increase signal strength and remove unwanted noise and distortion.

An **amplifier** increases a signal's strength (amplitude) and can extend a signal's range by boosting signal power to overcome attenuation. However, the effective length over which the signal travels can't be increased indefinitely; it's limited by two factors. The first is noise and interference introduced during transmission. An amplifier amplifies whatever signal is present at its input. If this signal includes noise along with the intended message, the noise as well as the message is amplified. In addition, amplifiers are never perfect. Some distortion or noise is generally introduced in the amplification process. A signal that's amplified many times contains noise from multiple transmission line segments and distortion introduced by each stage of amplification. Eventually, reduced signal quality and S/N ratio render the signal useless.

A **repeater** performs much the same function as an amplifier but operates on a different principle. Rather than amplify whatever signal is sent to it, a repeater extracts the data embedded in the signal it receives and retransmits a new signal containing the same data. An advantage of this device is that it doesn't retransmit noise or distortion. As long as the noise introduced since the last transmitting or repeating device doesn't cause misinterpretation of the data, the retransmitted data is the same as the original. Repeaters are typically required every 2 to 5 kilometers for coaxial cable and every 40 or 50 kilometers for single-mode fiber-optic cable.

WIRELESS TRANSMISSION

Wireless transmission shares many of the characteristics of wired media, including tradeoffs among data transfer capacity, signal frequency, bandwidth, attenuation, noise, distortion, and S/N ratio. Issues unique to wireless transmission include licensing, regulation, and security.

Radio Frequency Transmission

As mentioned, computer-related RF data transmission normally uses frequencies in the shortwave spectrum. Many other types of communication compete for the shortwave spectrum, including FM radio, broadcast TV, cell phones, land-based microwave transmission, and satellite relay microwave transmission (refer back to Figure 8.16).

The main advantage of RF transmission is high raw data transfer capacity (because of high frequency and bandwidth) and the mobility afforded by avoiding wired infrastructure. Another advantage in some situations is broadcast transmission—one sender transmitting the same message to many receivers at the same time. The drawbacks of wireless transmission include the following:

- Major regulatory and legal issues
- Cost of transmitting and receiving stations
- High demand for unused radio frequencies
- Susceptibility to many forms of external interference
- Security concerns—that is, anyone with the necessary receiving equipment can listen to the transmission

Wireless transmission frequencies are regulated heavily, and regulations vary in different parts of the world. The United States, Europe, Japan, and China have their own regulatory agencies. Different regulations apply to different transmission power levels.

In general, low-power transmission doesn't require a license, but the transmission equipment must be certified to meet regulations. Current wireless LAN technology uses unlicensed but regulated RF bands. High-power transmissions, such as long-distance microwave transmission, require certified transmitters and a license to use a specific frequency band. License costs are high because available bandwidth is a limited commodity.

NOTE

Long-distance wireless networks, such as satellite-based networks, are seldom implemented by a single-user organization because of the high cost of transmission equipment and licenses. Instead, companies are formed to purchase and maintain the required licenses and infrastructure. Users purchase capacity from these companies on a fee-per-use or bandwidth-per-time-interval basis.

The Institute of Electrical and Electronics Engineers (IEEE) creates many network and telecommunication standards. The 802.11 committee was formed in the early 1990s to develop standards for wireless LANs. In 1997, the committee published the 802.11 standard. Later versions of this standard include 802.11a (1999), 802.11b (1999), 802.11g (2003), and 802.11n (2009). Standards currently under development for higher throughput include 802.11ac and 802.11ad.

The following are some goals of the IEEE 802.11 standards:

- Meet the requirements of national and international regulatory bodies and pursue adoption by these bodies.
- Support communication by radio waves and infrared or visible light.
- Support stationary stations and mobile stations moving at vehicular speeds.

To expedite its work, the committee initially focused on RF transmission in the 2.4 GHz band because RF waves can travel several miles and penetrate walls and other obstacles. The 2.4 GHz band was chosen because most of this band is regulated but unallocated in the United States, Europe, and Japan.

The first **802.11** standard defined two RF transmission methods with a maximum raw data transfer rate of 2 Mbps. One method, frequency hopping spread spectrum (FHSS), was abandoned in later standards because its transmission speed couldn't be increased with FCC-mandated transmitter power limits. The other method, direct sequence spread spectrum (DSSS), was enhanced in the 802.11b standard to provide raw data transmission at 5.5 and 11 Mbps. The 802.11a standard uses a transmission method called orthogonal frequency division multiplexing (OFDM), which transmits across several channels simultaneously.

Error detection is a major problem in any RF data transmission scheme. RF transmissions are subject to many types of interference, and receivers encounter varying degrees of interference when in motion. Devices such as computers, TVs, microwave ovens, and cordless phones can interfere with wireless network communication. Another problem is multipath distortion, which occurs when an RF signal bounces off stationary objects (see Figure 8.19), causing the receiving antenna to receive multiple copies of the same signal at different times. Multipath distortion blurs or smears the signal content, causing bit detection errors. The problem is especially severe indoors, where there are many hard, flat surfaces that reflect RF waves.

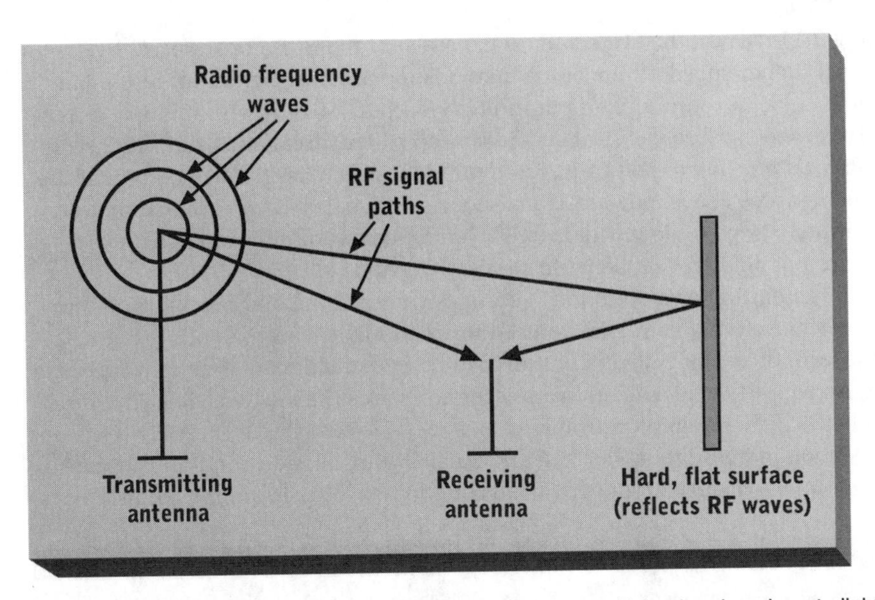

FIGURE 8.19 Multipath distortion caused by receiving multiple signal copies at slightly different times

Courtesy of Course Technology/Cengage Learning

The **802.11b** standard divides the 2.4 GHz band into 14 channels, each with 22 MHz of bandwidth. Transmissions in each band encode bits by using two-level, four-level, or eight-level phase-shift modulation, which yields raw data transfer rates of 22, 44, and 88 Mbps. A substantial portion of the raw data transfer rate is used for redundant bit transmission to improve error detection. The four standard transmission speeds (1, 2, 5.5, and 11 Mbps) use different bit-encoding and signal modulation methods, each with a different tradeoff in reliability, speed, and transmission distance. The fastest method uses eight-level modulation and encodes bits with minimal redundancy. Slower transmission uses two-level modulation with higher redundancy.

Transmitters monitor signal quality and error rates continuously. As noise, interference, and errors increase, or as the sender and receiver move farther apart, transmitters change modulation and bit-encoding methods to compensate. Transmitters step down through the 5.5, 2, and 1 Mbps transmission methods until reliable transmission is achieved and then move upward again as conditions improve.

The **802.11a** standard divides frequency bands in the 5.2, 5.7, and 5.8 GHz ranges into 12 channels. The standard transmission speeds are 6, 9, 12, 18, 24, 36, 48, and 54 Mbps. Bit transmission rates in each channel can be varied to account for different interference levels and error rates. 802.11a transmission is generally more reliable over short distances than 802.11b and 802.11g are because there are fewer sources of noise and interference. However, RF signals in the 5 GHz band attenuate more quickly than 2.4 GHz signals and are more easily absorbed by walls and other obstacles. Therefore, effective 802.11a transmission distance is shorter than for 802.11b and 802.11g.

The **802.11g** standard combines the frequencies and bit-encoding methods of 802.11b with the OFDM transmission method of 802.11a. Standard raw transmission speeds are the same as in 802.11a. The **802.11n** standard expands on the 802.11g standard by defining additional higher-speed bit-encoding methods, optional higher-bandwidth channels, and methods to achieve multiple-input multiple-output (MIMO) with multiple antennas. An 802.11n device can broadcast or receive on up to four frequencies in the 2.4 or 5 GHz bands. Raw data transmission rates up to 600 Mbps are possible by combining the highest bit-encoding rates and bandwidth across four frequencies. As a practical matter, these rates are usually impossible in the 2.4 GHz band because of interference and congestion; they're possible, but unlikely, in the 5 GHz band.

The 802.11n standard allows up to four antennas transmitting and receiving on four different frequencies. Antenna pairs can be used for redundant transmissions, usually described by the term **diversity**. A diversity transmitter sends redundant data transmissions across different frequencies, and a diversity receiver compares the received data streams and decodes only the higher quality transmission (usually the strongest). Diversity transmission and reception increase reliability under changing transmission conditions because interference tends to affect single frequency bands but does so by using more bandwidth.

NOTE

Development of the IEEE 802.11n standard was a long, difficult process because of considerable contention among committee members and a desire by many vendors to release products that complied with earlier drafts of the standards. This standard is also complicated by requirements for backward compatibility with earlier standards. As a result, available equipment varies widely in capabilities and its degree of adherence to the final standard.

Light Transmission

Wireless transmission using light frequencies can occur in the infrared, visible, or ultraviolet light spectra. In theory, raw transmission rates can be much higher than for RF transmission because of higher transmission frequency and available bandwidth. As a practical matter, however, atmospheric interference creates major barriers by attenuating light signals rapidly. Infrared transmissions are effectively limited to a few dozen meters, and laser transmissions in the visible spectrum are limited to a few hundred meters. The need for an unobstructed line of sight is also a barrier to widespread use.

Infrared transmission was once used as a LAN technology in large rooms with many cubicles, but it has been supplanted by the superior performance of RF transmission, which can penetrate walls and other obstacles. Laser transmission is used occasionally for point-to-point connections, as from one building rooftop to another, and when running fiber-optic cable is more expensive.

CHANNEL ORGANIZATION

A communication channel's configuration and organization affect its cost and efficiency. Configuration and organization issues include the number of transmission wires or bandwidth assigned to each channel, the assignment of wires or frequencies to carry specific signals, and the sharing (or not sharing) of channels between multiple senders and receivers.

Simplex, Half-Duplex, and Full-Duplex Modes

Figure 8.20 shows the most basic form of channel organization for electrical transmission through wires. A single communication channel requires two wires: a **signal wire**, which carries data, and a **return wire**, which completes an electrical circuit between sending and receiving devices. Optical transmission requires only one optical fiber per channel because a complete circuit isn't required. Some transmission modes might need two or more communication paths between sender and receiver, as shown in Figure 8.21. Multiple electrical transmission lines can share a single return wire to complete all electrical circuits.

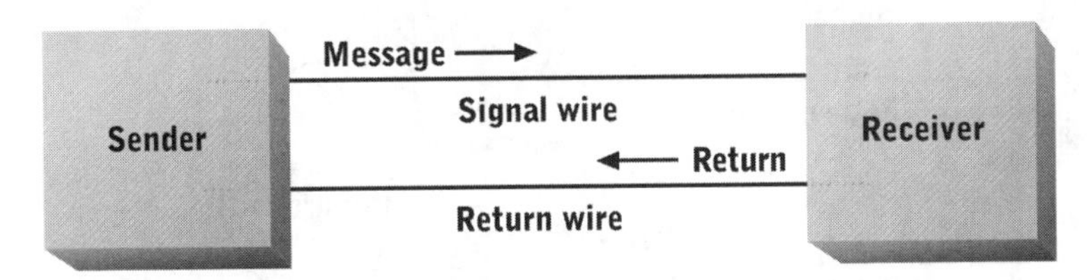

FIGURE 8.20 Configuration of a two-conductor electrical communication channel
Courtesy of Course Technology/Cengage Learning

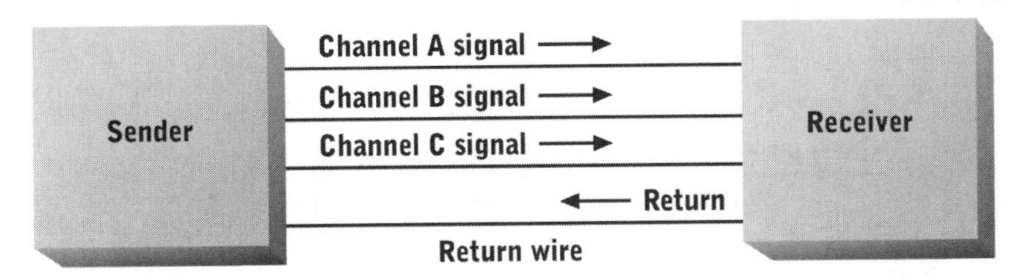

FIGURE 8.21 Multichannel communication with multiple signal wires and one common return wire
Courtesy of Course Technology/Cengage Learning

A single communication line can transmit messages in simplex or half-duplex mode. In **simplex mode**, messages flow in only one direction, as shown in Figure 8.22(a). This communication mode is useful when data needs to flow in only one direction and the chance of transmission error is small or irrelevant, as in ordinary radio broadcasts. However, if transmission errors are common or if error correction is important, simplex mode is inadequate. If the receiver detects transmission errors, there's no way to notify the sender or to request retransmission. A typical use of simplex mode is to send file updates or system status messages from a host processor to distributed data storage devices in a network. In these situations, the same message is transmitted to all devices on the network simultaneously, called **broadcast mode**.

Half-duplex mode uses a single shared channel, and each node takes turns using the transmission line to transmit and receive, as shown in Figure 8.22(b). The nodes must agree on which node is going to transmit first. After sending a message, the first node signals its intent to cease transmission by sending a special control message called a **line turnaround**. The receiver recognizes this message and subsequently assumes the role of

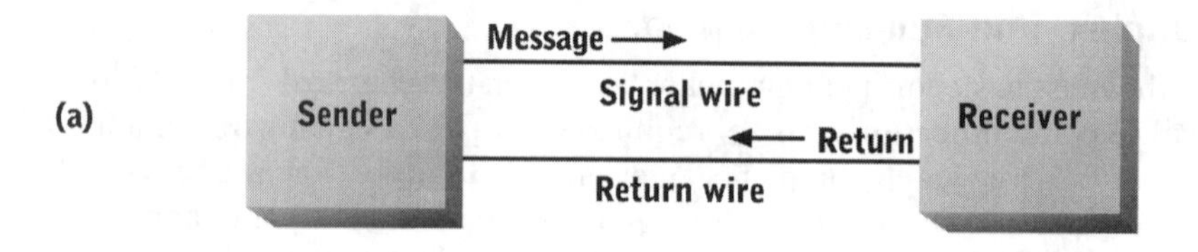

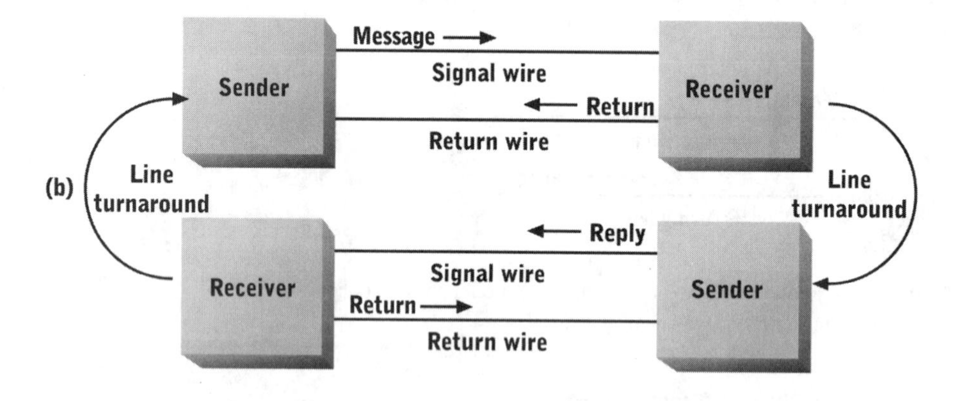

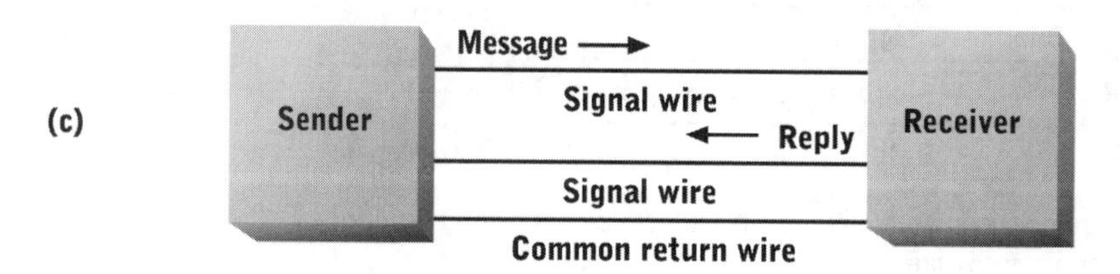

FIGURE 8.22 Configurations for simplex (a), half-duplex (b), and full-duplex (c) modes
Courtesy of Course Technology/Cengage Learning

sender. When its transmission is finished, it sends a line turnaround message, and the original transmitter again becomes the sender.

Half-duplex mode allows the receiver to request retransmitting a message if it detects errors. After receiving a message and a line turnaround, the former receiver briefly assumes the sender role and transmits a signal indicating correct or incorrect data receipt, followed by another line turnaround. If ASCII control characters are used, a **negative acknowledge (NAK)** control character is sent if errors are detected, and an **acknowledge (ACK)** control character is sent if no errors are detected. When a NAK is received, the recipient retransmits the most recent message.

Communication line cost is the same in simplex and half-duplex modes. However, the added reliability of half-duplex mode is achieved at a sacrifice in effective data transfer rate. When the receiver detects an error, it must wait until the sender transmits a line turnaround before it can send a NAK. If the error occurs near the beginning of a message, the transmission time used to send the remainder of the message is wasted. Also, because errors typically occur in bursts, many retransmissions of the same message might be required before the message is received correctly.

The inefficiencies of half-duplex mode can be avoided by using two transmission lines, as shown in Figure 8.22(c). This two-channel organization permits full-duplex communication. In **full-duplex mode**, the receiver can communicate with the sender at any time over the second transmission line, even while data is still being transmitted on the first line. If an error is sensed, the receiver can notify the sender immediately, which can halt the data transmission and retransmit. The speed of full-duplex transmissions is high, even if the channel is noisy. Errors are corrected promptly, with minimal disruption to message flow. This speed increase does involve the cost of a second transmission line, however.

Parallel and Serial Transmission

Parallel transmission uses a separate transmission line for each bit position (see Figure 8.23). A parallel channel's width, or number of lines, is typically one byte or one word plus a common return wire. Parallel communication is more expensive because of the cost of multiple transmission lines, but combining the capacity of multiple transmission lines results in a higher data transfer rate.

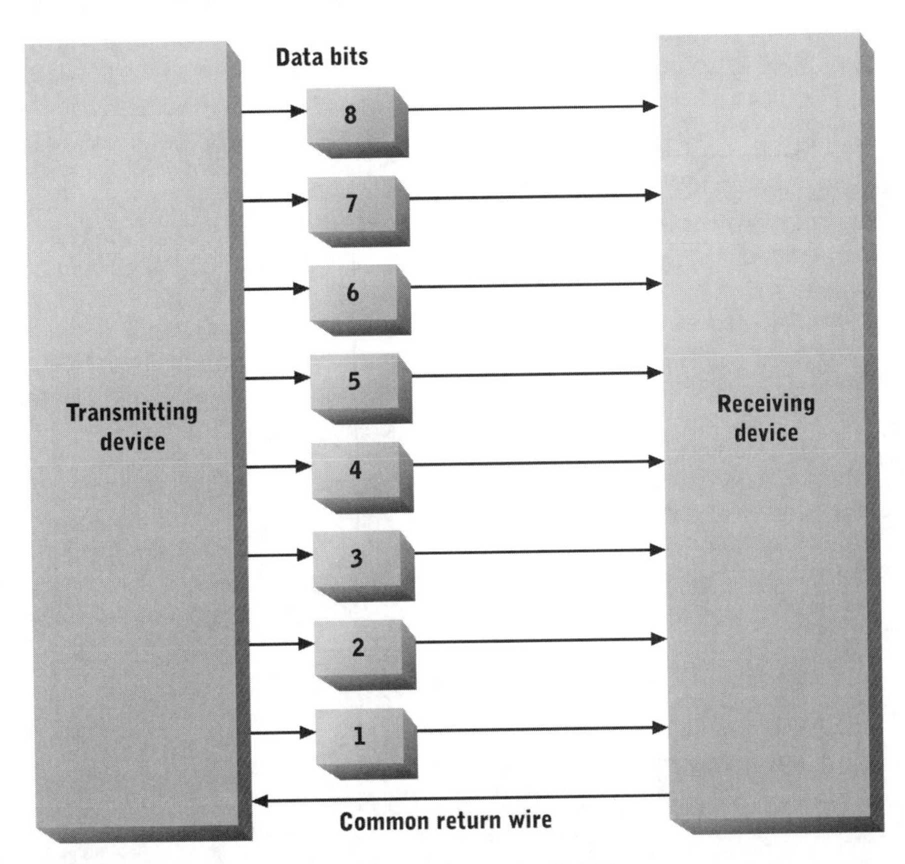

FIGURE 8.23 Parallel transmission of a data byte (8 bits)
Courtesy of Course Technology/Cengage Learning

The maximum distance over which data can be sent reliably via parallel transmission is limited. Because of slight differences in parallel transmission lines, data bits might arrive at the receiver at slightly different times. The timing difference between bits, called **skew**, increases with distance and transmission rate. Excessive skew causes bit values to be misinterpreted. To limit skew errors, high-speed parallel communication channels must be short, typically less than 1 meter.

Another problem with parallel channels is **crosstalk**, which is noise added to the signal in the wire from EMI generated by adjacent wires. Because of their limitations, parallel channels are used only over short distances where high data transfer rates are required—for example, the system bus and connections between memory caches and the CPU.

Serial transmission uses only a single transmission line and a return wire. Bits are sent sequentially through the transmission line, and the receiver reassembles them into larger data units, such as bytes (see Figure 8.24). Digital communication over distances of more than a few meters usually uses serial transmission to avoid skew and crosstalk and minimize wiring and cable cost. LAN, wide area network (WAN), and long-distance telecommunication lines use serial transmission.

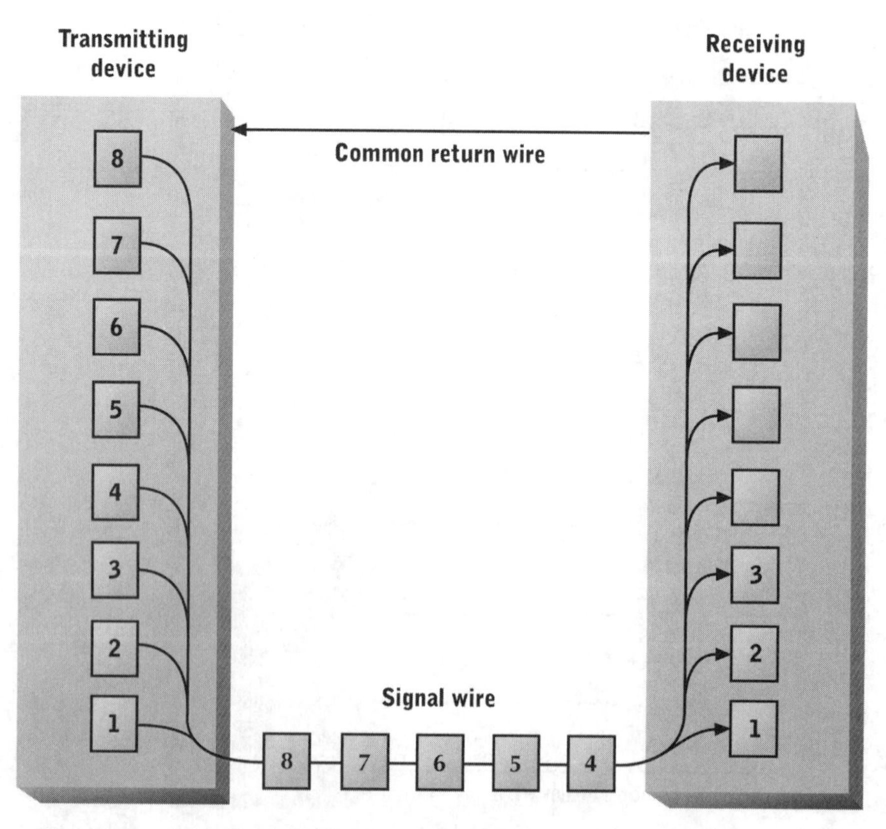

FIGURE 8.24 Serial transmission of a data byte (8 bits)
Courtesy of Course Technology/Cengage Learning

TECHNOLOGY FOCUS

Serial and Parallel Storage Connections

Until the mid-2000s, most computer systems used parallel cables to connect storage devices and device controllers. The most common examples included cables meeting the Advanced Technology Attachment (ATA) and Small Computer System Interface (SCSI) standards. Parallel connections were the only way to provide enough data transfer capacity to meet the needs of typical computer systems. However, a few important trends have combined to make serial cables the predominant form of storage device connection.

First, typical desktop computers and small servers continue to shrink. As a result, routing cables between the motherboard, device controllers, storage devices, and I/O devices is difficult. Because they contain many wires, parallel cables are bulky (see Figure 8.25) and difficult to route inside a typical small computer cabinet. Their size also restricts airflow, which can cause heat dissipation problems. In contrast, serial cables are thin and flexible. They occupy less space and are easier to route around computer components.

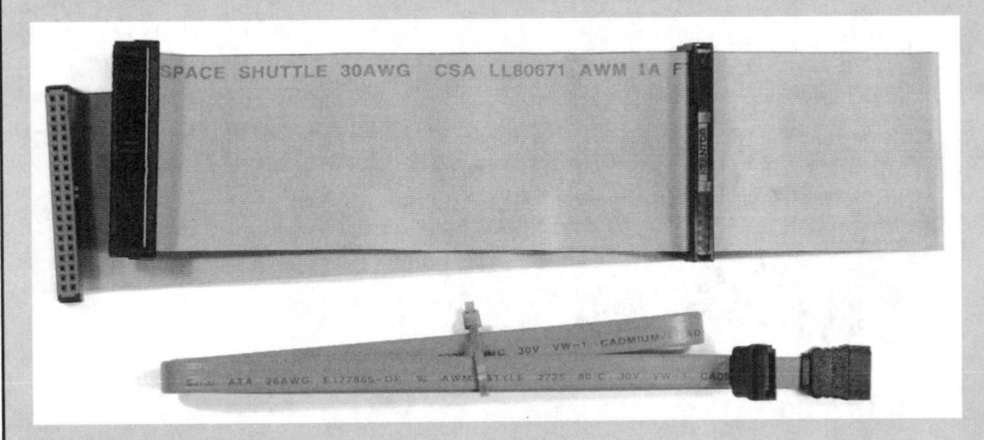

FIGURE 8.25 Parallel ATA (top) and Serial ATA (bottom) cables
Courtesy of Course Technology/Cengage Learning

Second, although desktops and small servers have grown smaller, the storage subsystems large-scale computers use have grown. Many transaction-processing and data-mining applications require terabytes of secondary storage, for example. Fulfilling this storage need requires hundreds of disk drives connected to dozens of storage controllers spread across multiple storage cabinets. Connecting all these devices with parallel cables is difficult because skew and crosstalk limit cable length to a few meters.

Finally, when electronic switching devices were slower, operating many in parallel was the only feasible way to achieve high data transfer rates. However, advances in semiconductor fabrication and electronic switching technology have yielded smaller and faster devices, which are the basic building blocks of computer components, such as network interface cards and storage device controllers. As a result, data transfer speeds of hundreds or thousands of megabits per second across a single cable pair are now possible.

(continued)

Recognizing these trends, standard-setting organizations for storage devices and cables have developed serial versions of older parallel standards. In the PC arena, storage device and computer manufacturers have widely adopted the **Serial Advanced Technology Attachment (SATA)** standard. SATA uses a software interface standard that's compatible with older Parallel ATA (PATA) standards but substitutes a seven-wire cable for PATA's 40-wire cable. A SATA cable uses two wire pairs for serial data transmission in each direction and three wires for shielding and ground. All desktop computers now use SATA storage devices and cables for higher performance and simpler low-cost assembly. SATA can also be used in servers because of its improved features, such as command queuing, hot-swapping, and better error detection.

In the server arena, serial standards, such as **Serial Attached SCSI (SAS)**, have replaced older parallel standards, too. Also, bulkier parallel cables carrying electrical signals have been replaced by optical serial cables. Figure 8.26 shows an older SCSI-3 cable for connecting a storage array to a server and a newer optical cable. SCSI-3 cables are bulky and limited to a few meters in length because of skew and crosstalk. Optical serial cables are much thinner and more flexible and can carry signals much longer distances, which enables connections between storage arrays and servers to span multiple racks or rooms.

FIGURE 8.26 Parallel SCSI-3 (top) and serial optical (bottom) cables
Courtesy of Course Technology/Cengage Learning

Channel Sharing

Most local phone service is based on a channel-sharing strategy called **circuit switching**. When a user makes a phone call, the full capacity of a channel between the user and the nearest phone-switching center is devoted to the call. The user has a continuous supply of data transfer capacity, whether it's needed or not, and the channel is unavailable to other users until the call is terminated. Circuit switching wastes channel capacity in most data transmission situations because the channel is held by one sender and receiver, even during inactive periods, such as silences during a phone conversation.

Circuit switching is inefficient for most data transmission settings because few users require high data transmission capacity continuously. Typically, transmission capacity is needed for short periods, or bursts. Channel-sharing techniques use available capacity efficiently by combining transmissions to and from multiple users. As long as many users don't need high capacity at the same time, the combined load on the channel averages to an acceptable level, and the cost of a single channel is shared across more users, thus reducing the cost per user.

Time-division multiplexing (TDM) describes techniques for splitting data transfer capacity into time slices and allocating them to multiple users. **Packet switching**, the most common type of TDM, divides messages from all users or applications into small pieces called **packets** (see Figure 8.27).

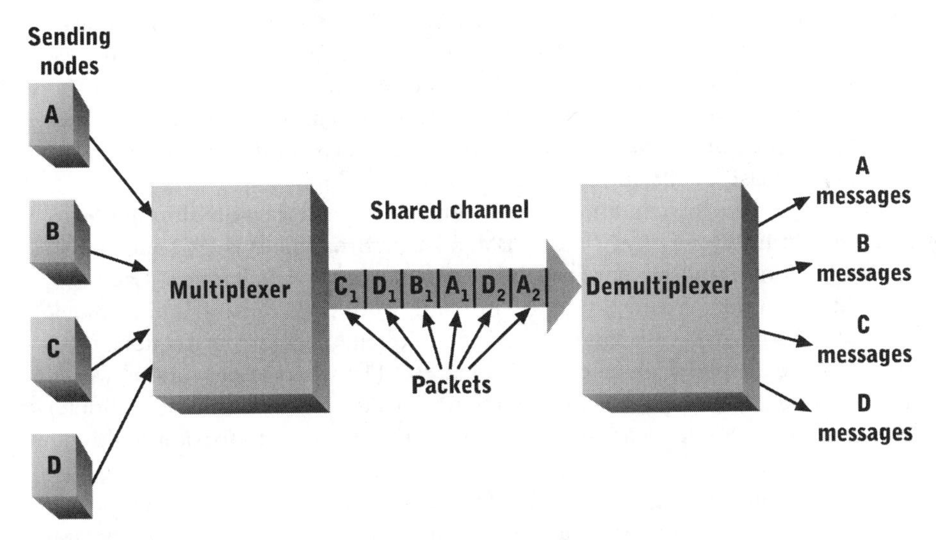

FIGURE 8.27 Packet switching—the most common form of TDM
Courtesy of Course Technology/Cengage Learning

Each packet contains a header identifying the sender, recipient, sequence number, and other information about its contents. Packets are then sent through the network to their destination as channel capacity becomes available. They can be held in a buffer temporarily, pending channel availability. If multiple channels are available, multiple packets can be sent simultaneously. The receiver reassembles packets into the original message in their intended sequence, even if they arrive in a different order. Transmission errors must be localized to a specific packet to request retransmitting the correct packet.

The primary advantage of packet switching is that the telecommunication service provider determines how to use available data transfer capacity and channels most effectively. In most situations, the provider can make rapid and automatic decisions that allocate available capacity efficiently to users. The result is substantially reduced telecommunication cost because of more efficient use of available capacity.

The main disadvantages of packet switching are varying delays in message transmission and the complexity of creating and routing packets. Because a user doesn't have a dedicated channel, the time required to send a message is unpredictable; it rises and falls with the total demand for the available data transfer capacity. Delays can be substantial when many users send messages simultaneously and available capacity is insufficient.

Despite its complexities, packet switching is the dominant form of intercomputer communication. Its cost and complexity are more than offset by more efficient use of available communication channels. Also, as in other areas of computer and communication technology, the inherent complexity doesn't translate into high costs because devices that can cope with the complexity are becoming cheaper and more powerful. Circuit switching is used only when data transfer delay and available data transfer capacity must be within precise and predictable limits, as when relaying a digital video broadcast from a sports arena to a broadcast facility in real time.

Another method for sharing communication channels is **frequency-division multiplexing (FDM)**, shown in Figure 8.28. Under FDM, a single **broadband** (high bandwidth) channel is partitioned into multiple **narrowband** (low bandwidth) subchannels. Each subchannel represents a different frequency range, or band, within the narrowband channel. Signals are transmitted in each subchannel at a fixed frequency or narrow frequency range. Common FDM examples include cable TV and 802.11n wireless networking with multiple antennas. Long-distance telecommunication providers also use FDM to multiplex single-mode optical fibers, an application commonly called **wavelength-division multiplexing (WDM)**.

FDM might require moving signals intended for one frequency band into another frequency band. For example, in most cable TV systems, channels 2 through 13 are carried across a coaxial cable at their original broadcast frequencies. Channels above 13 are remapped to frequencies other than their assigned broadcast frequencies so that they occupy contiguous narrowband channels within the coaxial broadband channel.

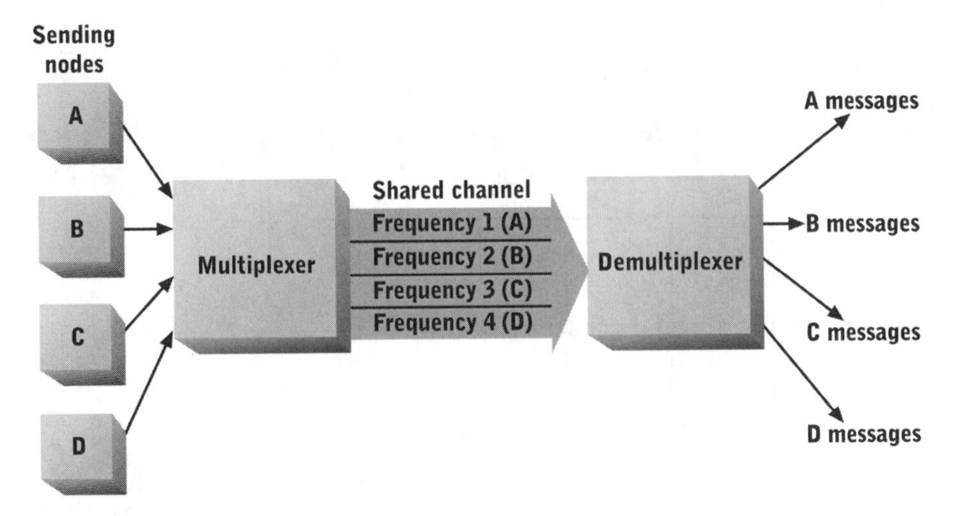

FIGURE 8.28 Channel sharing with FDM
Courtesy of Course Technology/Cengage Learning

Narrowband channels in a single broadband channel can use different signaling methods, communication protocols, and transmission speeds. They can be shared by using packet switching or other TDM methods. Multiple narrowband channels can also be combined into a parallel transmission channel.

TECHNOLOGY FOCUS

InfiniBand

InfiniBand is a data interconnection standard developed by the InfiniBand Trade Association, a consortium founded by Dell, Hewlett-Packard, IBM, Intel, Microsoft, and Sun Microsystems. This high-speed communication architecture is intended to interconnect devices such as servers, secondary storage appliances, and network switches. The goal is to replace the current jumble of competing interconnection standards with a unified standard and architecture that increases data transfer rates substantially.

InfiniBand is based on an interconnection architecture known as **switched fabric**, which interconnects devices with multiple data transmission pathways and a mesh of switches that resembles the interwoven threads of fabric (see Figure 8.29). A fabric switch connects any sender directly to any receiver and can support many simultaneous connections. Connections are created on request and deleted when they're no longer needed, freeing data communication capacity to support other connections. Switched

(continued)

fabric architecture isn't new, but the digital-switching technology that supports it has become cost effective since the late 1990s.

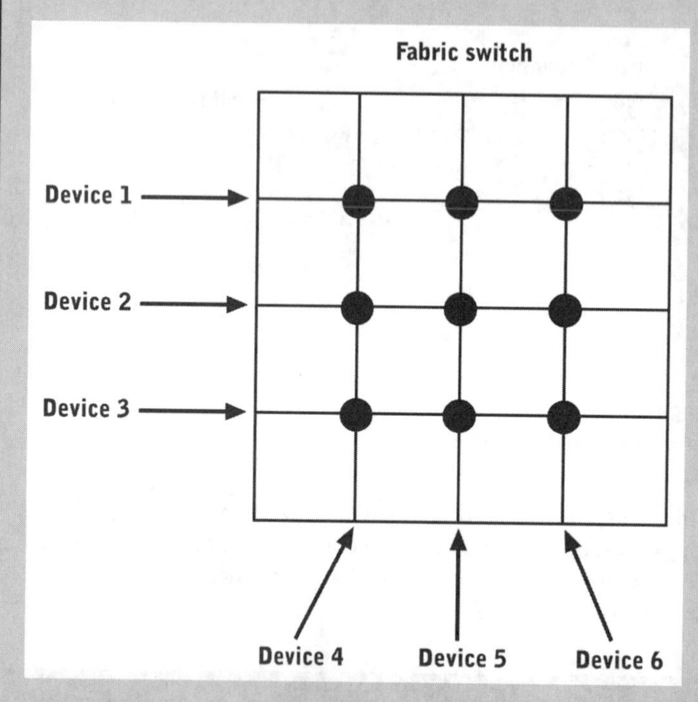

FIGURE 8.29 A 3 × 3 switched fabric
Courtesy of Course Technology/Cengage Learning

Each device is connected to an InfiniBand switch by a **host channel adapter (HCA)** or a **target channel adapter (TCA)**. HCAs are used by devices such as general-purpose servers that can initiate and respond to data transfer requests. An HCA has a direct connection to the host's primary storage via a device controller attached to the system bus or a special-purpose memory interface. TCAs are used by simpler devices, such as network switches and storage appliances.

InfiniBand devices are connected with copper wires or fiber-optic cables. The standard specifies cable connectors and operational characteristics but not physical cable construction. Conventional twisted-pair or coaxial cables can't meet the InfiniBand requirements. Some vendors use a bundle of twin-axial cables. Raw data transmission rates range from 2.5 to 30 Gbps per connection, and up to four connections can be used in parallel. Copper cables can be up to 25 meters long, and fiber-optic connections can extend up to 10 kilometers.

Figure 8.30 shows a typical architecture for a medium to large server farm used by large e-commerce Web sites. It includes multiple general-purpose servers configured for specific tasks, network storage servers, and network switches. Specializing each device's functions makes it easier to enlarge or reduce overall capacity, and device redundancy

(continued)

increases reliability and fault tolerance. However, many high-speed transmission lines and switches are necessary to interconnect all system components. These interconnections are InfiniBand's target market.

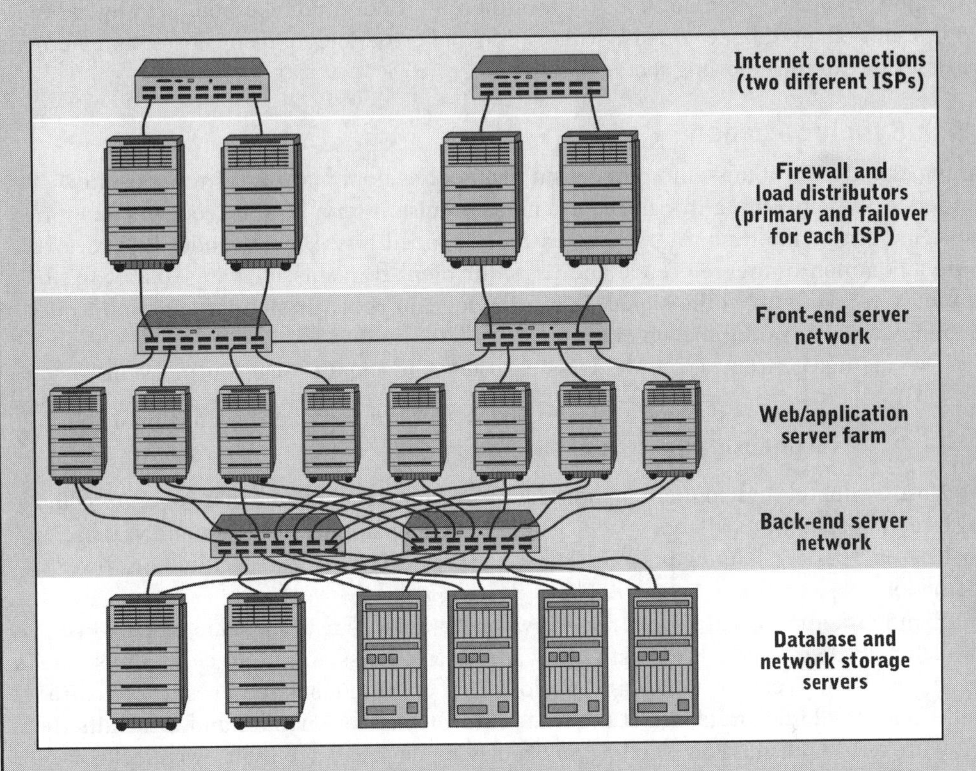

FIGURE 8.30 A server farm with network interconnections
Courtesy of Course Technology/Cengage Learning

InfiniBand has established a firm hold on high-speed connections in large server farms and supercomputer clusters. However, it competes with other interconnection standards, such as PCI, Gigabit Ethernet, and Fiber Channel. Many current InfiniBand products are switches and interconnection bridges for existing components that use other data communication standards.

COMMUNICATION COORDINATION

Sender and receiver must coordinate their approaches to various aspects of communication in a channel, including the start and end times of bits or signal events, error detection and correction, and encryption methods (or lack thereof). Encryption methods are beyond the scope of this chapter, but clock synchronization and error detection and correction methods are described in the following sections.

Clock Synchronization

An important part of any communication protocol is a common transmission rate. Senders place bits onto a transmission line at precise intervals, and receivers examine the signal at specific time intervals to extract encoded bits. Unless sender and receiver share a common timing reference and use equivalent transmission rates, data can't be extracted correctly from the signal. When sender and receiver share a common timing reference, they're said to be synchronized.

Two main synchronization problems can occur during message transmission:

- Keeping sender and receiver clocks synchronized during transmission
- Synchronizing the start of each message

Most computer and communication devices use internal clocks that generate voltage pulses at a specific rate when a specific input voltage is applied. Unfortunately, these clocks aren't perfect. Timing pulse rates can vary with temperature and minor power fluctuations.

Timing fluctuations aren't a problem when all devices use the same clock. Most parallel transmission standards assign a separate transmission line to carry the sender's clock pulse. The receiver monitors this clock line continuously and uses it to ensure that it reads and interprets bits at the same rate the sender encodes and transmits them.

With serial transmission, sending and receiving devices have their own clocks. Because the clocks are independent, there's no guarantee that their timing pulses are generated at exactly the same time. Most communication protocols transmit clock pulses from sender to receiver when a connection is first established. However, even if perfect synchronization is achieved at that time, clocks can drift out of synchronization later, and then bit interpretation errors by the receiver become more likely.

Figure 8.31 shows an example of errors resulting from unsynchronized clocks. The receiving device's clock pulses are generated at different times than the sender's clock pulses, resulting in misalignment of bit time boundaries. In this example, the receiver can't interpret several bits correctly because they contain two different signal levels.

Sender and receiver must occasionally transmit timing signals over the data transmission line to ensure that their clocks are synchronized. They must also agree on the boundaries of each message, and to do so, they must agree on message length. Two common approaches to synchronizing clocks and coordinating message boundaries are synchronous transmission and asynchronous transmission. These approaches are sometimes referred to as **character-framing methods** when messages consist of ASCII or Unicode characters.

Synchronous transmission ensures that sender and receiver clocks are always synchronized by sending continuous data streams. Messages are transmitted in fixed-size

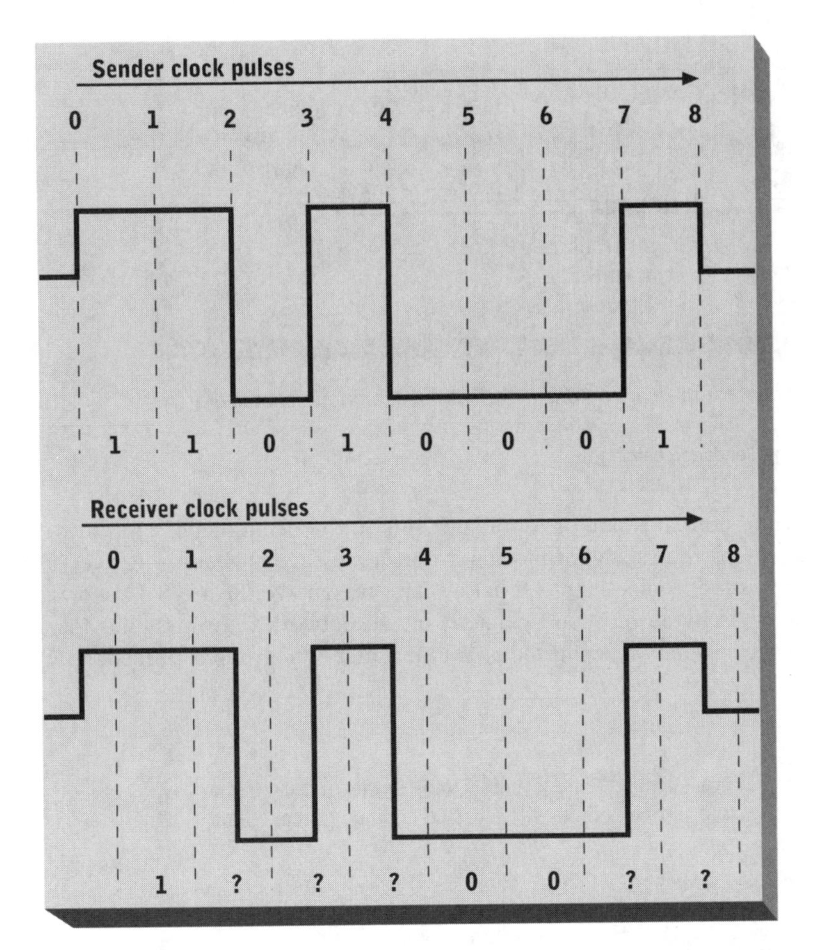

FIGURE 8.31 Communication errors resulting from unsynchronized clocks
Courtesy of Course Technology/Cengage Learning

byte groups called "blocks" (see Figure 8.32). Because block size is always the same, the receiver always knows where one block ends and another begins. If necessary, transmission blocks are separated by a continuous stream of **synchronous idle characters**, which have a predetermined pattern of signal transitions designed for easy clock synchronization. For example, the data start flag is an easily recognizable bit pattern that differs from the synchronous idle message and data end flag bit patterns. These differences enable the receiver to detect the beginning of a new transmission block.

In **asynchronous transmission**, messages are sent on an as-needed basis. They can be sent one after another, or there can be periods of inactivity between messages. From the receiver's standpoint, messages arrive at random or intermittent times. Clock drift is a major problem in asynchronous transmission. During idle periods, sender and receiver clocks can drift out of synchronization because no data or timing signals are being transmitted. When the sender does transmit data, the receiver must resynchronize its clock immediately to interpret incoming signals correctly.

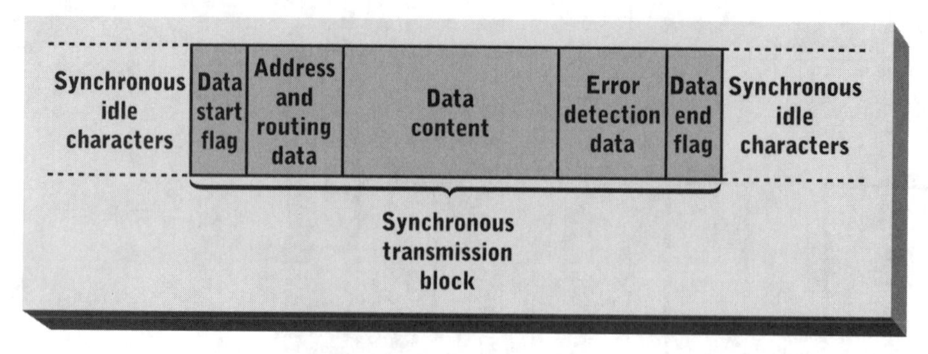

FIGURE 8.32 Typical format for messages transmitted with synchronous character-framing methods
Courtesy of Course Technology/Cengage Learning

Asynchronous transmission adds one or more **start bits** to the beginning of each message. The start bit "wakes up" the receiver, informs it that a message is about to arrive, and allows the receiver to synchronize its clock before data bits arrive. Figure 8.33 shows a message containing a single byte preceded by a start bit. In network transmissions, dozens or hundreds of bytes are usually transmitted as a unit. Each byte group is preceded by one or more start bits.

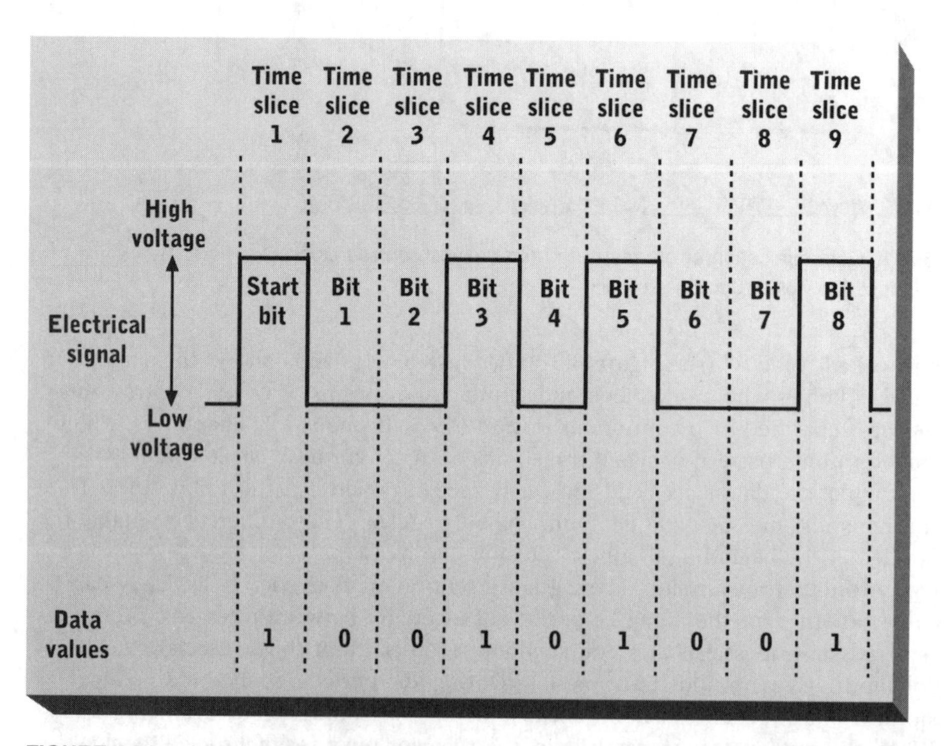

FIGURE 8.33 Asynchronous character framing for serial transmission, including a start bit
Courtesy of Course Technology/Cengage Learning

Synchronous transmission uses channel capacity more efficiently than asynchronous transmission because data is transmitted in large blocks, with fewer bits used to mark message boundaries and synchronize clocks. However, because transmission is continuous, synchronous transmission can't be used when multiple senders and receivers share the same channel. Asynchronous transmission is used in most networks, modem communication, and communication between computer systems and peripheral devices, such as keyboards and printers. Synchronous transmission is used only for high-speed communication between a dedicated sender and receiver, such as between two mirrored or clustered mainframe computers.

Error Detection and Correction

A crucial component of any communication protocol is a method for detecting and correcting errors in data transmission, reception, or interpretation. All widely used error-detection methods are based on some form of redundant transmission, meaning a redundant message or message component is transmitted with or immediately after the original message. The receiver compares the original message with the redundant transmission, and if the two match, the original message is assumed to have been transmitted, received, and interpreted correctly. If the two don't match, a transmission error is assumed to have occurred, and the receiver asks the sender to retransmit the message.

Error detection and correction methods vary in the following characteristics:

- Size and content of the redundant transmission
- Efficient use of the communication channel
- Probability that an error will be detected
- Probability that an error-free message will be identified as an error
- Complexity of the error-detection method

Size and content of the redundant transmission are inversely related to efficient use of the channel. For example, one possible error-detection method is to send three copies of each message and verify that they're identical. This method is easy to implement, but it uses the channel at only one-third of its capacity, or less if many errors are detected (correctly or incorrectly). Changing the method to send only two copies of each message increases maximum channel utilization to 50% of raw capacity.

For most methods and channels, the probability of detecting errors can be computed mathematically or statistically. The probability of not detecting a real error is called **Type I error**. The probability of incorrectly identifying good data as an error is called **Type II error**. For any error-detection method, Type I and Type II errors are inversely related—meaning a decrease in Type I error is accompanied by an increase in Type II error.

NOTE

Type II errors result in needless retransmission of data that was received correctly but incorrectly assumed to be in error. Increasing Type II error decreases the channel efficiency because a larger proportion of channel capacity is used to retransmit data needlessly.

In some types of communication channels, such as a system bus or a short-haul fiber-optic cable, the probability of a transmission or reception error is extremely remote. In other types of channels, such as high-speed analog modem channels, errors are common.

Different methods of error detection are suited to different channels and purposes. For example, error detection isn't normally used for digital voice transmissions because users aren't sensitive to occasional minor errors. At the other extreme, communication between bank computers and ATMs over copper wire uses extensive error checking because of the data's importance and the higher error rate in long-distance electrical transmissions.

Common methods of error detection include the following, discussed in the next sections:

- Parity checking (vertical redundancy checking)
- Block checking (longitudinal redundancy checking)
- Cyclic redundancy checking

Parity Checking

Character data is often checked for errors by using **parity checking**, also called **vertical redundancy checking**. In a character-oriented transmission, one **parity bit** is appended to each character. The parity bit value is a count of other bit values in the character.

Parity checking can be based on even or odd bit counts. With **odd parity**, the sender sets the parity bit to 0 if the count of 1-valued data bits in the character is odd. If the count of 1-valued data bits is even, the parity bit is set to 1 (see Figure 8.34). Under **even parity**, the sender sets the parity bit to 0 if the count of 1-valued data bits is even or to 1 if the count of 1-valued data bits is odd. The receiver counts the 1-valued data bits in each character as they arrive and then compares the count against the parity bit. If the count doesn't match the parity bit, the receiver asks the sender to retransmit the character.

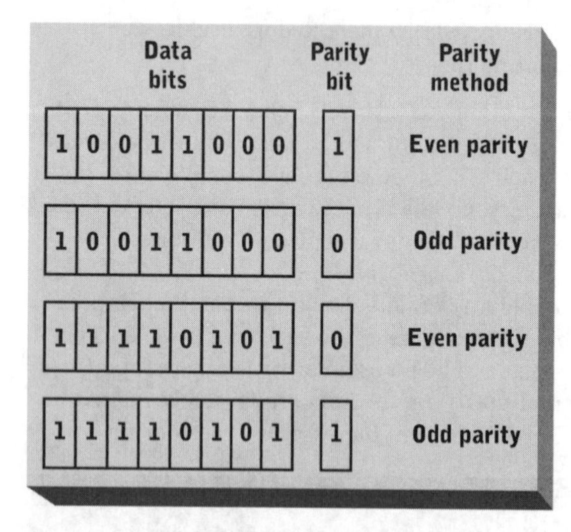

FIGURE 8.34 Sample parity bits
Courtesy of Course Technology/Cengage Learning

Parity checking has a high Type I error rate. For example, a transmission error that flipped the values of 2, 4, or 6 bits in an ASCII-7 character wouldn't be detected. In addition, this method is unreliable in channels that are subject to error bursts affecting many adjacent bits. It's more reliable in channels with rare errors that are usually confined to widely spaced bits.

Block Checking

Parity checking can be expanded to groups of characters or bytes by using **block checking**, also called **longitudinal redundancy checking (LRC)**. To implement this method, the sending device counts the number of 1-valued bits at each bit position in a block. It then combines parity bits for each position into a **block check character (BCC)** and adds it to the end of the block (see Figure 8.35). The receiver counts the 1-valued bits in each position and derives its own BCC to compare with the BCC the sender transmitted. If the BCCs aren't the same, the receiver requests retransmission of the entire block. In Figure 8.35, an even parity bit is computed for each bit position of an 8-byte block. The set of parity bits forms a BCC that's appended to the block for error detection.

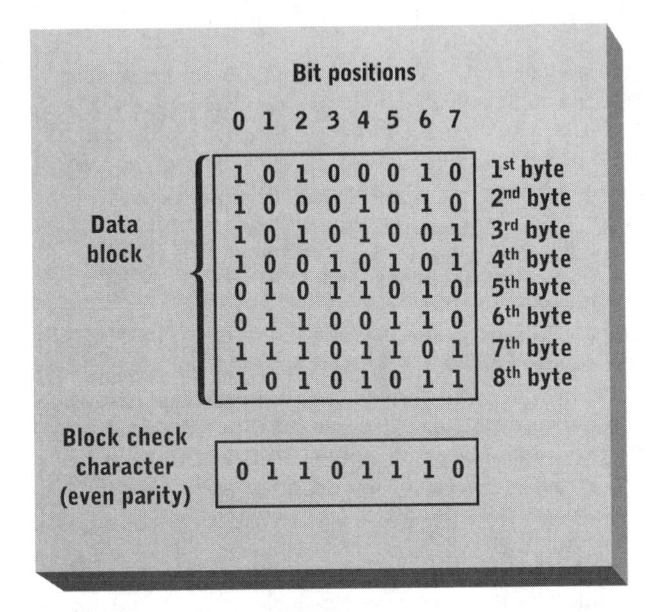

FIGURE 8.35 Block checking
Courtesy of Course Technology/Cengage Learning

This method is vulnerable to the same types of errors as parity checking. Type I error rates can be reduced by combining parity checking and block checking. However, even with this approach, some compensating errors might go undetected.

Cyclic Redundancy Checking

Cyclic redundancy checking (CRC) is the most widely used error-detection method. Like block checking, it produces a BCC for a group of characters or bytes. This CRC character, generated by a complex mathematical algorithm, is usually more than 8 bits and can be as large as 128 bits. CRC bit strings can be computed by software or special-purpose microprocessors incorporated into data communication and network hardware.

CRC has much lower Type I and Type II error rates than parity and block checking. Both error rates depend on the size of the transmitted data block and CRC bit string. CRC bit strings of 64 or 128 bits are commonly used in network packets and for magnetic tape data backups.

BUSINESS FOCUS

Upgrading Storage and Network Capacity

The Bradley Advertising Agency (BAA) specializes in designing print and video ads, mostly under contract to other advertising agencies. BAA employs 30 people; most are engaged in design and production work. Print ad copy is produced almost entirely by computer, using Adobe Illustrator and other software on 15 networked Dell and Apple desktop computers. BAA also uses many peripheral devices, including scanners, color printers, and a high-speed color copier. The company has two Dell rack-mounted servers connected to a Dell storage array with ten 146 GB SCSI disks. One server acting as the primary file server is also connected to an 80 GB DLT tape drive and a Blu-ray re-writable DVD drive.

Over the past five years, BAA has gradually migrated from standard-definition analog video cameras and videotape-editing equipment to high-definition video equipment for video production projects. Two high-performance workstations are dedicated to digital video editing and production, and BAA plans to add two more this year and two the following year. The files used to store print ads are normally 200 to 500 MB per project but can range as high as 5 GB, and BAA often has several dozen print projects in process at a time. Video production files are much larger than print files. For example, files for a typical standard-definition 30-second TV ad require 10 GB of storage; high-definition video files are six to eight times larger.

File-sharing performance was already slow when the first video-editing workstations were purchased. A consultant determined that the file server was using only about 20% of its data transfer capacity, and the network was the performance bottleneck. The existing network is an Ethernet LAN using twisted-pair cable installed in 1993. The LAN originally operated at 10 Mbps, but the network hub was upgraded to a 10/100 Mbps switch. All the desktop computers operate at 100 Mbps. The video-editing workstations and the server have network interfaces that support 100 Mbps or 1 Gbps speed, but they're limited to 100 Mbps by the current switch.

Network bottlenecks have become more common since the two video-editing workstations were acquired. Access to print ad files that took a few seconds a year or two ago can now take up to 5 minutes during busy periods. Recently, two 250 GB disks were added to each video-editing workstation so that video files can be stored locally during editing, but downloading and uploading a project to or from the server can take 15 minutes or more. Project files are sometimes moved between workstations several times a day if both machines are being used for the same project. Because BAA is growing rapidly, it expects disk storage requirements to double every year for the next three years.

BAA is studying alternatives to increase data storage, data communication, and network capacity, including the following:

- Upgrading the existing network to Gigabit Ethernet with existing cabling
- Installing a parallel 1 or 10 Gbps network for the server and video-editing workstations with new copper or fiber-optic cable
- Adding disks to the current storage cabinet (which has four empty slots) or adding a second storage cabinet
- Buying a network-attached storage server with at least 5 TB of disk capacity

(continued)

- Purchasing one or more InfiniBand storage devices and interfacing them to the existing file server and video-editing workstations via an InfiniBand switch

BAA is unsure whether its cabling will work with Gigabit Ethernet because it isn't certified as Category 5. It's also concerned that even if the existing cable works, the network will soon be saturated, and the cable will be unable to support a future move to 10 Gbps Ethernet. BAA is also concerned about investing in expensive InfiniBand technology.

Questions:

- If BAA's current cable is tested and determined to be reliable with Gigabit Ethernet, should the company use its existing wiring or install new wiring?
- If costs (including installation labor) for Category 6 and fiber-optic cable are $25,000 and $40,000, respectively, which cable should BAA choose if the existing network must be rewired for Gigabit Ethernet?
- What are the advantages of operating a separate high-speed network to support video-editing applications in addition to the existing network? What are the disadvantages? Which cable type is most suitable?
- Should BAA seriously consider moving to InfiniBand and stand-alone storage servers? Why or why not? If BAA should adopt InfiniBand, how quickly should it do so?

Summary

- A communication protocol is a set of rules and conventions covering many aspects of communication, including message content and format, bit encoding, signal transmission, transmission medium, and channel organization. It also includes procedures for coordinating the flow of data, including media access, clock synchronization, and error detection and correction.
- Data bits can be encoded into analog or digital signals, which can be transmitted via electrical, optical, or radio frequency sine waves. Bits are encoded in sine waves by modulating one or more wave characteristics. Possible modulation techniques include on-off keying and frequency, amplitude, and phase-shift modulation.
- Important characteristics of transmission media include raw data transfer rate, bandwidth, and susceptibility to noise, distortion, external interference, and attenuation. Bandwidth is the difference between the highest and lowest frequencies that can be propagated through a transmission medium. Higher-bandwidth channels can reliably carry composite signals with more data content over longer distances.
- The effective data transfer rate can be much lower than the raw data transfer rate because of attenuation, distortion, and noise. Attenuation is loss of signal power as it travels through the transmission medium. Distortion is caused by interactions between the signal and the transmission medium. Noise can be generated internally or added through external interference. The signal-to-noise (S/N) ratio is a measure of the difference between noise power and signal power.
- Electrical cables are of two main types—twisted-pair cable and coaxial cable. Both use copper but differ in construction and shielding. Twisted pair is inexpensive but limited in bandwidth, S/N ratio, and transmission speed because of limited shielding. Coaxial and twin-axial cables are more expensive but offer higher bandwidth, a higher S/N ratio, and lower distortion.
- Optical cables are of two types—multimode and single mode. Single-mode cable provides higher transmission rates than multimode cable but at a much higher cost. Optical cables have high bandwidth, little internally generated noise and distortion, and immunity to electromagnetic interference.
- Data can be transmitted wirelessly via radio waves and infrared or visible light. Short-distance RF transmission using IEEE 802.11 standards is widely used for LANs. Long-distance RF channels are in short supply because of their limited number and intense competition for licenses. RF channels are generally leased from telecommunication companies on an as-needed basis. Infrared and visible light transmission have limited applications because of line-of-sight requirements and atmospheric interference.
- Channel organization describes the number of lines dedicated to a channel and the assignment of specific signals to these channels. A simplex channel uses one optical fiber or copper wire pair to transmit data in one direction only. A half-duplex channel is identical to a simplex channel but can send a line turnaround message to reverse transmission direction. Full-duplex channels use two fibers or wire pairs to support simultaneous transmission in both directions.

- Parallel transmission uses multiple lines to send several bits simultaneously. Serial transmission uses a single line to send one bit at a time. Parallel transmission provides higher channel throughput, but it's unreliable over distances of more than a few meters because of skew and crosstalk. Serial transmission is reliable over much longer distances. Serial channels are cheaper to implement because fewer wires or wireless channels are used.
- Channels are often shared when no single user or application needs a continuous supply of data transfer capacity. In circuit switching, an entire channel is allocated to a single user for the duration of one data transfer operation. Packet switching, a time-division multiplexing method, allocates time on the channel by dividing many message streams into smaller units called packets and intermixing them during transmission. Frequency-division multiplexing divides a broadband channel into several narrowband channels.
- Sender and receiver must synchronize clocks to ensure that they use the same time periods and boundaries to encode and decode bit values. A single shared clock is the most reliable synchronization method, but it requires sending clock pulses continuously from sender to receiver. Asynchronous transmission relies on specific start and stop signals (usually a single bit) to indicate the beginning and end of a message unit. Synchronous transmission maintains a continuous flow of signals from sender to receiver to provide constant opportunities for clock synchronization.
- Error detection is always based on some form of redundant transmission. The receiver compares redundant copies of messages and requests retransmission if they don't match. Increasing the level of redundancy increases the chance of detecting errors but at the expense of reducing channel through put. Common error-detection schemes include parity checking (vertical redundancy checking), block checking (longitudinal redundancy checking), and cyclic redundancy checking.

At this point, you have a thorough understanding of how data is encoded and transmitted between computer systems. Data communication technology is the foundation of computer networks, but other software and hardware technologies are required. Chapter 9 discusses computer network hardware, protocols, and architecture, and Chapter 13 discusses system software that supports distributed resources and applications.

Key Terms

802.11
802.11a
802.11b
802.11g
802.11n
acknowledge (ACK)
amplifier
amplitude
amplitude modulation (AM)
amplitude-shift keying (ASK)
analog signal
asynchronous transmission
attenuation
bandwidth
binary signals
bit time
block check character (BCC)
block checking
broadband
broadcast mode
carrier wave
Category 5

Category 6
character-framing methods
circuit switching
coaxial cable
communication channel
communication protocol
composite signal
crosstalk
cycle
cyclic redundancy checking (CRC)
digital signal
discrete signal
distortion
diversity
effective data transfer rate
electromagnetic interference (EMI)
even parity
fiber-optic cable
frequency
frequency-division multiplexing (FDM)
frequency modulation (FM)
frequency-shift keying (FSK)
full-duplex mode
guided transmission
half-duplex mode
host channel adapter (HCA)
InfiniBand
line turnaround
longitudinal redundancy checking (LRC)
message
modulator-demodulator (modem)
multilevel coding
multimode graded-index cable
multimode step-index cable
narrowband
negative acknowledge (NAK)
noise
odd parity
on-off keying (OOK)
packets
packet switching
parallel transmission
parity bit
parity checking
phase
phase-shift keying (PSK)
phase-shift modulation
radio frequency (RF)
raw data transfer rate
repeater
return wire
Serial Advanced Technology Attachment (SATA)
Serial Attached SCSI (SAS)
serial transmission
signal
signal-to-noise (S/N) ratio
signal wire
simplex mode
sine wave
single-mode cable
skew
start bits
switched fabric
synchronous idle characters
synchronous transmission
target channel adapter (TCA)
time-division multiplexing (TDM)
transmission medium
twin-axial cable
twisted-pair cable
Type I error
Type II error
unguided transmission
vertical redundancy checking
wavelength-division multiplexing (WDM)
wired transmission
wireless transmission

Vocabulary Exercises

1. ________ transmission sends bits one at a time over a single transmission line or electrical circuit.
2. During half-duplex transmission, sender and receiver switch roles after a(n) ________ message is transmitted.
3. ________ encodes data by varying the distance between wave peaks in an analog signal.
4. A(n) ________ converts a digital signal to an analog signal so that digital data can be transmitted over analog phone lines.
5. Serial transmission standards, including ________ and ________, are replacing parallel transmission standards for connecting secondary storage devices and controllers.
6. The ________ of a sine wave is measured in hertz.
7. Most local phone service uses ________ switching to route messages from a wired home phone to the local phone-switching center.
8. Most networks use ________ switching to send messages from sender to receiver.
9. In ________, a bit is added to each character or byte, and the bit value is determined by counting the number of 1 bits.
10. A(n) ________ signal is a discrete signal that can encode only two possible values.
11. A(n) ________ wave transports encoded data through a transmission medium.
12. With parity checking, sender and receiver must agree whether error detection is based on ________ or ________.
13. A channel's ________ describes the mathematical relationship between noise power and signal power.
14. ________ is any change in a signal characteristic caused by components of the communication channel.
15. For any error-detection method, a decrease in ________ is accompanied by an increase in ________ error.
16. ________ can't affect optical signals but can affect electrical or RF signals.
17. A communication channel using electrical signals must have at least two wires—a(n) ________ and a(n) ________—to form a complete electrical circuit.
18. ________ measures a channel's theoretical capacity. ________ measures the actual capacity of a channel when a specific communication protocol is used.
19. Multiple messages can be transmitted on a single transmission line or circuit by ________ multiplexing or ________ multiplexing.
20. A(n) ________ signal can encode an infinite number of possible numeric values.
21. ________ is a measure of peak signal strength.
22. In asynchronous transmission, at least one ________ is added to the beginning of each message.
23. The term ________ describes encoding data as variations in one or more physical parameters of a signal.

24. In ________ transmission, blocks or characters arrive at unpredictable times, and no signal is transmitted during idle periods.
25. A medium's ________ is the difference between the highest and lowest frequencies that can be transmitted.
26. ________ mode implements two-way transmission with two separate communication channels; ________ mode implements two-way transmission with only one communication channel.
27. ________ encodes data by varying the magnitude of wave peaks in an analog signal.
28. ________ transmission uses multiple lines to send multiple bits simultaneously.
29. A(n) ________ extends a signal's range by retransmitting the signal without any noise or distortion from earlier transmission stages.
30. ________ generates a(n) ________ consisting of a single parity bit for each bit position in the group of characters or bytes.
31. In ________ transmission, signals are transmitted continuously, even when there's no data to send, to ensure clock synchronization.
32. ________ uses more than two signal characteristic levels to encode multiple bits in a single signal.
33. ________ encodes bit values with rapid pulses of electrical or optical power.
34. ________ is noise added to the signal from EMI generated by adjacent transmission lines in a parallel communication channel.
35. Frequency-division multiplexing of optical channels is sometimes called ________.
36. The length of a parallel communication channel is limited by ________, which can cause bits to arrive at slightly different times.
37. A receiver informs a sender that data was received correctly by sending a(n) ________ message. It informs the sender of a transmission or reception error by sending a(n) ________ message.
38. ________ is loss of signal strength as it travels through a transmission medium.
39. Messages transmitted by time-division multiplexing are divided into ________ before physical transmission.
40. ________ cable is an improved version of Category 5 cable that transmits data at high speeds more reliably.
41. Wireless LANs following the IEEE ________, ________, and ________ standards transmit in the 2.4 GHz band.

Review Questions

1. What are the components of a communication channel?
2. What are the components of a communication protocol?
3. Describe frequency modulation, amplitude modulation, phase-shift modulation, and on-off keying.

4. How does multilevel coding increase a channel's effective data transfer rate?
5. Describe the relationship between bandwidth, data transfer rate, and signal frequency.
6. How can noise and distortion be introduced into a transmission medium? How does a channel's S/N ratio affect the reliability of data transmission?
7. Compare twisted-pair, coaxial, twin-axial, multimode fiber-optic, and single-mode fiber-optic cable in terms of construction, susceptibility to EMI, cost, and bandwidth or transmission speed.
8. What are the advantages of wireless transmission using RF waves compared with infrared and visible light waves?
9. Describe simplex, half-duplex, and full-duplex transmission and compare them in terms of cost and effective data transfer rate.
10. Why is a channel's actual data transfer rate usually less than the theoretical maximum of the technology used to implement the channel?
11. Compare serial and parallel transmission in terms of channel cost, data transfer rate, and suitability for long-distance data communication. Why are standards for connecting secondary storage devices migrating from parallel to serial transmission?
12. What are the differences between synchronous and asynchronous data transmission?
13. What is character framing? Why is it generally not an issue in parallel data transmission?
14. Describe the differences between even and odd parity checking.
15. What's a block check character? How is it computed and used?
16. Compare frequency-division and time-division multiplexing. What physical characteristics of the communication channel does each technique require? Which provides higher data transmission capacity?
17. What's the difference between an amplifier and a repeater?
18. Compare IEEE 802.11b, 802.11g, and 802.11n wireless transmission in terms of raw data transfer rates, transmission frequencies, efficient use of available bandwidth, and susceptibility to noise and interference.

Problems and Exercises

1. Calculate the effective data transfer rate for a dedicated channel, given these characteristics: Data is transmitted in blocks of 48 bytes at 100 Mbps, an 8-bit BCC is used for error detection, and an error is detected, on average, once every 1000 transmitted blocks.
2. For data transmitted over twisted-pair cable, you have these characteristics: A signal's power on the channel is 75 dB, signal attenuation is 5 dB per 100 meters, and average noise power on the cable is 0.1 dB per meter. Answer the following questions:

   - What's the S/N ratio at the end of a 50-meter channel?
   - What's the S/N ratio at the end of a 1-kilometer channel?
   - What's the maximum channel length if the receiving device requires at least 20 dB of S/N ratio for reliable operation?

Research Problems

1. Parity checking is often used in computer memory to detect errors. Investigate current memory types, including parity and error-correcting code (ECC) memory. How is parity checking implemented in the memory module? How are errors detected, and what happens when an error is detected? What types of computer systems should use parity or ECC memory, and why?
2. Research InfiniBand Architecture (*www.infinibandta.org*) to investigate some issues not addressed in the chapter. What signal encoding, cables, and connectors are used? Is the channel serial or parallel? Is transmission synchronous or asynchronous? How are errors detected and corrected?
3. Investigate the IEEE FireWire data communication standard. How is data encoded, and what is the raw transmission speed? Describe how the standard enables quality of service guarantees when transmitting multimedia and other time-sensitive data.
4. Investigate the IEEE 806.16 (WiMAX) wireless networking standard. How do transmission speeds and distances compare with the IEEE 802.11 standards? What methods are used to detect and correct transmission errors? Will 802.16 networks eventually replace 802.11 networks? Why or why not?

CHAPTER 9

COMPUTER NETWORKS

CHAPTER GOALS

- Describe logical and physical network topologies
- Explain message forwarding across local and wide area networks
- Describe media access control protocols
- Describe network hardware devices
- Summarize the OSI network layers
- Describe the Internet architecture and network protocols
- Describe IEEE physical network standards, including Ethernet

Chapter 8 discussed data communication between a single sender and a single receiver, including data transfer between hardware devices in a computer system, between a computer and a peripheral device (such as a printer), and between two connected computers. Communication between computers in a network is more complex, as shown in Figure 9.1.

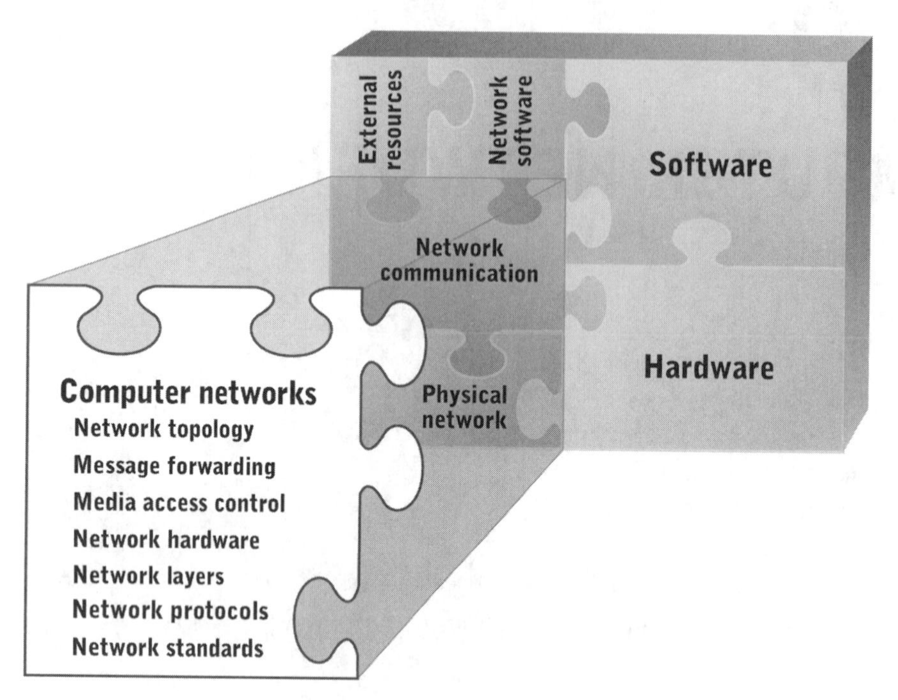

FIGURE 9.1 Topics covered in this chapter
Courtesy of Course Technology/Cengage Learning

NETWORK TOPOLOGY

The term **network topology** refers to the spatial organization of network devices, physical routing of network cabling, and flow of messages from one network node to another. An end node is a device such as a workstation, server, or printer that can be the source or destination of a message. Physically linking two nearby end nodes is usually straightforward. A point-to-point transmission line is laid over the shortest path and connected directly to both end nodes.

Figure 9.2(a) shows a **mesh topology**, in which every node pair is connected by a point-to-point link. This topology requires many transmission lines if the number of end nodes is large. To connect four end nodes, for example, six transmission lines and three connections per node are required. As the number of end nodes (n) increases, the number of point-to-point transmission lines rises quickly, as described by this formula:

$$1 + 2 + 3 + \cdots + (n - 1)$$

Mesh topology is impractical for all but very small networks. For larger networks, a segmented approach that shares links among end nodes is more efficient. Note the similarity of Figure 9.2(b) to a roadway system. End nodes are served by low-capacity links that connect to higher-capacity shared links, just as driveways connect houses to residential streets. Smaller shared links are connected to larger shared links in the same way residential streets connect to higher-capacity roads and highways.

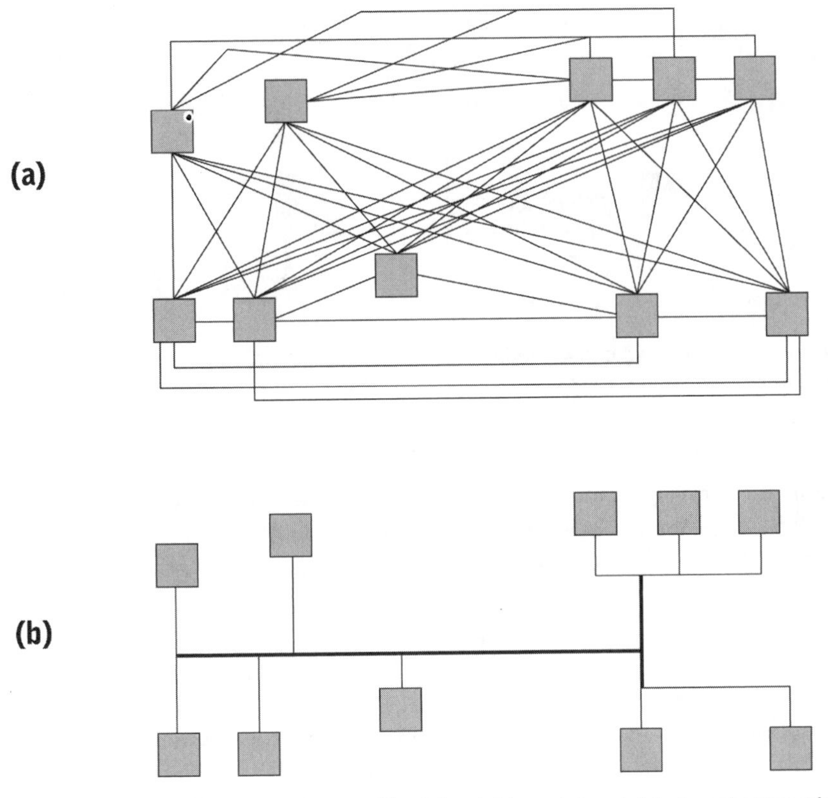

FIGURE 9.2 End nodes connected by (a) point-to-point and (b) shared connections
Courtesy of Course Technology/Cengage Learning

If the end nodes in Figure 9.2(b) were parts factories and assembly plants and the links between them were roads, how would you design a delivery service to move goods between them? Most large-scale delivery services use a **store and forward** system to connect source and destination nodes. Figure 9.3 shows the network in Figure 9.2(b) with three central nodes of small capacity and one of large capacity. In a physical delivery system, the central nodes are transfer points located at or near the junction of major roadways or air routes. Shipments move from end nodes to the nearest central node, where they are combined into larger shipments to other central nodes.

Computer networks and shipping networks are similar in that both connect many end nodes by using a system of interconnected transmission routes and central nodes. Both network types ensure reliable, rapid movement of shipments or messages between end nodes with minimal investment in network infrastructure. Both network types achieve reliability and cost efficiency by carefully organizing end nodes, transmission routes, and central nodes.

Network topology is an essential factor in computer network efficiency and reliability and can be referred to in physical or logical terms. **Physical topology** is the physical placement of cables and device connections to these cables. **Logical topology** is the path

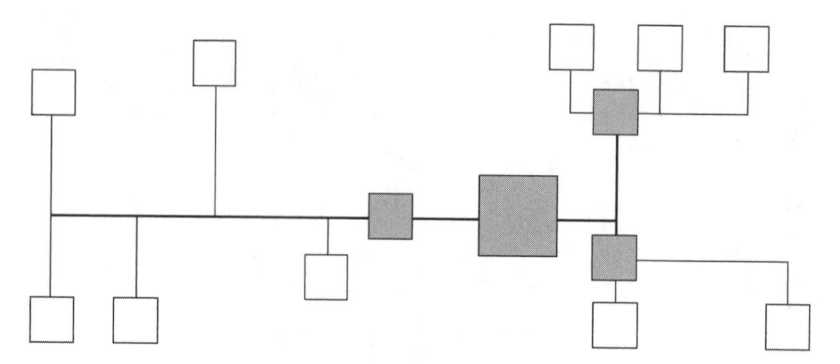

FIGURE 9.3 Shared connections and central nodes
Courtesy of Course Technology/Cengage Learning

messages traverse as they travel between end and central nodes. Logical topology can differ from the underlying physical topology, as later examples illustrate.

Physical topologies include the following:

- *Mesh*—This topology directly connects every node to every other node; practical only for very small networks.
- *Bus*—A **bus topology** directly connects every node to a single shared transmission line, as shown in Figure 9.4.
- *Ring*—A **ring topology** directly connects every node to two other nodes with a set of links forming a loop or ring (see Figure 9.4).
- *Star*—A **star topology** directly connects every node to a shared hub, switch, or router (see Figure 9.5).

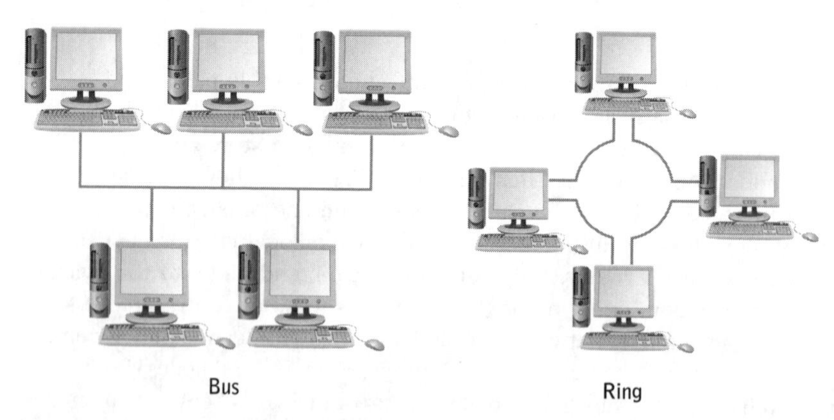

FIGURE 9.4 Bus and ring topologies
Courtesy of Course Technology/Cengage Learning

The characteristics differentiating these topologies include the length and routing of network cable, type of node connections, data transfer performance, susceptibility of the network to failure, and cost.

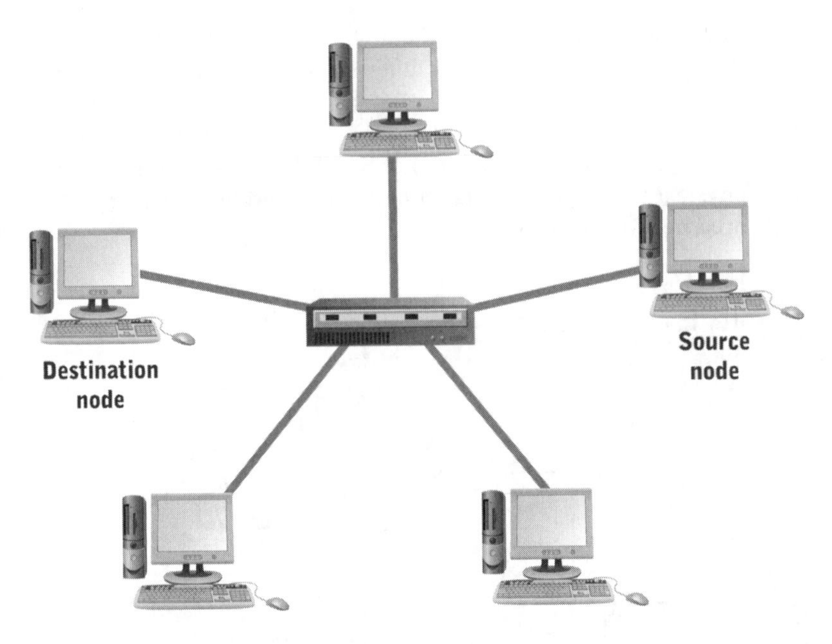

FIGURE 9.5 Star topology
Courtesy of Course Technology/Cengage Learning

Networks using bus and ring physical topologies were once common but are rarely used today. The historical advantages of bus networks (simplicity, reliability, and low-cost construction) have faded as network hardware has become cheaper and more reliable. Ring networks once had a performance advantage achieved at the expense of complex configuration. However, as network technology evolved, this advantage was eroded by faster network devices and network managers' desire to simplify network configuration. Star topologies now dominate physical network topology for wired networks. A bus topology is still common in wireless networks, described later in this chapter.

The advantage of a star topology is simple wiring. A transmission line connects each end node to the central node, which is typically in a central place in a building or a floor of a multistory building. The main disadvantage is that failure of the central node disables the entire network. Star networks now use two transmission lines (full-duplex mode, discussed in Chapter 8) between each end node and the central node, which enables messages to travel in both directions simultaneously.

MESSAGE ADDRESSING AND FORWARDING

This section discusses how messages sent by end nodes find their way through transmission lines and central nodes to their ultimate destination. Many types of central nodes can be used in a network, including hubs, switches, and routers. Although they have internal differences, they share a similar purpose—forwarding incoming messages from end nodes to the recipient end node or another central node that's one step closer to the recipient end node. To simplify the discussion of forwarding, the differences are ignored for now, and the generic term "central node" is used to describe any device that forwards messages. Later in "Network Hardware," you learn about different types of central nodes in more detail.

Figure 9.6 shows a typical campus wired network. Entire buildings, such as Jones Hall, or parts of buildings, such as the first three floors of Smith Hall, are organized into small networks wired as physical star topologies. Each network covering a floor or building is called a **local area network (LAN)**.

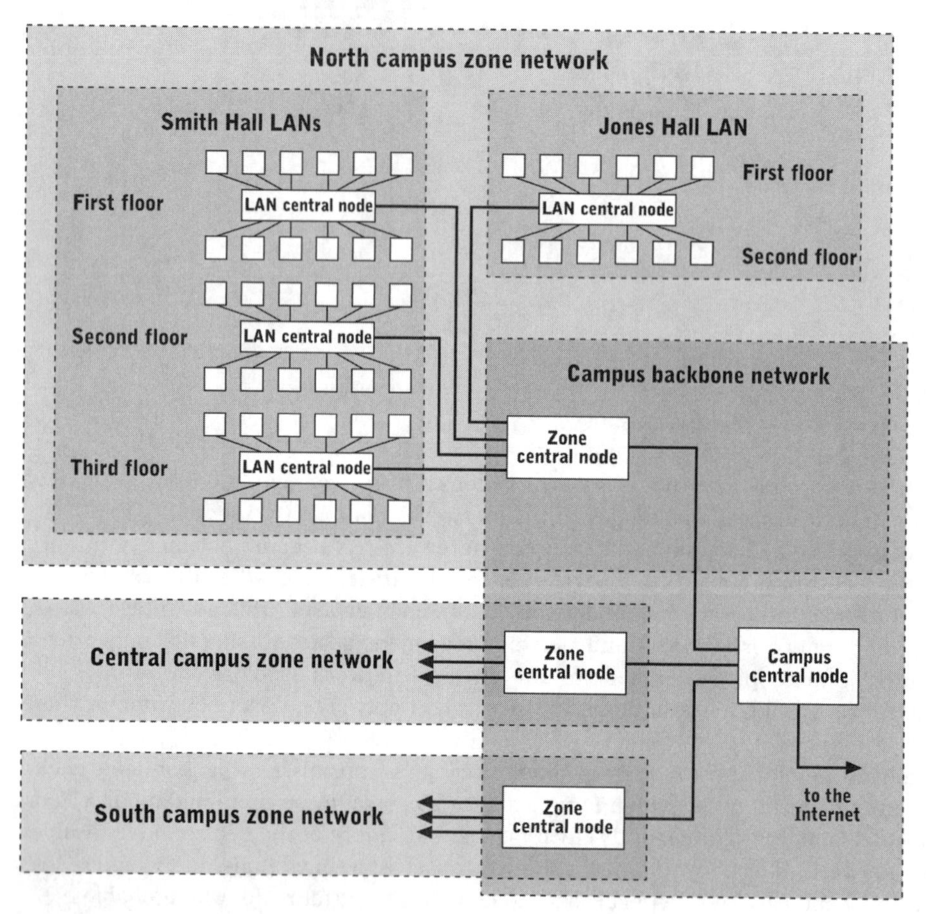

FIGURE 9.6 A typical campus wired network
Courtesy of Course Technology/Cengage Learning

A zone network connects all LAN central nodes in a group of buildings, a zone central node connects each zone network to the campus backbone network, and a campus central node connects the campus backbone network to the Internet. The entire campus network is a **wide area network (WAN)**, including end nodes, LANs, zone networks, the campus backbone network, and central nodes.

Messages can travel through many transmission lines and central nodes before reaching their destinations. Each message includes a destination address that a central node examines to determine where to forward the message. How does a central node "know" the addresses of other nodes in its network, and how does it forward a message to a distant address? Each central node maintains a table of node addresses and transmission lines or connection ports, called a **forwarding table**, and uses this table to make forwarding decisions. Each time an end node or a central node is powered on, it sends a message announcing its presence and its address to all nearby nodes. Central nodes listen for these messages and update their tables accordingly.

The procedure for forwarding a message between two end nodes in the same LAN depends on the logical network topology. For a logical bus, the central node examines the destination address to see whether it's local. If so, the central node forwards the message to all other local nodes. Each local node sees the message, but only the node with a matching destination address processes the message; all other nodes simply ignore it. If the destination address isn't local, the central node forwards the message to the next central node. The central node needs to distinguish only between local and nonlocal addresses to determine whether to forward the message to another central node.

Forwarding local messages in a logical star network is more complex. Because each node is attached to its own transmission line, the central node's forwarding tables keep track of which transmission line is connected to which address. A local message is forwarded from the source node's transmission line to the destination node's transmission line. The central node connects local sending and receiving nodes as needed for each incoming message.

Messages coming from outside the LAN to a local node are handled much like messages between local nodes. For logical bus networks, messages arriving from another central node are copied to the local bus. For a logical star network, the central node looks up the target address and copies the message to the target node's transmission line. Table 9.1 summarizes the forwarding decisions made by LAN central nodes.

TABLE 9.1 LAN central node routing decisions

Source and destination	Routing action
Local node to local node	For ring or bus, propagate message through local medium. For star, copy message from source node's transmission line to target node's transmission line.
Nonlocal node to local node	For ring or bus, copy message to local medium. For star, copy message from source node's transmission line to target node's transmission line.
Local node to nonlocal node	Forward message to the next central node.

The zone and campus central nodes in Figure 9.6 must make more sophisticated forwarding decisions than LAN central nodes do, and they need to know how to forward messages to any network address. One method of storing and accessing forwarding information on all addresses is to store a master directory of all known network addresses on one or more central servers and have central nodes get forwarding information by sending queries to these servers. This method works in theory, but in practice, it has many shortcomings, including the following:

- Every central node must know the location of one or more directory servers. If a server is moved, all central nodes must update their server location information.
- The master directory size is large, as is the volume of directory service requests. Many powerful computers are required to store the directory and answer queries.
- Every directory query travels over long distances, and WANs are clogged with directory queries and responses.
- Directory updates must be copied to each server, requiring multiple update messages (one for each server) or regular synchronization of directory content between servers. Either approach further clogs WANs.

Instead of the master directory method, most WANs use a distributed approach to maintaining directories. With this approach, each central node knows the addresses and physical locations of other nodes on its network and knows other nearby central nodes and the groups of addresses they control. Each central node also has one or more default destinations for addresses it doesn't know. Central nodes exchange information periodically, so they are kept updated on changes to network nodes and topology.

For example, a message is forwarded from an end node in Jones Hall to an end node in the south campus zone network in Figure 9.6. The Jones Hall LAN central node knows the addresses of all Jones Hall end nodes and the address of the north campus zone central node. When the central node receives a message to any node outside its LAN, it forwards the message to the north campus zone central node.

The north campus zone central node examines the destination address and compares it with addresses in its forwarding table. The LAN central nodes in Jones Hall and Smith Hall exchange forwarding information periodically with the north campus central node so that it knows the range of addresses each LAN central node services. Because the destination address doesn't match any of its known address ranges, the north campus central node forwards the message to the campus central node.

The campus central node knows that the destination address falls within a range of addresses controlled by the south campus central node, so it forwards the message to that central node. The south campus central node knows which addresses are controlled by which LAN central nodes and forwards the message to the correct LAN central node, which then forwards the message to its final destination.

NOTE

This description of distributed WAN forwarding is general and omits many details. There are many ways to implement it, but the method just described approximates the standard message forwarding used in most networks.

MEDIA ACCESS CONTROL

When multiple nodes share a common transmission medium, as in wireless LANs, they must coordinate their activities to avoid interfering with one another. If multiple nodes attempt to transmit across the same medium at the same time, their messages mix, producing noise or interference that's called a **collision**. Nodes sharing a common transmission medium follow a **Media Access Control (MAC)** protocol to determine how to share the medium efficiently. MAC protocols fall into two broad categories: those that allow collisions but detect and recover from them and those that attempt to avoid collisions altogether.

Carrier Sense Multiple Access/Collision Detection (CSMA/CD) is a MAC protocol described in the IEEE 802.3 standard and developed for early Ethernet networks, based on a wired bus topology. The basic strategy is not to prevent collisions but to detect and recover from them. Nodes using this protocol follow this procedure:

1. A node ready to transmit listens (carrier sense) until no traffic is detected.
2. The node then transmits its message.
3. The node listens during and immediately after its transmission. If abnormally high signal levels are heard (a collision is detected), the node ceases transmission.
4. If a collision is detected, the node waits for a random time interval and then retransmits its message.

Carrier Sense Multiple Access/Collision Avoidance (CSMA/CA) is a MAC protocol used in wireless networks. The name is a misnomer because collisions aren't avoided completely. Instead, CSMA/CA modifies CSMA/CD to prevent a unique problem in wireless transmission. In the node arrangement in Figure 9.7, say that both laptops can transmit to and from the access point reliably but are too far from one another to receive each others' transmissions. In this situation, carrier sense doesn't always prevent collisions because neither laptop can "hear" the other. Therefore, both might transmit at the same time because they think the RF channel is unused, and neither hears the interference pattern caused by the other's transmission.

FIGURE 9.7 Two laptops transmitting to a wireless access point
Courtesy of Course Technology/Cengage Learning

To prevent this problem, CSMA/CA uses a three-step carrier sense and transmission sequence:

1. A node that wants to transmit sends a **ready-to-send (RTS) signal**.
2. If the wireless access point detects no collision, it transmits a **clear-to-send (CTS) signal**.
3. After receiving the CTS signal, the node transmits its data.

Because all nodes can hear the wireless access point, they hear the CTS signal, and only the node that transmitted the RTS signal answers the CTS signal with a data transmission. If any other node wants to transmit, it waits at least one transmission interval after hearing a CTS before sending its own RTS. Collisions are still possible with CSMA/

CA but only with RTS signals. To avoid subsequent data transmission collisions, the wireless access point doesn't respond to colliding RTS signals. Lacking an immediate CTS response, each node assumes a collision has occurred and waits a random time interval before retransmitting an RTS signal.

NETWORK HARDWARE

Network topology, addressing and forwarding, and MAC functions are carried out by network hardware devices that include the following:

- Network interface cards (NICs) or network adapters
- Hubs
- Switches
- Routers
- Wireless access points

Table 9.2 summarizes the functions of each device. Network hardware is constantly evolving, and networking functions are frequently combined in a single device, particularly in newer hardware. When selecting devices, understanding their functional capabilities is important because device names might be misleading.

TABLE 9.2 Network hardware devices

Device	Function
NIC	Connects a node to the network Performs MAC and message-forwarding functions Acts as a bridge between the system bus and LAN
Hub	Acts as a central connection point for LAN wiring Implements the logical network topology
Switch	Forwards messages between nodes by creating and deleting point-to-point connections rapidly
Router	Connects two or more networks Forwards messages between networks as needed Makes intelligent choices between alternative routes
Wireless access point	Connects a wireless network to a wired network and forwards messages Performs MAC and security functions for wireless networks

Network Interface Cards

A device that connects a node, such as a computer or network printer, to a network transmission cable is called a **network interface card (NIC)** or **network adapter**. For a single computer, it can be a printed circuit board or card attached to a bus expansion port, or it can be integrated on the computer motherboard. An OS device driver controls the NIC and directs the hardware actions that move messages between the NIC and primary storage. For a peripheral device, such as a printer, a NIC is more complex because it can't rely on the processing and storage resources available in a computer.

In a bus network, the NIC scans the destination addresses of all messages and ignores those not addressed to it. In ring networks, it scans the destination addresses of all messages and retransmits those not addressed to it. The NIC in a star network accepts all incoming messages because the transmission line is shared only with the central node. When a correctly addressed message is received, the NIC stores it in a buffer and generates an interrupt on the system bus. It also performs MAC functions, including listening for transmission activity, detecting collisions, and retransmitting messages in CSMA/CD and CSMA/CA networks.

Hubs

A **hub** is a central connection point for nodes in a LAN. An evolutionary step in early bus and ring LANs was moving the physical topology from the LAN wiring to the hub. For example, early Ethernet LANs were wired with a shared coaxial cable (the bus), and end nodes connected to it with a transceiver. The introduction of Ethernet hubs eliminated the need for the coaxial cable and transceivers. The hub provided separate point-to-point connections between nodes and the hub by using less expensive cabling in a physical star topology and attached these connections to its internal shared bus (see Figure 9.8). As technology has progressed, LANs have abandoned bus and ring topologies entirely, and switches have replaced hubs.

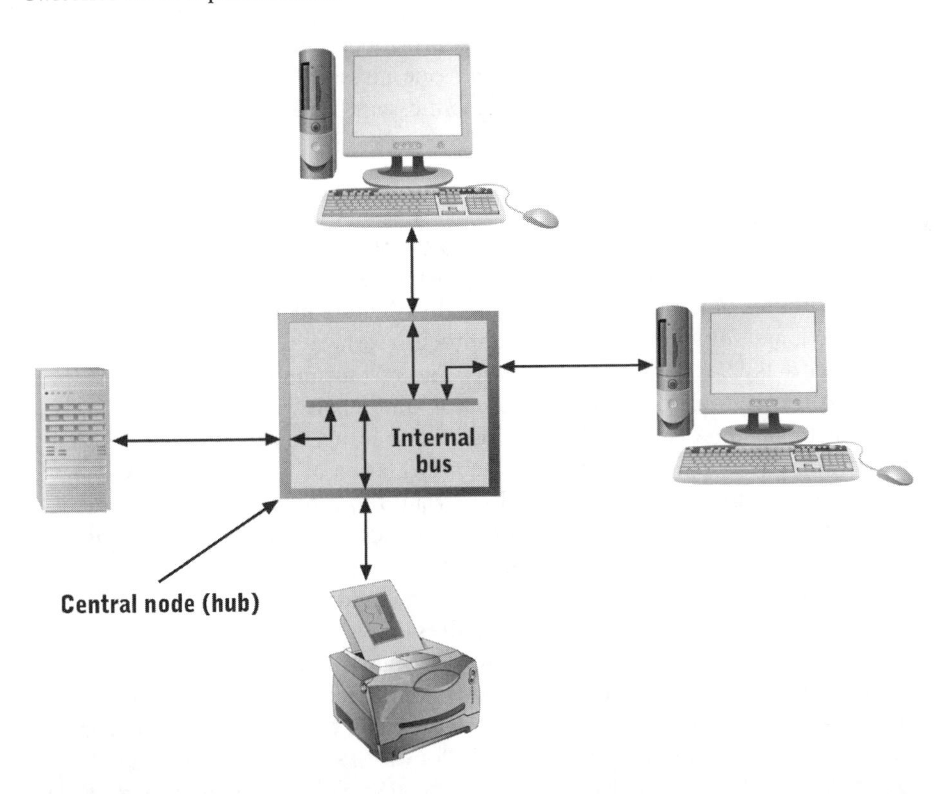

FIGURE 9.8 An Ethernet hub
Courtesy of Course Technology/Cengage Learning

Switches

Like a hub, a **switch** generally has a dozen or more input connections for computers and other network nodes. Unlike a hub, point-to-point connections with nodes aren't connected to a shared transmission medium. Instead, a switch examines the destination address of each incoming message and temporarily connects the sender's transmission line to the receiver's transmission line. It creates a temporary point-to-point connection for the time needed to transmit one message and breaks the connection as soon as the message has reached its destination.

Switches are more complex than hubs because they contain multiple connection points that can connect and disconnect rapidly and because they must "understand" message formats to extract addresses and control connection duration. Compared with a hub, a switch increases network performance for these reasons:

- Each connection has only one sending and one receiving node, thereby eliminating collisions and congestion.
- Multiple connections are supported, which enables multiple sender-receiver pairs to transmit at the same time.

In essence, each node occupies a separate LAN that's temporarily connected to another LAN as needed, eliminating collisions and retransmissions. A switch must deal with possible contention for transmission connections, as when two nodes try to send a message to a third node at the same time. Switches provide a memory buffer for each node connection that's large enough to hold at least one message. If a destination node is busy when a message destined for it is received, the message is held in the buffer until the connection is available.

Routers

A **router** intelligently forwards messages between two or more networks. It stores messages in a buffer, examines their contents, and applies decision rules to determine where to forward messages. Router buffers are large because they handle many messages that must be stored, examined, and forwarded. The routing process is more complex than switching, and there's a longer delay between message receipt and retransmission.

A router scans the network constantly to monitor traffic patterns and network node additions, modifications, and deletions. Routers use this information to build an internal "map" of the network. They periodically exchange information in their internal **routing tables** with other routers to learn about networks beyond those to which they're directly connected. Using this information, they can forward messages from local nodes to distant recipients and choose from multiple possible routes to a recipient.

Routers can be connected to more than two networks and can forward messages based on information other than the destination address. For example, an organization might have two internal networks—one dedicated to ordinary data traffic and another dedicated to audio or video streams. A router can be configured to examine messages arriving from an external network and forward them to one of the internal networks based on their content. A router might also be connected to two ISPs and route outgoing messages to a specific ISP based on criteria such as shortest path to recipient or maximal use of the least expensive ISP.

A stand-alone router is essentially a special-purpose computer with its own processor and storage. Routing and switching can be done in the same device, which blurs the distinction between them. In addition, a general-purpose computer with multiple NICs connected to different segments or networks can be configured as a router if the necessary software is installed. Routing software is usually a standard network OS component and might or might not be enabled by the server administrator.

Wireless Access Points

A wireless **access point (AP)** connects a wireless network to a wired network. Typically, wireless networks follow the IEEE 802.11a, 802.11b, 802.11g, or 802.11n standards, and wired networks follow one of the IEEE 802.3 (Ethernet) standards. Therefore, wireless APs serve as central points to carry out the central management functions of wireless protocols and as a switch connecting two different physical networks, translating between their protocols as needed. In their role as a central node for wireless networks, wireless APs manage media access, perform error detection, and implement security protocols.

OSI NETWORK LAYERS

In the late 1970s, the ISO developed a conceptual model for network hardware and software called the **Open Systems Interconnection (OSI) model**. The OSI model is useful as a general model of networks, a framework for comparing networks, and an architectural roadmap that enhances interoperability between different network architectures and products. It organizes network functions into seven layers (see Figure 9.9). Each layer uses the services of the layer immediately below it and is unaware of other layers' internal functions.

Application Layer

The **Application layer** includes communication protocols used by programs that make and respond to high-level requests for network services. Programs using Application layer protocols include end-user network utilities, such as Web browsers and e-mail clients; network services embedded in the OS, such as service routines that access remote files; and network service providers, such as File Transfer Protocol (FTP) and Web server programs.

The Application layer of one network node generates a service request and forwards it to the Application layer of another network node. The Application layers of the two network nodes "talk" to each other (shown by the top dashed arrow in Figure 9.9) by using all the lower-level OSI layers on both nodes as a communication channel (shown by solid arrows between layers in Figure 9.9). Other layers also communicate with their counterparts in this way and use the lower-level layers to transmit messages.

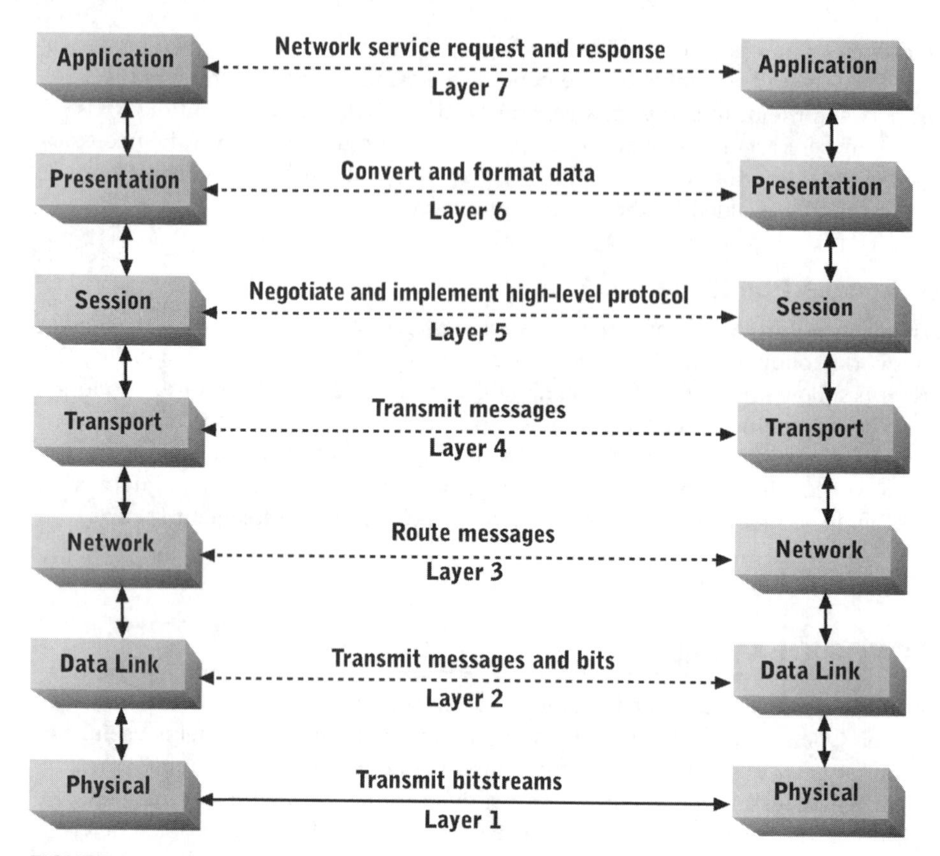

FIGURE 9.9 Layers of the OSI model
Courtesy of Course Technology/Cengage Learning

Presentation Layer

The **Presentation layer** ensures that data transmitted by one network node is interpreted correctly by the other network node. It's used mainly by applications that format data for user display. For example, a Web browser's Presentation layer formats compressed JPEG images for visual display and decides which fonts to use for text display. Other layered network models usually collapse Presentation-layer functions into the Application layer.

Session Layer

The **Session layer** establishes and manages communication sessions. When a session is first established, the Session layers negotiate protocol parameters, such as encryption and quality of service. After protocol parameters have been established, they monitor communication to detect and deal with any problems. For example, when an order is placed over the Web, the Web browser and server use Session-layer protocols to negotiate session parameters, such as encrypted transmission, and monitor the session for disruptions. If the communication channel is broken before the transaction is completed, the Web server node ensures that the transaction is abandoned—for instance, by canceling an unfinished order so that charges aren't applied to the customer's credit card.

Transport Layer

The **Transport layer** formats messages into packets suitable for transmission over the network. It places messages in a packet data area and adds required header and trailer information, including network addresses, error-detection data, and packet-sequencing data. Packets are given to the Network layer for delivery to the recipient. After receiving packets, the Transport layer examines them for errors and requests retransmission if necessary.

If a message doesn't fit in a single packet, the sending Transport layer divides the message content among several packets, adds a sequence number to each packet, and adds special coding in the first and last packets. The receiving Transport layer requests retransmission of any missing packets and reassembles data from multiple packets in the correct order.

Network Layer

The **Network layer** forwards messages to their correct destinations. As described earlier, the Network layers of sending nodes typically forward messages to the nearest central node. Network layers in the central node interact with one another to exchange forwarding information and update internal tables. Network layers in end nodes announce their presence to other end nodes and central nodes when they're initialized.

Data Link Layer

The **Data Link layer** is the interface between network software and hardware. In an end node, such as a computer or network printer, the NIC and device drivers implement the Data Link layer. Device drivers manage message transfer from secondary or primary storage to the NIC, which transmits messages as bitstreams across the physical network link. Data Link layer device drivers perform functions such as media access control and conversion of messages and addresses from one format to another—for example, from Internet formats to Ethernet formats.

Physical Layer

The **Physical layer** is where communication between devices actually takes place. It includes hardware devices that encode and decode bitstreams and the transmission lines that transport them.

INTERNET ARCHITECTURE

The U.S. Department of Defense (DOD) Advanced Research Projects Agency Network (ARPANET) developed network technology in the late 1960s and early 1970s to connect researchers working on defense projects. Many DOD researchers worked at universities, which soon incorporated ARPANET technology in their own networks. Networks based on ARPANET technology eventually evolved into the Internet. As a result of worldwide Internet adoption, core ARPANET protocols and their extensions have become worldwide standards.

NOTE

There's no universally accepted name for the layered protocol model that describes current Internet standards and technology. Some terms include DARPA model, Internet model, and **TCP/IP model**, which is the term used for the remainder of this chapter.

Transmission Control Protocol (TCP) and **Internet Protocol (IP)**, which are jointly called **TCP/IP**, are the core protocols of the Internet. Most services normally associated with the Internet are delivered via TCP/IP, including file transfer via File Transfer Protocol (FTP), e-mail distribution via Simple Mail Transfer Protocol (SMTP), and access to Web pages via Hypertext Transfer Protocol (HTTP). TCP/IP is the glue that binds private networks together to form the Internet and the World Wide Web.

The TCP/IP model predates the OSI model by almost a decade. Therefore, it's understandable that TCP/IP model layers and supporting protocols (see Figure 9.10) correspond only roughly to the OSI model. The following list describes the TCP/IP model layers briefly:

- *Application layer*—Roughly corresponds to the OSI Application and Presentation layers and includes many protocols. Only a few are shown in the figure: those defined previously as well as Domain Name System (DNS), Trivial File Transfer Protocol (TFTP), Dynamic Host Configuration Protocol (DHCP), and Network File System (NFS).
- *Transport layer*—Roughly equivalent to the OSI Session and Transport layers and includes TCP and User Datagram Protocol (UDP).
- *Internet layer*—Roughly corresponds to the OSI Network layer. IP is the primary protocol, although others, including Internet Control Message Protocol (ICMP), Address Resolution Protocol (ARP), and Reverse Address Resolution Protocol (RARP), play a supporting role.
- *Network Interface layer*—Roughly equivalent to the OSI Data Link layer. This layer connects Internet protocols to underlying network protocols in the Physical layer.
- *Physical layer*—Roughly equivalent to the OSI Physical layer. This layer contains physical network protocols, such as Ethernet and Asynchronous Transfer Mode (ATM).

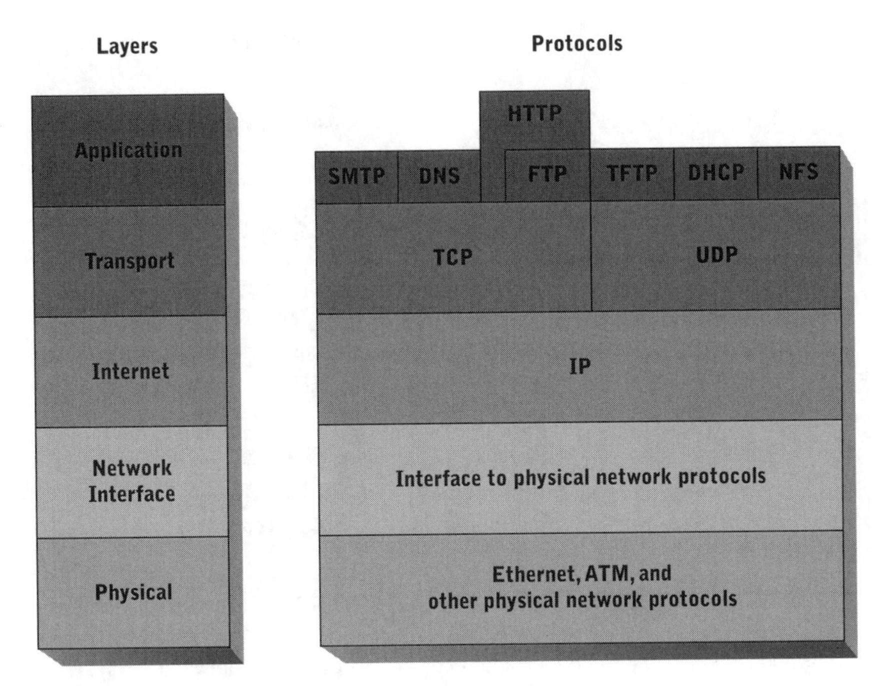

FIGURE 9.10 TCP/IP model layers and protocols
Courtesy of Course Technology/Cengage Learning

Internet Protocol

IP accepts messages called **datagrams** from Transport-layer protocols and forwards them to their destination. Because the Internet layer of the TCP/IP model is responsible for message forwarding, it includes rules for constructing valid addresses and routing packets from sender to receiver through central nodes as needed. IP assumes that a datagram traverses multiple networks via nodes called **gateways**, which are nodes connecting two or more networks or network segments that might be implemented physically as workstations, servers, or routers. Figure 9.11 shows the TCP/IP model layers involved in a route between two end nodes that includes two gateways. Note that the Transport and Application layers aren't implemented in the gateways.

Every IP node has a unique 32-bit or 128-bit address. The 32-bit addresses are defined in **Internet Protocol version 4 (IPv4)** and written in the form *ddd.ddd.ddd.ddd*, with *ddd* representing a decimal number between 0 and 255. IPv4 is slowly being replaced by IPv6, which uses 128-bit addresses written in the form *hhhh:hhhh:hhhh:hhhh:hhhh:hhhh:hhhh:hhhh*, with *hhhh* representing a sequence of four hexadecimal digits.

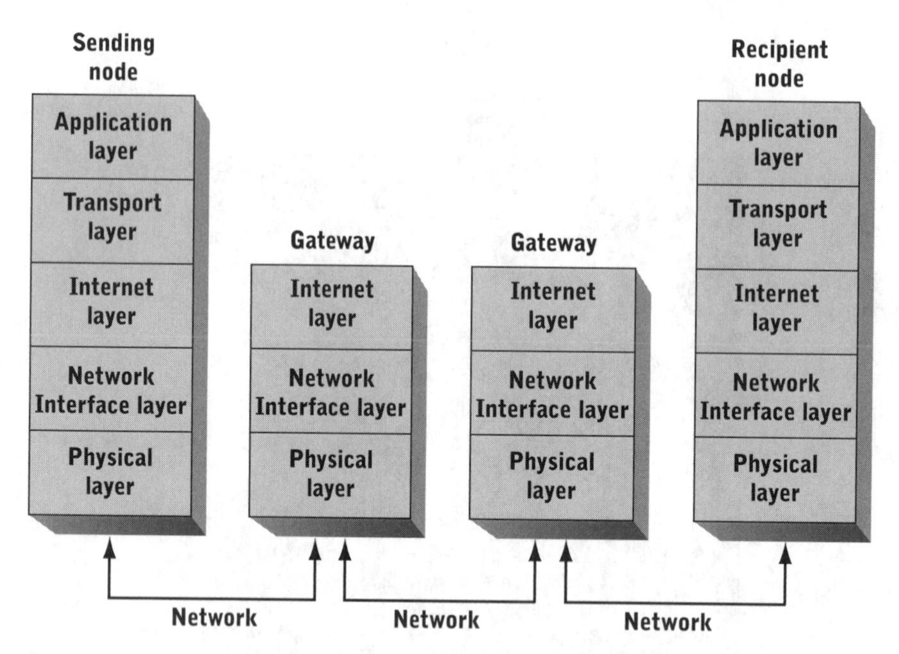

FIGURE 9.11 TCP/IP connection between sender and receiver, using multiple networks and gateways

Courtesy of Course Technology/Cengage Learning

NOTE

The written format of 32-bit IP addresses is called **dotted decimal notation**. The written format of 128-bit IP addresses is called **colon hexadecimal notation**, or "colon hex." Colon hexadecimal addresses containing a group of 0 digits can be abbreviated. For example, the address 2002:406A:3A9B:0000:0000:0000:C058:6301 can be written as 2002:406A:3A9B::C058:6301.

Each gateway maintains a routing table of local node IP addresses and their corresponding physical network addresses, such as Ethernet addresses. Each node and gateway knows the physical network address of at least one other gateway. If a received message contains an IP address that doesn't appear in a gateway's internal tables, the packet is forwarded to the default gateway.

IP tables can include entries for partial IP addresses. For example, most IPv4 node addresses in the University of New Mexico begin with 129.24. A node or gateway might not know the physical address for a specific node, such as 129.24.8.1, but its routing tables might contain the physical address of a gateway to any node with 129.24 in the first 16 address bits. Nodes and gateways announce their presence to nearby devices whenever they're started and exchange routing information periodically to keep their tables current.

IPv6

Internet Protocol version 6 (IPv6) is an updated version of IPv4 defined in the late 1990s and first deployed in the mid-2000s. It was developed to address a number of real and perceived problems with IPv4, including the following:

- Limited number of node addresses
- Poor fit to streaming multimedia
- Inability to multicast efficiently

IPv4's 32-bit addresses limited the total number of Internet nodes to around four billion. This number seemed more than enough in the late 1960s and was a good match to the limited networking and data communication technologies at that time. However, by the late 1990s, it was apparent that soon more than four billion devices would be connected to the Internet. IPv6's 128-bit addresses allowed enough expansion for the next century and offered more flexible ways to allocate groups of addresses to organizations and nodes.

The TCP/IP model and protocols were intended to transmit small data items quickly. When large data items, such as files, were to be transferred, it was assumed the transfer wasn't time sensitive. As computer and network technology improved in the 1980s and 1990s, Internet traffic increasingly included large streams of real-time audio and video data to support applications such as videoconferencing and Internet radio. IPv4 was poorly suited to transmitting this type of data, resulting in quality problems (see the VoIP Technology Focus for examples). IPv6 doesn't fully solve all the problems, but it does solve some.

The term **multicasting** describes transmission situations involving multiple senders and receivers. The most common is a single server transmitting the same data to multiple end nodes simultaneously (as in Internet radio and three-way calling). IPv4 had limited support for multicasting, but it was too inefficient for practical deployment. IPv6 expands support for efficient multicasting in one-to-many, many-to-one, and many-to-many modes.

Because IP is used in all Internet nodes and gateways, moving from IPv4 to IPv6 is a major undertaking. Many vendors didn't start providing IPv6 support until the late 2000s. End nodes can be updated by patching older OSs or switching to newer OSs designed to support IPv6. Upgrading gateways is more difficult. Most Internet gateways are routers implemented as dedicated hardware devices with embedded network software. Software in newer routers can be updated with remotely installed upgrades, but many older routers lack this capability, and others are simply incapable of supporting IPv6. As of 2010, IPv6 use is still far below IPv4, and full conversion to IPv6 will take many more years.

TCP

IP is an example of a **connectionless protocol**, in which the sender doesn't attempt to verify a recipient's existence or ask its permission before sending data. An end node sends an IP datagram to a gateway and assumes it will find its way to the intended recipient. IP doesn't support error detection or correction because datagrams are launched toward a recipient Internet layer without telling it that data is coming. There's no way for the recipient to know whether a datagram has been lost in transit or for the recipient node to send a negative acknowledgment or request retransmission.

TCP, on the other hand, is a **connection-oriented protocol**, so it has the framework needed to check for lost messages by establishing a connection between one sender and one recipient before transmitting messages. It can perform several connection management functions, including verifying receipt, verifying data integrity, controlling message flow, and securing message content. The sender and recipient TCP (Transport) layers maintain information about one another, including message routes, errors encountered, transmission delays, and status of ongoing data transfers. They can use this information to adjust their communication parameters to match changing network conditions.

TCP uses a positive acknowledgment (ACK) protocol to ensure data delivery. A recipient TCP layer sends an ACK signal to the sender TCP layer to acknowledge data receipt. The sender TCP layer waits for ACK signals for a time interval that's set based on the time required to establish a connection. This interval can be extended or shortened during the life of the connection to adapt to changing network conditions. If an ACK isn't received, the sending TCP layer assumes the message has been lost and retransmits it. If several consecutive messages are lost, the connection times out, and the sender assumes the connection has been lost.

TCP connections are established through a **socket**, which is the combination of an IP address and a port number. A **port** is a TCP connection with a unique integer number. Many ports are standardized to specific Internet services. For example, port 80 is the default TCP connection for HTTP (used by Web browsers and servers), and port 53 is the default TCP connection for Internet name service queries. The socket 129.24.8.1:53 is the TCP connection for Internet name services at the University of New Mexico.

UDP

User Datagram Protocol (UDP) is a connectionless protocol that provides less reliable transport services than TCP does. It lacks TCP's connection management features but does have one capability that TCP doesn't—supporting communication between multiple hosts and multiple senders. Like TCP, UDP addresses sockets, which enables targeting UDP communication to specific applications.

UDP is used mainly for communication that requires low processing overhead and doesn't require a guarantee of reliable delivery. Applications using UDP include broadcast applications, such as Internet radio; other streaming multimedia applications, such as video playback; and applications with multiple participating nodes, such as multipoint Internet phone service. UDP's lack of positive acknowledgment and other connection management messages reduces the data transfer requirements of these applications. Also, by using many small datagrams, loss of some messages can be tolerated because people aren't usually sensitive to the loss of small amounts of audio or video data in a continuous data stream.

TECHNOLOGY FOCUS

Voice over IP

Voice over IP (VoIP) is a family of technologies and standards for carrying voice messages and data over a single packet-switched network. As discussed in Chapter 8, packet-switching networks generally use available transmission capacity more efficiently than circuit-switching networks. With VoIP, consumers of phone services can place calls over the Internet or other networks by using their computers, VoIP cell phones, or other VoIP-enabled devices at a much lower cost than with a traditional public switched telephone network (PSTN). Skype and Vonage are examples of VoIP-based phone services. Many cable TV and integrated TV/phone companies, such as Comcast and Verizon, also offer VoIP services.

Because of its advantages, many organizations have migrated to VoIP, and more are considering doing so. However, any organization that wants to use VoIP must address many challenges, including the following:

- Choosing VoIP standards
- Acquiring and configuring compatible VoIP equipment and software
- Guaranteeing service quality
- Providing emergency phone connectivity

One challenge of VoIP deployment is the number of complex and competing standards. As with data networks, VoIP networks rely on a protocol suite—an integrated set of protocols, each of which performs a specific function, such as creating connections, packetizing the audio signal, routing packets, ensuring quality of service, and ensuring call security. The oldest and most widely used VoIP protocol suite is **H.323**, which also addresses video and data conferencing. It's an umbrella for many component protocols (see Figure 9.12). Supporting protocols include Registration, Admission, and Status (RAS), Real-Time Transport Protocol (RTP), RTP Control Protocol (RTCP), and H.225. Many of these protocols have multiple versions, which can cause compatibility problems for VoIP equipment and software that support different mixes of standards and versions.

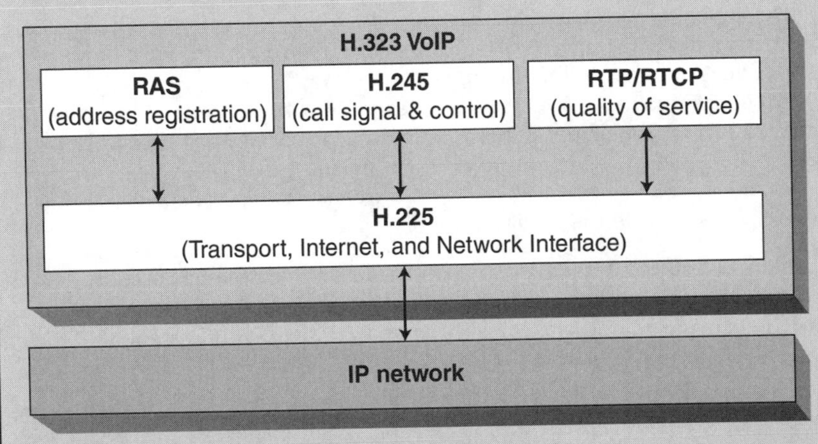

FIGURE 9.12 The H.323 VoIP protocol suite
Courtesy of Course Technology/Cengage Learning

(continued)

Further complicating VoIP deployment is that H.323 has several competitors, including Session Initiation Protocol (SIP), H.248, and Media Gateway Control Protocol (MGCP). Each protocol suite offers a different mix of capabilities, limitations, and interoperability with competing standards. Different vendors support different protocol suites, and equipment capability varies widely between vendors and often even in products from the same vendor. As a result, full interoperability has been the exception rather than the norm, although vendors are improving their products by supporting multiple standards and versions.

VoIP equipment and software can be loosely divided into two groups: end nodes and intermediate nodes. An end node is a device that a user uses to participate in a phone conversation. It can be a traditional phone with an external IP interface, a phone with an integrated IP interface, a computer, or a portable device combining phone and computer capabilities. An intermediate node performs many of the same functions as switches and routers in a data-oriented network, including routing and protocol conversion. Intermediate nodes also perform call and connection management functions that are unique to VoIP.

In the PSTN, a caller opens a connection to the local switching center, receives a dial tone, and then enters a series of numbers to identify the intended recipient. The local switching center uses a protocol suite that describes how a phone number is used to establish a two-way connection to the receiver across multiple networks and telephone switches. VoIP must perform similar call management functions, including the following:

- Establishing connections between VoIP end nodes on IP networks
- Establishing connections between VoIP end nodes and PSTN and other phones
- Performing protocol conversions as needed
- Managing available network capacity to ensure quality of service

Different VoIP protocol suites allocate these tasks to different hardware and software components. At one extreme, most call management functions can be embedded in end nodes. At the other extreme, end nodes can be simpler devices, with call management tasks performed by VoIP intermediate nodes, including specialized routers and switches. Again, the plethora of available standards yields many choices in end and intermediate nodes, with widely varying capabilities and interoperability. Some vendors, such as Cisco, market general-purpose equipment that can be customized to particular protocols and functions by adding software and hardware options.

Guaranteeing service quality isn't a simple matter in most IP networks. IPv4 can't guarantee minimal levels of throughput between a specific sender and receiver or guarantee that packets will be delivered in a timely fashion, in the correct order, or at all. IPv6 addresses some of these issues, but it will be years before it's fully deployed. Transmission quality problems include the following:

- *Packet loss*—VoIP packets might be lost during transmission because of network congestion, router overload, and intermittent device failures. Call participants perceive lost packets as "dropouts," which are missing audio segments of short or long duration.
- *Latency*—VoIP packets can be delayed in transit because of network congestion or having many routers in the chain of devices between end nodes. Call participants perceive periods of silence before the other participant's speech is heard. If the delays are too long, one party might assume the other is silent and start speaking, only to then hear both parties' overlapped speech.

(*continued*)

- *Jitter*—VoIP packets can be subject to latency that fluctuates over short periods. The user perceives jitter as periods of silence mixed with overlapped voice signals.

If the same organization controls both ends of a call and the network in between, intermediate nodes can detect VoIP packets and take actions to ensure their timely delivery. These actions might include routing packets over dedicated communication channels or assigning them higher priority during periods of heavy traffic. Little can be done to ensure service quality across a public IP network or when the same organization doesn't control both ends of the call.

Providing emergency phone service is difficult with VoIP for two reasons. First, ordinary phones are designed to operate during power outages, but most VoIP nodes are not. An ordinary phone receives all the electrical power it requires through the same wires that carry the analog voice signal. If this signal is carried over a separate circuit-switching network, phones can usually function during power blackouts. Because most VoIP end and intermediate nodes require external power, they don't function during power outages.

The second problem with emergency VoIP phone service is that many communities haven't implemented VoIP in emergency (911) call centers. Traditional phones and phone networks are the only means of reaching most of these call centers. To provide reliable emergency phone service, most organizations maintain at least some traditional phone equipment and connections.

Network Interface Layer

The TCP/IP model was originally intended to unite disparate network standards in a way that enabled any node connected to any type of network to communicate with any other node. When this model was developed, more than a dozen widely used network standards with incompatible address formats and forwarding methods were in use. IP provided a universal protocol specifying message and address formats and forwarding methods, but it didn't replace existing protocols. Instead, it was layered over existing protocols and designed to be translated easily into protocols used by the physical networks to which sending and receiving nodes were attached.

At the sending node, the Network Interface layer's role is to translate IP datagrams into a format that can be transported and forwarded over a specific physical network. Reverse translation occurs at the recipient node. A key concept in this translation is **encapsulation**, which embeds all or part of a datagram in a physical network's message format. Physical network protocols have different names for their message formats, including Ethernet frames or ATM cells. Because IP datagrams can be large and most physical network protocols specify a maximum message size, the Network Interface layer must often divide datagrams into pieces and encapsulate each piece in a separate frame or cell.

Another important concept in this layer is address resolution. Each physical network protocol has its own format for network addresses that differs from IPv4 and IPv6 address formats. In essence, an end node's network interface has two addresses: an IP address and an address assigned by the physical network protocol, such as an Ethernet address. The

TCP/IP model defines two protocols to translate between IP and physical network addresses. ARP converts IP addresses to physical network addresses, and RARP translates addresses in the opposite direction. In terms of the OSI model layers, IP addresses are Network layer (layer 3) addresses, and physical network addresses are Data Link layer (layer 2) addresses. A key distinction between switches and routers is that switches make forwarding decisions by using layer 2 addresses, and routers make forwarding decisions by using layer 3 addresses.

PHYSICAL NETWORK STANDARDS

The IEEE has drafted a number of network standards, collectively referred to as the **IEEE 802 standards**. These standards describe physical network hardware, transmission media, transmission methods, and protocols. Table 9.3 lists current standards and standards under development; discontinued and obsolete standards aren't included.

TABLE 9.3 IEEE 802 network standards

Standard	Description
802.1	Media Access Control (MAC)
802.2	Logical Link Control (LLC); inactive
802.3	CSMA/CD and Ethernet
802.11	Wireless LAN (Wi-Fi)
802.15	Wireless personal area network (WPAN)
802.16	Broadband wireless access (BWA)
802.17	Resilient packet ring (RPR)
802.18	Radio regulatory technical advisory group
802.19	Coexistence technical advisory group
802.20	Mobile broadband wireless access
802.21	Media independent handoff
802.22	Wireless regional area network

The 802.1 and 802.2 standards correspond roughly to the OSI Data Link layer. The 802.1 standard addresses issues such as media access and describes an architectural framework into which other standards fit. The 802.2 standard addresses issues such as routing, error control, and flow control. The methods defined in these standards are incorporated into most of the other 802 standards.

The most common standards for LANs are 802.3 (CSMA/CD in bus networks) and 802.11 (wireless LANs). Each standard has a number of subsidiary standards indicated by added letters, such as 802.11g, that cover additional implementation parameters. Many commercial networking products are based on the IEEE standards. For example, Ethernet is based on the 802.3 standard, and WiMAX is based on the 802.16 standard.

The IEEE standards development process is heavily influenced by companies and organizations in the telecommunication and networking industries. Standards are developed by committees whose membership is drawn from industry, government, and academia. These committees work closely with other standards and regulatory agencies, such as the American National Standards Institute (ANSI), the ISO, and the U.S. Federal Communications Commission and its international counterparts. The development of standards can be influenced as much by politics as by technological considerations.

The IEEE committees set criteria for evaluation and then invite proposals for adoption as standards. Proposals are usually submitted by vendors or vendor consortia based on products and technologies already under development. Occasionally, the process backfires. Adopted standards sometimes represent major compromises between very different proposals. Vendors might decide not to adjust their products under development to meet the compromise standard, thus generating an "orphan standard." A vendor might also choose to release products that don't adhere to any published standard and hope that its technology becomes a de facto standard. For example, Bluetooth version 1.2 is an IEEE standard, but later versions are not.

The marketplace ultimately decides which technologies and products succeed. The standard-setting process usually establishes an accepted target for implementation and, with a widely used standard, a degree of compatibility between competing products.

TECHNOLOGY FOCUS

WiMAX

Worldwide Interoperability for Microwave Access (WiMAX) is a group of wireless networking standards developed by the WiMAX Forum and codified in the IEEE 802.16 standard. It's neither a replacement for nor an extension to the IEEE 802.11 standards. Instead, it's targeted to different application areas—fixed and mobile Internet access spanning distances up to 50 kilometers (about 30 miles)—and a market between LANs and WANs, sometimes called **metropolitan area networks (MANs)**, which typically cover a town or city.

Table 9.4 summarizes characteristics of three current IEEE 802.16 standards. Higher transmission frequencies yield a higher raw data transmission rate (DTR) because of more available bandwidth, although later standards make more efficient use of lower frequencies and bandwidth with more sophisticated bit-encoding methods. A higher raw DTR is achieved over shorter distances with fixed transmission and receiving stations. As in 802.11 LANs, transmission shifts to lower DTRs with more robust error detection when stations are far apart or in motion. Like 802.11n, 802.16e includes multiple-antenna support that can yield higher DTRs or reduced error rates but not both at the same time. The 802.16m standard (WiMAX 2) is still under development, with ratification expected in 2011 or 2012. The expected raw DTR for mobile applications will be

(continued)

four times higher than for 802.11e, and high-power fixed-station DTRs will support multiple 120 Mbps connections.

TABLE 9.4 IEEE 802.16 (WiMAX) standards

Standard	Frequencies	Maximum raw DTR	Station types	Maximum range
802.16 (2001)	10–66 GHz	134 Mbps	Fixed	5 km
802.16 (2004)	2–11 GHz	75 Mbps	Fixed	5 km (low power); 50 km (high power)
802.16e (2005)	2–6 GHz	15 Mbps	Mobile	5 km

802.16 MAC protocols differ widely from 802.11 protocols. 802.16 uses a connection-oriented protocol in which stations establish a connection and negotiate transmission parameters before data is transmitted. A typical WiMAX deployment is a single high-power station with a tower-mounted antenna serving as an Internet access point for multiple lower-power fixed or mobile stations in nearby homes or vehicles. As each home or mobile station establishes a connection, the Internet access point station allocates transmission time slices, which prevents collisions and enables assigning higher quality of service levels to stations for purposes such as VoIP or H.323 videoconferencing.

The variety of transmission types defined in the 802.16 standards has been both a help and hindrance to adoption and deployment. Because RF spectrum availability and licensing above 10 GHz varies widely across the globe, 802.16 implementations have an advantage over competing technologies because they can be adapted to local conditions. On the other hand, the lack of a well-defined worldwide RF frequency range for high-throughput connections has fragmented the WiMAX market and made economies of scale and interoperability difficult to achieve. Later standard versions use unlicensed and open RF bands in most of the world, but they compete against a variety of other bandwidth users, including 802.11 LANs.

To date, most WiMAX adoption has occurred in suburban and rural areas with limited or no wired broadband connections. Adoption has been especially high in parts of Asia, where there's rapid growth but poor wired infrastructure. The biggest competitors to WiMAX are the data services provided by cell phone carriers, phone service providers, and cable TV companies. During the mid-2000s, it seemed that some cell phone carriers would adopt WiMAX as a supplement and eventual replacement for third-generation (3G) cell phone data services. As of 2010, however, it appears that most cell phone carriers will transition to a competing standard (Long Term Evolution [LTE]) that's more compatible with their existing 3G standards and infrastructure.

The rocky road traveled by 802.16 standards, vendors, and users is an excellent example of the ups and downs of the standard-setting process, market adoption, and the supporting roles of politics and economic competition. The 802.16 standards are based on solid technology that can deliver excellent data services, but a fragmented global regulatory environment coupled with major competition has slowed product development and adoption. WiMAX might still prove wildly popular, but so far, the lofty aspirations of many WiMAX vendors, investors, and adopters have yet to be realized.

Ethernet

Ethernet is a LAN technology, developed by Xerox in the early 1970s, that's closely related to the IEEE 802.3 standard. Digital Equipment Corporation (DEC), now part of Hewlett-Packard, was the driving force in developing commercial Ethernet products in the late 1970s. DEC, Intel, and Xerox jointly published a specification for Ethernet networks in 1980. The 802.3 standard, published a few years later, incorporated most aspects of this specification and has been extended many times.

Older Ethernet standards used a bus logical topology, Category 5 twisted-pair cable, and CSMA/CD. Figure 9.13 shows the 802.3 packet format. There's no provision for packet priorities or guarantees of quality of service, although incorporating these capabilities is a high priority for vendors and standard-setting organizations. The original Ethernet standard transmits data at 10 Mbps and was updated (802.3u) in the late 1990s to increase transmission speed to 100 Mbps. There were several competing and incompatible proposals for 100 Mbps Ethernet, including the never deployed 802.12 and 802.13 standards.

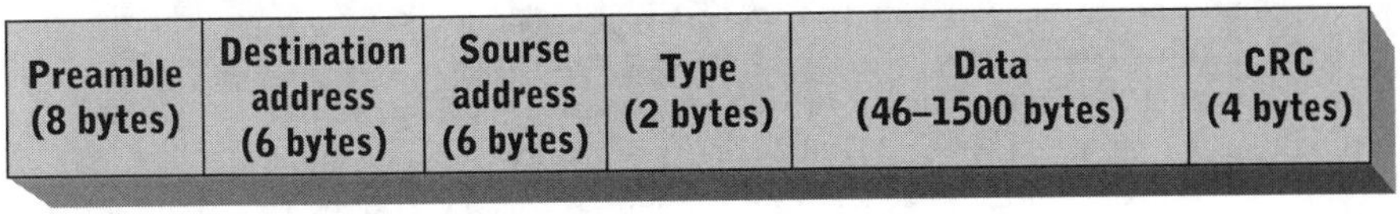

Preamble (8 bytes)	Destination address (6 bytes)	Sourse address (6 bytes)	Type (2 bytes)	Data (46–1500 bytes)	CRC (4 bytes)

FIGURE 9.13 Ethernet packet format
Courtesy of Course Technology/Cengage Learning

Gigabit Ethernet is based on the 802.3z standard (1998) and the 802.3ab standard (1999). As shown in Table 9.5, several physical implementations are defined, each representing a tradeoff between cost and maximum cable length.

TABLE 9.5 Gigabit Ethernet specifications

IEEE standard	Configuration name	Cable and laser type	Maximum length
802.3z	1000BaseSX	Short-wavelength laser over multimode fiber	550 meters
802.3z	1000BaseLX	Long-wavelength laser over multimode fiber	550 meters
802.3z	1000BaseLX	Long-wavelength laser over single-mode fiber	5 kilometers
802.3z	1000BaseCX	Category 5 twisted pair	25 meters
802.3ab	1000BaseT	Category 5 twisted pair	100 meters

10 Gigabit Ethernet is based on the 802.3ae (2002) and the 802.3ak (2004) standards. It supports 10 Gbps transmission speeds over multimode fiber-optic cables at LAN distances of 100 to 300 meters and WAN distances of 2 to 40 kilometers. Copper-based transmission is supported only over short distances. 10 Gigabit Ethernet is the first Ethernet standard to abandon CSMA/CD in favor of a full-duplex point-to-point switched architecture. 40 Gigabit Ethernet and 100 Gigabit Ethernet standards (802.3ba) are planned for release in 2010.

Ethernet has been a popular networking technology since the 1980s. Many competitors in the LAN marketplace, such as IBM token ring, have fallen by the wayside. In recent years, Ethernet has made major inroads into the WAN marketplace, supplanting older standards, such as Fiber Distributed Data Interface, and relegating others, such as Asynchronous Transmission Mode, to niche markets. If ongoing efforts to update the technology are successful, Ethernet should continue to dominate other network standards well into the 21st century.

BUSINESS FOCUS

Upgrading Network Capacity

Note: This Business Focus is a continuation of the one in Chapter 8.

The Bradley Advertising Agency (BAA) has decided to implement a network supporting both Gigabit and 10 Gigabit Ethernet networks to support general data communication and file access. The Gigabit Ethernet network will support general-purpose needs, and the 10 Gigabit Ethernet network will interconnect servers and video-editing workstations. BAA will purchase an additional server dedicated to high-speed, high-capacity storage for video editing. This server will have a 3 TB storage array, expandable to 20 TB. The reasons for the upgrade include the following:

- The existing twisted-pair cable was tested and found to meet Category 6 requirements.
- Gigabit Ethernet will meet the demand for general-purpose networking now and for the foreseeable future.
- 10 Gigabit Ethernet will meet current and near-term demands for video file sharing among video-editing workstations.
- Gigabit and 10 Gigabit Ethernet equipment is much cheaper than InfiniBand equipment.

BAA isn't sure how to integrate its microcomputers, video-editing workstations, and servers into the Gigabit and 10 Gigabit Ethernet networks and is considering three options:

- Purchase a new Ethernet switch with six 100/1000 Mbps ports and eight 10 Gigabit Ethernet ports. Leave the desktop computers connected to the current 10/100 switch and connect this switch to the new switch. Connect the video-editing workstations and all servers to 10 Gigabit Ethernet ports (upgrading NICs as needed). Use existing copper cabling for all desktop computer and video-editing workstation connections.
- Scrap the existing Ethernet switch and purchase two new ones—a 12-port 10 Gigabit Ethernet switch and a 24-port Gigabit Ethernet switch. Upgrade all desktop NICs to Gigabit Ethernet and connect them to a Gigabit Ethernet switch. Connect all servers and video-editing workstations to the 10 Gigabit Ethernet switch (upgrading NICs as needed). Use existing copper cabling for all desktop computer and video-editing workstation connections.
- Same as Option 2, except run new multimode fiber-optic cable from the 10 Gigabit Ethernet switch to the video-editing workstations.

(*continued*)

Questions:

- What are the costs of each option? Which option offers the best network performance? Which option provides the most operational flexibility?
- Which option should BAA choose to implement now? Why?
- BAA anticipates it will need to transmit high-quality video to and from client locations in the next three years. Video transfer modes will include transfer of large files and real-time video conferencing at HDTV resolution. Which option best enables BAA to adapt to this long-term need?

Summary

- Network topology refers to the spatial organization of network devices, the physical routing of network cabling, and the flow of messages from one network node to another. Topology can be physical (arrangement of node connections) or logical (message flow). Physical topologies include mesh, bus, ring, and star. The star topology is used in most modern wired networks, and wireless networks use the bus topology.
- LANs can be interconnected to form WANs. Specialized network hardware devices forward packets from source to destination. Message forwarding in a LAN is usually handled by the LAN hub or switch. Message forwarding across WANs uses a store and forward approach that's like the delivery and warehouse network of a package delivery service.
- A Media Access Control (MAC) protocol specifies rules for accessing a shared transmission medium. CSMA/CD is used in older Ethernet bus networks, and CSMA/CA is used in many wireless networks. CSMA/CD allows collisions to occur but provides a recovery mechanism. CSMA/CA prevents most collisions.
- Network hardware devices include NICs, hubs, switches, routers, and wireless APs. A NIC is the interface between a network node and the network transmission medium. A hub connects nodes to form a LAN. Switches are high-speed devices that forward messages based on physical network addresses. Routers can connect to more than two networks, forward messages based on IP addresses, and exchange information to improve forwarding decisions.
- The Open Systems Interconnection (OSI) model is a conceptual model that divides network architecture into seven layers: Application, Presentation, Session, Transport, Network, Data Link, and Physical. Each layer uses the services of the layer below and is unaware of other layers' implementations.
- TCP/IP is the core protocol suite on the Internet. IP provides connectionless packet transport across LANs and WANs. IP addresses are 32- or 128-bit values. TCP provides connection-oriented message transport to higher-level Internet service protocols, including HTTP and FTP. UDP provides connectionless service to support transporting multimedia data with minimal processing overhead. The Network Interface layer of the TCP/IP model enables transporting IP messages over physical network protocols, such as Ethernet and ATM.
- The IEEE 802 standards are developed by committees composed of members from industry, government, and academia and cover many types of networks. IEEE standards help ensure compatibility between products from competing vendors. Some standards, such as the Ethernet standard (802.3), are successful, and others are never implemented.
- Ethernet is the most widely deployed physical network standard. It has evolved through many versions, with Gigabit Ethernet and 10 Gigabit Ethernet standards currently used and 40 and 100 Gigabit Ethernet standards under development. Ethernet's support of multiple transmission speeds and media makes it suitable to both LANs and WANs.

Chapter 8 and this chapter have described all the hardware but only some of the software technology underlying computer networks. In Chapter 10, you examine tools used to build application software. In Chapters 11 and 12, you look at operating system technology, including resource allocation and file management systems. Then you return to computer networks in Chapter 13 and learn about distributed resources and applications.

Key Terms

10 Gigabit Ethernet
access point (AP)
Application layer
bus topology
Carrier Sense Multiple Access/Collision Avoidance (CSMA/CA)
Carrier Sense Multiple Access/Collision Detection (CSMA/CD)
clear-to-send (CTS) signal
collision
colon hexadecimal notation
connectionless protocol
connection-oriented protocol
Data Link layer
datagrams
dotted decimal notation
encapsulation
Ethernet
forwarding table
gateways
Gigabit Ethernet
H.323
hub
IEEE 802 standards
Internet Protocol (IP)
Internet Protocol version 4 (IPv4)
Internet Protocol version 6 (IPv6)
local area network (LAN)
logical topology
Media Access Control (MAC)
mesh topology
metropolitan area networks (MANs)
multicasting
network adapter
network interface card (NIC)
Network layer
network topology
Open Systems Interconnection (OSI) model
Physical layer
physical topology
port
Presentation layer
ready-to-send (RTS) signal
ring topology
router
routing tables
Session layer
socket
star topology
store and forward
switch
TCP/IP
TCP/IP model
Transmission Control Protocol (TCP)
Transport layer
User Datagram Protocol (UDP)
Voice over IP (VoIP)
wide area network (WAN)
Worldwide Interoperability for Microwave Access (WiMAX)

Vocabulary Exercises

1. The _____ standards define many aspects of physical networks.
2. The OSI _____ layer establishes and manages connections between clients and servers.
3. The _____ protocol is an updated version with larger addresses and improved support for multicasting and multimedia data.
4. The _____ topology is most common in wired networks, and the _____ topology is most common in wireless networks.
5. The OSI _____ layer forwards messages to their correct destinations.
6. A _____ is the combination of an IP address and a port number.
7. The OSI _____ layer refers to communication protocols used by programs, such as Web browsers, that generate requests for network services.
8. A network using a physical _____ topology connects all end nodes to a central node.
9. A physical connection between two different networks is implemented by using a(n) _____, _____, or _____.
10. A receiver can't detect loss of datagrams if a(n) _____ protocol is used.
11. In the TCP/IP model, a(n) _____ is the basic data transfer unit.
12. The original _____ standard transmits at 10 Mbps over twisted-pair cabling. Current standard versions support 1 and 10 Gbps transmission over twisted-pair and fiber-optic cable.
13. The _____ defines conceptual software and hardware layers for networks.
14. The _____ MAC protocol is used in wireless networks to prevent most collisions.
15. When two messages are transmitted at the same time on a shared medium, a(n) _____ has occurred.
16. With the _____ MAC protocol, collisions can occur, but they're detected and corrected.
17. A(n) _____ protocol defines the rules governing a network node's access to a transmission medium.
18. An end node's hardware interface to a network transmission medium is called a(n) _____.
19. The _____ protocol is used with broadcast and multimedia applications when processing overhead needs are low and reliable delivery doesn't need to be guaranteed.
20. The oldest and most widely used VoIP protocol suite is _____.

Review Questions

1. Describe the function of each layer of the TCP/IP model.
2. Compare 802.11 and WiMAX wireless networks in terms of transmission distances and frequencies, strategies for dealing with noise and interference, and how widely they're deployed.
3. How does a message from one LAN node find its way to a recipient on the same LAN? How does a message find its way to a recipient on another LAN?

4. Compare CSMA/CD and CSMA/CA in terms of how collisions are detected or avoided and their inclusion in physical network standards.
5. What is the function of a hub? How does it differ from a switch or router?
6. Describe the processes of encapsulation and address resolution. Why are they necessary features of the Internet?
7. Describe a connectionless and a connection-oriented protocol, and list one example of each.
8. How many bits are in an IP address? What is a TCP or UDP port? What is a TCP or UDP socket?
9. Describe past, current, and proposed Ethernet standards in terms of transmission speed, supported transmission media, and relevant IEEE standards.
10. What protocols are commonly used to implement VoIP? Are all VoIP protocols compatible with one another?
11. Describe the service quality problems that can occur in VoIP. Why are these problems so difficult to solve?

Research Problems

1. Investigate the videoconferencing equipment and software offerings of a major networking or telecommunication vendor, such as Cisco Systems (*www.cisco.com*), Tandberg (*www.tandberg.net*), Radvision (*www.radvision.com*), or Polycom (*www.polycom.com*). What related standards are supported in each company's product line? Which vendors and products support high-definition video, and what are the related standards? Describe available infrastructure products, including gatekeepers and multipoint control units (MCUs), and compare their capabilities with routers and switches in data networks.
2. You have been asked to design a network (wired, wireless, or a combination) for a home on which construction will start soon. The home is serviced by a cable TV provider and local phone company, and both provide Internet connectivity. It has two levels totaling 250 meters^2 (2700 feet^2). Five people will live in the home and use three laptop/netbook computers, a desktop computer, a multimedia server, an Internet-connected TV device (for example, TiVo or digital cable DVR), two multifunction printers, and at least five handheld 802.11g wireless devices, such as iPods and cell phones. Based on Internet connectivity options in your own area, recommend whether the home should get Internet services via DSL or cable TV. Also, recommend network infrastructure, including a router/firewall, wireless access points, wired switches, and combination devices. Should the homeowner have wired Internet connections installed during construction? If so, what wire types should be used, and in what rooms should connection points be placed?
3. Investigate the Ethernet connectivity devices offered by a major vendor, such as Linksys (*www.linksysbycisco.com*), NETGEAR (*www.netgear.com*), or Cisco (*www.cisco.com*). What range of features is available in LAN and WAN switches and routers? What devices are offered that don't clearly fall into the categories described in this chapter? What's the cost per port of Gigabit and 10 Gigabit switches and routers? What options are available for linking groups of switches and routers?

CHAPTER **10**

APPLICATION DEVELOPMENT

CHAPTER GOALS

- Describe the application development process and the role of methodologies, models, and tools
- Compare generations and types of programming languages
- Explain how assemblers, compilers, and interpreters translate source code instructions into executable code
- Describe link editing and contrast static and dynamic linking
- Explain the role of memory maps in symbolic debugging
- Describe integrated application development tools

Application development is a complex process that follows steps to translate users' needs into executable programs. There are automated tools to support each step, from user requirement statements through system models and program source code to executable code. Understanding the role and function of application development tools will make you a more efficient and effective analyst, designer, and programmer. Figure 10.1 shows the topics covered in this chapter.

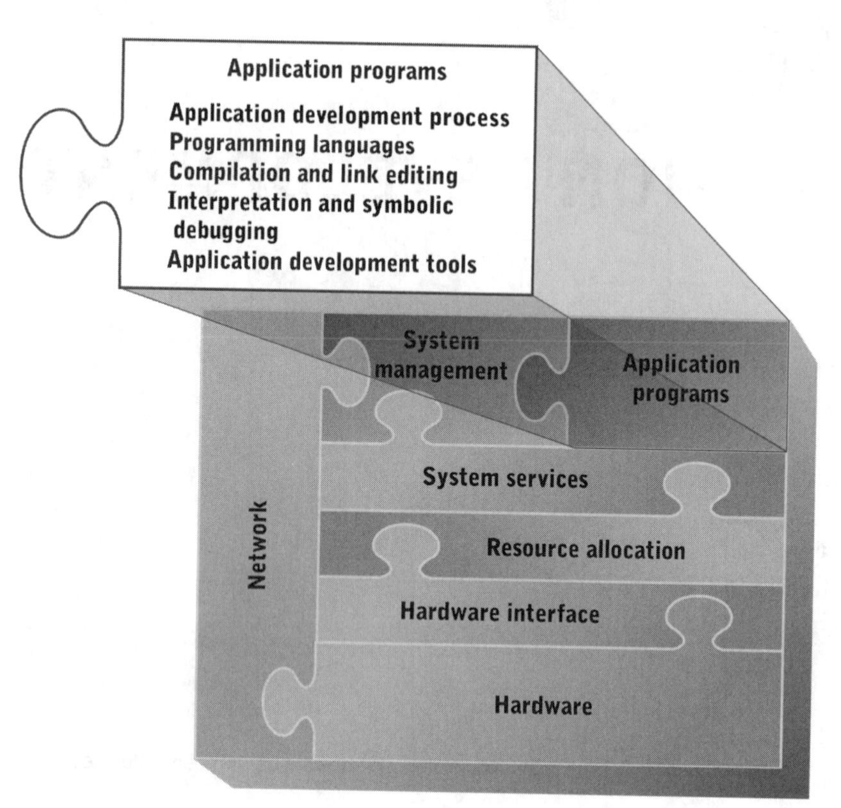

FIGURE 10.1 Topics covered in this chapter
Courtesy of Course Technology/Cengage Learning

THE APPLICATION DEVELOPMENT PROCESS

The process of designing and constructing software translates users' information-processing needs into CPU instructions that, when executed, address these needs. User needs are stated in general or abstract terms in natural human language, such as "I need to process accounts receivable and payable." Software programs are detailed, precise statements of formal logic written as sequences of CPU instructions. Application development involves two translations—from an abstract need statement to a detailed implementation that satisfies the need and from natural language to CPU instructions (see Figure 10.2).

Developing software requires a lot of effort and resources. For example, a user need stated in a single sentence might require millions or billions of CPU instructions to address. Developing and testing software require development methods and tools, highly trained specialists, and months or years of effort. Software has surpassed hardware to become the most costly component of most information systems.

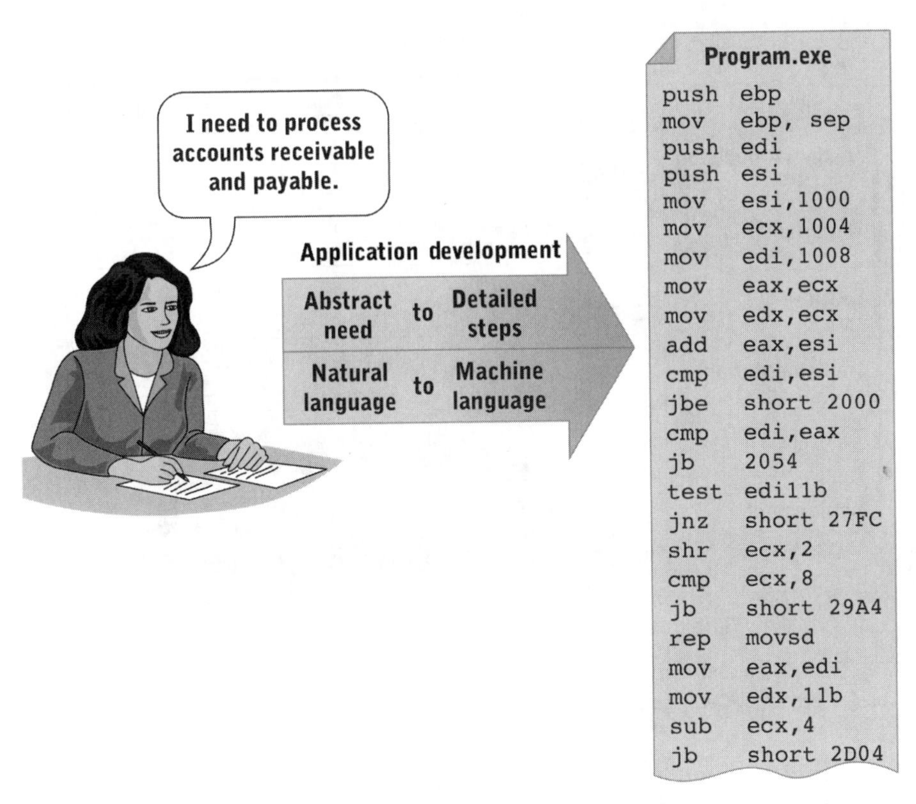

FIGURE 10.2 Application development translates users' needs to CPU instructions
Courtesy of Course Technology/Cengage Learning

Information system owners expect their investment in application development to produce software that meets their needs. Unfortunately, the complexity of this process creates many possibilities for error, resulting in software that doesn't work or fails to meet users' needs. The economic cost of application development errors can be much higher than the cost of developing the software. Reduced productivity, dissatisfied customers, and poor managerial decisions are just a few indirect costs of software that doesn't address users' needs completely or correctly.

Systems Development Methodologies and Models

In this chapter, you return to the systems development life cycle (SDLC) introduced in Chapter 1 but with a focus on application development. Figure 10.3 shows the disciplines of the Unified Process (UP) and their organization into iterations for a typical application development project. The UP is a system development methodology—a way of breaking the development process into smaller, more manageable pieces. In this chapter, you're concerned mainly with the requirements, design, and implementation disciplines of the UP.

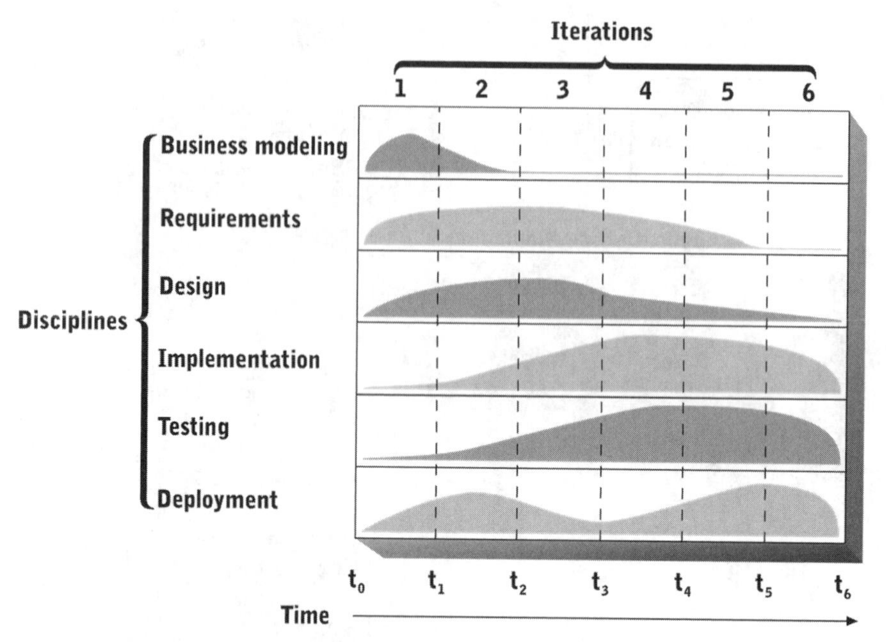

FIGURE 10.3 Disciplines and iterations in the Unified Process
Courtesy of Course Technology/Cengage Learning

Developers attempt to minimize errors by using proven development methodologies, which are integrated collections of models, tools, techniques, and processes. The UP methodology, for example, uses object-oriented analysis, design, and deployment models. Class diagrams and other types of diagrams document user and system requirements. Design and deployment discipline activities use additional model types. Many graphical models are supplemented with detailed text descriptions.

As part of the requirements discipline, members of the development team interact with users to understand and document users' needs. Much of the translation from abstract concepts to specific, detailed facts occurs during activities in this discipline. The components of a general statement, such as "I need to process accounts receivable and payable," are examined and documented in detail. What is an account payable or account receivable? What specific tasks are implied by the term "process"? What is the data content of system inflows and outflows? What data must be stored internally? What tasks are performed when and by which people or programs? The answers to these questions are summarized in a variety of forms, including text descriptions and diagrams that together compose a system requirements model.

System requirements models provide the detail needed to develop a system that meets users' needs. System developers perform design activities to create models that produce an architectural blueprint for system implementation. These models specify required features of system software, such as operating systems and database management systems; describe the required functions of application software subroutines, procedures, methods, objects, and programs in detail; and serve as a blueprint and roadmap for implementation. Figure 10.4 shows an example of a system requirements model; it's a class diagram, used to summarize a system's division into software and data as manageably sized components. Other UP system requirements models include use case, activity, and system sequence diagrams.

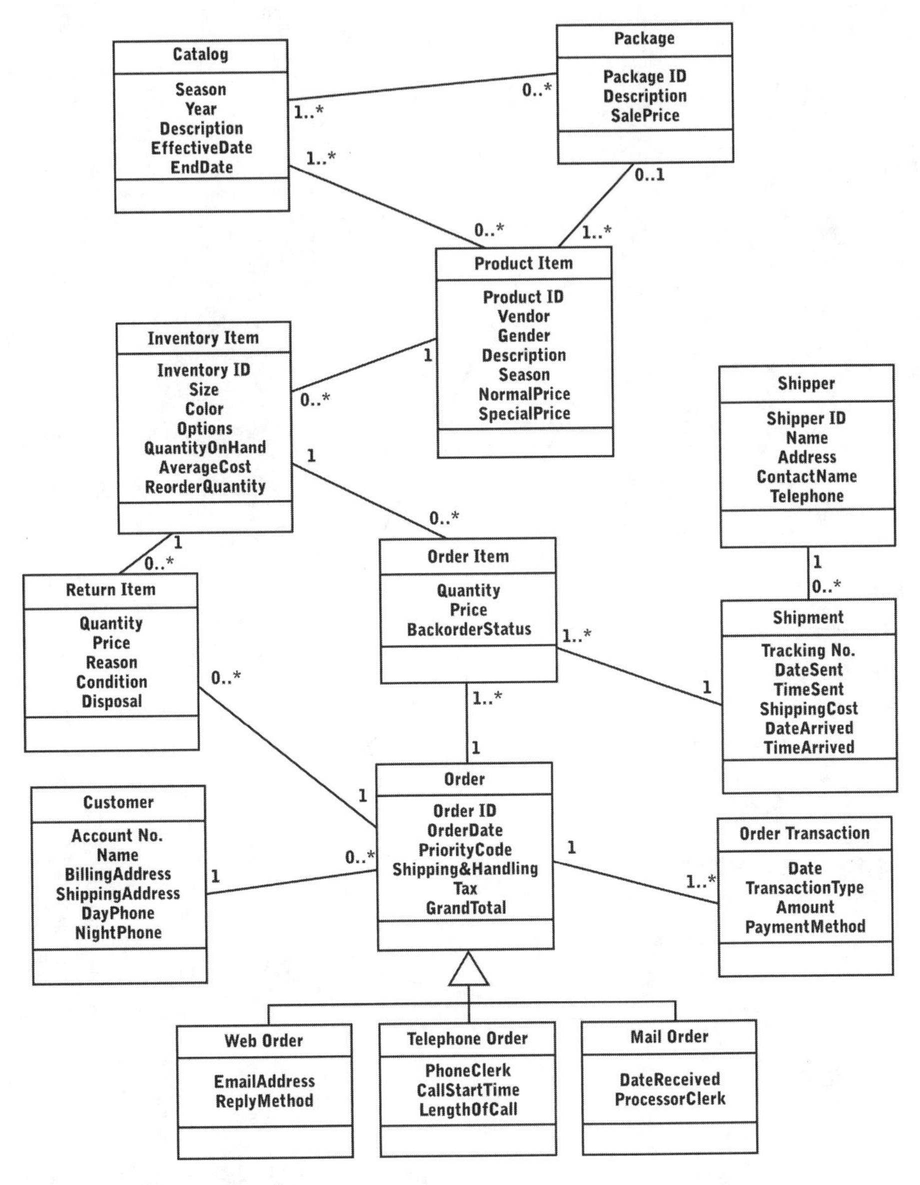

FIGURE 10.4 A class diagram
Courtesy of Course Technology/Cengage Learning

Design models specify detailed blueprints for software component construction and the interaction between software components and users. UP design models include package diagrams, interaction diagrams, and deployment diagrams. As shown in Figure 10.5, design models are developed based on system requirements models but add design and implementation details.

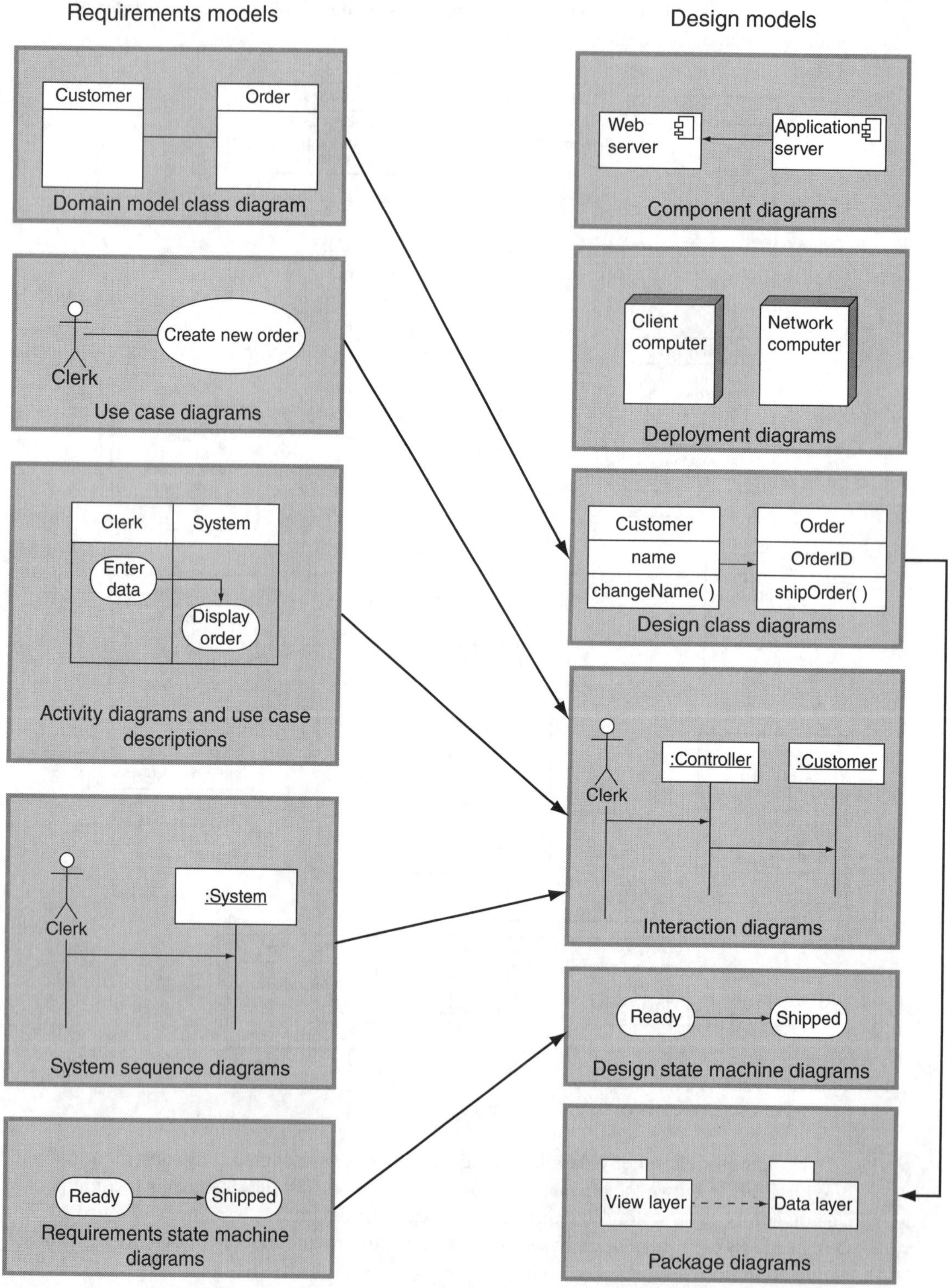

Requirements models

Customer
Order
Domain model class diagram

Clerk
Create new order
Use case diagrams

Clerk
System
Enter data
Display order
Activity diagrams and use case descriptions

Clerk
:System
System sequence diagrams

Ready
Shipped
Requirements state machine diagrams

Design models

Web server
Application server
Component diagrams

Client computer
Network computer
Deployment diagrams

Customer
name
changeName( )
Order
OrderID
shipOrder( )
Design class diagrams

Clerk
:Controller
:Customer
Interaction diagrams

Ready
Shipped
Design state machine diagrams

View layer
Data layer
Package diagrams

FIGURE 10.5 UP design models and their relationship to system requirements models

Implementation activities acquire or construct software and hardware components that match the design models. Typically, hardware and system software are purchased, and some or all application software is constructed, sometimes using purchased or previously developed software components. As acquisition and construction proceed, hardware and software are tested to ensure that they function correctly and meet users' needs as described in the system requirements model. The end result of the UP is a fully tested system that's ready to be put to productive use.

Tools

Many automated tools support application development. Some tools, such as word-processing and general-purpose drawing programs, can be used to build many types of models. Others, such as CASE tools (described later in "Application Development Tools"), are customized to specific models and development methodologies. Related tools are often integrated into a single suite that supports an entire system development methodology, in much the same way that word-processing, graphics, database, and spreadsheet programs are integrated into office suites.

These tools vary with the target system's deployment environment. They can be specific to an OS, a programming language, a database management system (DBMS), or a type of hardware. Wide variations in underlying methodology, supported models and techniques, and target deployment environment make selecting the right tool a critical and difficult undertaking. This task is further complicated by a dizzying array of descriptive terminology and varying levels of support for different parts of the development process.

Substituting automated application development tools for manual methods is a classic economic tradeoff between labor (such as analysts, designers, and programmers) and capital resources (sophisticated development tools and the hardware required to run them). In the early days of computers, hardware was so much more expensive than labor that it made economic sense to use labor-intensive processes for software development; hardware was simply too expensive to "waste" it doing anything but running application software developed by manual methods.

As hardware costs decreased, the economic balance shifted, leading to the introduction of automated tools to support application development. Programming languages, program translators, and OS service layers are the earliest examples of this shift. They enable computing hardware to support software development, thus reducing labor requirements. The current proliferation of application development tools is a continuation of this economic shift. Analysts, designers, and programmers now use a wide array of tools to automate application development tasks.

PROGRAMMING LANGUAGES

A programming language is used to instruct a computer to perform a task. Program instructions are sometimes called **code**, and the person who writes the instructions is called a programmer. Developers of programming languages strive to make software easier to develop by doing the following:

- Making the language easier for people to understand

- Developing languages and program development approaches that require writing fewer instructions to accomplish a task

Programming languages have evolved through multiple stages called "generations," as shown in Figure 10.6. Table 10.1 summarizes the characteristics of each generation; details are given in the following sections.

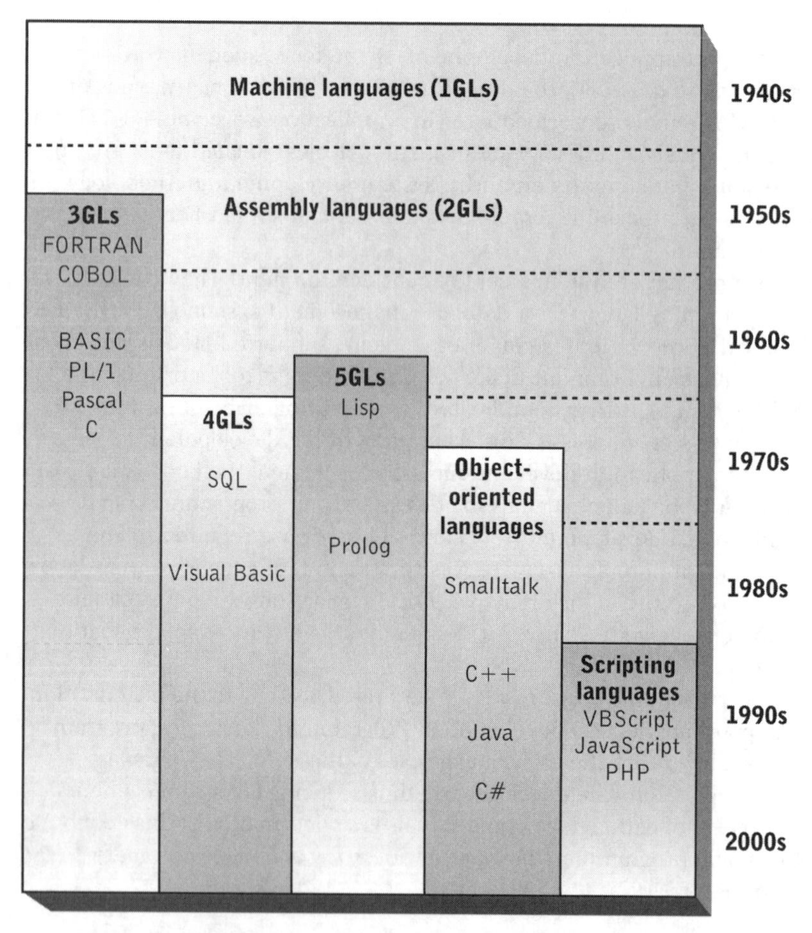

FIGURE 10.6 Programming language evolution
Courtesy of Course Technology/Cengage Learning

TABLE 10.1 Programming language characteristics

Generation	Description or characteristics
First	Binary instructions for a CPU
Second	Mnemonic instructions for a CPU
Third	Instruction explosion, machine independence, and usually standardization
Fourth	Higher instruction explosion, interactive and graphical I/O support, database support, limited nonprocedural programming, and proprietary standards
Fifth	High instruction explosion, nonprocedural programming, expert systems, and artificial intelligence applications
Object-oriented	High instruction explosion, support for modern code development, and code reuse methods, such as inheritance and message passing
Scripting	Very high instruction explosion; interpreted execution, used mainly for Web applications

First-Generation Languages

Binary CPU instructions, called **machine languages** or **first-generation languages (1GLs)**, are the earliest programming languages. People have difficulty remembering and manipulating long strings of binary digits, yet that's what early machine-language programmers did. They had to remember the binary codes representing each CPU instruction and specify all operands as binary numbers. Programming with binary numbers is tedious and error prone. As primary storage capacity and program size grew, developing error-free programs with only binary numbers became less feasible.

Second-Generation Languages

A **second-generation language (2GL)** is more commonly known as an **assembly language**. The instructions and programming examples in Chapter 4 were in a simplified assembly language. These languages use a short character sequence called a "mnemonic" to represent each CPU instruction, and programmers can define their own mnemonics to represent memory addresses of instructions and data items. Today, the term **variable** is often used to describe a mnemonic representing a data item's memory address and the term **label** to describe a mnemonic representing a program instruction's memory address. Short names are easier for people to manipulate than binary numbers. As a result, 2GLs became the most common programming languages during the early and mid-1950s. They're still used today in some types of system programming, such as writing device drivers.

Because the CPU processes only binary digits, any language other than a 1GL must be translated into a 1GL before a CPU can execute it. An **assembler** is a program that translates an assembly-language program into binary CPU instructions. It reads an assembly-language program file one instruction at a time. It then translates each mnemonic into its corresponding binary digit sequence and writes these digits to an output file ready to be loaded into memory and executed by the CPU. Assemblers are the earliest example of automated program development tools.

Although 2GLs made programming easier, they did nothing to minimize the number of instructions programmers had to write to perform a task. Each assembler instruction is translated into a CPU instruction, so a 2GL program still directs every CPU action. As memory capacity and program size grew, specifying every CPU action became more difficult. To increase programmer productivity, the one-to-one (1:1) correspondence between program instructions and CPU actions had to be broken.

Third-Generation Languages

Programming languages beyond the second generation enable programmers to specify many CPU actions with a single program instruction or statement. The one-to-many (1:N) relationship between later-generation programming statements and the CPU actions implementing them is called **instruction explosion**. For example, if 1000 machine instructions result from translating a 20-statement program, the degree of instruction explosion can be stated as a 50:1 ratio. Programming languages differ in the degree of instruction explosion, and even within a language, different types of statements vary in the degree of instruction explosion. In general, statements describing mathematical computation have low instruction explosion (10:1 or less), and statements describing I/O operations have high instruction explosion (100:1 or more).

A **third-generation language (3GL)** uses mnemonics to represent instructions, variables, and labels and has a degree of instruction explosion higher than 1:1. The first 3GL was FORTRAN. Later 3GLs include COBOL, BASIC, PL/1, Pascal, and C. A few 3GLs are still used today, but most were developed before graphical user interfaces (GUIs), DBMSs, and the Internet, which account for much of the decline in their popularity.

Like 2GL programs, 3GL programs must be translated into binary CPU instructions before the program is executed. Compilers, link editors, and interpreters are automated tools used to translate 3GL programs; they're covered in more detail later in this chapter. 3GL program translation is more complex than 2GL program translation, so 3GL translation tools are much larger and consume more computer resources than assemblers.

A single 3GL program can be translated to run on many different CPUs, a characteristic called machine independence or hardware independence (discussed in Chapter 2). Compilers, link editors, and interpreters translate machine-independent instructions into machine-language instructions for specific CPUs. Each CPU uses a different 3GL translation program that's customized to its instruction set.

Fourth-Generation Languages

In the 1970s and 1980s, **fourth-generation languages (4GLs)** were developed to address 3GL limitations. They shared many features, including the following:

- Much higher instruction explosion than in 3GLs
- Instructions or prewritten functions to implement interactive GUIs

- Instructions to interact directly with an internal or external relational database
- Capability to describe processing requirements without specifying solution procedures

The majority of 4GLs were proprietary, and many were optional components of DBMSs. Few of the original 4GLs survive today, although some current programming languages are their direct descendants. Two that have survived (and thrived) are Visual Basic and Structured Query Language (SQL). Visual Basic evolved from the 3GL BASIC and is adapted to modern software requirements. It has a variety of GUI capabilities and can interact directly with Microsoft Access databases. Although it's widely used, it lacks some features needed by "industrial strength" software.

SQL is specialized for adding, modifying, and retrieving data from relational databases and is usually used with DBMSs, such as Oracle or Microsoft SQL Server. SQL isn't a full-fledged programming language because it lacks some features of a general-purpose programming language, including many control structures and user interface capabilities. SQL programs are usually embedded in other programs, such as Web-based applications written in scripting languages. Figure 10.7 shows two programs that perform similar functions, one in C and the other in SQL. The C program has far more program statements, which indicates a lower degree of instruction explosion.

These programs also differ in their approach to solving a data retrieval problem. The C program describes a detailed, step-by-step procedure for extracting and displaying data that includes control structures and many database management functions, such as opening and closing files and testing for the end of file while reading records. In contrast, the SQL program contains no control structures, file manipulation, or output-formatting commands. It states what fields should be extracted, what conditions extracted data should satisfy, and how data should be grouped for display. SQL is a **nonprocedural language** because it describes a processing requirement without specifying a procedure for satisfying the requirement. The compiler or interpreter determines the most suitable procedure for retrieving data and generates or executes whatever CPU instructions are needed.

As a group, 4GLs are sometimes called "nonprocedural languages," although this term isn't completely accurate because most 4GLs support a mixture of procedural and nonprocedural instructions. A 4GL typically provides nonprocedural instructions for database manipulation and report generation. Some 4GLs also provide nonprocedural instructions for updating database contents interactively. Other types of information processing, such as complex computations, require procedural instructions.

Fifth-Generation Languages

A **fifth-generation language (5GL)** is a nonprocedural language suitable for developing software that mimics human intelligence. 5GLs first appeared in the late 1960s with Lisp but weren't widely used until the 1980s. Lisp and Prolog are the most common general-purpose 5GLs, but many proprietary 5GLs are used to build applications in medical diagnosis and credit application scoring, for example.

A 5GL program contains nonprocedural rules that mimic the rules people use to solve problems. A rule processor accepts a starting state as input and applies rules iteratively

```
SQL Example

open database banking;
select customer.acct_num, customer.name,balance+sum(transaction.
 amount)
from        customer,transaction
where       customer.acct_num=transaction.acct_num
group by    customer.acct_num,customer.name;


C Example

balance_report () {
   FILE  *cust_file, *trans_file;
   int   status,acct_num,a_num,balance,amount;
   char name [256];

   cust_file=fopen("customer","r");
   trans_file=fopen("transaction","r");
   status=scanf(cust_file, "%d%s%d\n", &acct_num,name, &balance);
   while (status != EOF {
      status=scanf (trans_file, "%d%d\n," &a_num, &amount);
      while (status != EOF){
         if (acct_num == a_num) {
            balance+=amount;
         }
         status=scanf(trans_file,"%d%d\n", &a_num,&amount);
      }
      printf("%d  %s  %d",acct_num,name,balance);
      trans_file=freopen("transaction","r");
      status=scanf(cust_file,"%d%s%d\n", &acct_num,name,&balance);
   }
   close(trans_file);
   close(cust_file);
   exit(0);
} /* end balance_report */
```


FIGURE 10.7 Equivalent programs in SQL and C
Courtesy of Course Technology/Cengage Learning

to achieve a solution. Figure 10.8 shows a sample Prolog program for solving the problem of moving a farmer, fox, chicken, and corn across a river on a boat that can hold only two. The rules describe valid moves and unsafe states; for example, the chicken and corn can't be left without the farmer because the chicken would eat the corn. The program prints messages to describe the rules it applies and moves it makes to solve the problem.

```
go (S,G) :- write ('start at state'),write (S),nl,path(S,G,[S]).

path (G,G,L) :- write ('SUCCESS, the path is:'),nl,ppt(L).

path (s(B,F,C,F,Cn),G,L) :-
move(s(B,F,C,F,Cn),s(Boat,Farmer,Chicken,Fox,Corn)),
   not(member(s(Boat,Farmer,Chicken,Fox,Corn),L)),
   not(unsafe(Boat,Farmer,Chicken,Fox,corn)),
   write('use this move to get to s('),
   write(Boat), write(','), write(Farmer), write(','), write(Chicken),
   write(','), write(Fox), write(','), write(Corn), write(')'), nl,
   path(s(Boat,Farmer,Chicken,Fox,Corn),
        G,
        [s(Boat,Farmer,Chicken,Fox,Corn) |L]), !.
/* s(Boat,Farmer,Chicken,Fox,Corn)  */

Move(s(e,e,e,Fox,corn),s(w,w,w,Fox,Corn)) :-
   Write('move farmer and chicken from east to west'), nl.

Move(s(w,w,w,Fox,Corn),s(e,e,e,Fox,Corn)) :-
   Write('move farmer and fox from west to east'), nl.

Move(s(e,e,Chicken,e,Corn),s(w,w,Chicken,e,Corn)) :-
   Write('move farmer and fox from east to west'), nl.

Move(s(w,w,Chicken,e,Corn),s(e,e,Chicken,e,Corn)) :-
   Write('move farmer and fox from west to east'), nl.

Move(s(e,e,Chicken,Fox,e),s(w,w,Chicken,Fox,w)) :-
   Write('move farmer and corn from east to west'), nl.

Move(s(w,w,Chicken,Fox,w),s(e,e,Chicken,Fox,e)) :-
   Write('move farmer and corn from west to east'), nl.

Move(s(e,e,Chicken,Fox,Corn),s(w,w,Chicken,Fox,Corn)) :-
   Write('move farmer only from east to west'), nl.

Move(s(w,w,Chicken,Fox,Corn),s(e,e,Chicken,Fox,Corn)) :-
   Write('move farmer only from west to east'), nl.

unsafe(X,X,Y,Y,_).
unsafe(X,X,Y,_,Y).

number(X,[X|T]).
number(X,[Y|T]) :- member(X,T).

ppt([]).
ppt([H|T]) :- ppt(T), write(H), nl.
```

FIGURE 10.8 A sample Prolog program
Courtesy of Course Technology/Cengage Learning

Object-Oriented Programming Languages

Both 3GLs and 4GLs have a similar underlying programming paradigm in which data and functions are distinct. Program instructions manipulate data and move it from source to destination. Data is processed by passing it between code modules, functions, or subroutines, each of which performs one well-defined task.

Software researchers in the late 1970s and early 1980s began to question the efficacy of this program/data dichotomy. They developed a new programming paradigm called **object-oriented programming (OOP)** that better addresses program reuse and long-term software maintenance. The fundamental difference between object-oriented and traditional programming paradigms is that OOP views data and programs as two parts of an integrated whole called an object (introduced in Chapter 3). Objects contain data and programs or procedures, called methods, that manipulate the data.

Objects reside in a specific location and wait for messages to arrive. A **message** is a request to execute a specific method and return a response. The response can include data, but the data included in a response is only a copy. The original data can be accessed and manipulated only by object methods. Figure 10.9 shows the relationships between data, methods, messages, and responses.

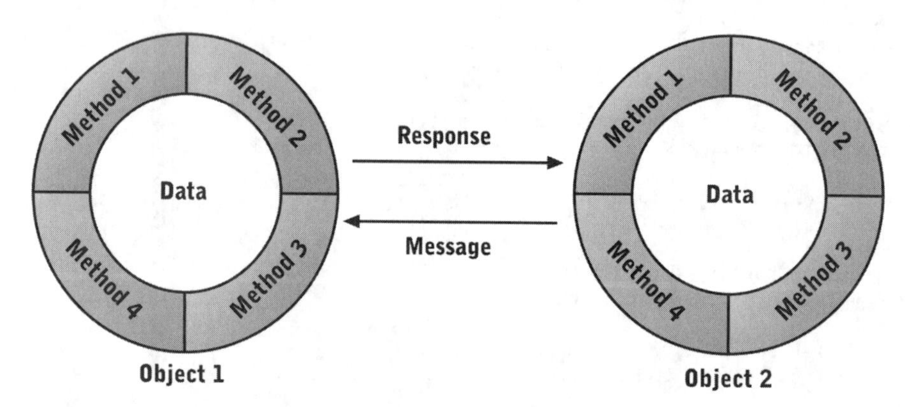

FIGURE 10.9 Objects, data, methods, messages, and responses in OOP
Courtesy of Course Technology/Cengage Learning

Smalltalk was the first commercial OOP language. C++ (an extension of C) and Java are more widely used now. Other languages, such as Visual Basic, have been extended with OOP concepts.

Object-oriented programming and design are uniquely suited to developing real-time programs, such as OSs and interactive user interfaces. Most current OSs and Web browsers are written in an OOP language. OOP languages and good object-oriented software design promote reusability and portability of source code. (Portability is explained later in "Programming Language Standards.") New objects can incorporate data and methods from existing objects through a feature called inheritance.

NOTE

Describing OOP languages by a name rather than a generation number is evidence that the concept of language generations is no longer relevant. OOP languages aren't a clear successor to 5GLs because they're neither nonprocedural nor used exclusively for developing AI and expert system applications. Some OOP languages, such as C++ and C#, are direct extensions of 3GLs, and others, such as Smalltalk and Java, are more closely related to 4GLs.

Scripting Languages

Scripting languages evolved from 4GLs, although most now incorporate OOP concepts. A **scripting language** enables programmers to develop applications that do most of their work by calling other applications and system software. These languages provide the usual control structures, mathematical operators, and data manipulation commands, but they extend these tools with the capability to call external programs, such as Web browsers, Web servers, and DBMSs. With scripting languages, programmers can assemble application software rapidly by "gluing" together the capabilities of many other programs. VBScript, JavaScript, and PHP are widely used languages for developing applications that use a Web browser as their main user interface.

Programming Language Standards

The American National Standards Institute (ANSI) and the International Organization for Standardization (ISO) set standards for some programming languages. Examples of ANSI standard languages include FORTRAN, COBOL, C, and C++. A programming language standard defines the following:

- Language syntax and grammar
- Machine behavior for each instruction or statement
- Test programs with expected warnings, errors, and execution behavior

Compilers, link editors, and interpreters are submitted to a standard-setting body for certification. Test programs are translated, and the behavior of the translator program and executable code is examined. The program's behavior is compared with the behavior defined by the standard to determine whether the translation program complies with the standard.

Standard programming languages guarantee program portability between other OSs and application programs, meaning a program can be moved to a new OS or CPU by recompiling and relinking or reinterpreting it with certified translation programs in the new environment. Portability is important because OSs and hardware change more frequently than most program source code.

NOTE

Setting standards for programming languages has become less relevant since the late 1980s. There are many reasons, including basic software research moving from academia into industry, software companies' reluctance to share proprietary technology, variability in applications and runtime environments, and increasing interdependence between application and system software. Current software standards are often more concerned with system software and interfaces between programs than with the languages used to build programs.

COMPILATION

Figure 10.10 shows application development with a program editor, compiler, and link editor. Input to the program editor comes from a programmer, a program template, or both, and its output is a partial or complete program, called **source code**, written in a programming language, such as Java or Visual Basic. Source code is normally stored in a file that's named to indicate both its function and programming language, such as ComputePayroll.cpp (with "cpp" as an abbreviation for C++).

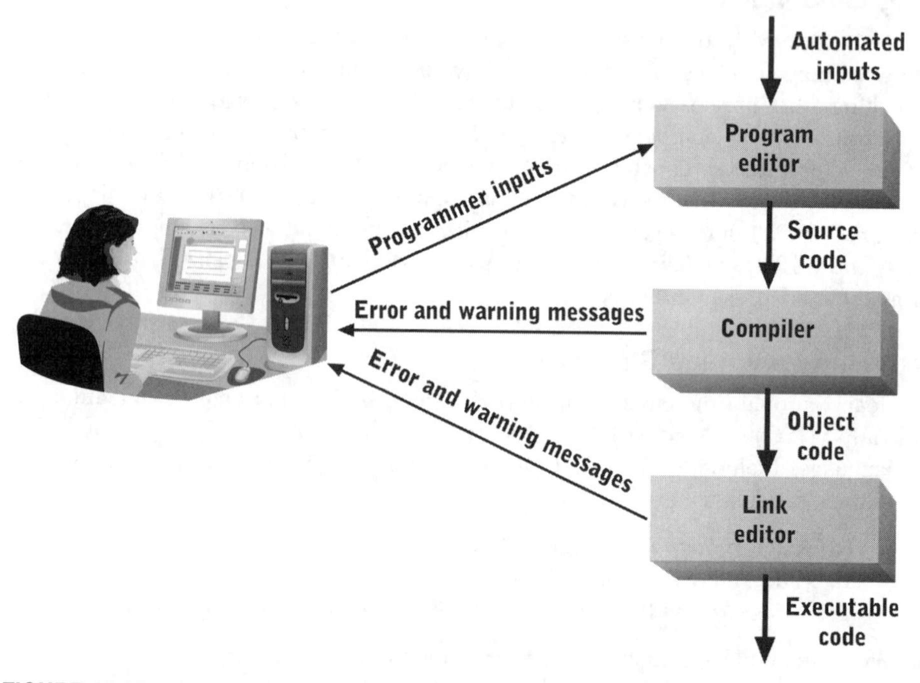

FIGURE 10.10 Application development with a program editor, compiler, and link editor
Courtesy of Course Technology/Cengage Learning

The compiler and link editor translate source code into executable code, although they perform different parts of the translation process. **Executable code** is stored in a file, such as an .exe file in Windows, and contains CPU instructions that are ready for an OS to load and execute. The **compiler** translates some source code instructions into executable code and others into library calls, which are further processed by the link editor. (Library calls and link editing are described later in "Link Editing.") Compiler output, called **object code**, contains a mixture of CPU instructions, library calls, and other information the link editor needs. It's usually stored in a file with an .o or .obj extension, such as ComputePayroll.obj.

The compiler reads and translates source code instructions. While reading source code, it performs the following tasks:

- Checks for syntax and other errors and issues warning or error messages, if needed

- Updates internal tables that store information about data items and program components, such as functions and subroutines
- Generates CPU instructions or library calls to carry out source code instructions

The actions the compiler takes vary, depending on the source code instruction type. Most programming languages have four source code instruction types, described in the following sections:

- Data declarations
- Data operations
- Control structures
- Function, procedure, or subroutine calls

Data Declarations

A **data declaration** defines the name and data type of program variables. For example, take a look at these C++ source code instructions:

```
float f_temperature;
float c_temperature;
```

These instructions declare two floating-point variables named `f_temperature` and `c_temperature`. When a compiler reads a data declaration, it allocates memory to store the data item. The amount of memory it allocates depends on the data type and the number of bytes used to represent this data type in the target CPU.

The compiler updates an internal table, called a **symbol table**, to keep track of data names, types, and assigned memory addresses. Information stored in this table includes the variable name, data type, and memory location. Table 10.2 shows some sample entries.

TABLE 10.2 Sample symbol table entries for data declarations

Name	Type	Length	Address
f_temperature	Single-precision floating-point	4	1000
c_temperature	Single-precision floating-point	4	1004

Data Operations

A **data operation** is any instruction, such as an assignment statement or a computation, that updates or computes a data value. The compiler translates data operation instructions into an equivalent sequence of data movement and data transformation instructions for the target CPU. It refers to entries in the symbol table to determine source and destination memory addresses for data movement instructions.

For example, here's a source code instruction that assigns a constant value to a program variable:

```
f_temperature = 212;
```

The compiler looks up `f_temperature` in the symbol table to find its memory address (1000) and generates this CPU instruction:

```
MOV 1000 <212>
```

In this instruction, `<212>` is the internal floating-point representation of the value 212. The compiler translates an instruction copying data from one variable to another

```
f_temperature = c_temperature;
```

into this CPU instruction:

```
MOV 1000 1004
```

Now look at a more complex example that combines computation and data movement:


```
c_temperature = (f_temperature – 32) ÷ 1.8;
```

The compiler translates this statement into the following sequence of CPU instructions:

```
MOV R1 1000        ; move f_temperature to register 1
MOV R2 <32>        ; store constant 32 to register 2
FSUB R1 R2 R3      ; subtract R2 from R1 and
                   ; store the result in register 3
MOV R2 <1.8>       ; store constant 1.8 in register 2
FDIV R3 R2 R3      ; divide register 3 by register 2 and
                   ; store the result in register 3
MOV R3 1004        ; copy register 3 to c_temperature
```

More complex formulas or more input variables and constants require longer CPU instruction sequences.

Control Structures

A **control structure** is a source code instruction that controls the execution of other source code instructions. Control structures include unconditional BRANCHes, such as a `goto` statement; conditional BRANCHes, such as an `if-then-else` statement; and loops, such as `while-do` and `repeat-until`.

The common thread in all control structures is the transfer of control between CPU instructions. A CPU BRANCH instruction requires a single operand containing the address of another instruction. Because all control structures require a CPU BRANCH instruction, the compiler must keep track of where CPU instructions for each source code instruction are located in memory. For example, examine these source code instructions:

```
      f_temperature = 0;
loop: f_temperature = f_temperature + 1;
      goto loop;
```

Here's the equivalent CPU instruction sequence:

```
2000 MOV 1000 <0>      ; copy 0 to f_temperature
2004 MOV R2 <1>        ; copy 1 to register 2
2008 MOV R1 1000       ; copy f_temperature to register 1
200C RADD R1 R2 R1     ; add R1 and R2 store result in R1
2010 MOV 1000 R1       ; copy result to f_temperature
2014 JMP 200C          ; loop back to add
```

The numbers to the left of each CPU instruction are its hexadecimal memory address. The name `loop` in the source code is a label. When the compiler reads a label, it stores the label in the symbol table along with the address of the first CPU instruction generated for the corresponding source code instruction. When the label is used as the target of a `goto` instruction, the compiler retrieves the corresponding memory address from the symbol table and uses it as the operand of the corresponding CPU BRANCH or JUMP instruction.

The `if-then-else`, `while-do`, and `repeat-until` control structures are based on a conditional BRANCH. For example, look at this source code instruction sequence:


```
f_temperature = 0;
while (f_temperature < 10)
{
    f_temperature = f_temperature + 1;
}
```

Here's the equivalent CPU instruction sequence:

```
2000 MOV 1000 <0>      ; copy 0 to f_temperature
2004 MOV R2 <1>        ; copy 1 to register 2
2008 MOV R3 <10>       ; copy 10 to register 3
200C MOV R1 1000       ; copy f_temperature to register 1
2010 RADD R1 R2 R1     ; add R1 and R2 store result in R1
2014 MOV 1000 R1       ; copy result to f_temperature
2018 FLT R1 R3         ; floating-point less than comparison
201C CJMP 2010         ; loop back to add if less-than is true
```

As before, each CPU instruction is preceded by its memory address. When the compiler reads the `while` statement, it temporarily stores the condition's contents and marks the memory address of the first instruction in the `while` loop (2010). It then translates each statement in the loop. When there are no more statements in the loop, as indicated by the closing brace (}), the compiler generates a CPU instruction to evaluate the condition (2018), followed by a conditional BRANCH statement to return to the first CPU instruction in the loop.

An `if-then-else` statement selects one of two instruction sequences by using both a conditional and unconditional BRANCH. For example, this source code instruction sequence

```
if (c_temperature >= -273.15)
    f_temperature = (c_temperature * 1.8) + 32;
else
    f_temperature = -999;
```

causes the compiler to generate the following CPU instruction sequence:

```
2000 MOV R1 1004          ; copy c_temperature to register 1
2004 MOV R2 <-273.15>     ; store constant -273.15 to register 2
2008 FGE R1 R2            ; floating-point > or =
200C CJMP 2018            ; branch to if block if true
                          ; begin else block
2010 MOV 1000 <-999>      ; copy constant -999 to f_temperature
2014 JMP 202C             ; branch past if block
                          ; begin if block (calculate f_temp)
2018 MOV R2 <1.8>         ; store constant 1.8 to register 2
201C FMUL R1 R2 R3        ; multiply registers 1 and 2 and
                          ; store the result in register 3
2020 MOV R2 <32>          ; store constant 32 in register 2
2024 FADD R2 R3 R3        ; add registers 2 and 3 and
                          ; store the result in register 3
2028 MOV 1000 R3          ; copy register 3 to f_temperature
202C                      ; next instruction after if-then-else
```


Function Calls

In most languages, a programmer can define a named instruction sequence called a **function**, **subroutine**, or **procedure** that's executed by a call instruction. A **call instruction** transfers control to the first instruction in the function, and the function transfers control to the instruction following the call by executing a **return instruction**.

NOTE

The terms "function," "subroutine," and "procedure" describe a named instruction sequence that receives control via a call instruction, receives and possibly modifies parameters, and returns control to the instruction after the call. For the remainder of the chapter, the term "function" is used, although everything said about functions also applies to procedures and subroutines.

Functions are declared much as program variables are, and the compiler adds descriptive information to the symbol table when it reads the declaration. For example, examine the following function:

```
float fahrenheit_to_celsius(float F)
{
    /* convert F to a celsius temperature */
    float C;
    C = (F - 32) / 1.8;
    return(C);
}
```

When the compiler reads the function declaration, it adds the function name `fahrenheit_to_celsius` to the symbol table and records the memory address of the function's first CPU instruction as well as other information, such as the function type (float) and the number and type of input parameters. When it reads a later source code line referring to the function, such as the following, it retrieves the corresponding symbol table entry:

```
c_temperature = fahrenheit_to_celsius(f_temperature);
```

The compiler first verifies that the types of the input variable (`f_temperature`) and result assignment variable (`c_temperature`) match the types of the function and function arguments. To implement call and return instructions, the compiler generates CPU instructions to do the following:

- Pass input parameters to the function.
- Transfer control to the function.
- Execute CPU instructions in the function.
- Pass output parameters back to the calling function.
- Transfer control back to the calling function at the statement immediately after the function call statement.

The flow of control to and from functions is similar to the flow of control to and from an OS interrupt handler, explained in Chapter 6. The calling function must be suspended, the called function must be executed, and the calling function must then be restored to its original state so that it can resume execution. As with interrupt handlers, register values are pushed onto the stack just before the called function begins executing and are popped from the stack just before the calling function resumes execution. The stack serves as a temporary holding area for the calling function so that it can be restored to its original state after the called function finishes execution.

Table 10.3 shows symbol table entries for the function `fahrenheit_to_celsius`, its internal parameters, and its variables in this function call:

```
c_temperature = fahrenheit_to_celsius(f_temperature);
```

TABLE 10.3 Symbol table entries

Name	Context	Type	Address
`fahrenheit_to_celsius`	Global	Single-precision floating-point, executable	3000
`F`	`fahrenheit_to_celsius`	Single-precision floating-point	2000
`C`	`fahrenheit_to_celsius`	Single-precision floating-point	2004
`f_temperature`	Global	Single-precision floating-point	1000
`c_temperature`	Global	Single-precision floating-point	1004

The compiler reads data names used in the function call and return assignment and looks up each name in the symbol table to extract the corresponding address. It also looks up the function parameter's name and extracts its addresses and lengths. The compiler then creates a MOV4 (move 4 bytes) instruction with this address and places it immediately before the PUSH instruction:

```
MOV4 2000 1000   ; copy fahrenheit temperature
PUSH             ; save calling process registers
JMP 3000         ; transfer control to function
```

The return from the function call is handled in a similar manner:

```
MOV4 1004 2004   ; copy function value to caller
POP              ; restore caller register values
```


The compiler generates code to copy the function result to a variable in the calling program. Control is then returned to the caller by executing a POP instruction.

LINK EDITING

The previous examples showed how a compiler translates assignment, computation, and control structure source code instructions into CPU instructions. It translates other types of source code instructions, such as file, network, and interactive I/O statements, into external function calls. The compiler also generates external function calls when it encounters call statements without corresponding function declarations. An **external function call**, sometimes called an **unresolved reference**, is a placeholder for missing executable code. It contains the name and type of the called function as well as the memory addresses and types of function parameters.

A **link editor** searches an object code file for external function calls. When one is found, it searches other object code files or compiler libraries to find executable code that implements the function. A **compiler library** is a file containing related executable functions and an index of the library contents. When the link editor finds executable code in a library or another object code file, it inserts the executable code into the original object code file and generates instructions to copy parameters and the function's return value. The result is a single file containing only executable code. If the link editor can't find executable code for an external function call, it generates an error message and produces no executable file.

At this point, you might be wondering why the compiler couldn't have generated all the executable code, eliminating the need for a link editor. Link editors provide two key benefits in program translation:

- A single executable program can be constructed from multiple object code files compiled at different times.
- A single compiler can generate executable programs that run in multiple OSs.

To understand a link editor's function better, say that each function in a program to convert temperatures between Fahrenheit and Celsius is being written by a different programmer. In other words, each programmer develops a function and stores the source code in a file. Each file is compiled separately, and none of the object code files contain a complete program. To combine these separate functions, the link editor patches together the call and return instructions in each object code file to create a single executable program (see Figure 10.11).

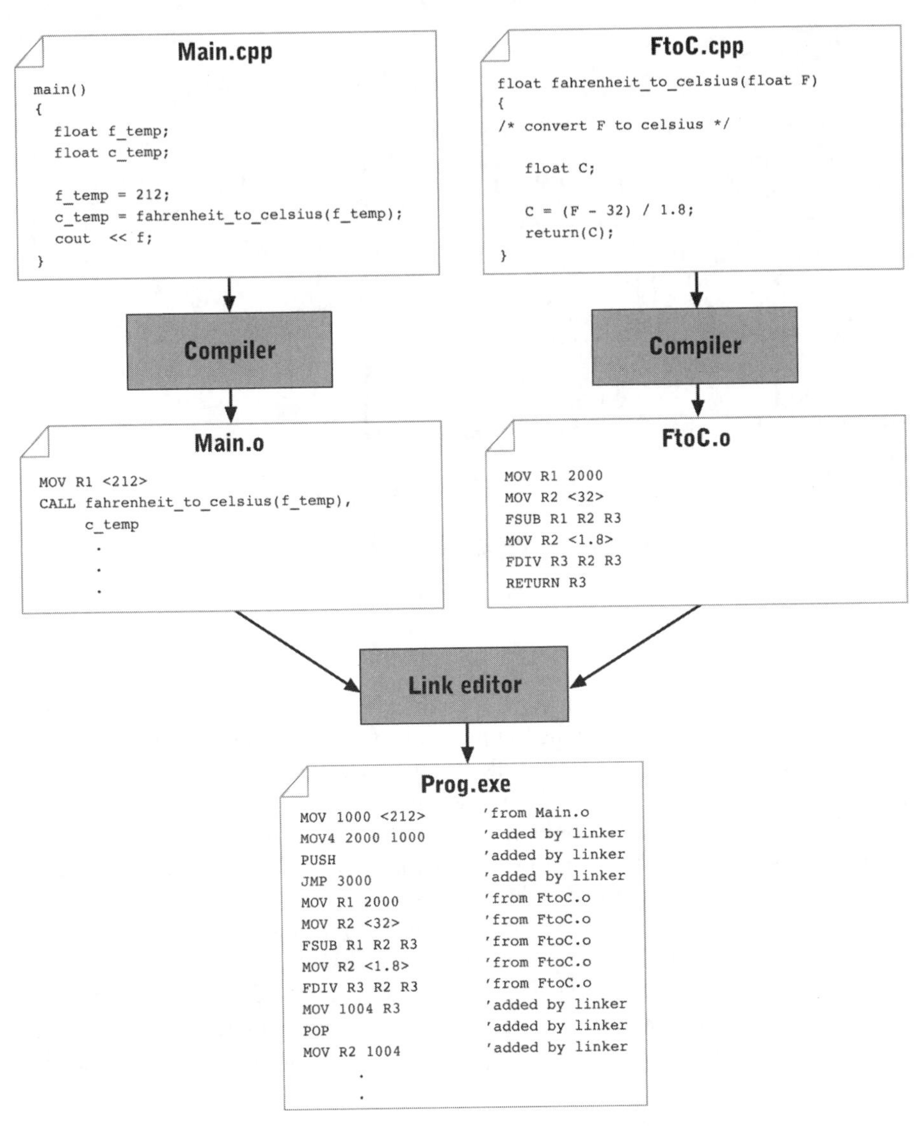

FIGURE 10.11 A link editor combines separately compiled functions into a single executable program

Courtesy of Course Technology/Cengage Learning

A link editor also makes it possible to use a single compiler for multiple OSs. For example, the C++ temperature conversion program in Figure 10.12 reads temperature input from the keyboard. (Lines beginning with // are comments.) Because the CPU instructions for interactive I/O vary by OS, if the compiler were responsible for generating CPU instructions for I/O operations, different compilers would be required to implement this program for each OS.

```
main()
{
      float f;
      float c;

      // print prompt
      cout << "Enter a fahrenheit temperature: ";
      // get keyboard input
      cin >> f;
      // compute celsius equivalent
      c = fahrenheit_to_celsius (f);
      // print celsius temperature
      cout << f << " degrees fahrenheit is ";
      cout << c << " degrees celsius." << endl;

      return(0);
}
```


FIGURE 10.12 A C++ program that reads input from the keyboard
Courtesy of Course Technology/Cengage Learning

Using external function calls and link editing enables a single compiler to serve multiple OSs. CPU instructions that implement I/O functions for each OS are stored in a compiler library. When object code files are linked, the programmer instructs the link editor to use the library for the target OS. The link editor replaces external function calls for interactive I/O with the CPU instructions for a specific OS. Target OSs might be completely different (such as Windows and Linux) or might be different versions of the same OS family (for example, Windows 7, Vista, and XP).

Link editing with compiler libraries can offer flexibility in other ways. For example, encryption algorithms vary in complexity and maximum length of the encryption key, and different versions of encryption algorithms with the same name and parameters can be stored in different libraries. An application programmer can choose the level of encryption to embed in an executable program by specifying which encryption library the link editor reads.

Dynamic and Static Linking

The linking process described previously is often called **static linking** or **early binding**. The term "static" is used because library and other subroutines can't be changed after they're inserted into executable code. **Dynamic linking**, or **late binding**, is performed during program loading or execution.

Calls to OS service routines are sometimes done with dynamic linking. For example, most Windows OS service routines are stored in **dynamic link libraries (DLLs)**. During program development, the link editor inserts code to call a special DLL interrupt handler and passes pointers to the service routine name and parameters in general-purpose registers. During program execution, the DLL interrupt handler accepts DLL service requests, locates the service routine by name, loads it into memory (if it isn't already there), copies input parameters, and then executes the service routine. When the service routine finishes, the DLL interrupt handler copies output parameters, terminates the service routine, and passes control back to the calling program.

Dynamic linking has two key advantages over static linking. The first advantage is smaller program executable files. Static linking incorporates copies of service routines into executable programs, which can increase executable program size by double, triple, or more, depending on the nature and number of functions included. Much of this additional storage is redundant because common subroutines (such as those to open and close files or windows) are used by many application programs. Dynamic linking avoids redundant storage of service routines in program executable files. Instead, each program contains only the code needed to call the DLL interrupt handler, which is typically only a few hundred kilobytes.

The second advantage of dynamic linking is flexibility. Portions of the OS can be updated by installing new DLLs. If application programs use dynamic linking, updating the OS updates all application programs with the new service-layer subroutines. The new subroutines are loaded and executed the next time an application program runs after an upgrade. In contrast, static linking locks an application program to a particular version of the service-layer functions. These functions can be updated only by providing an updated compiler library and relinking the application's object code.

The main advantage of static linking is execution speed. When a dynamically linked program calls a service-layer function, the OS must find the requested function, copy parameters, and transfer control. If the subroutine isn't in memory, the application program is delayed until the service routine is located and loaded into memory. All these steps are avoided if the application program is statically linked, resulting in much faster execution.

Static linking also improves the reliability and predictability of executable programs. When an application program is developed, it's usually tested with a specific set, or version, of OS service routines. Service routine updates are designed to be backward compatible with older versions, but there's always the risk that new routines dynamically linked into an older application program could result in unexpected behavior. Static linking ensures that this behavior can't happen. However, the reverse situation can also occur. That is, a new service routine version might fix bugs in an older version. Static linking prevents these fixes from being incorporated into existing programs automatically.

INTERPRETERS

Source and object code files are compiled and linked as a whole. In contrast, **interpretation** interleaves source code translation, link editing, and execution. An **interpreter** reads a single source code instruction, translates it into CPU instructions or a DLL call, and executes the instructions or DLL call immediately, before the next program statement is read. In essence, one program (the interpreter) translates and executes another (the source code) one statement at a time.

The main advantage of interpretation over compilation is that it offers the flexibility to incorporate new or updated code into an application program. This flexibility is a result of dynamic linking. Program behavior can be updated easily by installing new versions of dynamically linked code. It's also possible for programs to modify themselves—a feature required for certain types of applications, such as expert systems.

NOTE

Most desktop computers and many servers have more than enough memory and CPU resources to handle the additional demands of interpretation without noticeable performance degradation. The performance difference is an issue in some hardware environments, such as portable computing, and for many CPU-intensive applications, such as numerical modeling and computer animation.

The main disadvantage of interpretation compared with compiling and linking is increased memory and CPU requirements during program execution (see Table 10.4). With compiling and link editing, each program used to develop executable code resides in memory for only a short time. When a program is loaded and executed, all the software components that helped create it (program editor, compiler, and link editor) are no longer in memory. The memory requirements of a compiled and linked application program consist only of the memory required to store its executable code.

TABLE 10.4 Memory and CPU resources used during application execution

Resource	Interpretation	Compilation
Memory contents (during execution):		
Interpreter or compiler	Yes	No
Source code	Partial	No
Executable code	Yes	Yes
CPU instructions (during execution):		
Translation operations	Yes	No
Library linking	Yes	No
Application program	Yes	Yes

Symbolic Debugging

Compilers, link editors, and interpreters produce error messages to warn programmers of actual or potential mistakes in source code. However, even if a program is compiled and linked successfully, it might still produce errors when it runs. Diagnosing runtime errors based on a typical runtime error message is difficult because the error message refers to the executable code generated by the compiler and link editor instead of the programmer's source code. Variable names have been replaced by memory addresses, and symbolic source code instructions have been replaced by CPU instructions. Determining the source code instructions and variables that correspond to these memory addresses is a difficult task.

A compiler can be instructed to incorporate its symbol table into the object code file, and a link editor can be instructed to produce a **memory map**, or link map, based on the symbol table's contents. A memory map lists the memory location of every function and program variable, and a programmer can use it to trace error messages containing memory addresses to corresponding program statements and variables. Tracing these addresses to source code statements or data items requires a detailed memory map and a thorough understanding of machine code and the compiling and linking processes.

For example, look at the partial memory map in Figure 10.13 (the complete map is several hundred text lines) and the runtime error message in Figure 10.14. The error message lists the memory instruction that generated the error, and the memory map lists the starting address of each module in the program. The address in this error message falls between addresses in the third and fourth rows of the memory map. With this information, a programmer can determine that the error occurred somewhere in the `_main` function but can't tell exactly which instruction caused the error or determine the contents of program variables at the time the error occurred.


```
TemperatureConversion

Preferred load address is 00400000
  Location              Publics by Value                  Address   Lib:Object

0001:00000000  ?fahrenheit_to_celsuis@@YAMM@Z   00401000  Main.obj
0001:00000020  ?celsuis_to_fahrenheit@@YAMM@Z   00401020  Main.obj
0001:00000040  _main                            00401040  Main.obj
0001:000001b0  ??6ostream@@Z@Z                  004011b0  Main.obj
0001:000001d0  ?flush@@YAAAVostream@@AAV1@@Z    004011d0  Main.obj
0001:00009242  _ExitProcess@4                   0040a242  KERNEL32.dll
0002:000000f0  __real@8@3ffe8e38e38e38e39000    0040b0f0  Main.obj
0002:000000f8  __real@4@4004800000000000000000  0040b0f8  Main.obj
0002:00000100  __real@8@4004800000000000000000  0040b100  Main.obj
0002:00000108  __real@8@3fffe666666666666800    0040b108  Main.obj

entry point at           0001:00001ac2
```

FIGURE 10.13 A partial memory map
Courtesy of Course Technology/Cengage Learning

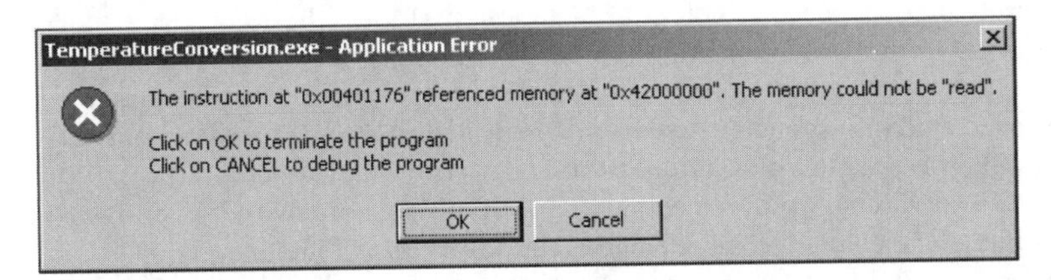

FIGURE 10.14 A typical runtime error message
Courtesy of Course Technology/Cengage Learning

A **symbolic debugger** is an automated tool for testing executable programs. It includes several features, including the capability to perform these tasks:

- Trace calls to specific source code statements or subroutines.
- Trace changes to variable contents.
- Execute source code instructions one at a time.
- Detect runtime errors and report them in terms of specific source code instructions and variables.

A symbolic debugger uses the symbol table, memory map, and source code files to trace memory addresses to specific source code statements and variables. It also inserts debugging checkpoints after each source code instruction so that program execution can be paused. An executable program containing symbol table entries and debugging checkpoints is sometimes called a **debugging version**. In contrast, a program's **production version**, or **release version**, omits the symbol table and debugging checkpoints to reduce program size and increase execution speed.

Interpreted programs have an advantage over compiled programs in detecting and reporting runtime errors because the symbol table and program source code are always available to the interpreter at runtime. If an error occurs, it can be reported in terms of the most recent source code line translated. Memory addresses can also be converted to source code names by looking them up in the symbol table. Most interpreters incorporate symbolic debugging capabilities, so they don't need a stand-alone symbolic debugging program.

TECHNOLOGY FOCUS

Java

Java is an OOP language and program execution environment developed by Sun Microsystems during the early and mid-1990s. It was first used as a plug-in to extend Web browser functionality and, as a result, has been described as a Web or network programming language. Although it does satisfy this description, Java is a full-featured programming language that supports almost any combination of hardware platform and OS.

Java's syntax and capabilities are similar to C++'s, so programmers familiar with either language can learn the other easily. Unlike C++, which is an extension of the C procedural language, Java was designed to be an OOP language. It supports object-oriented features, including encapsulation, inheritance, and polymorphism, but multiple inheritance isn't supported. It's designed to maximize application reliability and reusability of existing code.

What makes Java unique is a standardized target machine language for Java interpreters and compilers. Compilers and interpreters for other programming languages translate source code into executable CPU instructions and service routine calls to a specific OS. A Java compiler or interpreter translates Java source code into machine instructions and service routine calls for a hypothetical computer and OS called the **Java Virtual Machine (JVM)**. Instructions and library calls to the JVM are called Java "bytecode."

(continued)

The term "JVM" also describes an interpreter that translates JVM instructions into actual CPU instructions and OS service routine calls. The JVM simulates a Java computer with the CPU and OS of a conventional computer (see Figure 10.15). The JVM bytecode interpreter translates JVM machine instructions into CPU instructions for a real CPU, such as an Intel Core-i7 or IBM POWER7. Calls to JVM service routines are translated into equivalent calls to a real OS, such as Windows or Linux.

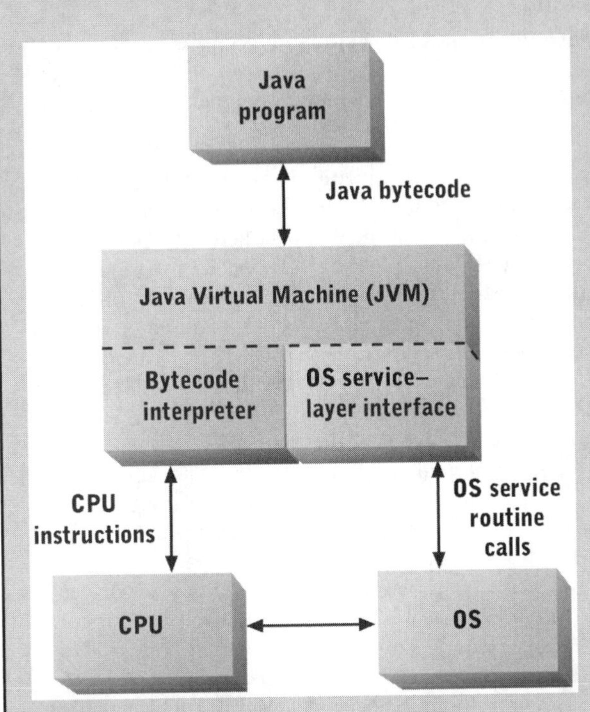

FIGURE 10.15 How Java and the JVM relate to the CPU and OS
Courtesy of Course Technology/Cengage Learning

Java programs come in three forms: stand-alone applications, applets, and servlets. Stand-alone Java applications run in the JVM and don't usually interact with other programs. A Java **applet** runs inside another program, such as a Web browser, and performs functions such as accepting user input and displaying forms and images. A Java **servlet** runs in a Web server and performs functions such as calculations, database access, and creation of Web pages that are transmitted to a Web browser for display. Applets and servlets run in a protected area called the **sandbox**, which provides extensive security controls to prevent them from accessing unauthorized resources or damaging the hardware, OS, or file system.

Native support for Java OS calls has been added to the Sun Solaris OS, which is Sun's version of UNIX. Other OS vendors haven't added native support for Java, although finding a general-purpose computer without an installed JVM is rare. Incorporating the

(continued)

JVM into the native OS reduces some of the inefficiency inherent in translating OS service routine calls. It also reduces the need for Java-enabled application software to install and use private and sometimes incompatible JVMs.

Java's popularity has far exceeded most initial expectations because of a number of factors, including the following:

- Sun's strategy of providing Java compilers and JVMs at little or no cost
- Incorporation of JVMs into Web browsers and servers
- The predominance of application software that uses a Web browser as the primary I/O device
- The capability of Java programs to run on almost any combination of hardware and OS

The most important drawback of Java is reduced execution speed resulting from using interpreted bytecode and OS translation. Emulation processes are inefficient users of hardware resources. **Native applications**—those compiled and linked for a particular CPU and OS—usually run 10 times faster than interpreted Java bytecode programs. Some JVMs use a just-in-time compiler to reduce the speed difference to approximately 3:1, but Java programs are always slower than native programs when running on anything other than a true Java machine, which doesn't exist currently. This inefficiency is an important consideration in some application environments, including portable computing devices, such as netbooks, tablet PCs, and Web-enabled cell phones, which typically use less powerful CPUs than other computer types.

Oracle acquired Sun Microsystems in 2009, and its intentions for Java's future aren't fully known. Oracle has incorporated Java technologies into its product offerings extensively since the late 1990s, including database servers and application development tools (see the JDeveloper Technology Focus later in this chapter). There has been some fear that Oracle might alter the Sun Microsystems tradition of open-source and user-driven development. However, current Java technologies are rooted firmly in the public domain because they're distributed under the GNU General Public License, which grants substantial rights to third parties to modify and redistribute Java programs, JVMs, and related tools. (You examine the GNU General Public License in "Research Problems" at the end of the chapter.)

APPLICATION DEVELOPMENT TOOLS

Programming languages, program translation tools, and debugging tools address only the implementation discipline of the Unified Process. Other tools are needed to perform activities in the other disciplines. Figure 10.16 summarizes the application development tools discussed, introduces new tools, and ties each tool to a specific part of the application development process.

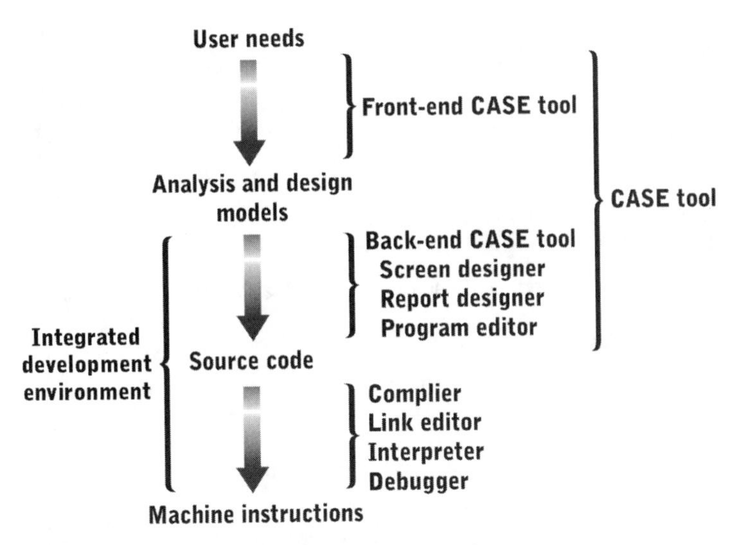

FIGURE 10.16 Application development tools
Courtesy of Course Technology/Cengage Learning

Integrated Development Environments

An **integrated development environment (IDE)** is a collection of automated support tools to speed development and testing, and it generally includes the following components:

- Intelligent program editor
- Compiler, interpreter, or both
- Link editor and a large library of classes or subroutines
- Interactive tool for prototyping and designing user interfaces
- Symbolic debugger
- Integrated GUI

The key feature of an IDE is the level of integration among tools, not the specific tools included. It leverages programmers' effort and skill so that programs can be written and tested in minutes, hours, or days; automates tedious and time-consuming tasks; streamlines other tasks; and maximizes reuse of existing source code. Using an IDE, a programmer can develop and test many program versions rapidly.

Editors assist programmers in writing syntactically correct code by verifying syntax as the code is being typed and highlighting errors (see Figure 10.17). They might also offer help in the form of completing program statements automatically or displaying pop-up messages or tool tips related to the current source code line (see Figure 10.18). Compiler and link editor errors displayed in one window are linked to source code instructions in another window. Some compilers supply suggested corrections or context-sensitive help in addition to error messages.

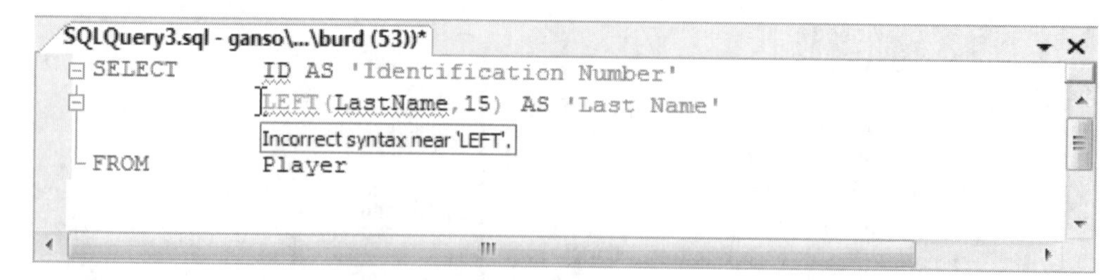

FIGURE 10.17 An intelligent program editor underlines suspected errors and displays a pop-up message

Courtesy of Course Technology/Cengage Learning

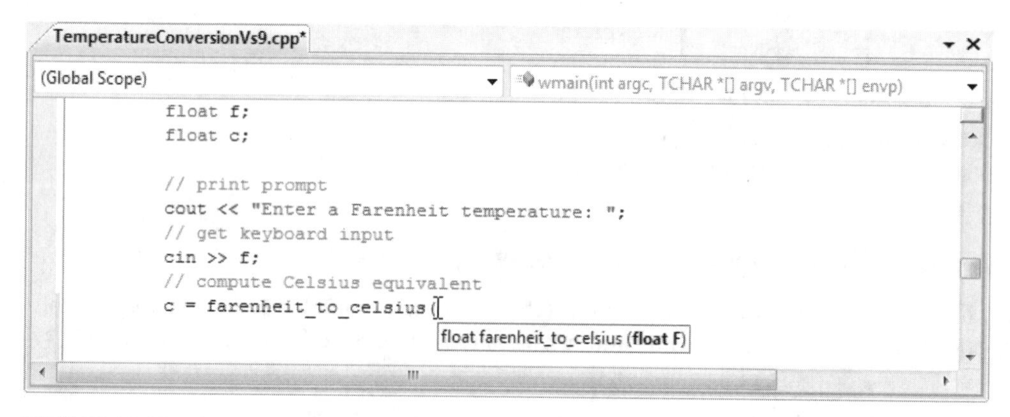

FIGURE 10.18 A pop-up tool tip prompting a programmer to correct a source code line

Courtesy of Course Technology/Cengage Learning

IDEs can also include program templates and skeletons stored in a library or generated based on input from the programmer or from other tools. They usually contain large libraries of predefined subroutines or classes that address user interface, database manipulation, network resource access, and other common functions. Library routines can be imported into program source code or added via function or method calls and linked into the executable program. With libraries, programmers can reuse source and executable code, thus speeding up application development.

NOTE

Complex tools, simulated execution environments, and large libraries require two to three times more CPU power, memory, and secondary storage than a typical desktop computer has. Programmers also need multiple large monitors to display all the windows used in the development process.

IDEs offer extensive symbolic debugging support. Method calls, function calls, and changes to program variables or object data can be traced during program execution.

Programs can be started and stopped at programmer-specified locations so that testing can be performed on specific code segments. Debugging and tracing can also be done on machine code and OS service calls. Testing and debugging occur in a simulated runtime environment so that program crashes can't disable the IDE or the entire workstation.

TECHNOLOGY FOCUS

Oracle JDeveloper

Oracle JDeveloper is an IDE for developing object-oriented Java software with Oracle and other DBMSs. A JDeveloper workspace is a container for application components, including program files, Web pages, system development models, user interface windows, and database connections. A programmer can interact with different views of a project, including software development models (such as class and package diagrams), files (see the upper-left pane in Figure 10.19), and classes and methods (see the lower-left pane in Figure 10.19).

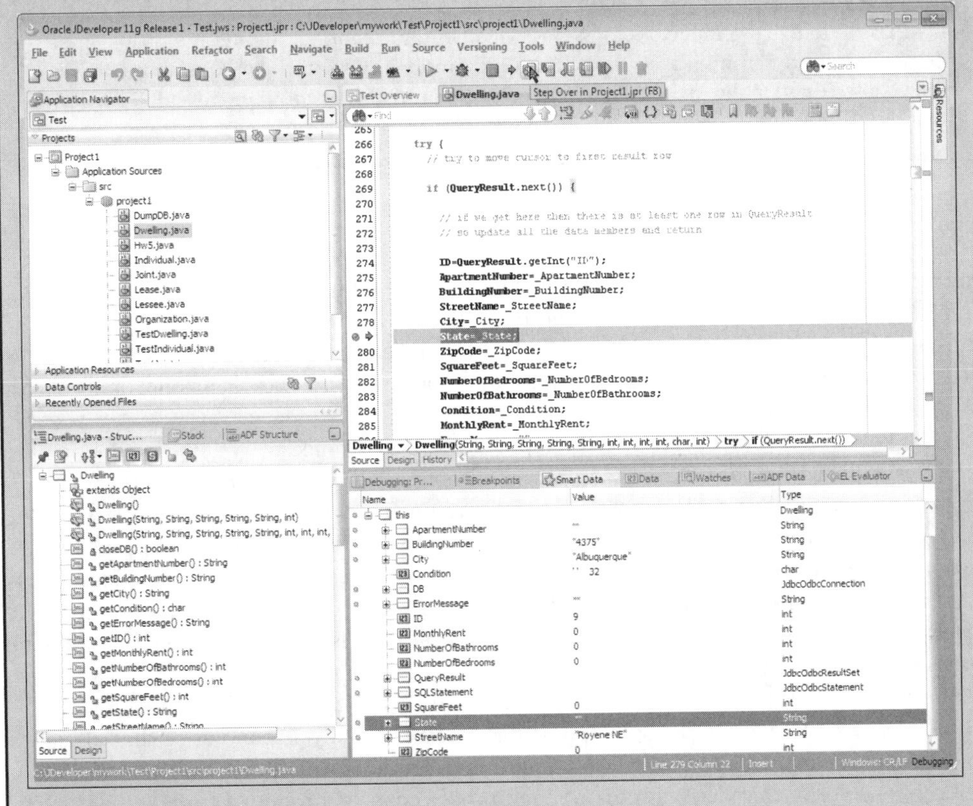

FIGURE 10.19 The Oracle JDeveloper interface
Courtesy of Course Technology/Cengage Learning

(continued)

JDeveloper contains many application development tools, including compilers, link editors, program text editors, symbolic debuggers, and interactive tools for designing user interface elements, such as menus, toolbars, icons, and forms. It also keeps track of which tools are used to create and modify programmer-defined objects and opens the correct tool automatically when a programmer selects an application component. Overlapping menus, toolbars, and icons provide access to tool features and commands.

Development tools are integrated automatically whenever possible. For example, a command to build a complete program automatically opens the link editor, user interface resource compilers, and program compilers for each component programming language in the right order. JDeveloper keeps track of file dependencies, so a change to one source code file causes all dependent files to be rebuilt automatically. It also manages data movement between tools and displays progress and results in an output window.

JDeveloper includes extensive symbolic debugging capabilities. For example, a programmer can pause program execution at specified source code lines. (In Figure 10.19, execution is paused before line 279.) Further execution can be directed on a line-by-line or function-by-function basis. A programmer can examine variable content while the program is paused and trace variable changes during execution (see the lower-right pane in Figure 10.19).

In addition, JDeveloper offers useful documentation containing many cross-references and hyperlinks to glossary terms, examples, and tutorials that can be accessed from a hard disk or the Web. A sophisticated documentation browser provides extensive search and indexing capabilities.

CASE Tools

The term **computer-assisted software engineering (CASE) tool** usually refers to a tool that supports the UP requirements and design disciplines. This term is a bit of a misnomer because any tool that assists system or application development fits the literal definition of a CASE tool. The singular form is also misleading because CASE tools are normally highly integrated tool suites.

The key feature of a CASE tool is support for a broad range of system development activities, with particular emphasis on model development. CASE tools support most or all system development activities up to, and sometimes including, program translation. They usually support a specific system development methodology, such as structured system development or the UP. A tool suite that primarily supports model development is sometimes called a **front-end CASE tool**. A tool suite that primarily supports application development based on specific analysis and design models is sometimes called a **back-end CASE tool** or code generator.

CASE tools can perform consistency checks and other types of error checking on models. For example, a CASE tool supporting structured analysis can verify that all data flows in a data flow diagram have corresponding definitions in the data dictionary. This tool can also verify that data outputs of each process can be produced from the data inflows based on the process logic specification.

When extensive back-end capabilities are included in CASE tools, much of the application development process can be automated. For example, a CASE tool supporting object-oriented development can generate class declarations in Java or C++ source code files, database schema descriptions, and component registration entries for specific database and component management systems.

The most comprehensive CASE tools automate the process of deploying working systems from analysis and design models. These tools generate program and other source code from models, compile and link the programs, create databases, and create, register, and install all components. They can also include comprehensive features for planning and managing development projects and for maintaining and modifying existing systems. Comprehensive CASE tools are usually expensive, require weeks or months of training, and support specific deployment environments.

BUSINESS FOCUS

Building the Next Generation of Application Software

Southwestern Gifts, Inc. (SGI), is a small catalog retailer of Southwestern art, jewelry, and gifts that distributes two catalogs per year, in May and October. About 40% of orders are received by phone and 10% by mail, and the rest via the SGI Web site. SGI employs up to 50 people, including 15 warehouse staff and 30 order-processing clerks during the Christmas shopping season, 4 accountants and bookkeepers to handle payroll and accounts payable, 3 buyers, a catalog designer, 4 IS staff, a warehouse manager, and the owner-operator.

SGI currently has a Hewlett-Packard (HP) ProLiant midrange computer running Linux, a small Dell server running Windows Server, and 20 desktop computers, all connected by a 100 Mbps Ethernet LAN with a single T1 (1.54 Mbps) Internet connection. Ten desktop computers are used only for order entry. The Dell server provides file and printer sharing for desktop computers and printer sharing for the HP computer. The other 10 desktop computers are used by the accountants, warehouse manager, buyers, catalog designer, and IS staff.

Application software on desktop computers includes Quicken (an accounting package), various productivity tools, such as Microsoft Office, and Adobe Illustrator (graphics design software). SGI purchased an order entry and inventory control package eight years ago from an out-of-state vendor. This package stores data in a MySQL relational DBMS and uses a proprietary IDE and operating environment called Data Entry, Retrieval, and Query System (DERQS).

DERQS provides forms-based data entry and data queries, a file definition tool, and a simple report generator. Screen forms are defined by using an interactive layout tool that compiles and stores screen layout and content in a screen definition file (SDF). Database tables are also defined with an interactive tool, and a simple query interface and report generator are included with the DBMS.

Application programs are written in a proprietary interpreted scripting language to display and manipulate screens, manipulate database content, and generate simple reports. Most of the existing application programs are DERQS scripts, but some are written in C and C++. DERQS includes a compiler library with functions that enable C programs to display DERQS screens and interact with DERQS files and the MySQL database. This library also works with C++ programs because C++ is a superset of C.

(continued)

Within a year after purchasing the order-entry and inventory control package, SGI contracted with the vendor to develop a Web-based order system for customers' use. The Web interface is targeted for Internet Explorer 5 (the most current version at that time). Web pages are hosted by Apache Web Server software running on the HP server, use HTML forms-based data input and JavaScript, and interact with back-end processing functions written in C.

One IS staff member specializes in DERQS data entry functions and script-based applications. A second IS staff member specializes in the DERQS query language and develops and maintains C and C++ applications. A third IS staff member maintains the Web site, including the online catalog and order-entry system. A fourth IS staff member manages the LAN and desktop computers.

SGI has seen the phone and mail order parts of its business decline steadily. Although it expects to keep these systems operating for several more years, SGI sees a need to improve its Web-based ordering system to attract younger, more tech-savvy customers. SGI also wants to minimize the number of software technologies it depends on and integrate purchasing, sales, inventory, and accounting functions more tightly. To achieve these goals, SGI plans to make several improvements to its information systems, including the following:

- Updating the Web-based ordering system to take advantage of more recent HTML standards, later IE versions, and other browsers, such as Firefox
- Implementing a direct interface to financial information in the MySQL database so that data can be uploaded to Quicken and Excel
- Enabling buyers to interact with the sales, purchasing, and accounts payable systems
- Using Web-based interfaces for existing and new application programs, with support for employees working from home or while traveling

SGI recently learned that the company that developed and supported DERQS has filed for bankruptcy. No company has shown an interest in purchasing the rights to DERQS, so it appears that no further upgrades or technical support will be available.

Questions:

- Should SGI develop any new software with DERQS? If not, what tools should it acquire for new system development?
- Should SGI reimplement the functions of its existing DERQS-based applications by using more up-to-date development tools?
- Should SGI consider replacing its disparate collection of tools and supporting software with an integrated suite from a large vendor, such as an Oracle DBMS with JDeveloper or Microsoft SQL Server with Visual Studio? What are the advantages and disadvantages of this replacement?

Summary

- Applications are developed by following the disciplines and iterations of the Unified Process. Application development follows a methodology, develops models, and uses automated tools. Most application development cycles include two sets of models: requirements models and design models.
- Executable software consists entirely of CPU instructions. Programmers write programs in higher-level programming languages that are translated into CPU instructions. Programming languages have evolved through many generations, and the most recent types are object-oriented and scripting languages. Later generations have incorporated improvements and capabilities such as instruction explosion, database access, support for GUIs, and nonprocedural programming.
- All programming language generations other than 1GL must be translated into CPU instructions before execution. A compiler translates an entire source code file, building a symbol table and an object code file as output. An interpreter translates, links, and executes source code programs one source code instruction at a time.
- Compiled and interpreted programs must be linked to libraries of executable functions or methods. A link editor statically links external reference calls in object code to library functions and combines them into a single file containing executable code. It also produces a memory map showing the organization of all software modules, which is useful for debugging. A link editor can also dynamically link external calls by statically linking them to an OS service routine that loads and executes DLL functions at runtime. Interpreters always use dynamic linking.
- Integrated suites of automated tools support the application development process. An IDE includes tools such as a program editor, screen and report designers, an intelligent compiler and link editor, a symbolic debugger, and extensive documentation. CASE tools provide a tool suite to support a wider range of development tasks. A front-end CASE tool supports developing requirements and design models. A back-end CASE tool generates program source code from models.

In the next two chapters, you look at operating systems in detail. Chapter 11 describes the internal architecture of an OS and discusses how it manages processes and memory. Chapter 12 concentrates on file and secondary storage device management.

Key Terms

applet
assembler
assembly language
back-end CASE tool
call instruction
code
compiler
compiler library
computer-assisted software engineering (CASE) tool
control structure
data declaration
data operation
debugging version
design models
dynamic link libraries (DLLs)

dynamic linking
early binding
executable code
external function call
fifth-generation language (5GL)
first-generation languages (1GLs)
fourth-generation languages (4GLs)
front-end CASE tool
function
instruction explosion
integrated development environment (IDE)
interpretation
interpreter
Java
Java Virtual Machine (JVM)
label
late binding
link editor
machine languages
memory map
message
native applications
nonprocedural language
object code
object-oriented programming (OOP)
procedure
production version
release version
return instruction
sandbox
scripting language
second-generation language (2GL)
servlet
source code
static linking
subroutine
symbol table
symbolic debugger
system requirements models
third-generation language (3GL)
unresolved reference
variable

Vocabulary Exercises

1. A compiler allocates storage space and makes an entry in the symbol table when it encounters a(n) ____________ in source code.
2. A(n) ____________ is produced as output during activities of the UP requirements discipline.
3. A link editor searches an object code file for ____________ .
4. A 4GL has a higher degree of ____________ than a 3GL does.
5. ____________ code contains CPU instructions and external function calls.
6. A(n) ____________ produces a(n) ____________ to show the location of functions or methods in executable code.
7. The compiler adds the names of data items and program functions to the ____________ as they're encountered in source code.
8. A 2GL is translated into executable code by a(n) ____________ .
9. A(n) ____________ translates an entire source code file before linking and execution. A(n) ____________ interleaves translation, link editing, and execution.
10. A Java ____________ runs in the ____________ of a Web browser.

11. ____________ in source code instructions are translated into machine instructions to evaluate conditions or transfer control from one program module to another.
12. A(n) ____________ uses the symbol table's contents to help programmers trace memory locations to program variables and instructions.
13. FORTRAN, COBOL, and C are examples of ____________ .
14. A(n) ____________ tool supports system model development. A(n) ____________ tool generates program source code from system models.
15. A link editor performs ____________ linking. An interpreter performs ____________ linking.
16. Java programs are compiled into object code for a hypothetical hardware and OS environment called the ____________ .
17. Widely used scripting languages include ____________ , ____________ , and ____________ .

Review Questions

1. Describe the relationships between application development methodologies, models, and tools.
2. Compare the generations and types of programming languages.
3. What is instruction explosion? What types of programming languages have the most instruction explosion? What types of programming languages have the least instruction explosion?
4. What are the differences between source code, object code, and executable code?
5. Compare assemblers, compilers, and interpreters.
6. What does a compiler do when it encounters data declarations in a source code file? Data (manipulation) operations? Control structures?
7. Compare the execution of compiled programs with interpreted programs in terms of CPU and memory utilization.
8. What is a link editor? What is a compiler library? How and why are they useful in program development?
9. What types of programming statements are likely to be translated into machine instructions by a compiler? What types are likely to be translated into library calls?
10. Compare error detection and correction capabilities in interpreters and compilers.
11. Compare static and dynamic linking.
12. What are the shortcomings of 3GLs in meeting the requirements of modern applications?
13. What are the main differences between OOP languages and traditional programming languages?
14. What components are normally part of an IDE? In what ways does an IDE improve programmer productivity?
15. What is a CASE tool? What's the relationship between a CASE tool and a system development methodology?
16. What's the difference between a front-end CASE tool and a back-end CASE tool?

Problems and Exercises

1. Develop assembly-language instructions to implement the following source code fragment:

```
a = 0;
i = 0;
while (i < 10) do
     a = a+i;
     i = i+1;
endwhile
```

Research Problems

1. Investigate a CASE tool, such as those offered by Computer Associates (*www.ca.com*), Oracle (*www.oracle.com*), Borland (*www.borland.com*), or IBM (*www.ibm.com*). On what system development methodologies is the tool based? What types of system models can be built with the tool? How is an analysis model translated into an implementation model? What programming languages, OSs, and DBMSs does the back-end CASE tool support? What deployment environments are supported?
2. Investigate an IDE, such as Microsoft Visual Studio or IBM WebSphere. Are application programs interpreted, compiled, or both? What program-editing tools are included? What tools are available to support runtime debugging? What DBMSs can be accessed by application programs?
3. Investigate the Free Software Foundation and the terms of its licenses—the GNU General Public License (GPL) in particular. What are the provisions of this license, and what is "copyleft"? Identify at least three current software packages, such as programming or application development tools, OSs, or DBMSs, distributed under the GPL's terms. Identify at least two large companies that incorporate these technologies into their own products and services sold for profit. What are the economic motivations for these companies to build their products and services around "free" software?

CHAPTER **11**

OPERATING SYSTEMS

CHAPTER GOALS

- Describe the functions and layers of an operating system
- List the resources allocated by the operating system and describe the allocation process
- Explain how an operating system manages processes and threads
- Compare CPU scheduling methods
- Explain how an operating system manages memory

Operating systems are critical components of information systems. Users and application programs rely on them to provide services and to manage software, data, and hardware resources. As an information systems professional, you'll often be asked to select, install, configure, and upgrade operating systems. Performing these tasks successfully requires a detailed understanding of how operating systems work. Figure 11.1 shows some topics you learn about in this chapter.

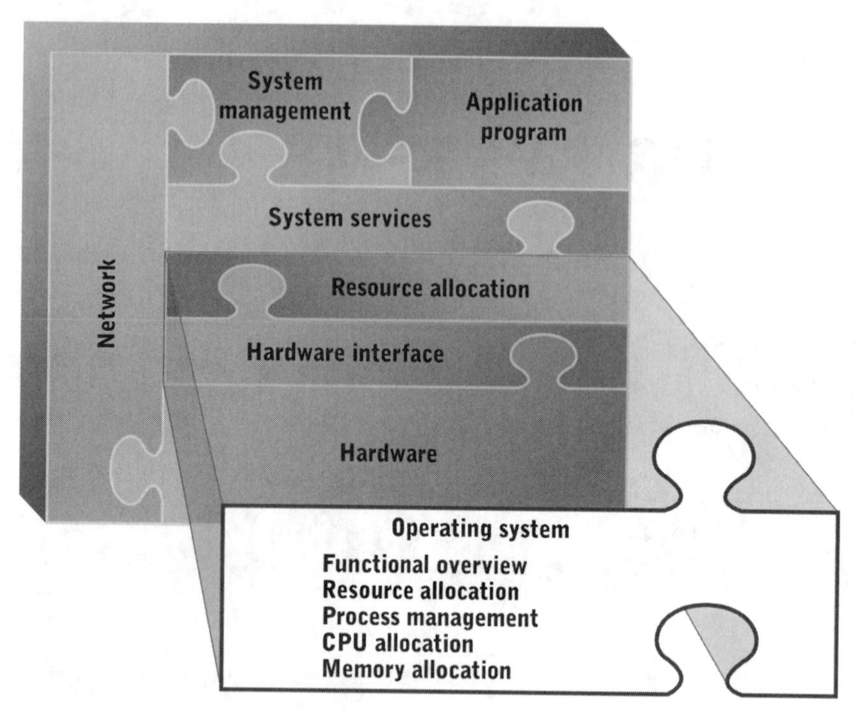

FIGURE 11.1 Topics covered in this chapter
Courtesy of Course Technology/Cengage Learning

OPERATING SYSTEM OVERVIEW

The operating system (OS) is an important part of all information systems. It plays a dual role as a high-level manager and a tireless low-level worker. As a manager, it oversees all hardware resources and allocates them to users and applications as needed. It also shifts resources to meet rapidly changing demands while enforcing resource security and sharing policies. With resources, the OS is similar to "Big Brother"—an omniscient government with unquestioned authority.

At the same time, the OS performs many low-level tasks on behalf of users and application programs, such as accessing files and directories, creating and moving windows, and accessing resources over a network. It provides these services for two reasons. First, it's efficient. If many application programs perform similar tasks, it makes sense to implement these tasks as reusable OS services available to all applications. This way, application programmers are free to address tasks unique to the type of application they're developing.

Second, providing so many low-level functions enables the OS to maintain control over hardware resources. For example, it provides file access services so that it can maintain control over secondary storage devices. With complete control, the OS can enforce systemwide security policies, prevent programs from accidentally overwriting one another's data, and allocate disk and I/O capacity efficiently to meet the needs of all users and programs.

Operating System Functions

OS functions are categorized as shown in Figure 11.2. You might want to refer back to this figure as you read this chapter and the next two to remind yourself how each low-level function fits into the "big picture." OS functions are divided into five main groups in this figure. In the leftmost column, hardware interface functions are listed by device type—CPU, primary storage, secondary storage, and I/O devices. The functions in the next three columns are also organized by these device types.

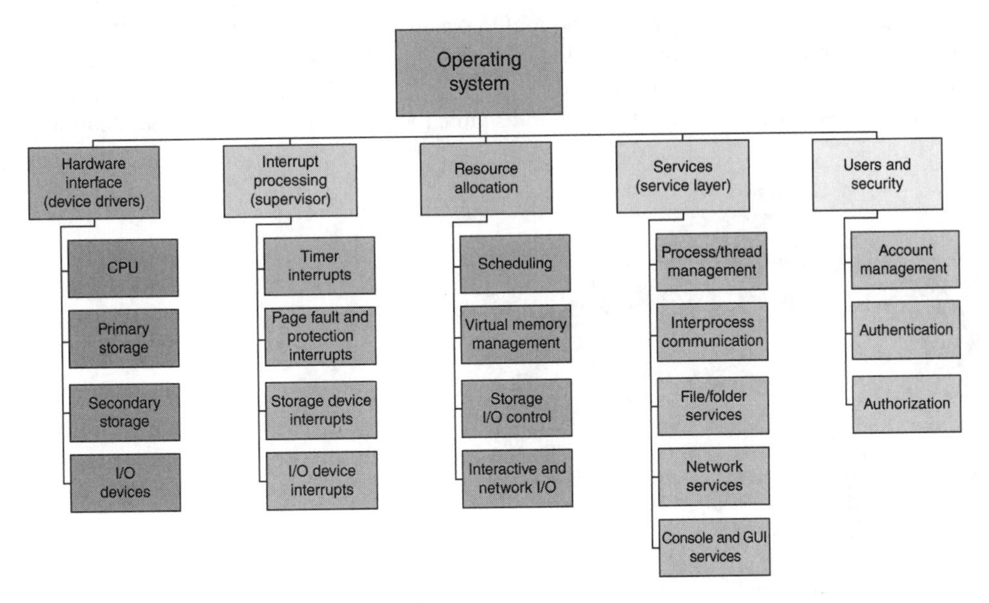

FIGURE 11.2 OS management functions
Courtesy of Course Technology/Cengage Learning

The functions in the Resource Allocation column are essentially a bridge between users, their processes, and the hardware resources used by these processes. As in any complex business or other organization, resource allocation is an important management function that provides resources to support many ongoing activities. OS resource allocation functions ensure that overall system objectives are achieved efficiently and effectively, as described in more detail later in this chapter.

As discussed in Chapter 6, interrupt processing is a way to coordinate I/O devices and improve system performance. This chapter expands the description of interrupt processing to include its role in coordinating hardware devices with their corresponding resource allocation functions. For example, you learn about the role of timers and other interrupts for CPU scheduling, process prioritization, and virtual memory management. The OS service layer, as you learned in Chapter 2, provides services to application programs and processes, such as file management, interfaces to external resources via networks, and user interface services (for example, GUI windowing functions). This chapter explores this layer further to include process and thread management services. Chapter 12 covers file-related services in depth, and Chapter 13 does the same with network-related services.

Finally, as shown in the rightmost column of Figure 11.2, this chapter introduces the OS's role in ensuring system security through user accounts, authentication of users, and authorizing access by users and processes to specific resources, such as files and I/O devices. As with service-layer functions, Chapter 12 explores file-related security issues in more depth, and Chapter 13 discusses these issues for network services.

Operating System Layers

Like other complex software, operating systems are organized internally into layers, as shown in Figure 11.3. Using layers makes the OS easier to maintain because functions in one layer can be modified without affecting other layers. The outermost layers provide services to application programs or directly to end users. The innermost layer encapsulates hardware resources, thus controlling and managing access by users and applications.

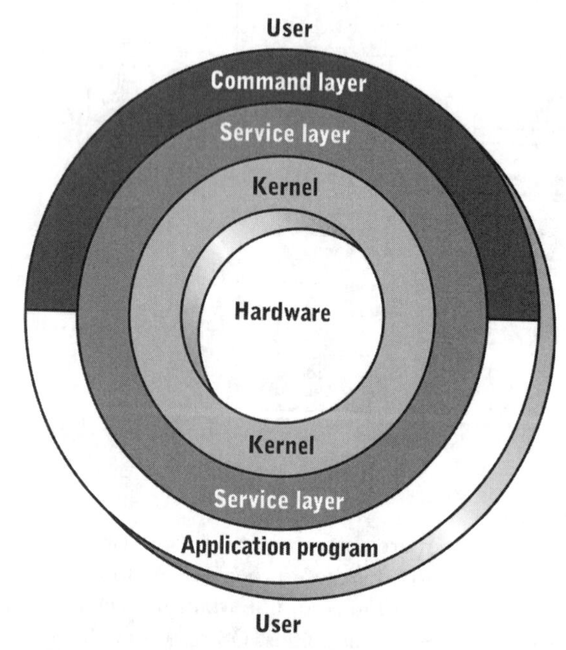

FIGURE 11.3 OS layers
Courtesy of Course Technology/Cengage Learning

The **command layer**, sometimes called the **shell**, is the user interface to the OS. Through this layer, a user or system administrator can run application and OS utility programs and manage system resources, such as files, folders, and I/O devices. Although end users tend to think of the command layer and the OS as one and the same, the command layer is only one part of the OS, and users can use different command layers to interact with the same OS.

Today's operating systems include a graphical user interface (GUI) that enables users to interact with visual representations (icons) of programs and other resources and manipulate them with actions such as dragging or clicking them. Having files, programs, and commands represented with icons is easy for most people to use. Xerox first used this form of interface in the late 1970s, and it has grown to dominate computing through decades of use and refinement and because of continuing advances in display technology and computer power.

The command layer can also be implemented with a text interface that accepts user input from the keyboard. For example, in MS-DOS, a user types commands, such as `DIR` to list a folder's contents or `COPY` to copy a file or folder from one storage location to another. A set of commands and syntax requirements is called a **command language** or sometimes a **job control language (JCL)**. Command languages tend to be difficult to learn and use, partly because of their similarity to programming languages. The user or programmer must know the language's syntax and semantics to issue commands. Examples of command languages include MS-DOS, IBM MVS JCL, Windows PowerShell, and UNIX Bourne shell. End users rarely use command languages now, but system administrators still use them often. Commands can be stored in a file to be reused, and they can be embedded in control structures, such as `if` statements and `while-do` loops, to automate repetitive tasks—for example, creating accounts and folders for a group of new users.

NOTE

Although Windows replaced MS-DOS in the 1980s and 1990s, a compatible command layer is included in most Windows versions. You can search for and run the cmd.exe file to open the text-based command prompt window that accepts MS-DOS commands. Similarly, UNIX and Linux OSs still include text-based command-layer shells, such as the Bourne (filename sh) and C (filename csh) shells.

The **service layer** contains thousands of reusable components that provide functions ranging from file and folder manipulation to accessing I/O devices (such as printers and scanners), starting and stopping programs, and creating, moving, and resizing GUI windows. Service components are packaged as functions that can be called from an application or other program. A request to execute a service-layer function is called a **service call**. Application programs, the command layer, and utility programs perform most of their functions by executing service calls.

The service layer is also an intermediary between programs, which request and use resources, and the kernel, which manages and provides access to resources. Think of a service call as an indirect request for system resources. For example, when an application program asks the OS to retrieve data from a file, it's indirectly asking for access to memory buffers, the system bus, the disk controller, and the disk holding the file. Similarly, an I/O service call is an implicit request for access to an I/O device and the resources needed to communicate with it.

The **kernel** is the OS portion that manages resources and interacts directly with computer hardware. It includes a resource allocation layer (described in the following section) and interface programs called device drivers for each hardware device in the computer (see Figure 11.4), such as keyboards, video displays, disk drives, DVD drives, and printers. Using device drivers makes the OS modular and flexible. As hardware is added or upgraded, device drivers are added or modified to reflect the changes (see Figure 11.5).

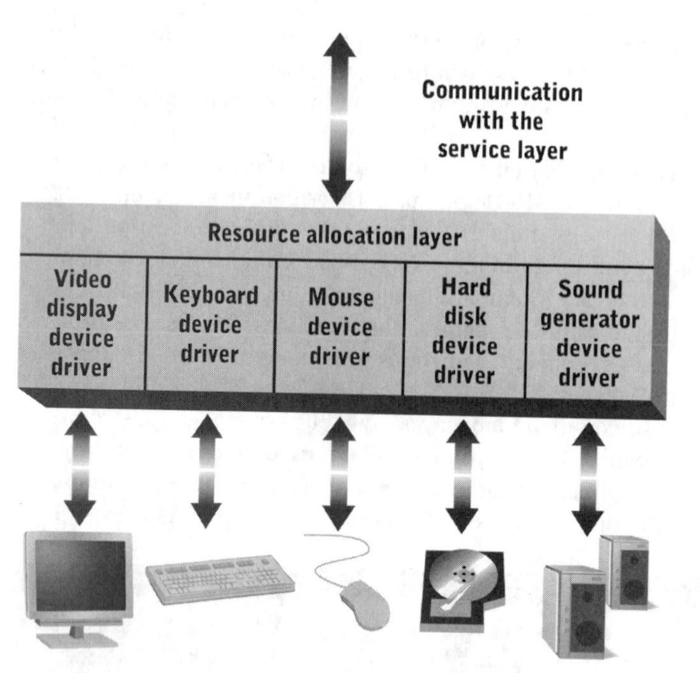

FIGURE 11.4 Components of the kernel
Courtesy of Course Technology/Cengage Learning

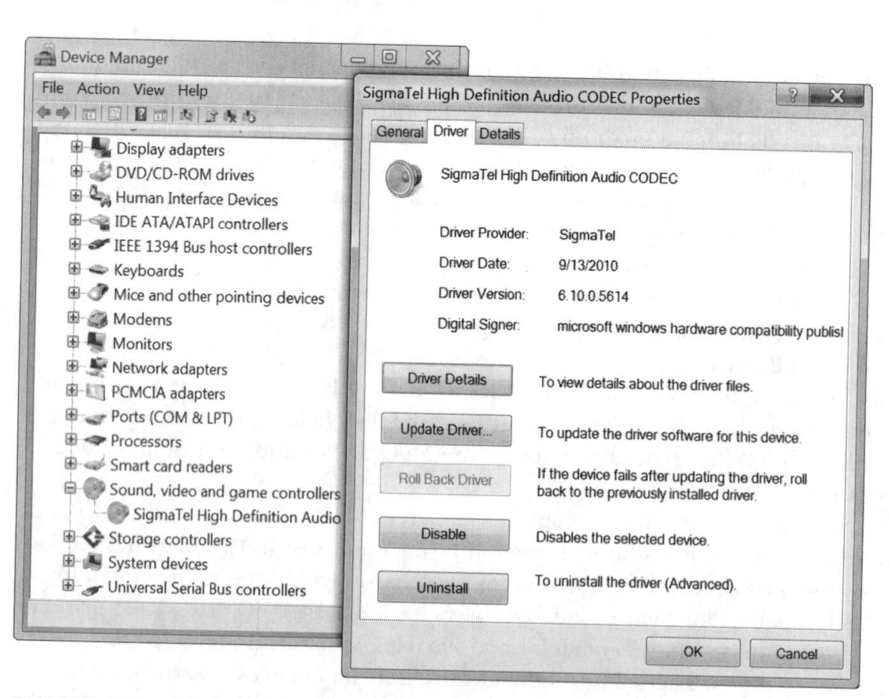

FIGURE 11.5 Device driver properties in Windows
Courtesy of Course Technology/Cengage Learning

RESOURCE ALLOCATION

Early computer systems were simple devices with few hardware resources. Because hardware resources were limited, complex resource allocation methods were unnecessary. An application program "took control" of the entire computer when it ran. Because only one application program ran at a time, there was no need for OS software to resolve competing demands for hardware resources.

Over time, computers became more powerful, and their hardware resources became more numerous and diverse. Extra processing power and storage capacity made it possible for multiple users and programs to use a computer at the same time. OS support for running multiple programs simultaneously is called **multitasking**. Multitasking OSs use elaborate procedures to share resources between processes and users and prevent them from interfering with one another.

Single-Tasking Resource Allocation

Resource allocation in a single-tasking OS involves only two running programs—an application and the OS. Single-tasking operating systems are small and efficient because they don't need complex resource allocation procedures. There's rarely any contention between the OS and application program for resources. The OS reserves whatever resources it requires when it boots, and all resources not allocated to it are available to an application program.

In a single-tasking environment, when an application program begins running, the OS grants it control of all unused hardware resources. The program controls these resources until it terminates, an error occurs, or it needs service from the OS. Errors and service calls are normally processed through interrupts (discussed in Chapter 6). When an interrupt is detected, control of the CPU is passed back to the OS. If an error has occurred, the OS can attempt to correct it so that the program is able to continue. If error correction isn't possible, the program is terminated, and the OS reclaims control of all hardware resources. In a single-tasking OS, no elaborate mechanisms are needed to prevent programs from accessing one another's resources because there's only one active program.

NOTE

MS-DOS, the most common single-tasking OS, was developed for the IBM PC and widely used until the early 1990s. It's still used sometimes in specialized applications that don't require multitasking capabilities, such as security systems, and for controlling hardware devices, such as elevators and irrigation systems.

Multitasking Resource Allocation

Multitasking operating systems are now the norm for general-purpose computers. They enable many programs or processes to run at the same time. Application and system software can be more flexible because large programs can be built from smaller independent modules or processes. Small modules are easier to develop and update, and they can be loaded, executed, suspended, and resumed as needed.

For example, a typical word-processing program often needs to format output and send it to a printer. With a multitasking OS, the program's printer formatting and interface functions can be a separate module that's loaded and executed when printing is required and suspended when it's not in use. The printing module can run independently of other word-processing modules, so the user can perform other tasks while a document is being printed.

System software also benefits from multitasking capabilities. Network interface cards, for example, typically have a dedicated OS module that detects and responds to incoming data. With multitasking, this module is loaded into memory when the OS boots and is placed in a suspended state until data arrives. Independent modules also manage other I/O devices, such as printers and video displays. Dividing OS tasks between independent modules makes the OS easier to build and upgrade and results in more efficient use of hardware resources. A multitasking OS manages hardware resources to achieve the following goals:

- Meet each program's resource needs.
- Prevent programs from interfering with one another.
- Use hardware and other resources efficiently.

Related goals include maximizing the amount of work performed, ensuring reliable program execution, protecting resource security, and minimizing resources the OS consumes.

Resource Allocation Tasks

To understand how resource allocation works, think of the resource allocation tasks that the road maintenance unit of a local government performs: maintaining a pool of repair personnel, equipment, and supplies; accepting repair requests from a variety of sources; scheduling resources to satisfy requests; and moving resources between sites to complete repairs. Managing resources to respond to repair orders is a complex task requiring detailed knowledge of what personnel, equipment, and supplies are available and when they're available. This unit also maintains detailed records of personnel skills, equipment capabilities, supplies on hand and on order, operational status, commitments, and schedules. These records are input data for future scheduling decisions and are modified as a result of these decisions. The road maintenance unit also follows well-defined policies and procedures to prioritize conflicting demands for service.

An OS's resource allocation functions are similar to those of the road maintenance unit. The OS keeps detailed records of available resources and knows which resources can satisfy which requests. It schedules resources based on allocation policies to meet current and anticipated demand, and it updates records constantly to reflect resource commitment and release by programs and users. Resource allocation goals and policies are defined as procedures or algorithms implemented in software. The records supporting these procedures are implemented as data structures. An OS's resource allocation functions are similar to any other program's; they're a set of algorithms and data structures organized to accomplish a particular purpose.

Computer scientists have long studied data structures and the most efficient ways to manipulate them, resulting in a large body of knowledge about which algorithms and data structures are most efficient for particular types of processing tasks. System programmers apply this body of knowledge to design and implement OS resource allocation procedures, taking care to use procedures that consume as few machine resources as possible. The reason is simple—every resource consumed by OS tasks is unavailable to application programs. You can compare it with management functions in a business: Every business has limited resources, and every resource devoted to a management function is unavailable for other business functions, such as producing and selling products.

The resources consumed by resource allocation procedures are sometimes referred to as **system overhead**. An important design goal for most OSs is to minimize system overhead while achieving acceptable levels of throughput and reliability. Note, however, that these goals conflict. Allocation decisions that make more resources available to programs typically require more complex allocation procedures. Ensuring higher levels of reliability also requires more extensive oversight procedures. However, increasing the complexity of allocation and oversight procedures consumes more hardware resources when these procedures are implemented in software. OS designers must determine the right balance between these competing objectives.

Real and Virtual Resources

A computer's physical devices and associated system software are called **real resources**. As allocated by the OS, the resources that are apparent to a program or user are called **virtual resources**. Operating systems make virtual resources appear to be equal to or greater than real resources. For example, a typical desktop computer usually has only one printer. However, if a user runs multiple programs, such as word-processing and spreadsheet applications, at the same time, each program "thinks" it has a printer all to itself. The single physical printer is a real resource. The printers each program "thinks" it controls exclusively are virtual resources.

Application design and programming are much simpler if the application isn't concerned with resource availability. Programs can be written under the assumption that whatever resources are requested will be provided. The programmer doesn't need to develop schemes to lock and unlock resources for exclusive use. For example, a word-processing program doesn't need to check whether the physical printer is being used by another program. It simply prints to its own virtual printer and leaves it to the OS to figure out how to print output from multiple programs without interference.

Providing virtual resources that meet or exceed real resources is accomplished by doing the following:

- Rapidly shifting resources unused by one program to other programs that need them
- Substituting one type of resource for another when possible and necessary

Although each program "thinks" it has control of all hardware resources in the computer, it's rare for one program to need them all at the same time. The OS shifts resources between programs as demand rises and falls. In this way, the sum of virtual resources for all active programs can substantially exceed the computer's real resources. Interrupt processing is an example of resource shifting. When a program makes an I/O request that can't be satisfied immediately, the OS suspends it and stores its register contents in memory (a stack push). Other programs receive control of the CPU while the I/O request is being processed. Later, the register contents are copied back from memory (a stack pop), and the program resumes execution and control of the CPU.

Certain types of resources can be substituted for each other. For example, memory is temporarily substituted for CPU registers while a program is suspended, pending completion of an I/O request. In the printing example, an application program "thinks" it's interacting directly with the printer. Instead, the OS stores output in a file temporarily. In essence, it substitutes secondary storage resources for the printer and related I/O resources. When the printer is available, the OS copies the data to the printer and deletes the file, thus reversing the temporary substitution.

TECHNOLOGY FOCUS

VMware ESX and ESXi

A **hypervisor** is an OS that enables dividing a single physical computer or cluster into multiple **virtual machines (VMs)**. Hypervisors are marketed or used by several software vendors, including VMware, IBM, Oracle, and Microsoft. Open-source hypervisors include Citrix XenServer and Kernel-Based Virtual Machine (a Linux version). VMware is the market leader in hypervisors, with approximately an 80% market share in 2009.

A hypervisor carries virtualization one step further than a normal OS. Instead of virtualizing each hardware resource for use by programs and processes, it virtualizes a subset of the physical computer's CPU, storage, and I/O resources to create multiple VMs. A guest OS is installed on each VM, and applications and services are installed in this OS.

A hypervisor can make the sum of physical resources allocated to all virtual machines appear greater than the underlying physical resources. This capability is a key reason for the widespread use of hypervisors for **server consolidation**, which uses VMs as small virtual servers hosted by a hypervisor running on a larger machine (see Figure 11.6). Because most servers don't use all their allocated resources all the time, a hypervisor can move resources between virtual servers as their needs rise and fall. As a result, total hardware resource requirements are reduced, compared with installing each server on a separate computer. Sharing a single hardware platform across multiple virtual servers can also simplify administrative tasks, such as installation, backup, and recovery, although it might complicate other tasks, such as hardware maintenance and upgrades.

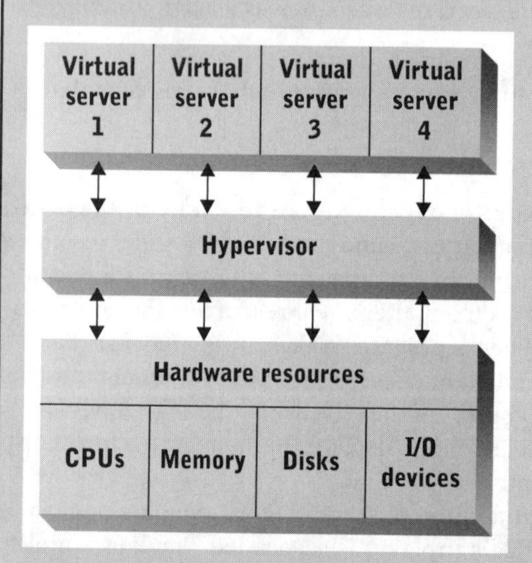

FIGURE 11.6 Virtual servers sharing a single computer system with a hypervisor
Courtesy of Course Technology/Cengage Learning

(continued)

Virtualization software varies in several characteristics, including how many traditional OS functions are provided and whether the software is installed directly on the hardware or in a host OS (the OS running on the physical computer). At one end of the spectrum are **bare-metal hypervisors** that are installed much like OSs but provide only minimal OS-type functions, such as the capability to start and stop VMs and manage physical resources on behalf of VMs. At the other end of the spectrum are **virtualization environments** that are installed as an application in another OS.

VMware offers two hypervisors, ESX and ESXi, intended for server consolidation and a virtualization environment, VMware Workstation, which is installed as an application in Windows or Linux. ESXi is a bare-metal hypervisor with minimal OS services. Its small hardware footprint makes it efficient, thus maximizing the hardware resources available to VMs. ESX includes a portion of the Linux kernel, which enables it to interact directly with some Linux applications, but it consumes more hardware resources than ESXi. VMware Workstation is used mainly to enable desktop computers to run multiple OSs simultaneously.

A critical performance factor in any hypervisor or virtualization environment is I/O processing and communication between the hypervisor and guest OSs. Interactive I/O devices, such as keyboards, mice, and video displays, typically use highly optimized drivers for quick response time. Virtualization introduces another software layer between device drivers and I/O hardware that can reduce performance to unacceptable levels. To address this issue, VMware includes VMware Tools, which provides customized device drivers for interactive I/O devices that streamline the I/O interface between the hypervisor and guest OS. VMware Tools also allows direct communication between the hypervisor and guest OS to enable functions such as shutting down and restarting a VM.

PROCESS MANAGEMENT

A **process** is a unit of executing software that's managed independently by the OS and can request and receive hardware resources and OS services. It can be a stand-alone entity or part of a group of processes cooperating to achieve a common purpose. Processes can communicate with other processes executing on the same computer or with processes executing on other computers.

Process Control Data Structures

The OS keeps track of each process by creating and updating a data structure called a **process control block (PCB)** for each active process. It creates a PCB when a process is created, updates the PCB as process status changes, and deletes the PCB when the process terminates. Using information stored in the PCB, the OS can perform a number of functions, including allocating resources, securing resource access, and protecting active processes from interference by other active processes.

PCB content varies across operating systems and can also vary within a single OS, depending on whether certain functions, such as resource accounting and auditing, are enabled. The following data items are typically included:

- A unique process identification number
- The current state of the process—for example, executing or suspended
- Events the process is waiting for

- Resources allocated exclusively to the process, including memory, files, and I/O devices
- Machine resources consumed—for example, CPU seconds consumed and bytes of data transferred to and from disk
- Process ownership and access privileges
- Scheduling priority or data for determining scheduling priority

PCBs are normally organized into a larger data structure, such as a linked list. The list of PCBs, sometimes called a **process queue** or **process list**, is often searched by OS components. For example, a user control process might search the process queue for all processes owned by a certain user so that they can be terminated when the user logs off. The speed of searches and updates is an important characteristic of PCBs and the process list. The process list and PCB contents change frequently as processes are created and terminated and resources are allocated and released. Using efficient data structures and processing algorithms minimizes system overhead.

Processes can create, or **spawn**, other processes and communicate with them. The original process is called the **parent process**, and the newly created process is called the **child process**. Parent processes can spawn multiple child processes, and the child processes of a single parent are collectively called **sibling processes**. Child processes can, in turn, spawn children of their own. A group of processes descended from a common ancestor, including the common ancestor itself, is called a **process family**.

Application programs often spawn a child process to run general-purpose utility programs that aren't part of the OS service layer. For example, a Web browser can spawn a process to view specialized content, such as a spreadsheet or Adobe Acrobat document. The process can be another program in the same application, a utility program supplied with an application suite, or a utility program supplied with the OS. Subdividing application programs into multiple cooperating processes can improve execution speed, especially if the computer has multiple CPUs, because multiple processes can execute simultaneously.

Even if only one CPU is available, subdividing application programs into cooperating processes increases execution speed because of more efficient resource allocation. For example, a user of a word-processing program wants to edit one document while generating an Acrobat version of another. If the word-processing program spawns a separate Acrobat formatting process, both tasks can be performed concurrently. Because the two tasks use different mixes of computer resources (heavy I/O and light CPU for the editing process and the reverse for the Acrobat formatting process), the OS can allocate resources efficiently so that both processes move toward completion quickly.

Threads

The benefits of subdividing large processes into smaller ones are subject to the law of diminishing returns. As the number of active processes increases, the system overhead required to track and manage them also increases. The process list grows, and OS processes that search and update the list become less efficient. System overhead increases, and the resources available to execute application processes are reduced.

In many operating systems, processes can subdivide themselves into more easily managed subunits called threads. A **thread** is a portion of a process that can be scheduled and executed independently. Process threads can execute concurrently on a single processor or simultaneously on multiple processors. Threads share all resources allocated to their parent processes, including primary storage, files, and I/O devices.

The advantage of organizing an application program into a family of threads instead of multiple processes is that system overhead for resource allocation and process management is reduced. Because all threads of a process share storage and I/O resources, the OS can track allocation of these resources on a per-process basis, which reduces the number of PCBs and speeds up searching and updating the process queue.

The OS keeps track of thread-specific information in a **thread control block (TCB)**. Each PCB contains pointers to its related TCBs, and all active TCBs are organized into a data structure called a **run queue** or **thread list**. As with a process list, a run queue is implemented with efficient data structures that can be searched or updated quickly.

A process or program that divides itself into multiple threads is said to be **multithreaded**. An OS that supports threads creates a thread for each process automatically, but subdividing processes into multiple threads isn't automatic. The programmer or compiler must specifically identify code segments that can execute independently and create a thread for each segment. Desktop and server OSs have supported multithreaded processes since the 1990s.

CPU ALLOCATION

Threads progress toward completion only when they have CPU cycles in which to execute their instructions. A multitasking OS can execute dozens, hundreds, or thousands of threads in the same time frame. Most computers have only one or two CPUs, so threads must share CPUs.

NOTE

The term "thread" is used throughout this chapter, but all references to threads also apply to processes in OSs that don't support threads.

The OS makes rapid decisions about which threads receive CPU control and for how long control is retained. Typically, a thread controls a CPU for no more than a few milliseconds before it relinquishes control and the OS gives another thread a turn. Figure 11.7 shows three threads sharing a single CPU by using small time slices. This method of CPU sharing is called **concurrent execution** or **interleaved execution**.

	Time slice 1	Time slice 2	Time slice 3	Time slice 4	Time slice 5	Time slice 6	Time slice 7
Thread 1	Running	Idle	Idle	Idle	Idle	Running	Idle
Thread 2	Idle	Running	Idle	Running	Idle	Idle	Idle
Thread 3	Idle	Idle	Running	Idle	Running	Idle	Running

FIGURE 11.7 Concurrent (interleaved) thread execution on a single CPU
Courtesy of Course Technology/Cengage Learning

Thread States

An active thread can be in only one of the following states:

- Ready
- Running
- Blocked

Figure 11.8 shows the three thread states and the events that move a thread from one state to another. Threads in the **ready state** are idle, pending availability of a CPU. Many threads can be in the ready state at one time. When a CPU becomes available, the OS chooses a ready thread to execute on that CPU.

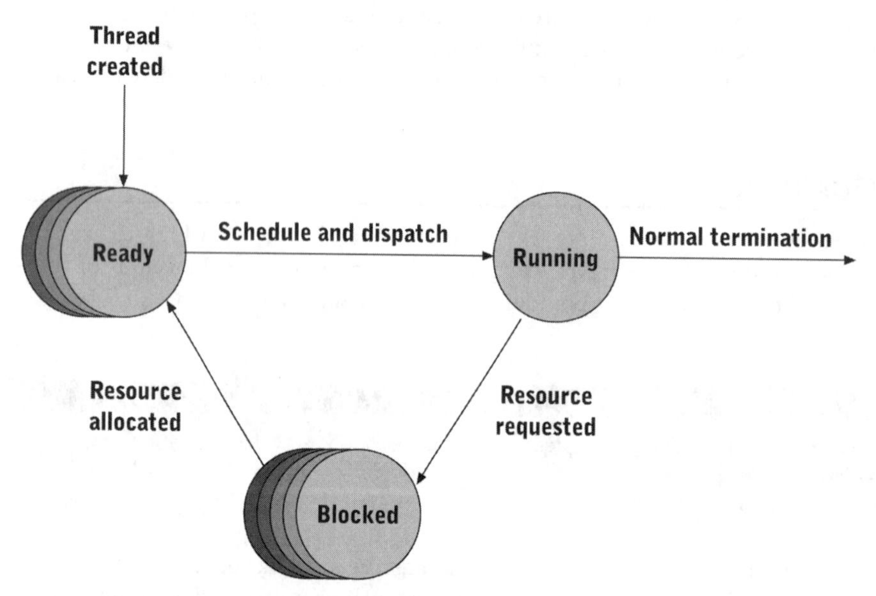

FIGURE 11.8 Thread movement between states
Courtesy of Course Technology/Cengage Learning

The act of giving control of a CPU to a ready thread is called **dispatching**. The OS dispatches a thread by loading the instruction pointer with an instruction address that belongs to the dispatched thread. During the next fetch and execution cycle, the thread takes control of a CPU and its related resources, such as general-purpose registers.

After it's dispatched, a thread has entered the **running state** and retains control of a CPU until one of these events:

- The thread or its parent process terminates (halts) normally.
- An interrupt occurs.

When a process or thread terminates itself, control returns to the OS. The way in which the OS regains CPU control varies across operating systems and CPUs. The most common method is for the process or thread to execute an exit service call, which triggers a software interrupt.

Interrupts can occur for a variety of reasons, including the following:

- Executing a service call, such as a file I/O request
- A hardware-generated interrupt indicating an error, such as overflow, or a critical condition, such as a power failure alarm from an uninterruptible power supply
- An interrupt generated by a peripheral device, such as a NIC when a packet arrives

Recall from Chapter 6 that when any interrupt is received, the CPU automatically suspends the currently executing thread, pushes current register values onto the stack, and transfers control to the OS master interrupt handler. The suspended thread remains on the stack until interrupt processing is completed. During this time, the thread is in a **blocked state**. After the interrupt has been processed, the OS can leave the suspended thread in the blocked state, move it to the ready state, or return it to the running state.

Interrupt Processing

A blocked thread is waiting for an event to occur, such as allocation of a requested resource or correction of an error condition. Error conditions might or might not be correctable. If they can't be corrected—for example, overflows or memory protection faults—the thread is halted. If the error can be corrected, the thread remains in the blocked state until the error condition is resolved.

For example, if a thread attempts to access a file on a removable disk and the disk isn't in the drive, the disk drive controller generates an interrupt to signal the error. The CPU automatically suspends the thread and passes CPU control to an OS interrupt handler. In an OS such as Windows, the interrupt handler displays an error message asking that a disk be inserted. If the user complies, the error condition is resolved, and the thread leaves the blocked state. If the user doesn't comply, the error isn't corrected, and the requesting thread is terminated.

Most service calls directly or indirectly require resource allocation. For example, a service call requesting additional memory for a thread is a direct request for resource allocation. A request to open a file is an indirect request for memory buffers to hold file data as it's transferred to or from secondary storage. If the request is for read/write access, it's an indirect request for exclusive access to the file. A subsequent request to read data from the file is an indirect request for access to the secondary storage device, its controller, and the I/O channels connecting the secondary storage device to its memory buffers.

Some service calls, such as requests for additional memory, can usually be performed immediately, but many service calls, such as requests for input from a file or device, require time to carry out. In this case, the thread remains in the blocked state until the request is satisfied.

The following event sequence is typical when a process or thread reads from a file:

1. The thread requests file input by issuing a software interrupt.
2. The thread is pushed onto the stack automatically, and the corresponding interrupt handler is called.
3. The interrupt handler executes. It checks to see whether the requested data is already in a memory buffer. If it is, the interrupt handler transfers the data

from the memory buffer to the data area of the requesting process. The interrupt handler exits, and the OS places the suspended thread in the ready or running state.

4. If the requested data isn't already in a memory buffer, the interrupt handler generates a read command and sends it to the secondary storage controller. The interrupt handler then exits without waiting for the data to be returned. The OS places the suspended thread in a blocked state, pending arrival of the requested data.
5. When the data is read from disk, the secondary storage controller sends an interrupt to the CPU. The interrupt handler transfers the data from the secondary storage controller to a memory buffer over the system bus and then copies it to the data area of the requesting process. The interrupt handler exits, and the OS places the blocked thread in the ready or running state.

Scheduling

The decision-making process the OS uses to determine which ready thread moves to the running state is called **scheduling**, and the OS portion that makes scheduling decisions is called the **scheduler**. Operating systems vary widely in their scheduling methods, although the following methods are typical:

- Preemptive scheduling
- Priority-based scheduling
- Real-time scheduling

Preemptive Scheduling

In **preemptive scheduling**, a thread can be removed involuntarily from the running state. A running thread controls a CPU by controlling the instruction pointer's content. CPU control is lost whenever an interrupt is received and the CPU pushes the current thread onto the stack and transfers control to the OS. As you learned in Chapter 6, the portion of the OS that receives control is called the supervisor, which performs two important functions:

- Calling the correct interrupt handler
- Transferring control to the scheduler

The supervisor uses the value in the interrupt register as an index to the interrupt table. It extracts the corresponding address, pushes itself on the stack, and transfers control to the interrupt handler by placing its address in the instruction pointer. When the interrupt handler finishes, it passes control back to the supervisor by popping the stack. The supervisor then passes control to the scheduler by executing an unconditional BRANCH instruction.

The scheduler performs four tasks:

- Updating the status of any thread affected by the last interrupt
- Deciding which thread to dispatch to the CPU
- Updating thread control information and the stack to reflect the scheduling decision
- Dispatching the selected thread

Processing an interrupt usually changes the state of at least one thread. If the interrupt resulted from a request for a resource that couldn't be provided immediately, the thread currently on top of the stack must be moved from the running state to the blocked state. If processing the interrupt cleared an error condition or supplied a resource some thread was waiting for, the thread must be moved from the blocked state to the ready state. The scheduler updates the corresponding TCBs to reflect these state changes. Figure 11.9 shows the processing sequence for a service request and delayed satisfaction of that request. The processing steps on the left occur after Thread 1 makes an I/O service call. The processing steps on the right occur after the I/O device finishes the I/O operation.

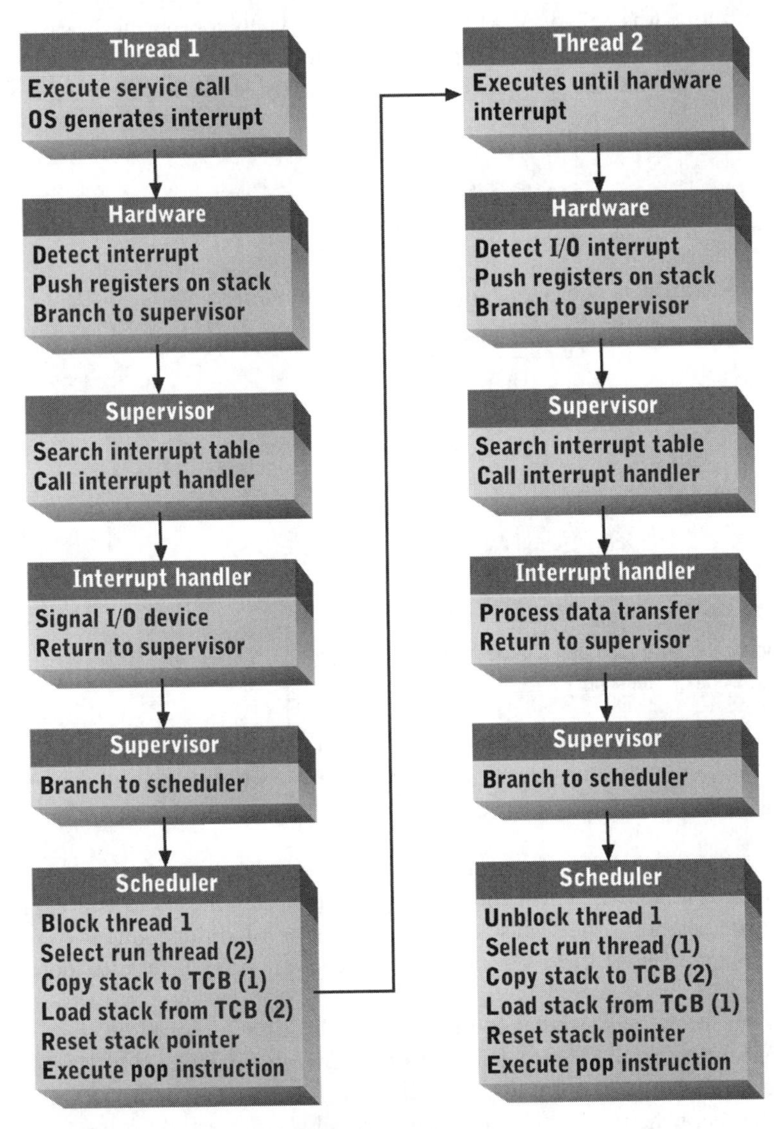

FIGURE 11.9 Interrupt processing
Courtesy of Course Technology/Cengage Learning

After the scheduler decides which thread will run next, it updates the thread's TCB and manipulates the stack contents, if necessary. If the suspended thread is to become the running thread, the scheduler pops the stack and the thread resumes execution. If another thread is selected as the running thread, the scheduler must save the entire stack contents to the corresponding TCB because elements below the topmost element represent suspended subroutines or functions of the interrupted thread.

For example, a thread named "main" begins execution and calls subroutine A. Subroutine A calls subroutine B, and an interrupt is received before any subroutine completes execution. After the interrupt is received and control is passed to the supervisor, there are three thread states on the stack (see Figure 11.10). The stack contents are main, which was pushed when subroutine A was called; A, which was pushed when subroutine B was called; and B, which was pushed when the interrupt was detected. The entire stack must be saved to the thread's TCB so that it can be restored when the thread is dispatched at some future point.

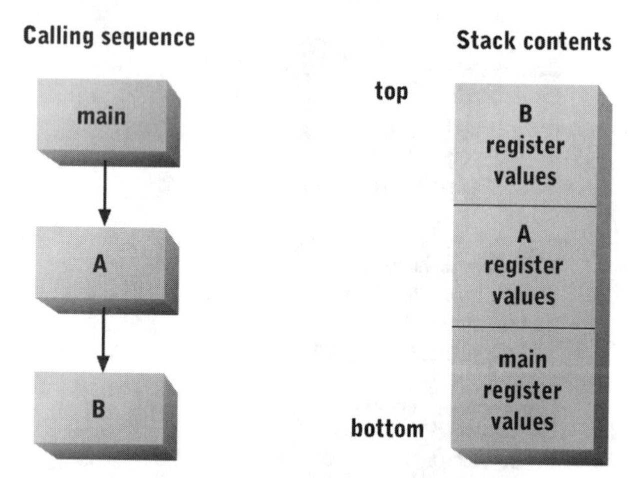

FIGURE 11.10 Subroutine calls (left) and associated stack content (right)
Courtesy of Course Technology/Cengage Learning

After the suspended thread's stack is saved to its TCB, the OS copies the saved stack of the next running thread from its TCB to the stack. The scheduler then transfers control to the thread by issuing a pop instruction.

Timer Interrupts Because interrupt arrival is unpredictable, the time a thread remains in the ready state is also unpredictable. Theoretically, it's possible for a thread to control the CPU indefinitely if it makes no service calls, generates no error conditions, and receives no external interrupts. Under these conditions, a thread stuck in an infinite loop would execute forever.

Most CPUs generate an interrupt periodically to give the scheduler an opportunity to suspend the currently executing thread. A **timer interrupt** is generated at regular intervals of between several dozen and several thousand CPU cycles. As with other interrupts, the currently executing thread is pushed onto the stack, and control is transferred to the supervisor. Because a timer interrupt isn't a "real" interrupt, there's no interrupt handler to call, so the supervisor passes control to the scheduler.

Timer interrupts are an important CPU hardware feature for multitasking operating systems. They guarantee that no thread can hold the CPU for long periods because the scheduler has regular opportunities to place another thread in the running state.

Priority-Based Scheduling

Priority-based scheduling determines which ready thread should be dispatched to the CPU according to one or more of the following methods:

- First come first served
- Explicit priority
- Shortest time remaining

With **first come first served (FCFS)** scheduling, the scheduler always dispatches the ready thread that has been waiting the longest. The simplest way to determine which thread has been waiting the longest is to order the run queue by waiting time. Each time a thread is moved into the ready state, the scheduler places its TCB at the back of the run queue. The scheduler then searches the run queue from front to back until it finds a ready thread, and this thread is the next to run. Implementing the run queue in this fashion is simple, so FCFS scheduling has low system overhead.

Explicit priority scheduling assigns a priority level to each process or thread. A priority level can be assigned to a newly created thread based on the default priority of the process owner, a priority stated explicitly in an OS command, or other methods. Many operating systems automatically assign high priority levels to their own internal threads under the assumption that OS threads must be completed quickly to maintain total system throughput. This scheduling method can use priority levels in two ways:

- Always dispatch the highest-priority ready thread.
- Assign larger time slices to high-priority threads.

The first strategy ensures that high-priority threads are always dispatched before lower-priority threads. However, scheduling decisions based solely on initial priority level can result in extremely long or infinite idle time for long-running threads with low priority. These threads are always moved to the back of the run queue and can't get enough CPU time to be completed. To prevent this problem, many operating systems increase the priority level of "old" threads automatically. Priority-based scheduling with variable-length time slices also addresses this problem, although this method is seldom used.

Shortest time remaining (STR) scheduling chooses the next thread to be dispatched based on the expected amount of CPU time needed to complete the process. It can be implemented directly by tracking time to completion for each thread and ordering the run queue with STR threads first. STR scheduling can be used indirectly by increasing a thread's explicit priority as it nears completion. In either case, the scheduler must know how much CPU time is required for each thread to execute and how much time has already been used. For the scheduler to have this information, the required CPU time must be provided to the scheduler when the thread is created and then stored in the TCB. The TCB must also store the amount of CPU time a thread uses, and this information must be updated each time the thread leaves the ready state. Creating, updating, and using time-to-completion information increases system overhead, so STR scheduling is rarely used.

Real-Time Scheduling

Real-time scheduling guarantees a minimum amount of CPU time to a thread if the thread makes an explicit real-time scheduling request when it's created. It guarantees a thread enough resources to complete its function in a specified time and is often used in transaction processing, data acquisition, and automated process control. For example, in a transaction-processing application, an ATM typically expects a response to an account balance inquiry within a specified time.

Data acquisition and process control programs use data that arrives at a constant rate from hardware devices. Data acquisition programs, such as those used to record radio astronomy observations, copy incoming data to a storage device, such as tape or disk. Data analysis is performed later by a separate process. Process control programs actively process the data as it arrives. For example, a process control program in a chemical manufacturing plant processes data from sensors in pipelines and reaction vessels. Data inputs flow from data collection devices or process sensors into memory buffers. A data acquisition or process control program must extract data from buffers for analysis quickly enough to keep them from overflowing. The program must be allocated enough resources to ensure that it can process or store incoming data at least as quickly as it arrives.

When a thread with a real-time scheduling requirement is created, it informs the OS of the maximum CPU time needed to complete one thread cycle and the frequency of cycles. A **thread cycle** can execute instructions to process a single transaction, retrieve and store data from an I/O device, or retrieve and analyze one set of process variables. For example, data acquisition equipment might send data every second, and the thread might need a maximum of 150 milliseconds to finish processing the data.

The thread signals the OS at the start of each thread cycle, and the scheduler starts a timer, tracking thread progress by updating CPU time used in the TCB each time the thread enters and leaves the ready state. The scheduler checks the timer, CPU time used, and maximum thread cycle time whenever it makes a scheduling decision to ensure that the thread completes its cycle before the timer expires. The scheduler can manage multiple real-time threads, although real-time scheduling overhead can quickly consume available CPU resources.

TECHNOLOGY FOCUS

Windows Scheduling

Current Windows desktop and server operating systems support multitasking, multi-threaded processes, and preemptive priority-based scheduling and dispatching. Scheduling, dispatching, and interrupt handling are performed by an OS component called the microkernel, which acts as both supervisor and scheduler.

Each process the microkernel manages has a base priority level described by an integer value from 0 to 31. Higher values represent higher scheduling priority. Priority levels are grouped into four categories called base priority classes: Idle (priority levels 0–6), Normal (priority levels 6–10), High (priority levels 11–15), and Real-time (priority levels 16–31). A process is assigned a base priority class when it's first created.

(continued)

Most user processes start in the Normal base priority class. In desktop computers and workstations, the microkernel raises the base priority class to High when a process is moved from the background to the foreground—in other words, when the process is receiving input from the keyboard and mouse. The Real-time base priority class doesn't meet the definition of real-time scheduling given previously; it simply describes the highest category of priority levels.

Threads inherit the base priority class of their parent processes, and thread priority classes are named Idle, Lowest, Below Normal, Normal, Above Normal, Highest, and Time Critical. Table 11.1 summarizes the meaning of thread priority classes in the base priority classes for processes.

TABLE 11.1 Thread priority classes and levels

Base priority class	Thread priority class	Priority level
Real-time	Time Critical	31
	Highest	26
	Above Normal	25
	Normal	24
	Below Normal	23
	Lowest	22
	Idle	16
High	Time Critical	15
	Highest	15
	Above Normal	14
	Normal	13
	Below Normal	12
	Lowest	11
	Idle	1
Normal	Time Critical	15
	Highest	10
	Above Normal	9
	Normal	8
	Below Normal	7
	Lowest	6
	Idle	1
Idle	Time Critical	15
	Highest	6
	Above Normal	5
	Normal	4
	Below Normal	3
	Lowest	2
	Idle	1

A thread's current priority level is called its "dynamic priority." The initial value of a thread's dynamic priority is the same as the base priority class of the thread and parent process. The microkernel can alter dynamic priority to improve system performance and to reflect high or low demand for system resources by a thread.

(*continued*)

Figure 11.11 shows a Windows utility called Process Viewer (pview.exe), which shows information about active processes, including each process's base priority class and the initial priority class and dynamic priority of its threads. This example shows information for a process named jdeveloper and six of its threads. The base priority class of jdeveloper is Normal, and the initial priority class for the selected thread (4) is Above Normal. The dynamic priority for this thread started at 9, but the current value shown at the bottom is 12. At some point after the thread was created, the microkernel adjusted its dynamic priority upward.

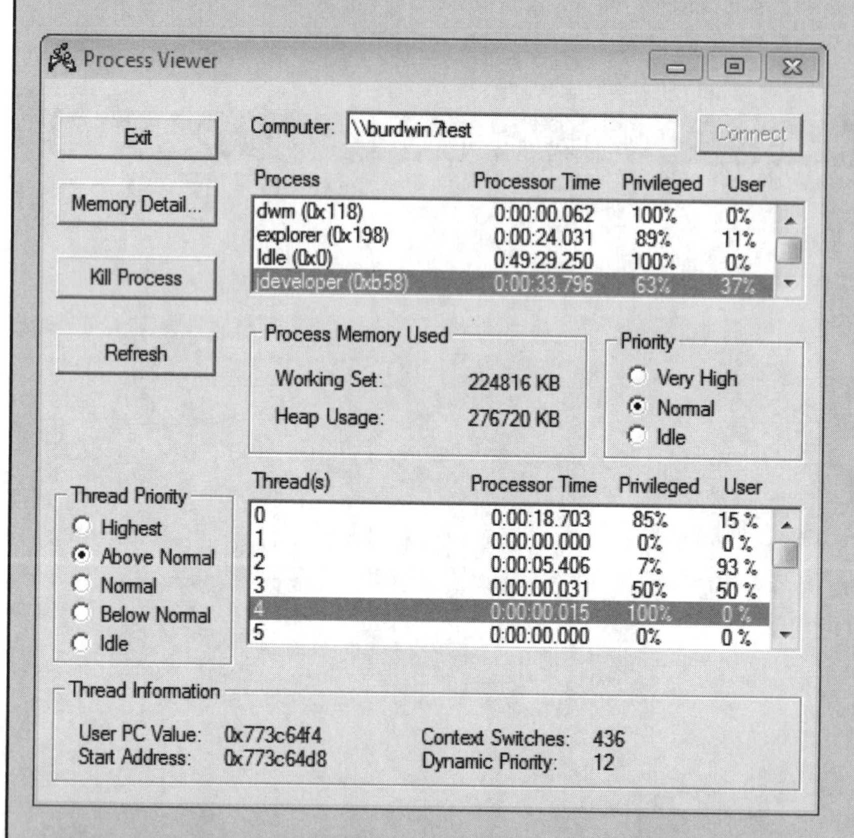

FIGURE 11.11 Windows Process Viewer
Courtesy of Course Technology/Cengage Learning

The microkernel uses a strict priority-based scheduling algorithm. A higher-priority thread in the ready state is always executed before a lower-priority thread. When a new thread is created with a higher priority level than the currently executing thread, the microkernel moves the currently executing thread to the ready state, where it remains until all higher-priority threads terminate or become blocked.

BUSINESS FOCUS

Choosing a Server Operating System

Microsoft has steadily gained market share in server operating systems, starting with Windows NT and continuing with various versions of Windows Server. Windows Server faces competition from several operating systems, including OS/390 (IBM) and OpenVMS (Hewlett-Packard). UNIX is also a major competitor, with Linux and products from hardware vendors such as Solaris (Sun Microsystems), AIX (IBM), and HP-UX UNIX (Hewlett-Packard).

Early Windows versions, such as Windows 3.1 and Windows 95, lacked many features needed for a server or multiuser OS. Beginning with Windows NT, Microsoft has added server-oriented features, such as the following:

- A reliable, robust file management system (NTFS)
- Access controls for key system resources, including files and directories
- Preemptive priority-based scheduling and multithreading
- Memory protection and virtual memory management
- Interprocess communication and Internet-oriented networking support
- Server clustering with load sharing and failover

Windows Server includes many server-oriented services and functions beyond basic server OS capabilities, including the following:

- Core Internet services, such as DNS, DHCP, and routing
- User and business-oriented Internet services, such as e-mail, FTP, HTTP, and e-commerce
- File and printer sharing
- Support for component-based software with the COM+ and .NET component frameworks

By including these functions as part of Windows Server, Microsoft has forced other vendors to bundle software that had been sold separately with their server operating systems.

UNIX, another widely used server OS, has a long history, beginning in the early 1970s as an experimental OS developed at AT&T Bell Laboratories. At first, AT&T didn't commercialize UNIX and allowed others to use both the OS and its source code free. Many universities experimented with UNIX and made changes and enhancements to it. Eventually, commercial vendors began to offer UNIX operating systems with proprietary extensions, such as GUIs, advanced networking support, and database management systems.

At the same time, nonprofit organizations enhanced UNIX and developed useful applications and services that ran in UNIX. These enhancements and extensions were always distributed free and with complete source code (called "open source") so that others could make improvements or modify the software as they saw fit. Currently, the driving force behind UNIX development and standardization is the Open Group (*www.opengroup.org*), which has owned the UNIX trademark since 1993. Until the late 1990s, the Open Group's UNIX standards competed with others from the IEEE and ISO, and these competing standards had major incompatibilities that created problems for software developers. In the late 1990s, these groups collaborated to produce the Single UNIX Specification. Version 3 is the most widely adopted, and systems conforming to it are described as "UNIX 03." A newer version released in 2008 is called POSIX:2008; despite the name, it's the successor to version 3 of the Single UNIX Specification.

(*continued*)

Linux is a UNIX variant developed by Linus Torvalds in the early 1990s. It combines many early open-source enhancements with new ones to create a unified package that's generally easier to install and administer than other UNIX variants. Many Linux applications and services are based on nonproprietary standards, such as MIT's X Window GUI and the CORBA component standard. (CORBA is discussed in Chapter 13.) Organizations can deploy Linux and Linux-based applications and services without paying the per-user or per-server licensing fees required with Windows and proprietary UNIX variants.

Linux, UNIX, and all current UNIX variants provide essentially the same basic OS functions as Windows Server. In addition, a variety of applications and services are available, ranging from personal productivity software to industrial-strength database management systems and computer-aided design and manufacturing software. The main differences from Windows Server are as follows:

- Many commonly used services and software (such as Web services) aren't provided as part of the server OS. A wide variety of for-profit and nonprofit organizations develop and distribute this software for Linux and UNIX.
- Most Linux/UNIX application and server software is provided with source code, which enables users to make modifications for their own purposes.
- Application and server software can't be distributed in binary (executable) form because Linux and UNIX run on so many different CPUs and computer systems. Software is normally distributed in source code format, with installation scripts that recompile and relink it for a specific hardware platform.
- Linux/UNIX application and server software are usually based on open standards, controlled by nonprofit organizations with wide representation from hardware and software vendors, academia, and user communities.

The differences in standards, bundling, binary compatibility, and number of software sources yield a complex set of tradeoffs that must be considered when deciding which OS to acquire. For example, many people see the open-source nature of Linux and UNIX as a security and reliability advantage, assuming that open-source software undergoes rigorous testing by a large user and developer community and enables faster incorporation of security patches and improvements. On the other hand, many people consider Windows Server to be much easier to use and configure because of binary software compatibility, an extensive set of similar configuration tools, and single-vendor support.

Describe the strengths and weaknesses of Windows Server and UNIX or Linux in the following settings. Which OS would you recommend for each company?

- A small electrical supply wholesaler with 25 employees, 15 user workstations, and an application and services suite that includes basic accounting and inventory control functions, file and printer sharing, and a small Web site not used for e-commerce
- An engineering design and consulting partnership with 25 employees, 15 user workstations, and a mix of applications and services, including basic accounting functions, bid preparation, CAD and drafting, construction management, file and printer sharing, and a small Web site not used for e-commerce
- A large catalog seller of musical equipment with hundreds of employees; three warehouses in New Jersey, St. Louis, and Portland; and a mix of applications and services, including all accounting functions, catalog preparation and distribution, phone and Web-based sales, and inventory control and logistics

MEMORY ALLOCATION

Memory allocation is the assignment of specific memory addresses to system software, application programs, and data. Processes and threads need memory to hold instructions and data during execution. The OS allocates memory when processes and threads are created and responds to requests for additional memory during execution. It also allocates memory to itself and for other needs, such as buffers and caches.

Note that most of what's said about memory allocation in this section also applies to managing secondary storage devices. Keep this in mind as you read this section because many of the terms and concepts are used again in Chapter 12.

Physical Memory Organization

Any computer's main memory can be regarded as a sequence of contiguous, or adjacent, memory cells, as shown in Figure 11.12. Addresses of these memory locations are assigned sequentially so that available addresses proceed from zero (low memory) to the maximum available address (high memory).

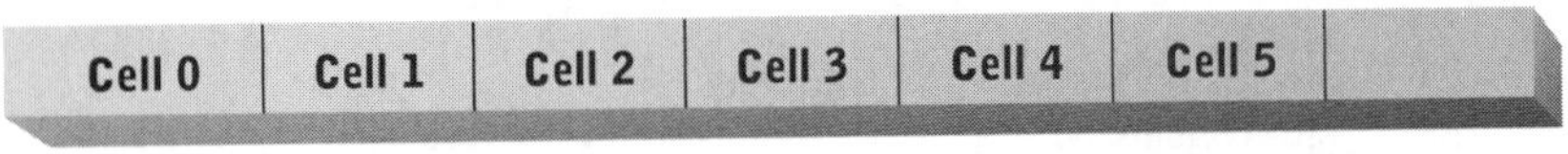

Cell 0 | Cell 1 | Cell 2 | Cell 3 | Cell 4 | Cell 5

FIGURE 11.12 The sequential physical organization of memory cells
Courtesy of Course Technology/Cengage Learning

Data values and instructions generally occupy multiple bytes of storage. For example, a single-precision integer typically occupies 4 consecutive bytes, or 32 bits, of storage. When written in a program, the bits and bytes of a numeric value are typically ordered from highest-position weight on the left to lowest-position weight on the right. If a 32-bit integer is written as follows, it's assumed that the left-position weight is the largest (2^{31}) and the right-position weight is the smallest (2^0):

00110010 11010011 01001100 01101001

When considered as a byte sequence, the leftmost byte is called the **most significant byte**, and the rightmost byte is called the **least significant byte**. In many CPU and memory architectures, a data item's least significant byte is placed at the lower memory address. For example, the 32-bit value $02468ACE_{16}$ would be stored in memory with CE at the lowest memory address, 8A at the next address, 46 at the address after that, and 02 at the highest address.

NOTE

Other CPU and memory architectures store bytes in the reverse order. **Big endian** describes architectures that store the most significant byte at the lowest memory address. **Little endian** describes architectures that store the least significant byte at the lowest memory address.

A CPU's or computer's **addressable memory** is the highest numbered storage byte that can be represented. Addressable memory is usually determined by the number of bits used to represent an address. For example, if 32 bits are used to represent a memory address, addressable memory is $2^{32} = 4{,}294{,}967{,}296$ bytes, or $FA000000_{16}$ bytes, or 4 GB. A computer system's **physical memory** is the actual number of memory bytes that are physically installed in the machine. Physical memory can be smaller than addressable memory but never larger.

Single-Tasking Memory Allocation

Memory allocation in a single-tasking OS is fairly simple. A memory allocation diagram, or memory map, shows physical memory as an array of addresses, with the lowest memory address at the bottom and the highest memory address at the top. Figure 11.13 shows a memory map for a single-tasking OS. The bulk of the OS normally occupies lower memory addresses, and the application program is loaded immediately above it. For the application program in this figure, memory allocation is **contiguous**, meaning all portions of the program and OS are loaded into sequential locations in memory.

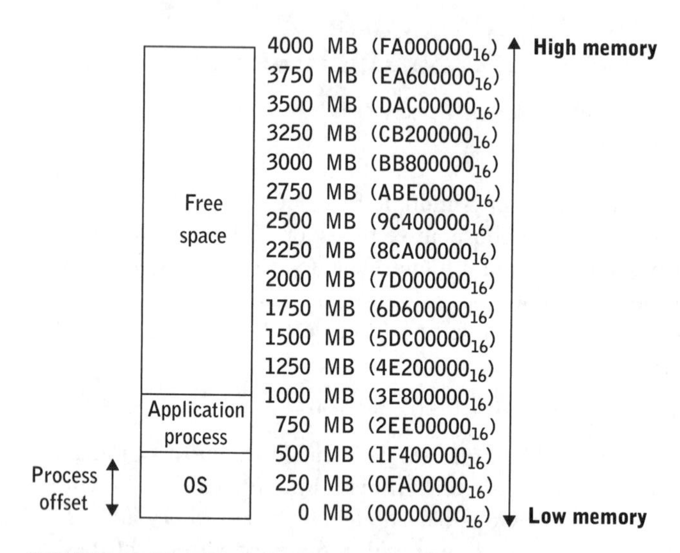

FIGURE 11.13 Contiguous memory allocation in a single-tasking OS
Courtesy of Course Technology/Cengage Learning

In Figure 11.13, the OS occupies the lowest range of memory addresses, and the application process's first instruction is one address higher than the last byte of the OS. The difference between the first address in physical memory and the address of the first process instruction is called the **process offset**. In this figure, the offset is $1F400000_{16} - 0 = 1F400000_{16}$.

A programmer could explicitly include the process offset in any memory address operands when writing or compiling a program. For example, if the process offset is $1F400000_{16}$ and the 10th program instruction is a BRANCH to the first instruction, the 10th instruction's operand must be $1F400000_{16}$. The programmer must know the process offset to specify any memory address operands in BRANCH, MOVE, and other instructions. This method is called **absolute addressing**, which uses memory address operands that refer to actual physical memory locations.

However, a programmer doesn't usually know exactly where a program will be loaded in memory when it's executed or the exact value of the process offset. Even if the programmer does know the process offset when writing the program, the offset value might change later for many reasons, including the following:

- Upgrading the OS, thus changing its size and the process offset
- Reconfiguring the OS, such as allocating more I/O buffers or adding device drivers
- Loading and running multiple processes

If a process uses absolute addressing, it must be rewritten every time the offset is changed.

Because a process offset usually can't be known in advance, processes are written as though their first instruction will always be loaded in memory address 0. The CPU automatically converts memory address operands into physical memory addresses as the program runs. To do so, the OS calculates and stores the process offset in a register when the program is first loaded into memory. During program execution, the CPU automatically adds the process offset to all memory address operands before accessing memory. This method of computing physical memory addresses automatically is called **indirect addressing** or **relative addressing**. The register holding the offset value is called an **offset register**.

When a CPU instruction uses a memory address as an operand, the operand is called a "memory reference." Examples of CPU instructions containing memory references include LOAD, STORE, and BRANCH. The process of determining the physical memory address that corresponds to a memory reference is called **address mapping** or **address resolution**. Address resolution is a simple process when memory is allocated contiguously; each memory reference is mapped to its equivalent physical memory address by adding the offset value. For the application process shown in Figure 11.13, all memory references are mapped to physical memory addresses by adding the value $1F400000_{16}$ (500 MB) to the reference.

Multitasking Memory Allocation

Memory allocation is more complex when the OS supports multitasking. The OS finds free memory regions in which to load new processes and reclaims memory when processes terminate. The goals of multitasking memory allocation are as follows:

- Allow as many active processes as possible.
- Respond quickly to changing memory demands of processes.
- Prevent unauthorized changes to a process's memory regions.
- Perform memory allocation and addressing as efficiently as possible.

Multitasking operating systems use partitioned memory, which divides memory into equally sized regions, each capable of holding all or part of a process or thread. Figure 11.14 shows 4 GB of physical memory divided into 250 MB partitions, each of which can hold an OS component, a process, or nothing at all (free space).

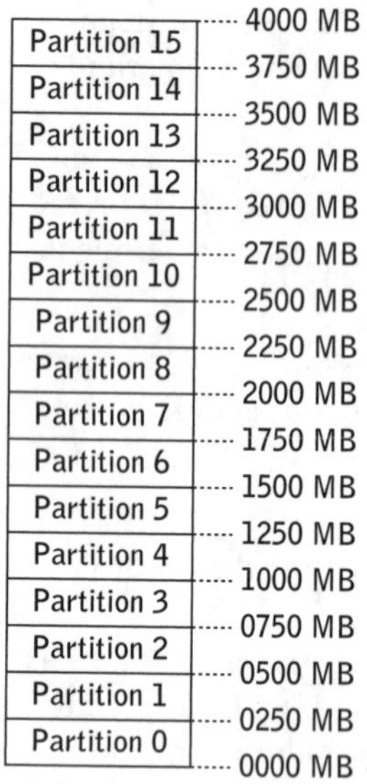

Partition 15
Partition 14
Partition 13
Partition 12
Partition 11
Partition 10
Partition 9
Partition 8
Partition 7
Partition 6
Partition 5
Partition 4
Partition 3
Partition 2
Partition 1
Partition 0

4000 MB
3750 MB
3500 MB
3250 MB
3000 MB
2750 MB
2500 MB
2250 MB
2000 MB
1750 MB
1500 MB
1250 MB
1000 MB
0750 MB
0500 MB
0250 MB
0000 MB

FIGURE 11.14 Dividing main memory into fixed-size partitions
Courtesy of Course Technology/Cengage Learning

Figure 11.15 shows three processes and the OS loaded into fixed-size memory partitions. Process 1 occupies three memory partitions, and Process 2 occupies two memory partitions. Each process uses all memory in its allocated partitions. Process 3 occupies four complete partitions and part of a fifth. Partitions 14 and 15 are unallocated.

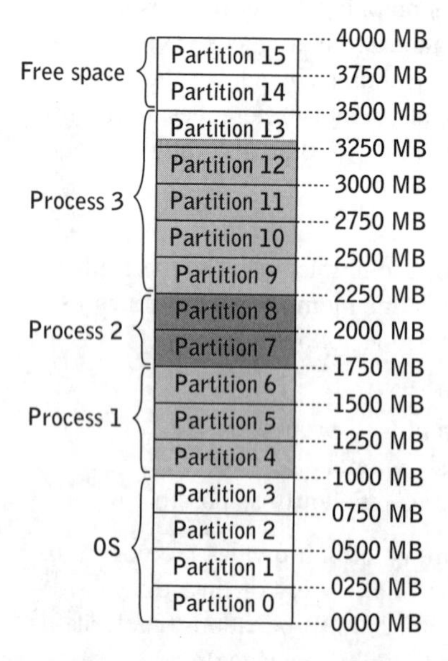

FIGURE 11.15 Several processes and the OS loaded into 250 MB fixed-size memory partitions
Courtesy of Course Technology/Cengage Learning

The OS maintains a memory partition table and updates it each time a partition is allocated or freed (see Table 11.2). When a process is ready to be loaded, the OS searches the table to find enough contiguous free partitions. If they're found, the table is updated and the process is loaded into free partitions. When a process terminates, the table is updated to free the memory partitions.

TABLE 11.2 Fixed-size memory partitions maintained by the OS

Partition	Starting address	Status	Allocated to
0	0 MB	Allocated	OS
1	250 MB	Allocated	OS
2	500 MB	Allocated	OS
3	750 MB	Allocated	OS
4	1000 MB	Allocated	Process 1
5	1250 MB	Allocated	Process 1
6	1500 MB	Allocated	Process 1
7	1750 MB	Allocated	Process 2
8	2000 MB	Allocated	Process 2
9	2250 MB	Allocated	Process 3
10	2500 MB	Allocated	Process 3
11	2750 MB	Allocated	Process 3
12	3000 MB	Allocated	Process 3
13	3250 MB	Allocated	Process 3
14	3500 MB	Free	
15	3750 MB	Free	

Contiguous program loading, coupled with fixed-size memory partitions, usually results in wasted memory space. For example, Process 3 uses only part of its last partition. The remainder is wasted because the OS can't allocate less than a full partition. Wasted space can be reduced by reducing partition size. In general, the smaller the partition size, the less wasted space. However, smaller partitions require a larger memory partition table, which increases OS memory requirements and the time needed to search and update the table.

Memory Fragmentation

As processes are created, executed, and terminated, memory allocation changes accordingly. Over time, memory partition allocation and deallocation leads to an increasing number of small free partitions separated by allocated partitions. Figure 11.16 shows changes in memory allocation over time as processes are created and terminated.

Figure 11.16(a) shows a starting point with four processes in memory. Figure 11.16(b) shows memory allocation after Process 3 has terminated and Process 5 has been loaded into memory. A new free space area has been created because Process 5 is smaller than Process 3. In Figure 11.16(c), Process 2 has terminated, and its former partition is now free.

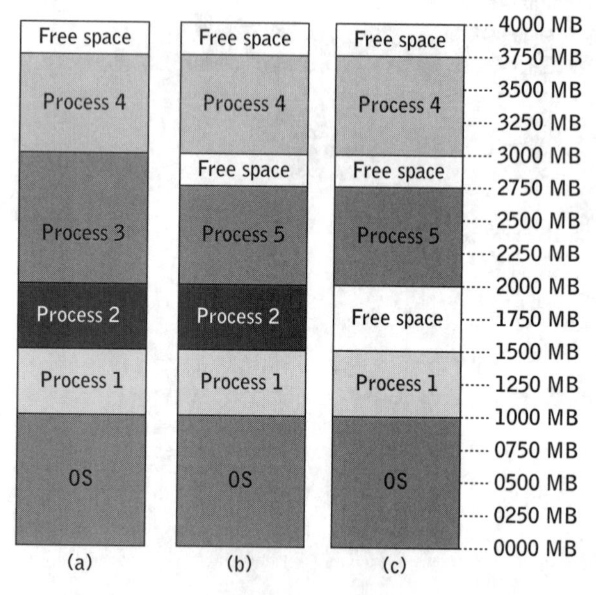

FIGURE 11.16 Changes in memory allocation as processes execute and terminate
Courtesy of Course Technology/Cengage Learning

There's now 1 GB of free memory but only two contiguous free 250 MB partitions (in the space between 1500 and 2000 MB). Although there might be other processes waiting to execute that require 1 GB or less, only those requiring 500 MB or less can be loaded because of fragmentation of available free space.

Fragmentation occurs when memory partitions allocated to a single process or purpose are scattered throughout physical memory, as shown for free space in Figure 11.16(c) and Figure 11.17(a). Over time, free space becomes more fragmented, and larger processes have increasing difficulty finding enough contiguous partitions. One way to address the problem of fragmented free space is to relocate all programs in memory periodically in a process called **compaction**. After compaction, all free partitions form a contiguous block in upper memory, as shown in Figure 11.17(b). Compaction is a time-consuming process because entire programs are moved in memory and many partition table entries are updated. The overhead required for compaction is generally larger than the overhead required for a more common strategy—noncontiguous memory allocation.

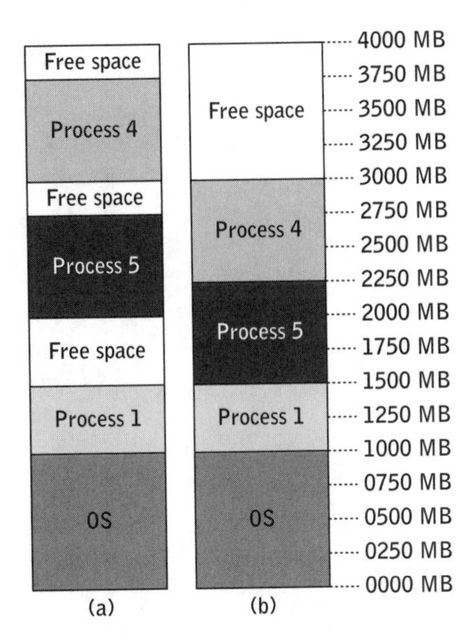

FIGURE 11.17 Compaction combines multiple free space fragments (a) into a single contiguous partition (b)

Courtesy of Course Technology/Cengage Learning

NOTE

Compaction of allocated space is done during disk fragmentation. Scattered sectors allocated to each file are moved to sequential sectors in the same or adjacent cylinders. File compaction isn't required because sectors are allocated to files noncontiguously, but compaction does improve read/write performance by minimizing track-to-track seek time and rotational delay.

Noncontiguous Memory Allocation

In an OS that supports noncontiguous memory allocation, portions of a process can be allocated to free partitions anywhere in memory. **Noncontiguous memory allocation** uses small fixed-size partitions, usually no larger than 64 KB, although 250 MB partitions are assumed to simplify the discussion and figures in this section.

In Figure 11.17(a), imagine that a new process (Process 6) awaits loading and needs 1 GB of memory. Under contiguous memory allocation, it can't be loaded unless memory is compacted first. With noncontiguous memory allocation, Process 6 can be divided into four 250 MB partitions that are allocated to available free memory partitions, shown as white boxes in Figure 11.18.

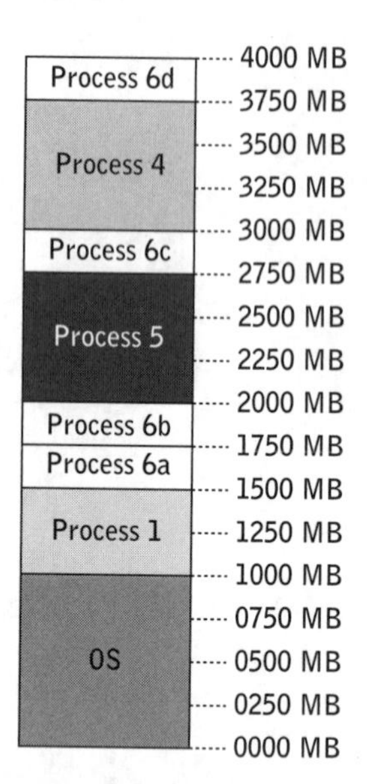

FIGURE 11.18 Process 6 is allocated to available free memory partitions
Courtesy of Course Technology/Cengage Learning

Noncontiguous memory allocation is more flexible than contiguous memory allocation, but flexibility comes at a price. Process 6 was compiled and linked assuming that it would occupy contiguous memory locations starting at address 0. However, memory references in Process 6 must now cross noncontiguous partitions. For example, if a BRANCH instruction in partition 7 references an instruction in partition 15, what offset value should be used to resolve the memory reference to a physical address?

In noncontiguous memory allocation, each process partition has its own offset value. To resolve addresses correctly, the OS must keep track of each program partition and its offset and make adjustments to memory references that cross partition boundaries. The partition tables and address calculations that support noncontiguous memory allocation are more complex than for contiguous memory allocation. Table content and address resolution are described in the next section.

Virtual Memory Management

The only portions of a process that must be in memory at any point during execution are the next instruction to be fetched and any operands stored in memory. Only a few bytes of any process must reside in memory at any one time. Most operating systems minimize the amount of process code and data stored in memory at one time, which frees large quantities of memory for use by other processes and substantially increases the number of processes that can execute concurrently.

Virtual memory management divides a program into partitions called pages. Each **page** is a small fixed-size portion of a program, normally between 1 and 4 KB. Memory is also divided into pages of the same size. Each memory page is called a **page frame**. During program execution, one or more pages are allocated to page frames, and the rest are held in secondary storage. As pages in secondary storage are needed for current processing, the

OS copies them into page frames. If necessary, pages currently in memory are written to secondary storage to make room for pages being loaded.

Each memory reference is checked to see whether the page it refers to is currently in memory. A reference to a page held in memory is called a **page hit**, and a reference to a page held in secondary storage is called a **page fault. Page tables** store information about page locations, allocated page frames, and secondary storage space. Each active process has a page table or portion of a page table dedicated to it. Table contents include page numbers, a status field indicating whether the page is currently held in memory, the page frame number in main memory or secondary storage, and a status field indicating whether the page has been modified since being swapped into memory. Table 11.3 shows a sample page table.

TABLE 11.3 Portion of a process's page table

Page number	Memory status	Frame number	Modification status
1	In memory	214	No
2	On disk	101	N/A
3	On disk	44	N/A
4	In memory	110	Yes
5	On disk	252	N/A

Because page size is fixed, memory references can easily be converted to the corresponding page number and offset in the page. The page number can be determined by dividing the memory address by the page size. The whole portion of the result is the page number, and the remainder is the offset into this page. For example, if page size is 1 KB, a reference to address 1500 is equivalent to an offset of 476 (1500 - 1024) into page number 2 (1500/1024 + 1). If the table is stored sequentially, the corresponding entry in the process's page table can be computed as an offset into the table.

If the reference is to an address in a page held in memory, the corresponding memory address is an offset into the memory page indicated in the table. Using Table 11.3 as an example, a reference to address 700 (offset of 700 into page 1) resolves to an offset of 700 into memory frame 214. Again, fixed page size is used to calculate the corresponding memory address as 219836 ($214 \times 1024 + 700$). Note that these calculations are similar to those for addressing array contents, described in Chapter 3.

A secondary storage region, called the **swap space**, **swap file**, or **page file**, is reserved for the task of storing pages not held in memory. The swap space is divided into page frames in the same manner as memory. A memory reference to a page held in the swap space results in this page being loaded into a page frame in memory. As with address resolution, page location in the swap space can be computed by multiplying the page number by the page size.

If all page frames are allocated, a page currently in memory, called the **victim**, must be written to the swap space before the reference page is loaded into a page frame. Some common methods for selecting the victim are as follows:

- Least recently used
- Least frequently used

Both methods require the OS to maintain information about accesses to pages in memory. Searching and updating this information is part of the system overhead associated with virtual memory management.

When a victim has been selected, it might or might not be copied back to the swap space. The sample data in Table 11.3 shows entries indicating whether a page has been modified since it was swapped into memory. If a page hasn't been modified, the copy held in the swap space is identical to the copy in memory, which means the page doesn't have to be copied to the swap space if it has been selected as the victim.

Memory Protection

Memory protection refers to protecting memory allocated to one program from unauthorized access by another program. It can apply to interference between programs in a multitasking environment or interference between a program and the OS in a single-tasking or multitasking environment. If memory isn't protected, errors in one program can generate errors in another. If the program being interfered with is the OS, the result might be a system crash.

In the simplest form of memory protection, the OS checks each write to a memory location to ensure that the address being written is allocated to the program performing the write operation. Complicating factors include various forms of indirect addressing, virtual memory management, and cooperating processes, which are two or more processes that "want" to share a memory region. Memory protection adds overhead to each write operation.

Memory Management Hardware

Early types of multitasking, memory protection, and virtual memory management were implemented exclusively by the OS, which imposed severe performance penalties. Consider, for example, the overhead required to map program memory references by using virtual memory management. Each reference requires the OS to search one or more tables to locate the correct page and determine its corresponding memory or disk location. A memory reference that should consume only one or a few CPU cycles consumes many additional cycles for paging, swapping, and address-mapping functions.

The benefits of advanced memory addressing and allocation schemes are offset by reduced performance when they're implemented in software. For this reason, CPUs and computer systems now incorporate advanced memory allocation and address resolution functions in hardware. For example, all Intel microprocessors since the 80386 have included hardware support for virtual and protected memory management.

TECHNOLOGY FOCUS

Intel Core Memory Management

Intel Core CPUs dedicate six registers named CS, DS, ES, FS, GS, and SS to hold data structures called segment descriptors. ("Segment" is an Intel term for a memory region

(continued)

containing instructions or data.) Segment descriptions contain information about a segment, including the following:

- The physical memory address of the segment's first byte (its base address)
- The segment size
- The segment type
- Access restrictions
- Privilege level

A memory address used in an operand consists of a reference to a segment register and an offset value. For example, the operand DS:018C refers to byte 018C (hexadecimal) in the segment the DS segment descriptor points to. The CPU converts a memory address operand to a physical memory address by adding the segment's base address and the offset.

Segment size ranges from 1 byte to 4 GB. Segment types include data segments (further divided into the types "true" data and stack data) and code segments (executable instructions). The CS register always contains a segment descriptor for a code segment. The DS, ES, FS, and GS registers always contain segment descriptors for a data segment or a null descriptor. The SS registers always contain the segment descriptor of a stack.

Core CPUs protect process memory by checking instruction operands that reference segment descriptors. The simplest check is a limit check on the offset value that prevents processes from reading or writing data outside their own segments. For each memory reference, the CPU compares the offset value to the size parameter stored in the segment descriptor. If the offset is larger than the segment size, an error interrupt is generated.

Access types must also be specified for each segment. Data segments can be marked as read-only (RO) or read-write (RW). Code segments can be marked as execute-only (EO) or execute-read (ER). A segment's access type restricts the types of instructions that can execute by using memory references in the segment. For example, the CPU generates an error interrupt if an instruction attempts to write to a segment marked RO or read from a segment marked EO. Write operations to code segments always generate an error interrupt. The CPU also checks segment types when loading segment descriptors into a register. Segment descriptors for data segments can't be loaded into the CS register, and segment descriptors for code segments can't be loaded into the DS, ES, FS, GS, or SS registers.

Core CPUs use two different methods to prevent access to code segments not allocated to a process. The first specifies a privilege level for each process and segment. Privilege levels are numbered 0 through 3, with 0 being the most privileged. Every process is assigned a privilege level and can't access a segment with a privilege level numbered lower than its own. This method protects OS components from interference by application programs. Segments allocated to the kernel typically have a privilege level of 0, and segments allocated to other OS components typically have a privilege level of 1 or 2. Application processes usually have a privilege level of 3.

Memory segments can be hidden from a process. The CPU maintains a global descriptor table (GDT) containing descriptors for all segments. If an application process uses the GDT, it can "see" all segments. The OS can execute an instruction that allocates a local descriptor table (LDT) to a process and defines the LDT's content before issuing the instruction. Typically, the OS populates the LDT only with descriptors allocated to the process. It's impossible for a process using an LDT to "know of" any segments other than its own.

(continued)

Core CPUs implement virtual memory management tables in hardware. Page size is normally 4 KB. The CPU uses two types of tables—page directories and page tables. A page directory is a table of pointers to page tables. By convention, the entries in the page directory correspond to the segment's descriptors in a process's LDT (see Figure 11.19). Each entry in a page directory points to one entry in a page table. Each entry in a page table contains descriptive information about one 4 KB page.

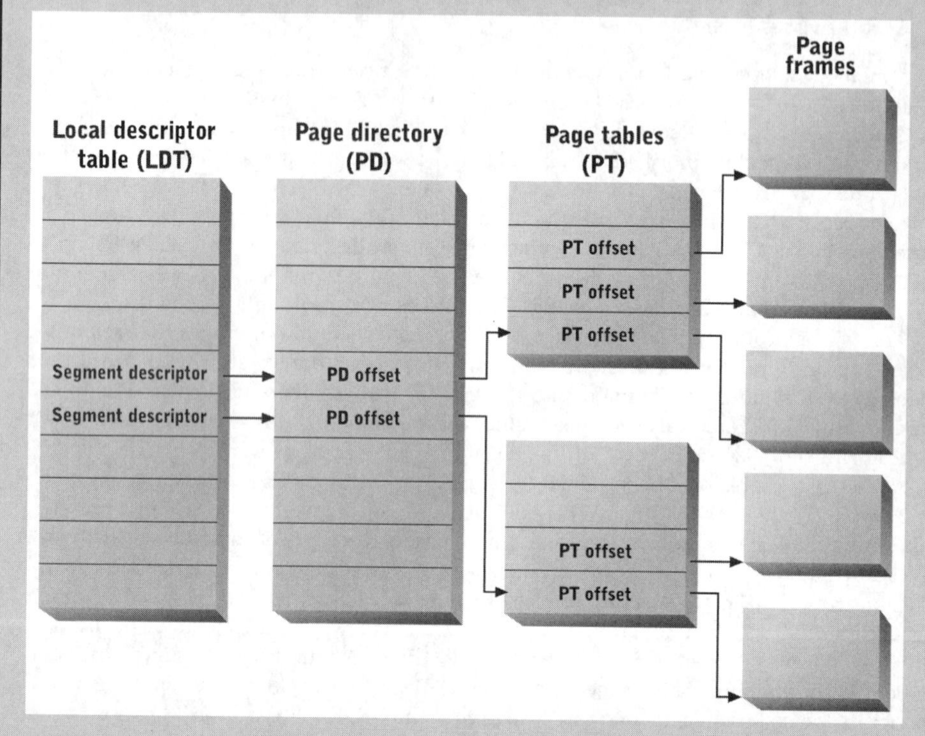

FIGURE 11.19 Relationship between segment and page (virtual memory) tables
Courtesy of Course Technology/Cengage Learning

A page table entry includes the page's physical memory address if it's loaded into memory. It also contains bit field flags, including a flag indicating whether the page is in memory, a flag indicating whether the page has been accessed since it was loaded, and a flag indicating whether the content of a page in memory has been written.

Virtual memory management responsibility is split between the CPU and the OS. When virtual memory management is enabled, the CPU converts addresses containing a segment descriptor and offset into addresses containing a page directory offset, a page table offset, and an offset into the 4 KB page. The CPU automatically performs a lookup in the page directory and page table based on these offsets to find the page's base address (see Figure 11.20). It then adds the page offset to generate a physical address. If the page table entry indicates that the page isn't in memory, the CPU generates an interrupt. It also sets the access and write flags in the page table as page contents are read or written.

(continued)

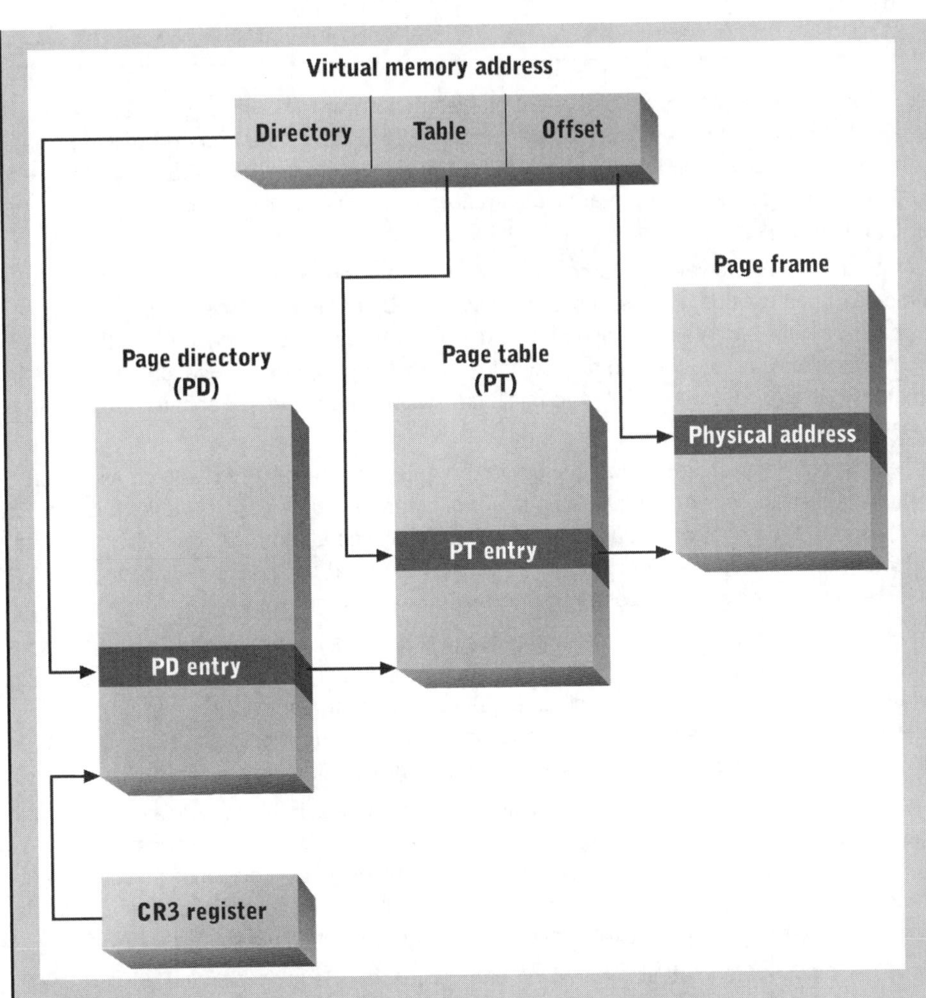

FIGURE 11.20 Address resolution with virtual memory management
Courtesy of Course Technology/Cengage Learning

The OS maintains its own table that maps pages to page frames in the swap space and clears the access and write flags when a page is loaded into memory. It also sets and clears "in memory" flags in the page table each time pages are loaded or swapped and provides an interrupt handler to perform page swaps in response to page fault interrupts.

Early Microsoft and UNIX operating systems were unable to implement usable multitasking and virtual memory management on Intel-based PCs because early Intel CPUs didn't have these features. Intel introduced hardware support for memory protection and virtual memory management with the 80386 CPU in 1987, and these features have continued into current Core processors. This hardware support enabled important improvements in Microsoft operating systems and provided enough CPU power to support UNIX on PCs.

Summary

- The OS is the most important component of system software. Its primary purpose is to manage hardware resources and provide support services to users and application programs. In functional terms, it manages the CPU, memory, processes, secondary storage (files), I/O devices, and users. In architectural terms, it consists of the kernel, service layer, and command layer.
- The OS allocates hardware resources to user processes on demand. In multitasking environments, it coordinates access to shared resources by many processes while ensuring that each process receives the resources it needs. The resource allocation function requires extensive recordkeeping and complex procedures to make allocation decisions. An important design goal for any OS is to minimize system overhead yet ensure that resource allocation goals are met.
- Application software is simpler to develop if programs are unaware of resource allocation functions. Operating systems make a large set of virtual resources appear to be available to each program. The sum of virtual resources generally exceeds the real resources existing in a computer system. The OS implements virtual resources by rapidly reallocating real resources to programs and substituting one resource for another.
- The OS stores information about each process in a process control block (PCB). PCB contents are updated continuously as resources are allocated to processes. In some operating systems, processes can create executable subunits called threads, which can be scheduled independently. Threads share all resources with their parent process. Execution speed is increased when multiple threads execute concurrently or simultaneously.
- An active thread is always in one of three states—ready, running, or blocked. In a computer with a single CPU, only one thread can be in the running state at a time. Ready threads are waiting for access to a CPU. Blocked threads are waiting for some event to occur, such as completing a service request or correcting an error condition. Processes can be scheduled by many methods, including first come first served, explicit priority, and real-time scheduling.
- Memory is divided into fixed-size partitions and processes are allocated one or more memory partitions to store instructions and data. The OS maintains tables to track partition allocations and free space. Memory references are mapped to physical addresses through page table lookups and address calculations.
- Processes are created as though they occupy contiguous primary storage locations starting at the first location. They aren't usually placed in low memory because system software or other programs reside there. Indirect addressing reconciles the difference between where a program "thinks" it's located in memory and where it's actually located. The actual address of a program's first instruction is stored in an offset register, and this address is added to each address reference the program makes. Indirect addressing enables multiple programs to reside in memory and allows programs to be moved during execution.
- Current operating systems implement virtual memory management. Portions of processes, called pages, are allocated to small memory partitions called page frames. Pages are swapped between memory and secondary storage as needed. Complex memory management procedures incur substantial overhead. To reduce system overhead, CPUs now implement many of these functions in hardware.

This chapter gave you an overview of internal OS architecture and described resource, CPU, and memory allocation. In Chapter 12, you examine secondary storage allocation and file management and see that many of the issues in memory allocation must also be addressed in secondary storage allocation. In Chapter 13, you turn your attention to external resources and explore how one OS cooperates with other operating systems to give local users access to external resources.

Key Terms

absolute addressing
address mapping
address resolution
addressable memory
bare-metal hypervisors
big endian
blocked state
child process
command language
command layer
compaction
concurrent execution
contiguous
dispatching
explicit priority
first come first served (FCFS)
fragmentation
hypervisor
indirect addressing
interleaved execution
job control language (JCL)
kernel
least significant byte
little endian
memory allocation
memory protection
most significant byte
multitasking
multithreaded
noncontiguous memory allocation
offset register
page
page fault
page file
page frame
page hit
page tables
parent process
physical memory
preemptive scheduling
priority-based scheduling
process
process control block (PCB)
process family
process list
process offset
process queue
ready state
real resources
real-time scheduling
relative addressing
run queue
running state
scheduler
scheduling
server consolidation
service call
service layer
shell
shortest time remaining (STR)
sibling processes
spawn
swap file
swap space

system overhead
thread
thread control block (TCB)
thread cycle
thread list
timer interrupt
victim
virtualization environments
virtual machines (VMs)
virtual memory management
virtual resources

Vocabulary Exercises

1. A(n) ________ OS supports multiple active processes or users.
2. In virtual memory management, a memory page's location is determined by searching a(n) ________.
3. A(n) ________ occurs when a process or thread references a memory page not currently held in memory.
4. Dispatching a thread moves it from the ________ state to the ________ state.
5. The CPU periodically generates a(n) ________ to give the scheduler an opportunity to allocate the CPU to another ready process.
6. A(n) ________ is an OS that enables dividing a single physical computer or cluster into multiple virtual machines.
7. In the ________ scheduling method, threads are dispatched in order of their arrival.
8. A(n) ________ process contains subunits that can be executed concurrently or simultaneously.
9. ________ scheduling guarantees that a thread receives enough resources to complete one ________ in a maximum time interval.
10. Hardware resources consumed by an OS's resource allocation functions are called ________.
11. ________ scheduling refers to any type of scheduling in which a running thread can lose control of the CPU to another thread.
12. The act of selecting a running thread and loading its register contents is called ________ and is performed by the ________.
13. To achieve efficient use of memory and a large number of concurrently executing processes, most OSs use ________ memory management.
14. When a thread makes an I/O service request, it's placed in the ________ state until processing of the request is finished.
15. Memory pages not held in primary storage are held in the ________ of a secondary storage device.
16. On a computer with a single CPU, multitasking is achieved by ________ execution of multiple processes.
17. With ________, all portions of a process must be loaded into sequential physical memory partitions.
18. The ________, ________, and ________ are the main layers of an OS.

19. A(n) ______ is the unit of memory read from or written to the swap space.
20. A(n) ______ resource is apparent to a process or user, although it might not physically exist.
21. Under a(n) ______ memory allocation scheme, portions of a single process might be physically located in scattered partitions of main memory.
22. In virtual memory management, memory references by a process must be converted to an offset in a(n) ______.
23. Information about a process's execution state, such as register values and process status, are stored in a(n) ______.
24. With the ______ scheduling method, threads requiring the least CPU time are dispatched first.
25. A(n) ______ causes the currently executing process to be ______ and control to be passed to the ______.
26. The process of converting an address operand into a physical address in a memory partition or page frame is called ______.
27. A(n) ______ is an executable subunit of a process that's scheduled independently but shares memory and I/O resources.
28. The ______ endian storage format places the ______ byte of a word in the lowest memory address. The ______ endian storage format places the ______ byte of a word in the lowest memory address.
29. In ______, program memory references correspond to physical memory locations. In ______, the CPU must calculate the physical memory location that corresponds to a program memory reference.
30. In indirect addressing, the content of a(n) ______ is added to calculate the corresponding physical memory address.

Review Questions

1. Describe the functions of the kernel, service, and command layers of an OS.
2. What's the difference between a real resource and a virtual resource?
3. What are the goals of an OS resource allocation function? Describe the conflicts between them.
4. What characteristics or capabilities differentiate a bare-metal hypervisor from a virtualization environment?
5. How and why does a thread move from the ready state to the running state? How and why does a thread move from the running state to the blocked state? How and why does a thread move from the blocked state to the ready state?
6. What is a process control block, and what is it used for?
7. What is a thread? What resources does it share with other threads in the same process?
8. Briefly describe the most common methods for making priority-based scheduling decisions.

9. What complexities are introduced by real-time scheduling requirements?
10. Describe the operation of virtual memory management.
11. What is memory protection, and why is it needed? What factors complicate it?
12. What is absolute addressing? What is indirect addressing?
13. What are the costs and benefits of indirect addressing?

Research Problems

1. Viruses and other malware often exploit bugs known as buffer overflows in widely used software. One method of preventing these exploits in software running on current Windows versions is Data Execution Prevention (DEP). Investigate buffer overflows and DEP. How does DEP prevent malicious attacks via buffer overflows? Is it always effective? Why might a user or administrator disable DEP?
2. Microsoft includes the hypervisor Hyper-V in Windows Server 2008 and later. Compare the capabilities of the most recent versions of Hyper-V and VMware ESX and ESXi. Which product has the best features for server consolidation? How has VMware responded to Microsoft bundling a free hypervisor with its server OSs?

CHAPTER **12**

FILE AND SECONDARY STORAGE MANAGEMENT

CHAPTER GOALS

- Describe the components and functions of a file management system
- Compare the logical and physical organization of files and directories
- Explain how secondary storage locations are allocated to files and describe the data structures used to record these allocations
- Describe file manipulation operations
- List access controls that can be applied to files and directories
- Describe file migration, backup, and recovery methods
- Explain methods for ensuring fault tolerance
- Compare storage consolidation methods, such as storage area networks and network-attached storage

File and secondary storage management are important system software functions because stored programs and data are important user and organizational resources. The collection of system software that performs file and secondary storage management and access functions is known as a **file management system (FMS)**. The FMS is usually part of the OS, although it's sometimes supplemented by additional software, such as database management systems. Figure 12.1 shows the FMS functions described in this chapter.

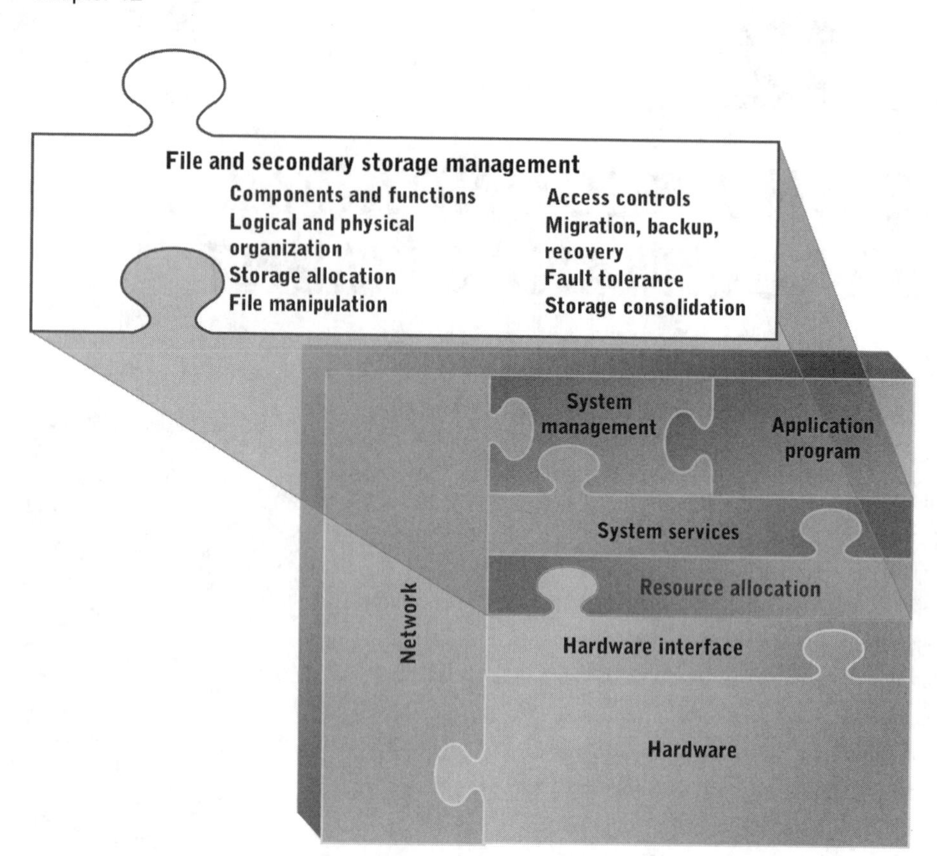

FIGURE 12.1 Topics covered in this chapter
Courtesy of Course Technology/Cengage Learning

FUNCTIONS AND COMPONENTS OF FILE MANAGEMENT SYSTEMS

An FMS is implemented in the following layers, which are similar to those in an OS (see Figure 12.2):

- Command layer or application program
- File control
- Storage I/O control
- Secondary storage devices

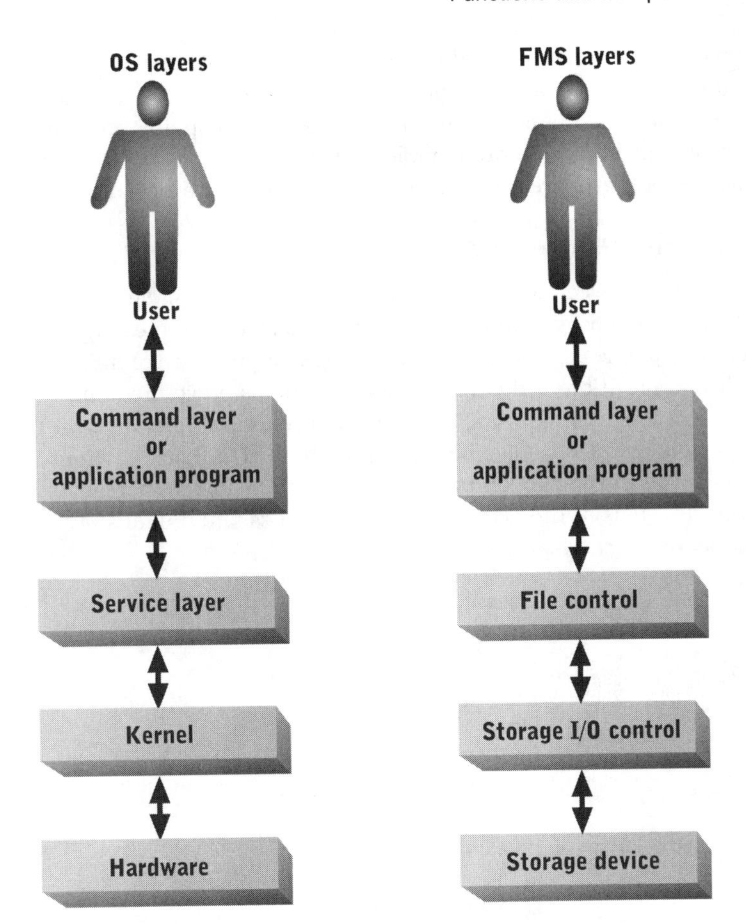

FIGURE 12.2 OS layers compared with FMS layers
Courtesy of Course Technology/Cengage Learning

Storage devices physically store bits, bytes, and blocks in a storage medium, and storage device controllers interact with the bus and OS device drivers to transfer data between storage devices and memory. As discussed in Chapter 6, the device controller presents a logical view of the storage device or media to the device driver. A logical view is a linear sequence of storage locations called a linear address space.

The **storage I/O control layer** is the part of the kernel that accesses storage locations and manages data movement between storage devices and memory. Software modules in this layer include the following:

- Device drivers for each storage device or device controller
- Interrupt handlers
- Buffers and cache managers

The **file control layer** provides service functions for manipulating files and directories. It processes service calls from the command layer or application program and issues commands to the storage I/O control layer to interact with hardware. It also maintains the directory and storage allocation data structures used to find files and their associated physical storage locations.

An FMS provides command-layer functions and utility programs for users and system administrators to manage files, directories, and secondary storage devices. Via the command layer, users perform common file management functions, such as copying, moving, and renaming files. Utility programs address more complex functions, such as creating text files, formatting storage devices, and creating backup copies of files and directories.

Logical and Physical Storage Views

The file control layer is the bridge between logical and physical views of secondary storage. Users and applications view secondary storage logically as a collection of files organized in directories and storage volumes. On a desktop or laptop computer, a **volume** is usually an entire physical disk, a partition of the disk, or a removable storage medium, such as a DVD or flash drive. On larger computers, a volume can span multiple physical disks. The physical view of secondary storage is a collection of physical storage locations organized as a linear address space. A typical computer has up to a few dozen storage volumes, thousands of directories, tens of thousands to millions of files, and billions of physical secondary storage locations (see Figure 12.3).

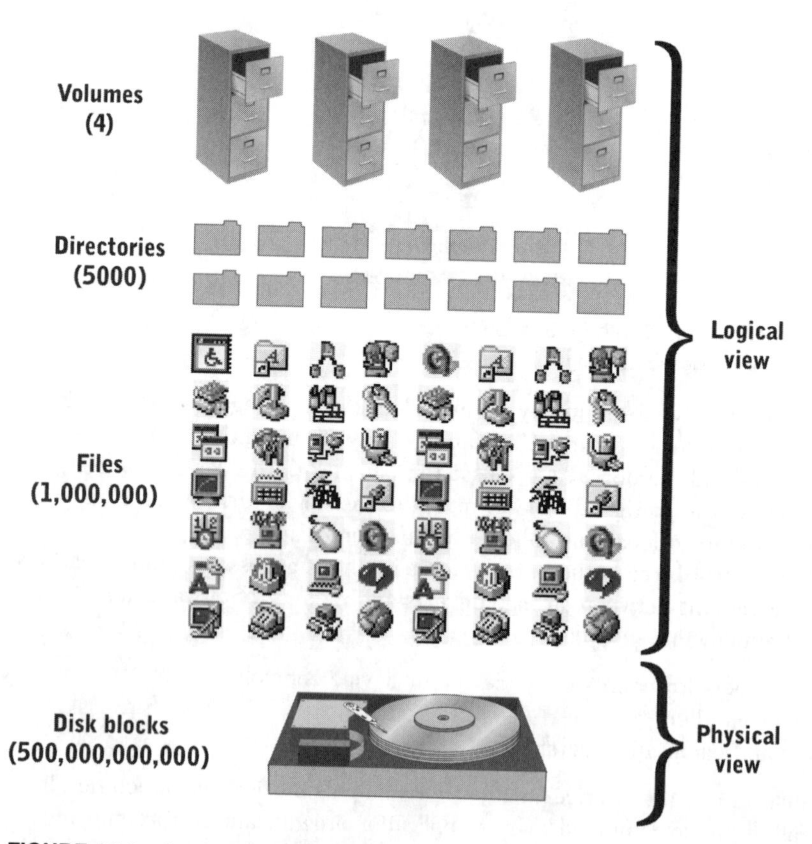

FIGURE 12.3 Logical and physical secondary storage for a typical small server
Courtesy of Course Technology/Cengage Learning

Figure 12.4 shows the logical structure of a typical data file. It's subdivided into records, and each record is composed of multiple **fields**. As discussed in Chapter 3, a record usually contains information about a single person, such as a customer or employee; a thing, such as a product in inventory; or an event, such as a transaction. A field contains a single data item describing the record.

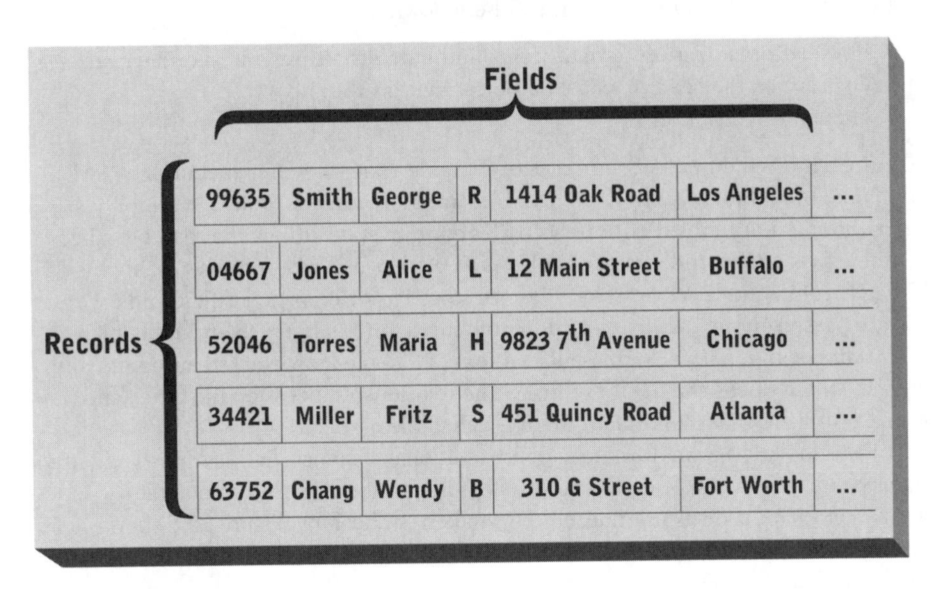

FIGURE 12.4 Logical structure of a data file
Courtesy of Course Technology/Cengage Learning

The logical file structure is independent of the physical device on which it's stored. A number of physical structure characteristics, such as the following, are simplified or ignored in the corresponding logical file structure:

- Physical storage allocation
- Data access methods
- Data-encoding methods

Physical storage allocation considerations include placement of fields and records in a file and distribution of a file across storage locations, media, or devices. Physical data access factors include whether file content is accessed sequentially, with an index, or by some other method. Data-encoding issues include the data structures and coding methods for representing each field. Related issues include data encryption and data compression.

File Content and Type

A file can store many different data types, including text, numbers, complex data structures, and executable instructions, but designing an FMS that accounts for all possible variations in file content and organization is difficult. This type of FMS would be complex and have a huge number of file-oriented commands and service routines. To avoid these problems, most FMSs support only a few file types directly, including the following:

- Executable programs
- OS commands
- Text or unformatted binary data

Among other things, file type determines the following:

- Physical organization of data items and data structures in secondary storage
- Operations that can or can't be performed on the file
- Filename restrictions

For example, the Windows FMS supports executable code stored in executable (EXE) files and dynamic link library (DLL) files. EXE files are stored in a format that simplifies loading them into memory for execution. DLL files contain subroutines that can be called from executable files and an index that enables the OS to locate subroutines quickly by name. OS batch command (BAT or CMD) files are stored as ordinary text files, and text commands are passed automatically to a command interpreter when a user double-clicks a BAT or CMD file icon. If a user double-clicks a text (TXT) file icon, the OS automatically starts Notepad, which opens the file for editing. The relationship between file types and the programs or OS utilities that manipulate them is called **file association**.

Current FMSs include a framework to support additional file types. In this framework, users and programs can register new file types and install programs to perform common file-manipulation operations, such as printing, editing, and error checking. Figure 12.5 shows a partial listing of file types registered on a Windows computer. Right-clicking a file opens a context menu for that file type.

Normally, the file type is declared when a file is created. In some FMSs, such as UNIX, the file type is stored in the directory. In other FMSs, the file type is declared with a naming convention, such as a file extension. For example, in Windows, executable filenames must end in .exe, dynamic link library filenames must end in .dll, and text filenames end in .txt by default. Windows registered file types also use file extensions, such as .docx for WordPad or Word document files (see the Name column in Figure 12.5) and .dvr for recorded TV programs.

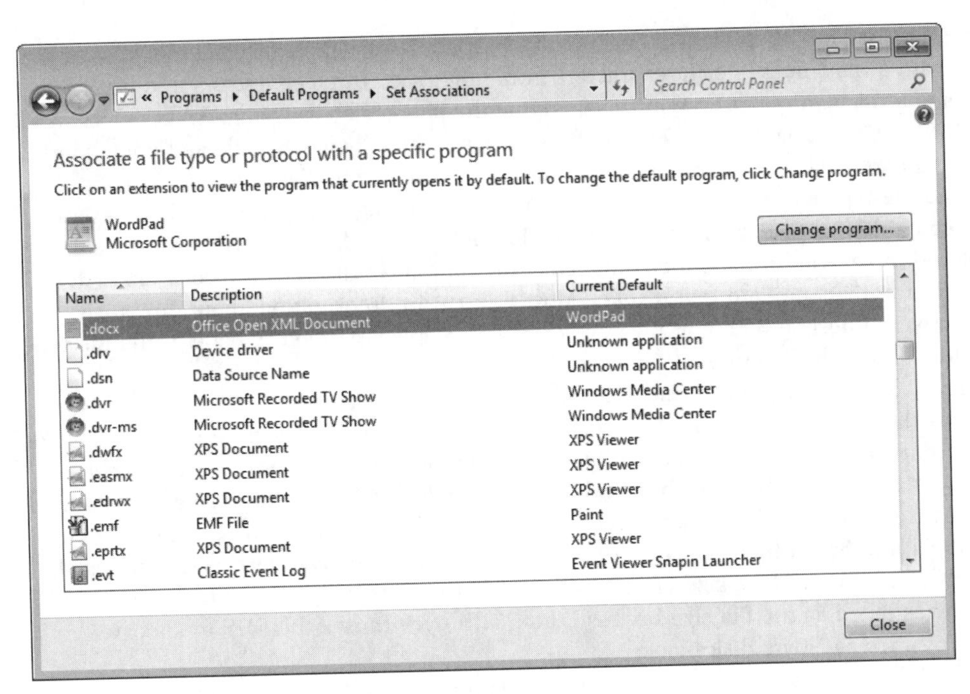

FIGURE 12.5 Registered Windows file types and associated programs
Courtesy of Course Technology/Cengage Learning

DIRECTORY CONTENT AND STRUCTURE

A **directory** contains information about files and other directories. Directory information is normally stored in a table, although more complex data structures are used sometimes. Users can view most directory information with graphical or command-line interfaces, such as Windows Explorer or the UNIX/Linux `ls` command. Some directory information, such as disk location, is hidden from users because it's used only by the FMS. Typical directory contents include the following:

- Name
- File type
- Location
- Size
- Ownership
- Access controls
- Time stamps

NOTE

In Windows operating systems, the term "folder" is often used instead of "directory."

All files and directories must have a unique name in their own directory, but names can be reused in other directories. Operating systems vary in requirements for valid names. Many older OSs restrict both the length and characters used. For example, MS-DOS uses a two-part name with a maximum of eight characters in the first part, three characters in the second part, and a mandatory dot (.) symbol separating the parts, such as "Sample.doc." (This format is sometimes called "8.3.") Embedded spaces and most nonalphabetic and nonnumeric characters aren't allowed. Current OSs are far less restrictive in filename requirements.

The file type can be stored implicitly with a filenaming convention, such as .exe for an executable file, or directly by using a coded field in the directory. A file's type can be displayed in several formats, including icons in a graphical display or special characters added to the filename in a text display. Most FMSs store information about directories the same way as information about files. An FMS uses a special directory code in the file type field to distinguish directories from files.

The location field content varies considerably across FMSs. In simpler FMSs, it usually contains the disk address of the file's first disk block. In more complex FMSs, it contains or points to a data structure, such as an index, containing the addresses of all disk blocks.

The size field can contain the number of bytes allocated to the file, the actual number of bytes stored in the file, or both. The number of allocated bytes is usually larger than the actual number of bytes because typically, a portion of the last disk block isn't used. For example, with a block size of 512 bytes, a file that's actually 513 bytes might have 1024 bytes listed as the file size because the 513th byte causes the FMS to allocate another entire 512-byte disk block.

The file owner field contains the account name or identification number of the file creator or the last account to take ownership. Most complex FMSs grant special permissions to the file owner. Access controls, described in more detail later in this chapter, list the accounts that have been granted or denied rights to the file. Rights vary depending on the OS, but the following are typical:

- *List*—An account or a group can view a file in a particular directory listing or list a directory's contents.
- *Read*—An account or a group can view a file's content, which also implies the right to copy it.
- *Modify*—An account or a group can add new content, alter or delete existing content, rename the file, move the file to a new directory, or delete the file.
- *Change*—An account or a group can alter permissions for other accounts or groups; this right is usually granted to the owner by default.

Most FMSs store one or more time stamps for each file in the directory, which can include date and time data for the following:

- File creation
- Most recent read
- Most recent write
- Most recent backup

Hierarchical Directory Structure

In a **hierarchical directory structure**, directories can contain other directories, but a directory can't be contained in more than one parent. This structure is sometimes called a **tree directory structure** because directory diagrams resemble upside-down trees. The left pane of

Figure 12.6 shows a portion of a hierarchical directory structure in Windows Explorer. There's a root directory for each secondary storage device (drives C and T in this figure). Each root directory can contain other directories, files, or both, as can directories below the root directory. The number of recursively descending directory levels is theoretically unlimited.

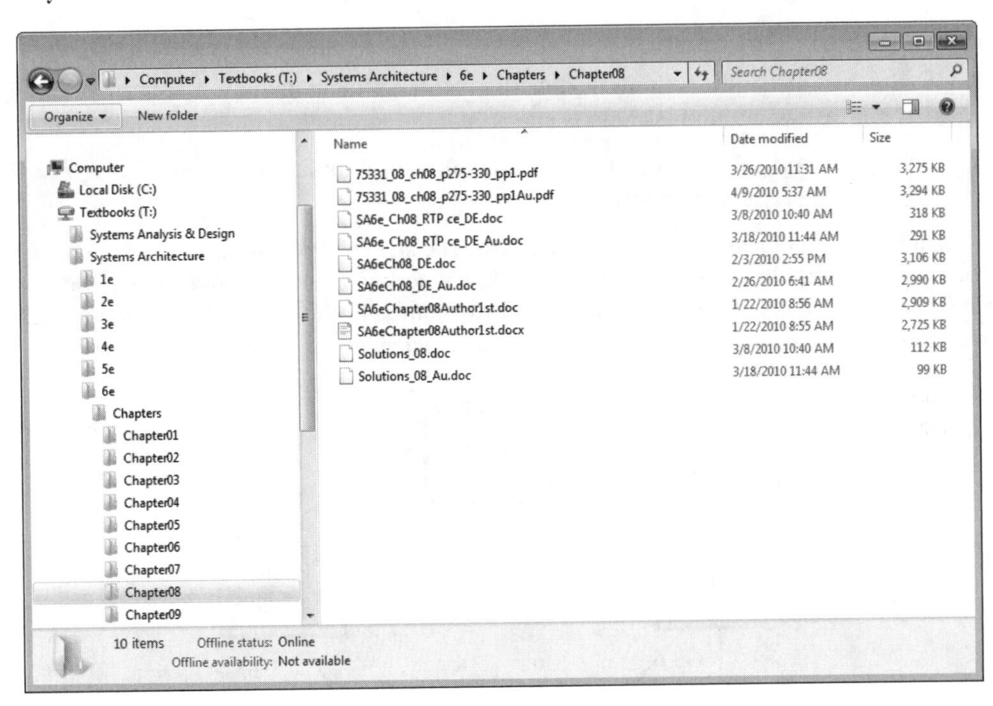

FIGURE 12.6 A hierarchical directory structure
Courtesy of Course Technology/Cengage Learning

The right pane of Figure 12.6 shows the contents of the Chapter08 directory. For each process or user, the OS maintains a pointer to the directory that's currently being accessed. This directory is called the **current directory** or **working directory**. In a multiuser OS, each user normally has a default working directory, or **home directory**. When a user logs on interactively or runs a batch process, his or her home directory is the default current directory. The user or process can change the current directory by issuing a command or making a service call.

In the hierarchy of directories, names of access paths can be specified in two ways. A **complete path**, also called a **fully qualified reference**, begins at the root directory and proceeds through all directories along a path to the file being accessed. Directory names are separated by a special character, such as "\" in Windows or "/" in UNIX/Linux. The fully qualified reference to the current directory is shown without \ characters in the address text box at the top of Figure 12.6:

```
T:\Systems Architecture\6e\Chapters\Chapter08
```

The fully qualified reference to the last file in this directory is as follows:

```
T:\Systems Architecture\6e\Chapters\Chapter08\Solutions_08_Au.doc
```

A **relative path** begins at the current directory's level and extends downward to a specific file. For example, if Chapters is the current directory, the name .\Chapter08\Solutions_08_Au.doc is a relative path to the Solutions_08_Au.doc file in the Chapter08 subdirectory of the current directory, which is indicated with a period.

Graph Directory Structure

A **graph directory structure** is more flexible than a hierarchical directory structure because it relaxes two of the restrictions enforced in a hierarchical directory structure:

- Files and subdirectories can be contained in multiple directories.
- Directory links can form a cycle.

Figure 12.7 shows a graph directory structure. The shaded directory is a pointer to a directory stored elsewhere, as indicated by the dashed arrow. Pointers from one directory to another are called **links** in UNIX and **shortcuts** in Windows. In the figure, the Public Reports directory is referenced from two parent directories: Financial Statements and Accounting.

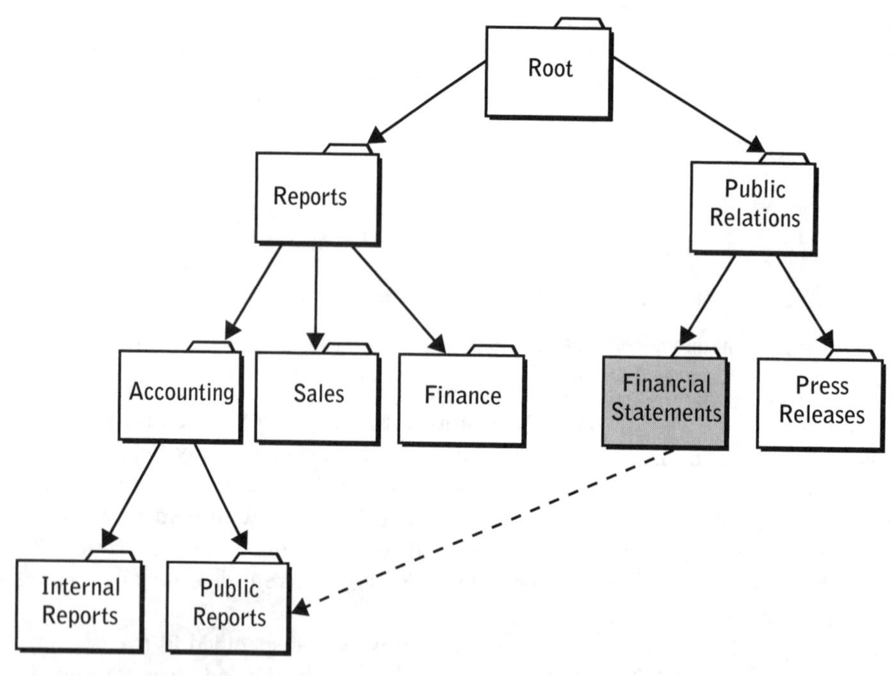

FIGURE 12.7 A graph directory structure
Courtesy of Course Technology/Cengage Learning

Links require special processing to avoid multiple listings and recursive loops in programs, such as backup and synchronization utilities, that list a directory's contents and all its subdirectories. Without special processing, a backup utility would back up all files in the Public Reports directory twice. Also, a program can enter an infinite loop if a directory structure contains a link between a subdirectory and one of its parent directories. To prevent these problems, a program can build a list of directories already accessed and never reread a directory already on this list. It can also be programmed to ignore links and shortcuts, as in most Windows applications.

STORAGE ALLOCATION

The storage I/O control layer allocates secondary storage locations to files and directories. Because there are so many directories, files, devices, and secondary storage locations, storage allocation data structures and the functions for maintaining their content are complex. Data structures and procedures for managing secondary storage allocation to files and directories are similar to those for allocating memory to processes, described in Chapter 11. The most important differences are the number of storage locations, which is much larger for secondary storage, and the frequency of allocation changes, which is much lower for secondary storage.

Allocation Units

An **allocation unit** is the smallest number of secondary storage bytes that can be allocated to a file. Allocation units can't be smaller than the unit of data transfer (normally a block) between the storage device and device controller. The typical block size for secondary storage devices ranges from 512 bytes to 4 KB, in multiples of 512 bytes.

Allocation unit size can be a multiple of block size. For example, a disk might use a 512-byte block size, but the system administrator might decide to use 16 KB allocation units, each containing 32 blocks. Allocation unit size is usually set when an OS or storage device is installed. Some OSs, such as DOS and early Windows versions, set this size automatically based on storage device or media capacity. In other OSs, such as UNIX and later Windows versions, the system administrator selects the unit size when a storage volume is created. After allocation unit size has been set, changing it is difficult.

Allocation unit size is a tradeoff among these factors:

- Efficient use of secondary storage space for files
- Size of storage allocation data structures
- Efficiency of storage allocation procedures

Smaller allocation units result in more efficient use of storage space. For example, a file containing a single byte must be stored in a full allocation unit. If allocation unit size is 512 bytes, 511 bytes are empty and wasted. If the allocation unit size is 16 KB and only 1 byte is stored, 16,383 bytes are empty and wasted. If a storage device holds many small files, a large allocation unit can waste a lot of storage capacity.

The advantage of larger allocation units is that storage allocation data structures can be smaller. As allocation unit size increases, the number of allocation units decreases. For example, in a 100 GB disk, if the unit size is set to 512 bytes, there are 209,715,200 ($100 \times 1024^3 \div 512$) allocation units in the device. If the unit size is set to 16 KB, there are 6,553,600 ($100 \times 1024^3 \div 16{,}384$) allocation units in the device.

Storage Allocation Tables

A **storage allocation table** is a data structure that records which allocation units are free and which belong to files. It contains one entry for each allocation unit. Smaller allocation units increase the number of entries in the table. As this table grows larger, the time required to search and update it increases, which slows down any processing function that creates or deletes a file or changes its size.

Storage allocation table format and content vary across FMSs. Simple OSs, such as early Windows versions, used simple data structures—tables and linked lists, for example. More complex OSs, such as UNIX and later Windows versions, use complex data structures, including bitmaps and B+ trees. This chapter sticks to simpler data structures for examples.

Figure 12.8 shows a hypothetical storage device with 36 allocation units; those shaded the same color belong to the same file or to free space. Table 12.1 shows directory entries for the files stored in this figure. Free allocation units are assigned to SysFree, a hidden system file. A file's allocation units can be stored in any order on the device. The linear address of each file's first allocation unit is stored in the directory, and addresses of other file allocation units are stored in the storage allocation table.

0 1 2 3 4 5
6 7 8 9 10 11
12 13 14 15 16 17
18 19 20 21 22 23
24 25 26 27 28 29
30 31 32 33 34 35

FIGURE 12.8 Storage blocks allocated to three files
Courtesy of Course Technology/Cengage Learning

TABLE 12.1 Directory content for the files in Figure 12.8

Filename	Owner name	First allocation unit	Length (in allocation units)
File1	Smith	0	14
File2	Jones	3	3
File3	Smith	5	9
SysFree	System	2	10

Figure 12.9 shows a storage allocation table for the files in Figure 12.8. Its format and content are nearly identical to the **File Allocation Table (FAT)** file system used in early Windows versions. This table contains an entry for each allocation unit, and a file's allocation units are chained together in sequential order by a series of pointers. The table entry for each allocation unit contains a pointer to the next allocated unit in the file. For example, the entry for unit 0 (the first unit allocated to File1) contains a pointer to the table entry for unit 1 (the second unit allocated to File1). The pointers form a linked list that ties together the entries of all allocation units assigned to a file. The entry for a file's last allocation unit contains a special code ("End," in this example) to indicate that it's the last unit. All unallocated units are linked into a single chain, which simplifies finding storage units to allocate to new or expanded files.

Unit	Pointer	Unit	Pointer	Unit	Pointer	Unit	Pointer
0	01	9	10	18	22	27	32
1	08	10	13	19	20	28	29
2	7	11	12	20	21	29	30
3	04	12	18	21	23	30	34
4	06	13	15	22	28	31	End
5	14	14	16	23	24	32	33
6	End	15	17	24	31	33	35
7	11	16	19	25	26	34	End
8	09	17	25	26	27	35	End

FIGURE 12.9 A storage allocation table matching the storage allocations in Figure 12.8
Courtesy of Course Technology/Cengage Learning

Sequential access to a file's allocation units is efficient when the storage allocation table uses linked lists. However, random access is much less efficient, particularly if the file is large. Some FMSs define a separate file type for random access files and store an index or similar data structure in the file's first allocation unit. The contents of this index are redundant with the storage allocation table's contents, but the index makes locating and accessing specific units of a file more efficient.

Blocking and Buffering

Some application programs access files by logical records. A **logical record** is a collection of data items, or fields, that an application program accesses as a single unit. A **physical record** is the unit of storage transferred between the device controller and memory in a single operation. For disks and other devices using fixed-size data transfer units, a physical record is equivalent to a block. For storage devices with variable-size data transfer units, such as tape drives, block size might differ from physical record size or be undefined.

If the logical record size is less than the physical record size, a single physical record might contain multiple logical records, as shown in Figure 12.10(a). If the logical record size is larger than the physical record size, multiple physical records are needed to hold a single logical record, as shown in Figure 12.10(b). Logical record grouping in physical records is called **blocking**, which is described by a numeric ratio of logical records to physical records called the **blocking factor**. The blocking factor in Figure 12.10(a) is 4:3, and the blocking factor in Figure 12.10(b) is 2:3. If a physical record contains just one logical record, the file is said to be **unblocked**.

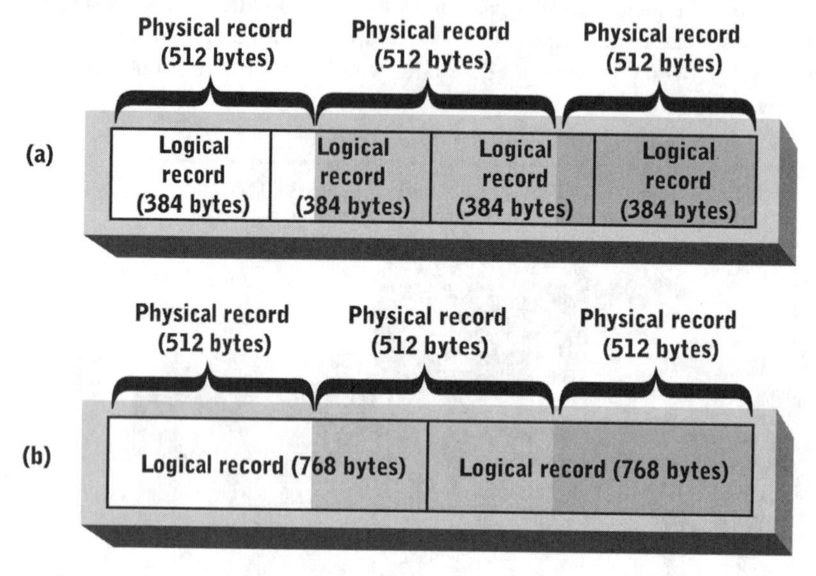

FIGURE 12.10 Blocking logical records
Courtesy of Course Technology/Cengage Learning

Read and write operations to or from an unblocked file can be carried out with simple, efficient algorithms because of the one-to-one correspondence between logical and physical records. File I/O is more complex when logical and physical records have different sizes because the FMS must coordinate physical record I/O and extract logical records on behalf of the requesting program.

The FMS uses buffers in primary storage to store data temporarily as it moves between programs and secondary storage devices. Buffers are allocated automatically when the file is first accessed and managed by the OS on behalf of application programs. Each buffer is the size of one allocation unit. As physical records are read from secondary storage, they're stored in buffers. The FMS extracts logical records from the buffers and copies them to the data area of the application program (see Figure 12.11). This procedure is reversed for write operations.

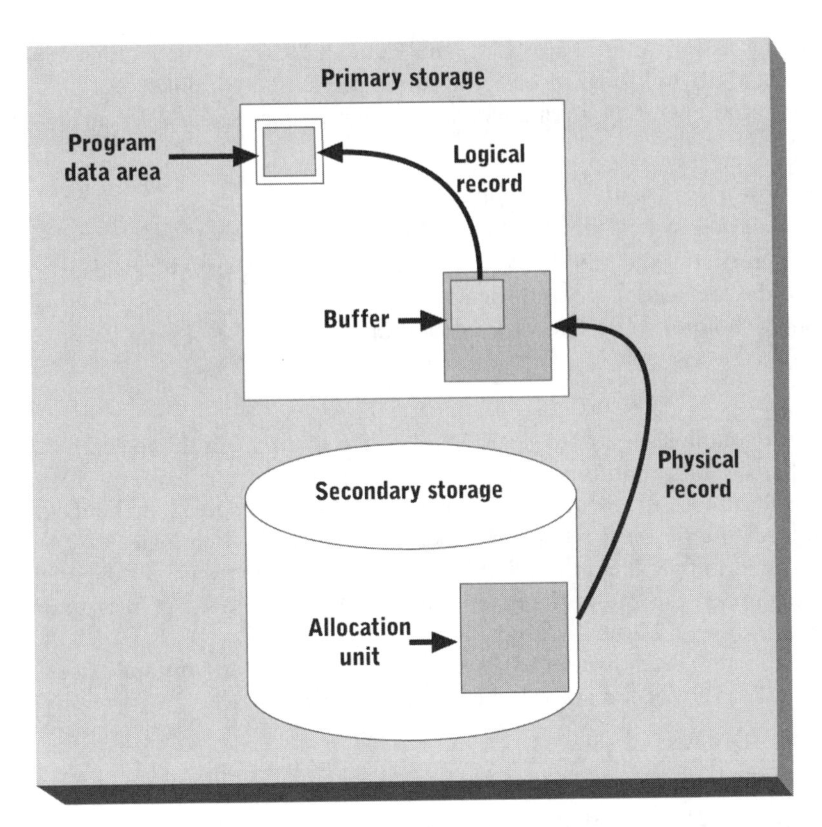

FIGURE 12.11 Input from secondary storage to an application program using a buffer
Courtesy of Course Technology/Cengage Learning

A buffer is a temporary holding area for extracting logical records from physical records. As discussed in Chapter 6, buffering also improves I/O performance if enough buffers are used. For high blocking factors, a small number of buffers can improve performance dramatically. For example, if each physical record contains 10 logical records, reading a single physical record provides enough data for 10 logical read operations. Reading the first logical record results in copying 10 logical records, which is one physical record, into the buffer. The next nine sequential read operations can be satisfied from the buffer without further input from secondary storage.

Low blocking factors, with larger logical records, require more buffers to achieve substantial performance improvements. The FMS usually allocates enough buffers to hold at least one physical record. If enough buffers can't be allocated to hold an entire physical record, a logical record must be moved to the program's data area in a series of physical read operations and buffer-copying operations.

An Example of Storage Allocation and File I/O

The following example is typical of storage allocation and file I/O procedures in a simple FMS. Assume the directory entries in Table 12.1 represent actual files stored on a disk drive, and the allocation unit size used by the disk drive and FMS is 512 bytes. Allocation units are assigned to files as shown in Figure 12.8, and Figure 12.9 shows the storage allocation table. In addition, 55-byte logical records are stored sequentially in File3.

In response to any read operation an application program performs, the FMS carries out the following tasks:

1. Determine which allocation unit contains the requested record.
2. Load this allocation unit into the buffer if it's not already there.
3. Copy the portion of the allocation unit containing the requested logical record to the application program's data area in memory.
4. Increment a pointer to the current position in the file.

Sequential Access

The first allocation unit contains the first logical record or the first part of the logical record, if it's larger than an allocation unit. When the first logical record is read, the FMS looks up File3's first allocation unit (5) in the directory, issues a read request to the disk controller, and then loads the physical record into the buffer. Next, the first logical record's first byte is stored in byte 0, and the FMS copies the first 55 bytes from the buffer to the application program. At the end of the operation, the FMS sets the file pointer to 55 to point to the start of the second logical record.

During subsequent read operations, the FMS calculates the allocation unit and offset for each logical record. For the second record, the calculation is as follows:

$$\frac{(\text{Pointer} - 1) \times \text{Record size}}{\text{Block size}} = \frac{(2 - 1) \times 55}{512} = 0, \text{ remainder } 55$$

The second record begins in allocation unit 0 at byte 55. Because this block is already in the buffer, the FMS copies 55 bytes starting at offset 55 and adds 55 to the file pointer.

Direct Access

Say that the first read operation requests the 37th logical record. Using the same formula, the calculation is as follows:

$$\frac{(\text{Pointer} - 1) \times \text{Record size}}{\text{Block size}} = \frac{(37 - 1) \times 55}{512} = 3, \text{ remainder } 499$$

The 37th record begins in the fourth allocation unit allocated to File3 at byte 499. Allocation unit 3 is the fourth allocation because allocation unit 0 is the first allocation unit.

To find File3's third allocation unit, the FMS follows the chain of entries in the storage allocation table. The first allocated unit (5) is recorded in the directory. File3's second allocation unit is the pointer field of entry 5 in the storage allocation table, which is allocation unit 14. File3's third allocation unit is the pointer field of entry 14 in the table, which is allocation unit 16. File3's fourth allocation unit is the pointer field of entry 16 in the table, which is allocation unit 19. Therefore, the 37th record is stored in allocation unit 19, starting at byte 499.

The FMS loads allocation unit 19 into the buffer and begins transferring 55 bytes to the application program, but it reaches the end of the buffer before it finishes copying 55 bytes. The FMS recognizes this condition, looks in the storage allocation table to determine File3's next allocation unit (20), issues a read request for allocation unit 20, loads it into the buffer, and copies the first 42 bytes to the application program.

FILE MANIPULATION

All FMSs provide service functions in the file control layer to enable application programs to create, copy, move, delete, read, and write files, but the exact functions vary widely across FMSs. Application programs interact directly with the FMS through the OS service layer, and users interact with the FMS indirectly through the command layer.

File Open and Close Operations

The FMS must perform several tasks, collectively called a **file open operation**, before an application program can read or write a file's contents. The application program executes a file open service call to inform the FMS that it intends to read or write a file. In response to the service call, the FMS performs the following steps:

1. Locate the file in the directory structure and read its directory entry.
2. Search an internal table of open files to see whether the file is already open.
3. Ensure that the process has enough privileges to access the file.
4. Allocate one or more buffers.
5. Update an internal table of open files.

The FMS maintains an open file table to prevent application programs from interfering with each other's file I/O activities. If a file is already open for read-only access and another program tries to open it, the FMS normally opens the file but allocates separate buffers to each program. If another program attempts to open the file for writing, the request is normally denied because buffer content is difficult to manage if one program can change content that another program has already read. In some complex FMSs, multiple programs can read or write a file by using complex schemes for locking physical records.

When an application program finishes reading or writing a file, it executes a file close service call. The FMS completes the **file close operation** by performing these steps:

1. Flush the program's file I/O buffers to secondary storage.
2. Deallocate buffer memory.
3. Update the file's directory entry time stamps.
4. Update the open file table.

If the program that issued the request is the only program accessing the file, the FMS deletes the file's entry in the open file table. If other programs are still accessing the file, the FMS deletes the program from the list of programs actively using the file.

Delete and Undelete Operations

In most FMSs, files aren't immediately removed from secondary storage when they're deleted. Instead, the file's storage allocation units are marked as free, and its directory entry is marked as unused. As new files are created or existing files are expanded, the deleted file's allocation units are reassigned to other files and overwritten with new content. Some portion of the deleted file's content remains in secondary storage until all its allocation units have been reassigned and overwritten. When a new file is created in the deleted file's former directory, the deleted file's directory entry is overwritten. Handling file deletion in this way is efficient, but it has two important consequences:

- Files can be undeleted by reconstructing directory and storage allocation table contents.
- File content can be visible to intruders who can bypass the storage allocation table and read allocation units directly.

A user might be able to recover a deleted file by performing an **undelete operation**. For example, if File2 in Figure 12.8 and Table 12.1 is deleted, the second row of the directory table is marked as deleted, and allocation units 3, 4, and 6 are added to the chain of allocation units for SysFree. If the user runs a file recovery utility before performing any other file operations, File2 can be recovered based on the information still in the directory entry and the chained contents of the storage allocation table in Figure 12.9.

NOTE

In some environments, such as law enforcement and defense research, users need to know that a deleted file can never be recovered. In these environments, an FMS is used that can be configured so that directory entries and allocation units are overwritten immediately with blanks. This feature increases security but slows the deletion process, particularly for large files.

ACCESS CONTROLS

Because data is an important organizational resource, an FMS helps prevent loss, corruption, and unauthorized access to files. An FMS relies on the OS to identify and authenticate users and their processes. In OSs that enforce access controls, each user has a unique account name or ID number and must authenticate his or her identity through passwords or other means. After a user's identity is authenticated, the user's account name or ID number is passed to the FMS with every service request.

By default, users are the owners of files they create and can grant or deny access privileges to other users or groups. Different FMSs have different access controls. For example, UNIX defines three access control types:

- *Read*—A user or process can view a file's contents.
- *Write*—A user or process can alter a file's contents or delete it altogether.
- *Execute*—A user or process can execute a file, assuming it contains an executable program or OS commands.

A file owner can reserve access privileges for himself or herself, thereby denying these access privileges to all other users except the system administrator. The file owner can also grant any access privilege to other members of a group or to all users. For example, a file owner might grant read and write access to one workgroup but only read access to other users. A file owner can also revoke his or her own access privileges. For example, a file owner might deny write access to prevent deleting an important file accidentally.

Most FMSs use similar access controls for files and directories. Users are the owners of their home directories and any directories they create below the home directory in a hierarchical directory structure. Access controls for reading (listing directory contents) and writing (altering directory contents) are defined.

Access controls are enforced automatically in FMS service routines that access and manipulate files and directories. Although access controls are a necessary part of file manipulation, they impose additional processing overhead. In some FMSs, the system administrator might choose from several levels of enforcement for file access controls to balance overall FMS performance with the need for file security.

An FMS restricts access to secondary storage devices, storage allocation tables, and root directories to prevent users and processes from bypassing security controls built into FMS service routines. File and directory accesses can also be logged for the system administrator to review later. Enforcing access controls reduces the speed of many file access operations because of the extra processing required.

TECHNOLOGY FOCUS

Windows NTFS

As mentioned, early Microsoft OSs used the FAT file system. When Microsoft developed Windows NT, which evolved into later Windows versions, including Server, Vista, and 7, it decided to develop a new file system: **New Technology File System (NTFS)**. NTFS was targeted to high-performance and mission-critical applications requiring features such as the following:

- High-speed directory and file operations
- Capability to handle large disks, files, and directories
- Secure file and disk content
- Reliability and fault tolerance

NTFS organizes secondary storage as a set of volumes containing storage allocation units called clusters, which can be 512, 1024, 2048, or 4096 bytes. Each cluster is identified by a 64-bit logical cluster number (LCN) in a linear address space. A volume can be as large as 4096×2^{64} bytes.

A volume's master directory is stored in a data structure called the Master File Table (MFT), which contains a sequential set of file records, one for each file in the volume. All volume contents are stored as files, including user files, the MFT itself, and other volume management files, such as the root directory, storage allocation table, bootstrap program, and bad (corrupted) cluster file. The first 16 MFT entries, numbered 0 through 15, are reserved for the MFT and volume management files. All subsequent MFT entries, numbered 16 and higher, store records about user files.

Conceptually, a file is an object with a collection of attributes, including name, global access restrictions (such as read only), and a security descriptor that identifies the owner and owner-defined access controls. A file's data content is just another attribute, although it's usually much larger than other attributes. Each attribute type is assigned a numeric code, and file attributes are stored in ascending numeric code order in the file's

(continued)

MFT record. MFT record size is 1, 2, or 4 KB and is determined by the OS when a volume is formatted.

Each file attribute contains a header and a data value. The header contains the attribute name, a resident flag, the header length, and the length of the attribute value. Attributes can be resident or nonresident. A resident attribute is a short value, stored in an MFT record immediately after the header, for holding information such as access restrictions or filename (see Figure 12.12a).

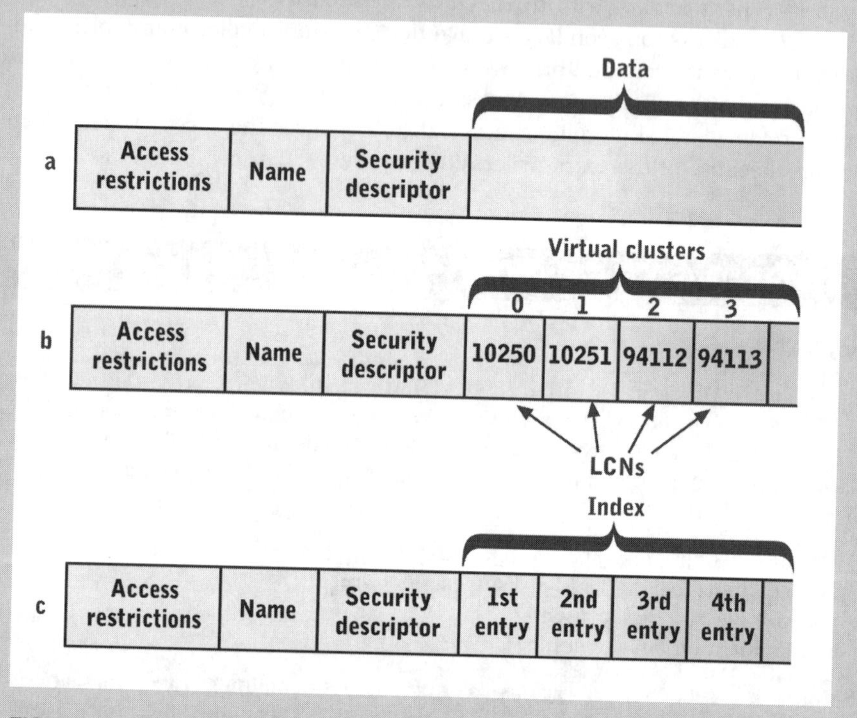

FIGURE 12.12 MFT records for a small file (a), large file (b), and small directory (c)
Courtesy of Course Technology/Cengage Learning

A nonresident attribute is stored in clusters elsewhere in the storage device, and these cluster addresses are stored in the file's MFT record. A file's data content is the most common nonresident attribute, although some usually resident attributes, such as the security descriptor, can grow too large to fit in an MFT record and are also stored in clusters. The clusters assigned to a nonresident attribute are called virtual cluster numbers (VCNs). The area immediately after the nonresident attribute's header stores a sequence of logical cluster numbers (LCNs), as shown in Figure 12.12b. The VCN corresponds to an LCN's position in this sequence. For example, LCN 94112 is VCN 2 in this figure.

The volume root directory and all user-defined directories are stored in the same manner as files—as sequences of attributes in an MFT record. A directory's data attribute

(continued)

contains an index of files it contains, as shown in Figure 12.12c. The index is sorted by filename and contains the file number (equivalent to the MFT record number), time stamps, and size. By duplicating this information in the index, directories can be listed more quickly. The index of small directories is stored sequentially in the MFT record, and larger directories are stored as B+ trees.

File security is implemented through the OS's object manager capabilities. The OS manages files, I/O devices, and many system services as objects. Accessing an object automatically invokes a security subsystem that compares the object's security descriptor against the security descriptor of the accessing process or user. An MFT security descriptor has the same structure and content as security descriptors for other object types.

NTFS has several fault-tolerance features, including redundant storage of critical volume information, mapping of bad clusters, logging of disk changes, and optional RAID (discussed later in "Fault Tolerance"). The MFT is always stored at the beginning of a volume, but a partial second copy is stored in the middle of the disk in case a block assigned to the primary MFT becomes corrupted or unreadable. The FMS detects unreadable blocks during formatting and subsequent read and write operations. Clusters containing bad blocks are marked as unreadable in a separate bad cluster file.

NTFS uses a delayed, or "lazy," write protocol. Disk blocks are cached in memory, and a background process performs cache flushing whenever the disk isn't otherwise busy. Although the actual write to disk might be delayed, the program performing the write operation gets immediate confirmation of its completion. Write operations affecting volume structure, such as file creation, file deletion, and directory modification, are written to the cache and also written immediately to a log file stored on disk. If a system crash happens, the log file's contents are always current and can be used to restore the volume structure to a consistent state.

FILE MIGRATION, BACKUP, AND RECOVERY

Most file management systems have utilities and embedded features, such as the following, to protect files against damage or loss:

- File migration (version control)
- Automatic and manual file backup
- File recovery

Smaller-scale FMSs, such as those in LAN and PC operating systems, typically don't support file migration but do support backup and recovery.

File Migration

When a user alters a file, the original file version is usually overwritten by the new version. However, there are advantages to maintaining the original file version, including allowing "undo" operations and having the original available as a backup. Many commonly used application programs, such as word processors and text editors, save original file copies automatically. For example, Microsoft Word can be configured to create a backup of the original version automatically each time it saves a document file. The most recently saved file is stored with a .doc file extension, and the backup of the original file is stored with a .wbk (for "Word backup") file extension.

Many transaction-processing programs also preserve original versions of input files. For example, the original version of a bank's master account file is usually copied before processing daily batch transactions, such as checks, interest, and monthly fees. The transaction-processing program then reads the master and transaction files and updates the master with the transactions (see Figure 12.13). The original master account file is commonly called the **parent**, and the copy that's updated to reflect new transactions is called the **child**. After another set of transactions is processed, the parent becomes the **grandparent**, the child becomes the parent, and the new copy of the master account file becomes the new child.

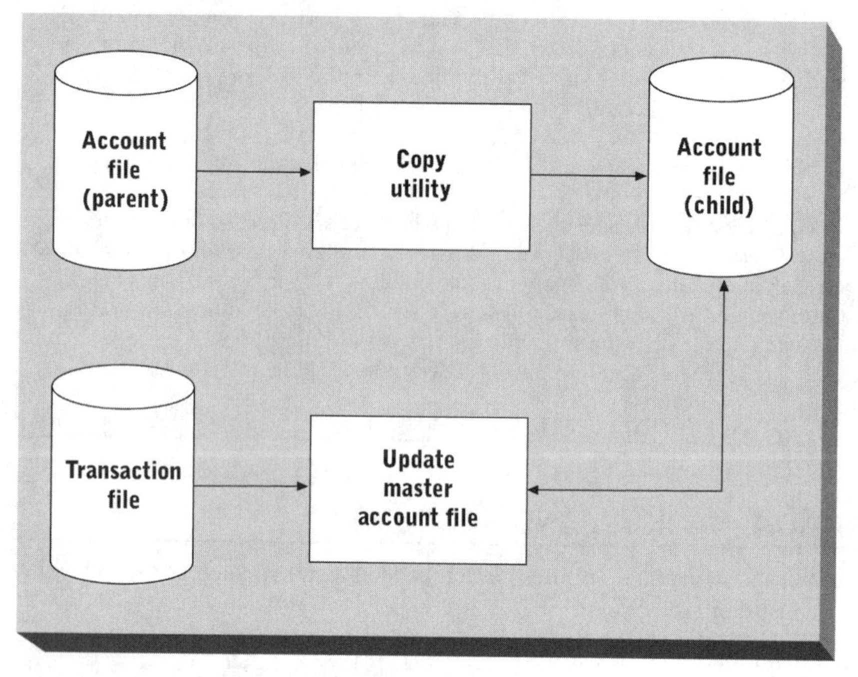

FIGURE 12.13 Updating batch transactions
Courtesy of Course Technology/Cengage Learning

In mainframe FMSs, such as IBM's OS/390, and some third-party file management software, the process of naming and storing original versions of altered files is automated. In a process called **versioning**, a file's original version is archived automatically whenever the file is modified. In these systems, attaching a version number to each filename is typical. When a file is created, it's assigned a version number of 1. The first time it's altered, the modified copy is assigned a version number of 2, and so forth. The FMS generates and stores copies and tracks version numbers. A user or an application program can access a file's older version by referring to its version number.

As files are modified, older versions accumulate on secondary storage and can consume this resource rapidly, especially when files are large or altered frequently. To compensate, the FMS automatically performs **file migration**, a file management technique that balances each file version's storage cost with anticipated user demand for this version. As a file version becomes outdated, the probability of access decreases. The FMS migrates (moves) older file versions from local disks to remote storage media and eventually to offline backup storage, such as tape.

A simpler file migration method is based on file age rather than file version. The FMS periodically checks each file's last access time stamp and migrates older files. The file still appears in directory listings and is reloaded automatically from remote or backup storage if it's accessed. This method is often used in FMSs that manage infrequently accessed files for many users and can be coupled with robotic devices to access offline tape or removable optical disc volumes.

File Backup

Most FMSs include general-purpose utilities to create backup copies of files and directories on removable or remote storage media. Backups protect against data loss caused by storage device or media failure, accidental erasure, malicious attack, and so forth. Backup utilities can be run manually or automatically at set periods and protect file content, directory content, and storage allocation tables.

Backup copies should be stored on a separate storage device to prevent losing both the original and backup in case of total device failure. For backups created on removable media, the media should be stored in a different physical location. Large computer centers typically store backup copies in a separate building or site to minimize the probability of a disaster destroying both copies, such as a fire that destroys an entire building. Backup copies can be transmitted to remote storage facilities via a high-speed network, eliminating the cost, delay, and risks of physically transporting storage media.

Types of periodic backups include the following:

- Full backup
- Incremental backup
- Differential backup

When a **full backup** is performed, the FMS copies all files and directories for an entire storage volume. As each file is backed up, the backup utility usually modifies the file's directory entry to set the time of the last backup to the current time. A full backup can also include storage allocation tables, partition tables, and other important disk management data structures. This backup type is time consuming because of the large number of files copied and the slow write speeds of most backup storage devices. It's usually done during off-peak hours at long time intervals, such as weekly.

An **incremental backup** archives only files that have been modified since the previous incremental or full backup. To make these operations possible, the FMS must keep track of when backups are performed and when files are modified. The backup utility compares the most recent update and backup time for each file and directory. Only files and directories modified since their last backup are copied to backup storage, and their backup times are reset to the current time. Incremental backups are usually much faster than full backups because many files aren't modified frequently.

A **differential backup** is a variation on an incremental backup in which backup times aren't reset as files are copied. As a result, subsequent differential backups take longer because they include files that wouldn't be backed up if the backup time stamp had been reset during the previous backup. The main advantage of this backup type is faster recovery time after storage device failure. With a differential backup, data is fully recovered by restoring the most recent full backup and then the most recent incremental backup. With incremental backups, full recovery is accomplished by restoring the most recent full backup and then restoring *all* subsequent incremental backups in the order they were created.

Most large-scale FMSs use all the backup types. For example, for smaller data sets, incremental backups might be created at the end of each business day with full backups created each weekend. For larger or more important data sets, differential backups might be created at the end of each business day with full backups created twice per week.

Transaction Logging

Transaction logging, also called **journaling**, is a form of automated file backup. The term "transaction" in this context shouldn't be confused with the more generic meaning of a business transaction, such as a customer purchase. To an FMS, a **transaction** is any change to file contents or attributes, such as an added record, a modified field, or changed access controls. In an FMS that supports transaction logging, all changes to file content and attributes are recorded automatically in a log file in addition to being written to the file's I/O buffer. Log entries are written immediately or frequently to a separate physical storage device.

Transaction logging provides a high degree of protection against data loss caused by program or hardware failure. When an entire computer system fails, the contents of file I/O buffers are lost. If these buffers weren't written to physical storage before the failure, file content becomes corrupted and content changes are lost. With transaction logging, the FMS can recover most or all of the lost changes and repair corrupted files. When the system is restarted, the transaction log's contents are reviewed and compared with the file content on disk. Lost updates are identified and written to the files.

Transaction logging imposes a performance penalty because every file change requires two write operations: one to the data file and another to the log file. It's typically used only when the costs of data loss are high, as for large-scale e-commerce sites.

File Recovery

Backup procedures and utilities must be supplemented by reliable recovery procedures to form a complete file protection mechanism. Typically, recovery procedures have both automated and manual components. For example, transaction log replay and subsequent file repair is usually fully automated. Recovery procedures based on full or incremental backups stored on removable media usually rely on manual procedures to some degree.

The FMS maintains backup logs to aid in locating backup copies of lost or damaged files. Recovery utilities can search these logs for particular files or groups of files. Backup logs record the storage device or medium holding backup copies. At the time of the backup, the backup utility writes an ID number or code to the backup storage medium and sometimes to a label that's manually applied to the medium. The system administrator can locate the medium and mount this external label on the correct device. The recovery utility reads the embedded ID number or code to verify that the correct medium has been mounted before beginning recovery operations.

Recovery procedures for a crashed system or physically damaged storage device are usually more sophisticated and highly automated. Damage might have occurred to files, directories, storage allocation tables, and other important disk management data structures. The recovery utility reconstructs as much of the directory and storage allocation data structures as possible and makes a consistency check to ensure the following:

- All storage locations appear in the storage allocation table and other data structures.
- All files have correct directory entries.
- All storage locations of a file can be accessed through the storage allocation table.
- All storage locations can be read and/or written.

Consistency checking and repair procedures consume anywhere from a few minutes to several hours, but they reduce the need to do large amounts of data recovery from backup copies and minimize the amount of current data that's lost. They can also eliminate the need to reinstall system and application software.

FAULT TOLERANCE

As applied to FMSs, **fault tolerance** describes methods of securing file content against hardware failure. Magnetic and optical drives are complex devices with many mechanical parts. To improve performance, manufacturers use high spin rates and small distances between read/write heads and recording media. The result is devices that provide high performance but at the cost of occasional catastrophic failure.

Common causes of disk failure include head crashes, which are contacts between a read/write head and a spinning platter, and burned-out motors and bearings. Repairing a failed disk drive is prohibitively expensive because of the nature of manufacturing methods. Most enterprise disks have 5-year warranties, and manufacturer claims for mean time between failures (MTBF) range as high as 180 years, but the large number of disks in use guarantees that some failures will occur.

File backup, recovery, and transaction logging are forms of protection against disk failure, but they require time to carry out, and data is unavailable to users during recovery operations. In many processing environments, occasional downtime for file recovery is acceptable. In others, such as banking, retail sales, e-commerce, and production monitoring and control, downtime is unacceptable or expensive. In general, any business or organization performing continuous updates and queries against files and databases is a candidate for advanced methods of fault tolerance, such as mirroring or RAID.

Mirroring

Disk mirroring is a fault-tolerance technique in which all disk write operations are made simultaneously or concurrently to two storage devices. In some cases, the two devices might be in different cabinets, rooms, or buildings. If one device fails, the other device contains a duplicate of all data. Data is available continuously because either device can respond to a read request.

The FMS can perform disk mirroring by being configured to make duplicate writes to duplicate storage devices, but this method can reduce system performance substantially.

Software-based mirroring is required if duplicate disks aren't attached to the same disk controller. When duplicate disks are in the same cabinet, mirroring is usually implemented in hardware by the device controller, which reduces CPU and system bus overhead. Multiple disk drives are attached to the controller, and the controller duplicates write operations automatically to each drive. Read operations are split between the drives to improve performance. Special utilities are used to configure the disk controller for mirroring and to initialize a new duplicate drive if a failure occurs.

Disk mirroring provides a high degree of protection against data loss with no performance penalty if it's implemented in hardware. The primary disadvantages of mirroring are doubling storage device cost and the higher cost of disk controllers that perform mirroring.

RAID

Redundant array of independent disks (RAID) is a disk storage technique that improves performance and fault tolerance. The original RAID version, now known as RAID 0, was developed at the University of California, Berkeley, in the late 1980s. RAID has evolved considerably since then, and many products are now available commercially. A flurry of RAID development in the early 1990s resulted in many incompatible approaches and products, and the RAID Advisory Board (RAB) was formed in 1992 to define standard methods of implementing RAID levels through 5 (see Table 12.2). The RAB disbanded in the mid-2000s, so RAID techniques developed since then (such as RAID 6 and 10) can vary across software and hardware providers. RAID levels 2 through 4 are rarely used.

TABLE 12.2 RAID levels

Level	Description
0	Data striping without redundancy
1	Mirroring
2	Data bit striping with multiple error checksums
3	Data byte striping with parity check data stored on a separate disk
4	Data block striping with parity check data stored on a separate disk
5	Data block striping with parity check data stored on multiple disks
6	Data block striping with two sets of parity check data stored on multiple disks
10	Data striping combined with mirroring

RAID 1 is disk mirroring, described in the previous section. All other RAID levels use some form of **data striping**, which breaks a unit of data into smaller segments and stores these segments on multiple disks. For example, a 16 KB block of data can be divided into four segments, with each 4 KB segment written in parallel to a separate disk (see Figure 12.14). A subsequent read of the original 16 KB block accesses all four disks in parallel.

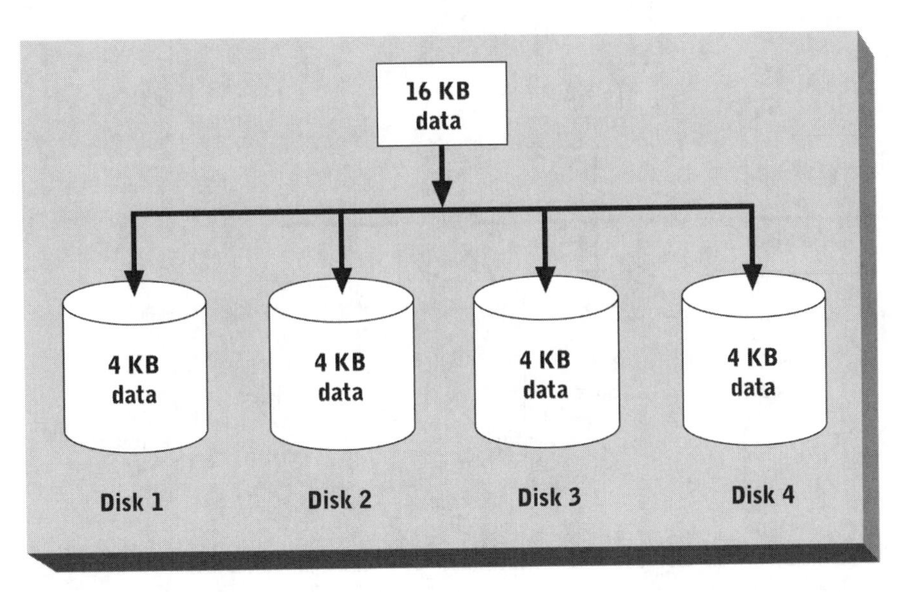

FIGURE 12.14 Data striping across four disks
Courtesy of Course Technology/Cengage Learning

Data striping improves read performance by breaking a large read operation into smaller parallel read operations. The elapsed time to perform an entire read or write operation is reduced because multiple disks can perform small parallel operations faster than a single disk can perform one large operation. However, overhead is incurred to disassemble and reassemble data segments and to issue read or write requests to multiple disks.

RAID 1 through 6 achieve fault tolerance by generating and storing redundant data during each write operation. RAID 3 through 6 generate parity bits for data bytes or blocks and store them on one or more of the RAID disks. If a single disk fails, no data is lost because missing bits can be reconstructed from the data and parity bits on the remaining disks, as described in Chapter 8.

In RAID 3 and 4, parity information is stored on a dedicated disk (see Figure 12.15). In RAID 5 and 6, parity information is distributed across all disks in a round-robin fashion. RAID 5 and 6 have slightly better performance during the period between a disk failure and regeneration of its content to a new disk. With RAID 3 and 4, failure of any disk except the parity disk requires recomputing lost data bits from the remaining data and the parity data for all read operations. With RAID 5 and 6, some read operations don't require parity computations. For example, with five disks, RAID 5 stores 20% of the parity information on each disk. If a single disk fails, 80% of subsequent read operations require parity computations, but the other 20% don't.

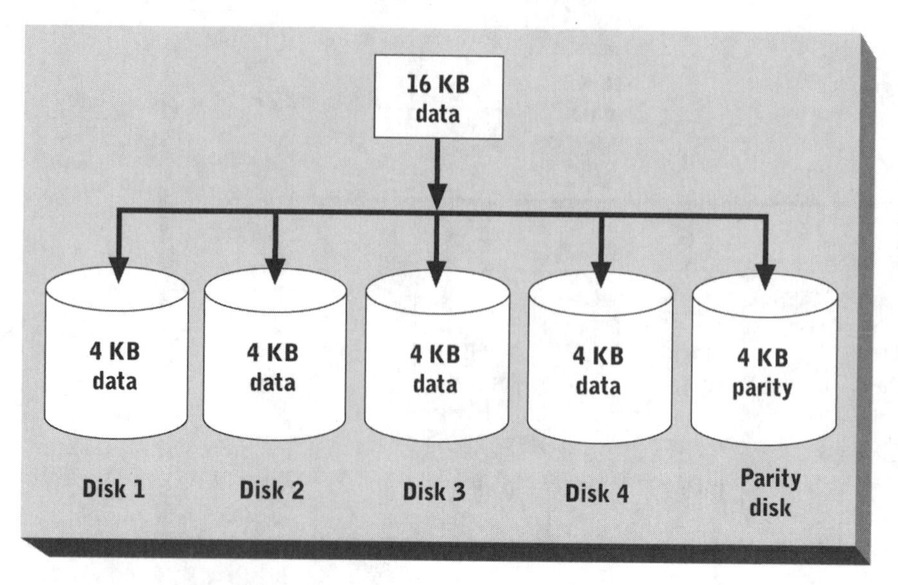

FIGURE 12.15 A RAID 3 or 4 write operation to multiple disks
Courtesy of Course Technology/Cengage Learning

Storing parity data reduces a disk array's usable capacity. The proportional reduction depends on the number of disks in the array. If five disks are used, 20% of the available disk space stores redundant parity information, and the effective data storage capacity is 80% of the raw storage capacity. Using more disks would decrease the portion of space used for parity bits but with a slight increase in the probability of data loss in the event of multiple disk failures.

Multiple RAID levels can be layered to combine their best features. The most common example is RAID 1+0, also called RAID 10. RAID 10 mirrors each disk (RAID 1) and then stripes data (RAID 0) across multiple mirrored pairs (see Figure 12.16). Because striping requires at least two disks and each disk is mirrored, RAID 10 requires at least four disks. For RAID 10 with four disks, read and write performance is improved by up to 100%, and the system can recover from the failure of any single disk or two disks if they're in different mirror pairs.

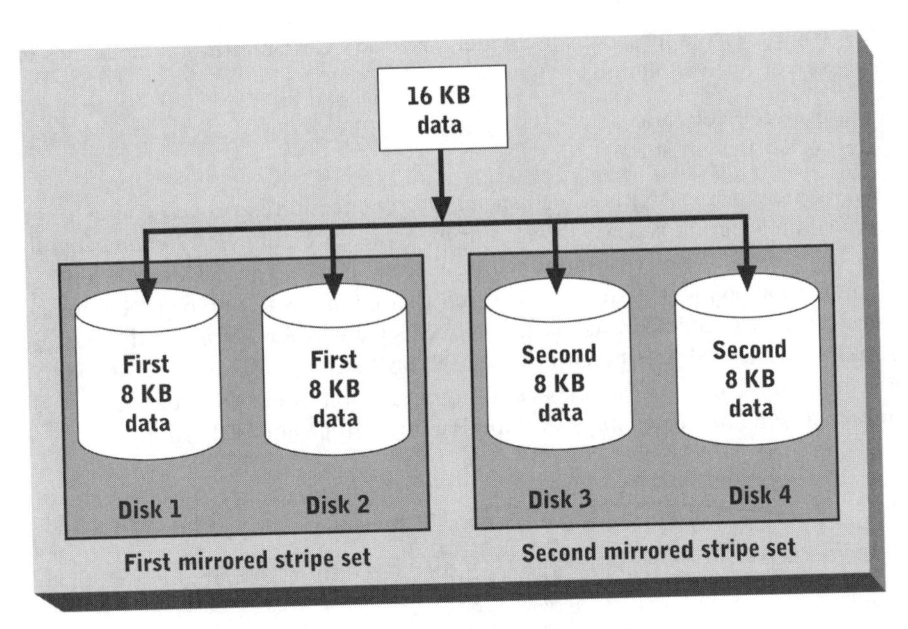

FIGURE 12.16 16 KB stored in a four-disk RAID 10 array
Courtesy of Course Technology/Cengage Learning

RAID can be implemented in software or hardware. Hardware systems are common for a number of reasons, including the following:

- The ability to configure all RAID components in a single cabinet
- Hardware fault tolerance extended with redundant power supplies and disk controllers
- Reduced load on the host CPU
- Reduced system software complexity

A RAID storage device looks like a single large disk drive to an OS. A dedicated controller performs all RAID-related processing, including segmenting read and write operations, generating parity data, reconstructing missing data if a disk fails, and repopulating data to a replacement disk. Hardware-based RAID systems for LAN and small WAN servers are typically based on a SCSI standard. RAID systems for larger computer systems usually use other high-capacity communication channels, such as Fibre Channel.

STORAGE CONSOLIDATION

The traditional model of storage access by application software relies on an approach commonly called **direct-attached storage (DAS)**. DAS describes any architecture in which software running on a CPU accesses secondary storage devices in the same computer. It's an efficient approach to storage access when a single computer interacts with a single storage subsystem. However, DAS can be expensive and inefficient for organizations with dozens or hundreds of servers and terabytes of shared data. In this environment, storage overlap between servers can be substantial, resulting in high costs and redundant updates of multiple data copies. Two approaches are widely used to overcome the inefficiencies of DAS in multiple-server environments:

- Storage area network
- Network-attached storage

A **storage area network (SAN)** is a high-speed interconnection between general-purpose servers and a separate storage server. For example, in Figure 12.17 each general-purpose server has a device controller attached to its system bus. The device controller attaches via an external connection to a SAN switch that connects to one or more storage servers. A storage server accepts storage access requests from other servers and accesses embedded storage devices on their behalf. As in DAS, SAN storage accesses are at the level of disk sectors in a logical address space. Communication in a SAN is based on a high-speed protocol, such as Fibre Channel, InfiniBand, or 10 Gigabit Ethernet.

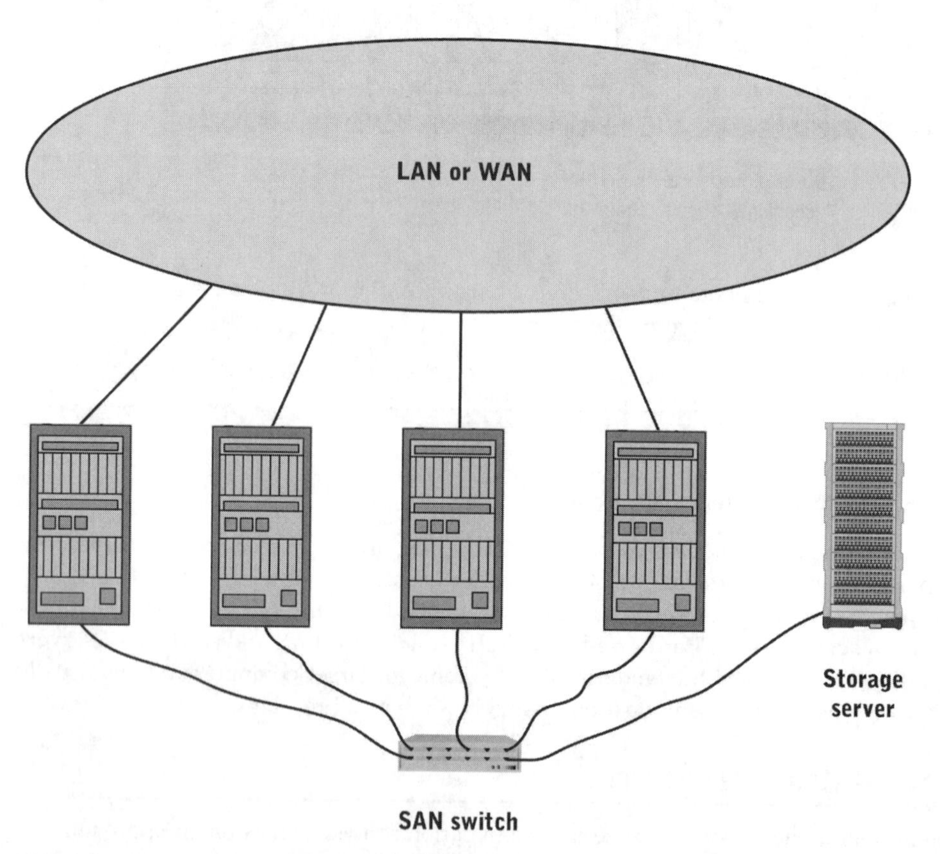

FIGURE 12.17 A server cluster with a storage area network
Courtesy of Course Technology/Cengage Learning

The storage server in a SAN isn't a complete general-purpose computer, although it might have a CPU and a limited-purpose OS. In essence, it acts as a disk controller for each server, translating accesses in a logical address space into physical accesses to one or more disk drives. However, the storage server must also handle the complexities of shared resource access. For example, if one general-purpose server asks to read a storage location while another server is writing to the same location, the storage server must queue the read request until the write request is completed.

The term **network-attached storage (NAS)** describes any architecture with a dedicated storage server attached to a general-purpose network to handle storage access requests from other servers. Figure 12.18 shows an NAS server with four application and Web servers attached via a LAN or WAN. An NAS server can be a general-purpose server customized for storage applications or a limited-purpose server, sometimes called a "server appliance." In either case, an NAS server has all the hardware attributes of a complete computer system, including a CPU, memory, system bus, storage subsystem, and network I/O devices. It also has an OS that can manage its hardware resources and respond to storage service requests from other servers.

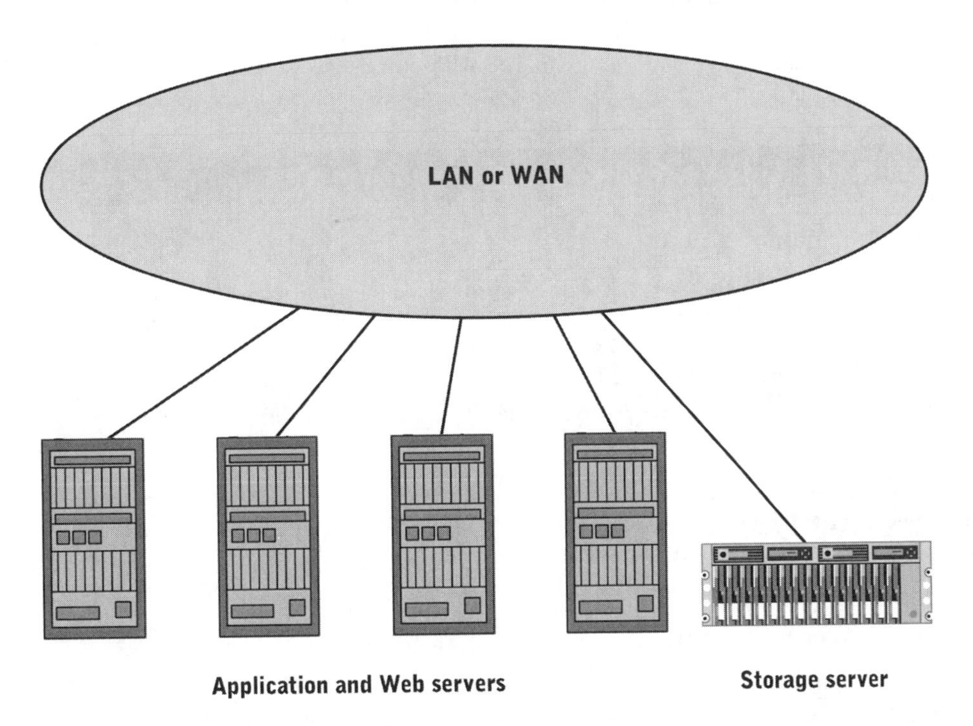

FIGURE 12.18 Network-attached storage
Courtesy of Course Technology/Cengage Learning

A key distinction between SAN devices and NAS servers is the type of storage access requests that are serviced. A SAN device accepts low-level requests to locations in the storage subsystem's logical address space. This type of access is sometimes called "block-oriented access" or "sector-oriented access." In contrast, an NAS server accepts access requests to files, which can encompass many storage blocks. This type of access is usually called "file-oriented access." In NAS, a storage server OS manages one or more file systems shared by other servers or clients. In a SAN, a storage server provides a pool of physical storage locations. Other servers use these locations to store one or more file systems.

SANs and NASs have unique advantages, although many organizations use a combination of both approaches. SANs are most common in multiserver environments with mainframes or supercomputers and substantial overlap in server storage needs. Clusters commonly use SANs to share data and system software among many identical computers. SANs are expensive to purchase and administer, but they avoid the costs of duplicate storage and storage administration.

NAS is commonly used when geographically dispersed servers need access to a common file system. One example is a shared file system accessed by servers and clients spread across a university or corporate campus. NAS is much cheaper to acquire and administer than a SAN, but the cost savings come at the price of lower performance. The reason is that NAS connections are ordinary LAN or WAN connections, which are usually slower and more congested than SAN interfaces, which are high-speed connections dedicated to storage access. NAS servers also have the additional task of file system management, which slows their response time compared with storage servers in a SAN.

TECHNOLOGY FOCUS

Google File System

Traditional FMSs embedded in widely used OSs, such as Linux and Windows, are designed to support many kinds of applications. Most organizations achieve acceptable levels of application performance and support with these FMSs. However, some organizations have applications that aren't well matched to traditional FMSs, which leads to a difficult decision—tolerate the suboptimal performance and application support or invest the resources to develop a customized FMS that better matches application requirements.

In its early years, Google recognized that its FMS needs were quite different from other organizations and poorly matched to traditional FMS capabilities. Because Google executives anticipated rapid growth in data storage requirements, they were concerned that suboptimal FMS performance for their applications would be a continuing problem. They decided to develop **Google File System (GFS)**, an FMS specifically matched to their needs. Key assumptions underlying GFS led to important differences from traditional FMSs:

- Scalability to petabyte storage
- Large files, ranging from hundreds of megabytes to dozens of gigabytes
- Data storage on a distributed collection of commodity servers
- Simultaneous file access by multiple distributed applications

Because traditional FMSs support files of widely varying sizes, their allocation units are generally small (512 KB to 4 MB) for efficient storage allocation and I/O with smaller files. However, most files stored in GFS are 100 MB or larger. In addition, most Google

(continued)

applications read entire files sequentially. These characteristics, combined with the need to scale to petabyte storage, led GFS designers to choose 64 MB as the standard unit of storage allocation (called a "chunk"). As a result, storage allocation tables are smaller and can fit entirely in memory, thus improving file server performance substantially.

Large allocation unit size can result in poor I/O performance when applications interact with many small files or small portions of larger files. For example, with a 64 MB block size, an application such as a Web server retrieving kilobyte-sized pages would access large quantities of data needlessly. However, few Google applications perform I/O on random locations in files, so large storage allocation units don't incur the performance penalties associated with a poor match between allocation unit size and typical patterns of application storage access.

Given Google's growth expectations, it was clear from the beginning that relying on a few large file servers wasn't feasible. Instead, data storage would require thousands or more servers that needed to be distributed geographically and built with inexpensive components to control cost. Therefore, failure of disks, disk arrays, entire servers, and network connections was an important design issue. GFS has to tolerate these failures without a major impact on performance or data availability.

To address this need, GFS uses a unique approach to distribution. A single server cluster (the GFS master) stores all directory, storage allocation, and replication information, and a distributed network of "chunk servers" running Linux stores file content (see Figure 12.19). When an application opens a file, it interacts with the GFS master, which transfers much of the file's storage allocation and replication data to the application. Most of the application's subsequent file accesses go directly to the chunk servers holding the necessary data. Instead of implementing fault tolerance at the disk array level, GFS routinely duplicates entire files across multiple chunk servers.

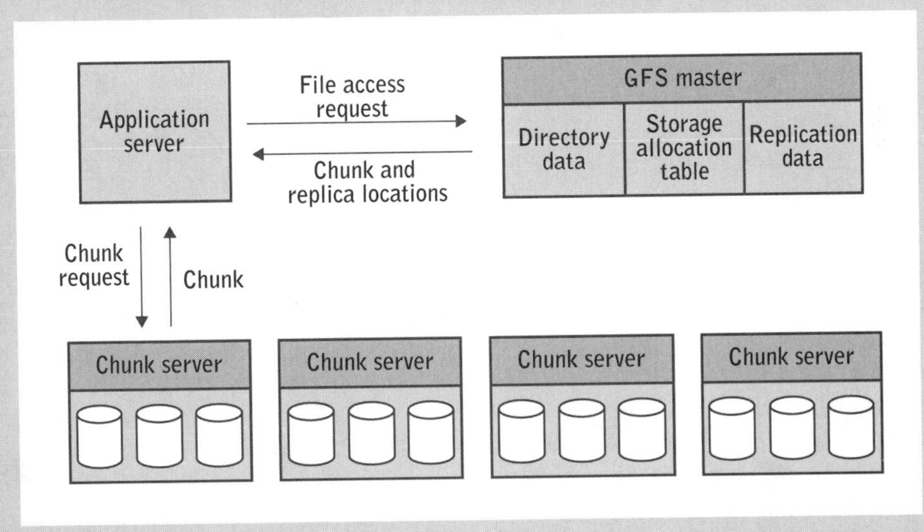

FIGURE 12.19 Interaction between an application server and the GFS master and chunk servers

Courtesy of Course Technology/Cengage Learning

(continued)

Unlike traditional FMSs, applications that interact with GFS "see" storage allocation and replication information and interact directly with underlying allocation units via the chunk servers. This approach reduces the I/O load on the GFS master at the expense of more complex application programming. In the world of general-purpose OSs and application software, hiding FMS complexity and internals from the applications is the norm. However, because Google develops all its own applications, it can choose to make more FMS details available to applications with little concern about increased application expense and complexity.

The use of a single GFS master is a potential performance bottleneck and a single point of failure. Most large-scale FMSs use a multimaster approach that's more fault tolerant but requires complex processes for distributing and managing updates to directory, storage allocation, and replication information. The choice of a single GFS master was for expediency. In the late 1990s, Google wanted to develop and deploy GFS as rapidly as possible. Google estimates that bypassing the complexity of a "true" multimaster FMS reduced development and deployment time by at least a year. However, as Google's data storage needs continue to grow, having a single GFS master looms larger as a possible impediment to performance and fault tolerance.

Summary

- A file management system (FMS), usually part of the OS, manages all aspects of user and program access to secondary storage. The FMS presents a logical view of stored data and programs to users as files organized into directories and storage volumes. As users and programs manipulate the logical view, the FMS translates these operations into commands to physical storage devices. FMSs support only a few file types directly, including executable programs, executable OS command files, and user data files. In current FMSs, users and programs can define new file types and have programs for manipulating these files associated with the file types.
- Directories enable users to organize the many files stored in a typical computer. Each storage device or volume has a root directory. In hierarchical directory structures, directories can contain other directories, creating a tree structure in which each file belongs to only one directory. In graph directory structures, files can belong to more than one directory, and there's the possibility of loops or cycles in the directory structure. Directories store descriptive information about files and directories, such as name, owner, file type, access controls, and time stamps.
- Secondary storage devices are divided into allocation units, typically a few kilobytes in size. The FMS assigns allocation units to files and directories as they're created or expanded and reclaims allocation units as files and directories shrink or are deleted. The FMS uses data structures called storage allocation tables to track the assignment of allocation units to files and directories. Entries in these tables can be linked lists in simple FMSs; in more complex FMSs, they can be indexes or other complex data structures.
- The FMS allocates buffers to support program file I/O. A program executes a file open service call before reading or writing a file, which causes the FMS to find the file, verify access privileges, allocate buffers, and update the open file table. Multiple programs can open the same file for reading, but in most FMSs, only one program can open a file for writing. When a program is finished with a file, it executes a file close service call, which causes the FMS to flush buffer content to the storage device, release the buffers, update the file's time stamps, and update the open file table.
- The FMS enforces access controls when accessing files on behalf of a user or program. File owners and system administrators can grant or deny access privileges for reading, writing, and executing files. When a user or program attempts to access a file, the FMS checks the account name or ID number against the access controls stored in the file's directory entry to determine whether access is permitted. Enforcing access controls provides security but increases FMS overhead.
- FMSs include utilities to make backup copies of files and directories and to recover them if needed. Backups can be performed manually or be fully automated, and backup types include full, incremental, and differential. Some FMSs support automatic storage and backup of old file versions with file migration. As file contents change, the original versions are archived and migrated from local disks to remote storage media and eventually to offline storage as they become further out of date.
- Fault tolerance describes methods of securing file content against hardware failure. One method is disk mirroring, in which all disk write operations are duplicated to two storage

devices. Servers and some workstations use RAID technology to improve performance and fault tolerance. RAID improves performance by striping data across multiple disks and reading or writing small blocks of data in parallel. It improves fault tolerance through disk mirroring, storing redundant parity data on one or more disks, or both. There are multiple levels of RAID, each providing a unique mix of fault tolerance and performance improvement.

- Organizations that store large amounts of data typically use some form of consolidated storage. A storage area network (SAN) is a high-speed interconnection between general-purpose servers and one or more storage servers. Network-attached storage (NAS) servers are dedicated to managing one or more file systems and are accessed by other servers and clients over a LAN or WAN.

This chapter and Chapter 11 have covered allocation and management of the CPU, primary storage, and secondary storage. In Chapter 13, you examine how users and applications interact with external resources, including files, I/O devices, and programs.

Key Terms

allocation unit
blocking
blocking factor
child
complete path
current directory
data striping
differential backup
direct-attached storage (DAS)
directory
disk mirroring
fault tolerance
fields
File Allocation Table (FAT)
file association
file close operation
file control layer
file management system (FMS)
file migration
file open operation
full backup
fully qualified reference
Google File System (GFS)
grandparent
graph directory structure
hierarchical directory structure
home directory
incremental backup
journaling
links
logical record
network-attached storage (NAS)
New Technology File System (NTFS)
parent
physical record
redundant array of independent disks (RAID)
relative path
shortcuts
storage allocation table
storage area network (SAN)
storage I/O control layer
transaction
transaction logging
tree directory structure
unblocked
undelete operation
versioning
volume
working directory

Vocabulary Exercises

1. A(n) _______ is the unit of file I/O accessed by an application program as a single unit. A(n) _______ is the unit of storage transferred between the device controller and memory in a file I/O operation.
2. The term _______ describes the ratio of logical records to physical records.
3. A(n) _______ operation releases allocated buffers and flushes their content to secondary storage.
4. A(n) _______ operation allocates buffers for file I/O and updates a table of files in use.
5. The content of a logically, but not physically, deleted file can be recovered in a(n) _______ operation.
6. _______ describes tracking old file versions and moving them to archival and offline storage devices.
7. In _______, changes to files are written to a log file as they're made.
8. In an FMS, the _______ layer processes service calls from the command layer or application program. The _______ layer manages movement of data between storage devices and memory.
9. A(n) _______ specifies all directories leading to a specific file. A(n) _______ specifies file location based on the current or working directory.
10. In a(n) _______ directory structure, a file can be located in only one directory. This restriction doesn't apply in a(n) _______ directory structure.
11. MS-DOS and some Windows versions record storage allocation information in the _______ file system.
12. An FMS can implement _______ with disk mirroring or _______.
13. A(n) _______ records the assignment of storage locations to files.
14. When an old version of a master file is saved, the current version can be called the _______, the previous version is the _______, and the version before that is the _______.
15. RAID 10 combines disk mirroring and _______ to achieve fault tolerance and improve performance.
16. In a(n) _______, multiple servers share access to the same storage server over a high-speed dedicated network.
17. In _______, a storage server manages one or more file systems and responds to file I/O requests sent across a LAN or WAN.

Review Questions

1. List the FMS layers and describe their functions.
2. What's the difference between the logical and physical structure of a file? What are the advantages of not having an application program interact directly with the physical file structure?

3. What file types does a file management system usually support?
4. What is an allocation unit? What are the advantages of using small allocation units? What are the disadvantages?
5. Describe the use of buffers in file I/O operations. When are buffers allocated? When are they released?
6. Describe a hierarchical directory structure. What are its advantages and disadvantages compared with a graph directory structure?
7. How is file deletion normally accomplished? What security problems might result from this method?
8. What levels of access rights can exist for a file?
9. What is transaction logging or journaling? Describe the performance penalty it imposes on file update operations.
10. Describe the levels of RAID. What are their comparative advantages and disadvantages?
11. Compare storage area networks and network-attached storage. Which is more common in environments where many servers in the same location access the same data?

Problems and Exercises

1. Modify the directory content in Table 12.1 and the storage allocation table in Figure 12.9 to store a new file containing seven allocation units.
2. Assume the first character of the filename in a deleted file's directory entry is overwritten with an ASCII 0 to mark the file as deleted. Write a step-by-step procedure for undeleting a deleted file, assuming no file operations have been performed since the deletion.
3. You have six 1 TB disks for assembling a RAID storage array and are considering RAID levels 0, 5, 6, and 10 for the array. For each level, what's the effective data storage capacity, and how many disks can fail before data is lost permanently? Which level would you recommend for an array holding data for a Web site that's changed infrequently and backed up daily? Which would you recommend for an array holding transaction data for an e-commerce site that's backed up daily?

Research Problems

1. Google has two customized FMSs: GFS, described in the chapter, and a newer file system, called BigTable, that supports applications such as Google Earth and Google Finance. Investigate BigTable to determine how it's implemented and what application services it provides. What are the unique characteristics of applications using BigTable, and how is it optimized for these applications?
2. Investigate the SAN and NAS products of a major computer vendor, such as IBM, Hewlett-Packard, or Dell. What are the approximate costs of configuring each device to store 2 TB of data and respond to requests from eight other servers? Which one has higher storage access performance? Which one is easier to configure and administer? Which is best suited to file sharing for fewer than a dozen computers in a home or small business network? Which is best suited to handle storage duties in a high-performance computing cluster?

CHAPTER **13**

INTERNET AND DISTRIBUTED APPLICATION SERVICES

CHAPTER GOALS

- Describe types of distributed software architecture and discuss their advantages compared with centralized applications
- Explain how operating systems and network protocol stacks cooperate so that users and programs can access remote resources
- Explain the role and function of directory services and the LDAP standard
- Describe low-level protocols for interprocess communication across networks
- Describe standard Internet protocols for accessing distributed resources
- Discuss component-based application development and describe the protocols and standards that support it
- Describe cloud computing models and compare their economic benefits and risks

Users of computers and information systems interact with a variety of resources located on computer systems all over the world. In this chapter, you examine the complex set of network protocols, infrastructure, and services that make it possible for users to interact with geographically dispersed resources. Figure 13.1 shows the topics covered in this chapter.

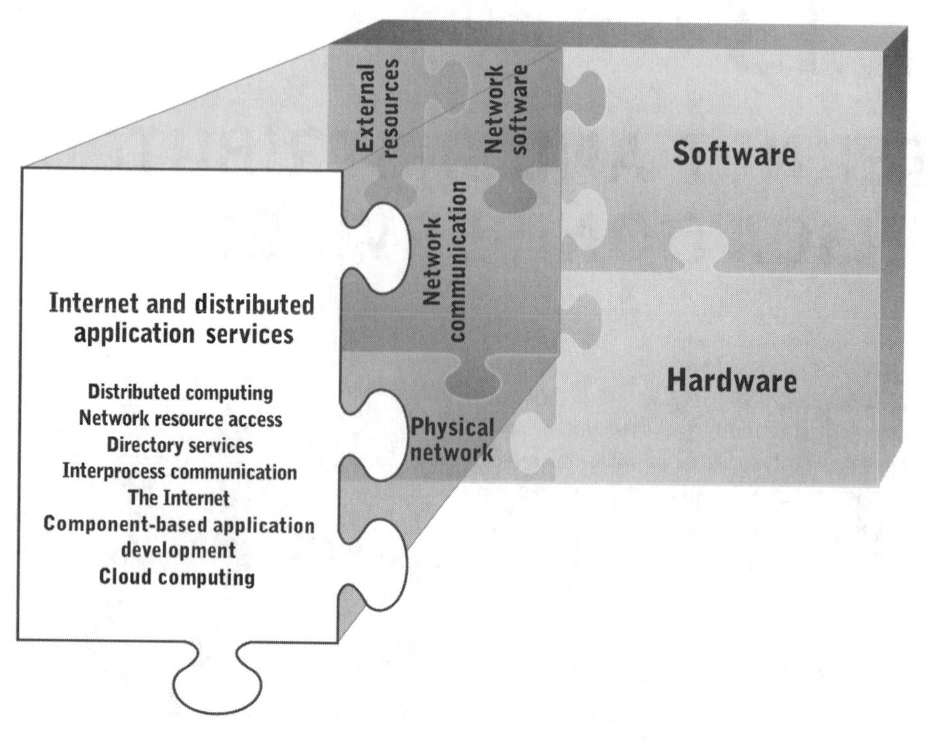

FIGURE 13.1 Topics covered in this chapter
Courtesy of Course Technology/Cengage Learning

DISTRIBUTED SOFTWARE ARCHITECTURE

Information systems are composed of software components distributed across many computer systems and geographic locations. For example, an organization's corporate financial data might be stored on a mainframe computer in its central office. Midrange computers in regional offices might generate accounting reports periodically based on data stored on the mainframe, and desktop computers in branch offices might access and view these reports as well as query and update the central database. Similar tasks might be performed with laptop computers or PDAs using wireless networks. Distributing parts of an information system across many computer systems and locations is called **distributed computing** or **distributed processing**.

Client/Server Architecture

Client/server architecture is a method of organizing software to provide and access distributed information and computing resources. It divides software into two classes: client and server. A server manages system resources and provides access to these resources through a well-defined communication interface. A **client** uses the communication interface to request resources, and the server responds to these requests. Servers are typically available at all times to respond to clients. In contrast, clients can be offline or idle except when needed to access server resources.

The client/server architectural model can be applied in many ways. A simple example is how workstations access a shared printer on a LAN, as shown in Figure 13.2. An application program on a workstation sends a document to a server, which dispatches it to a management process for the specified printer. The server acknowledges the client request and notifies the client when the document is sent to the printer.

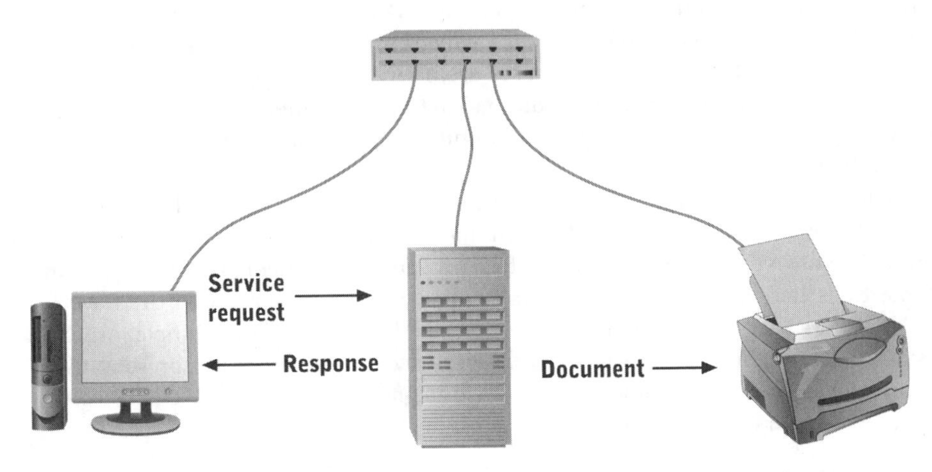

FIGURE 13.2 Network printing services implemented with client/server architecture
Courtesy of Course Technology/Cengage Learning

N-Layer Client/Server Architecture

A variation of client/server architecture, called **three-layer architecture** or **three-tier architecture**, divides application software into the following client and server processes called layers or tiers (see Figure 13.3):

- The **data layer** manages stored data, usually in databases.
- The **business logic layer** carries out the rules and procedures of business processing.
- The **view layer** accepts user input and formats and displays processing results.

The view layer acts as a client of the business logic layer, which in turn acts as a client of the data layer.

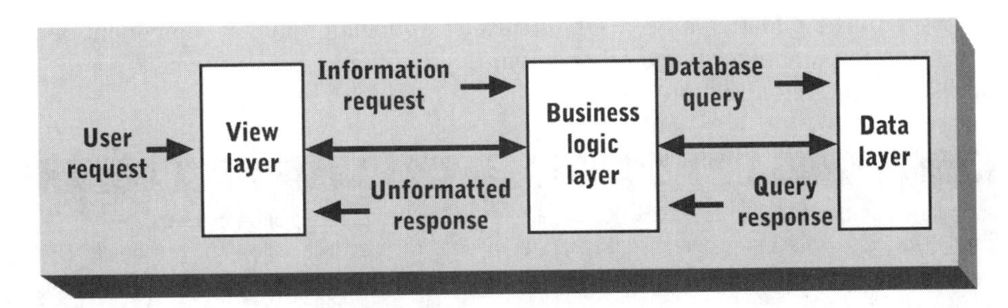

FIGURE 13.3 Three-layer architecture
Courtesy of Course Technology/Cengage Learning

Three-layer architecture simplifies distributing or replicating application software across a network. Interactions between layers are service requests and responses, and layers can be placed in different processes on the same computer or on different computers. Layers that overload a single computer's capacity can be replicated on multiple machines.

Layers can be added when processing requirements or data resources are complex. Architectures having more than three layers are called **n-layer architectures** or **n-tier architectures**. For example, corporate databases are often stored by using multiple database management systems (DBMSs). An additional data layer is interposed between the business logic layer and the DBMSs to present a unified view of data resources from all corporate databases.

Layers can also be replicated in different forms. A single business logic layer might interact with multiple view layers from different applications. For example, business logic for validating inventory items might be part of an inventory-ordering application and part of a customer merchandise return application. Similarly, a single application can have multiple view layers that interact with a single business logic layer. For example, order entry business logic might interact with a view layer used by phone sales representatives that runs on simple VDTs. This same business logic might interact with a Web-based view layer used by customers.

Middleware

The connections between layers of client/server and multitier application software can be complex. Every layer can use a unique combination of programming language, OS, and hardware. Well-defined interfaces and communication protocols enable the layers to function as an integrated whole.

The term **middleware** describes software that "glues" together parts of a client/server or multitier application. It's a wide-ranging system software category because applications' connection and communication requirements vary. For example, a simple client/server application might only need to transmit messages between a single client and server in predetermined locations. A more complex multitier application might need additional middleware, such as Web servers that support embedded client-side programs, database servers that support stored procedures, and network OSs that share files among clients on multiple computers.

As OSs have evolved, they have incorporated more middleware functions. For example, software to implement network protocols, such as TCP/IP, wasn't embedded in microcomputer OSs until the mid-1990s. Web server software, once an optional component of most server OSs, is now standard, and other server OS functions, such as e-mail and document distribution, often rely on it.

NOTE

Other middleware can be obtained as optional OS utilities or as separate packages. For example, the Microsoft BackOffice suite is optional middleware that extends the capabilities of Windows Server. Novell GroupWise and Oracle Tuxedo are examples of middleware packages that work with multiple OSs.

Peer-to-Peer Architecture

In **peer-to-peer (P2P) architecture**, the roles of client and server are combined into a single application or group of related applications. For example, a family has several Internet-connected computers in separate locations containing photos and home movies. Each family member can install a P2P file-sharing application, such as BitTorrent, and configure a shared file space that enables each computer to access files on the other computers. In essence, each computer acts as a client when accessing files on other computers and a server when providing files to other computers.

Advantages of P2P architecture include improved scalability and reducing the number of computer and network connections needed to support an application. For example, family photos and movies can also be shared via a client/server architecture, as when a family posts pictures and videos to a server-based sharing site, such as Facebook or YouTube. Using these sites reduces the amount of software and files stored on users' computers but adds a server and associated network connections. The server and its network connections are central points of failure, and server capacity must be sufficient to handle many users. With P2P architecture, there are fewer potential bottlenecks and points of failure. In addition, the system's overall capacity scales up with the number of participating computers.

Some applications combine client/server and P2P architectures. For example, when an Internet phone application, such as Skype, starts on a user computer, it connects to a server to register its IP address and the connected user account. When users call one another, they interact with this server to get connection information. After the connection is established, further communication between users is P2P.

P2P architecture has a bad reputation resulting from P2P applications that enable illicit file sharing of copyrighted material. Because many of these applications bypass centralized servers, the illicit activity is spread across many users and computers, which makes it harder for copyright holders to stop this practice, compared with client/server applications. However, there are many legitimate uses of P2P architecture. Because today's computers and devices such as cell phones have more than enough power to run both client/server and P2P applications, P2P architecture is here to stay, and its use will likely increase over time.

NETWORK RESOURCE ACCESS

An operating system's primary role is to manage hardware, software, and data resources. As part of this role, it accepts and processes resource access requests from users and applications via the service layer. OSs enable users to interact with resources of the local computer and remote computers. To provide distributed access, the OS must be able to distinguish between local and remote resources and interact with remote operating systems. This section covers the OS components that perform these functions.

Protocol Stacks

An OS implements network I/O and services as a complex set of software layers. The TCP/IP model, covered in Chapter 9, defines the number and functions of these software layers.

Software implementing the Transport, Internet, and Network Interface layers of this model is commonly called a **protocol stack**. Figure 13.4 shows one possible protocol stack for a workstation. Stacks 1 and 2 have different Application and Transport layers but share the other three layers. A client application, such as a Web browser, uses stack 1, and Skype uses stack 2.

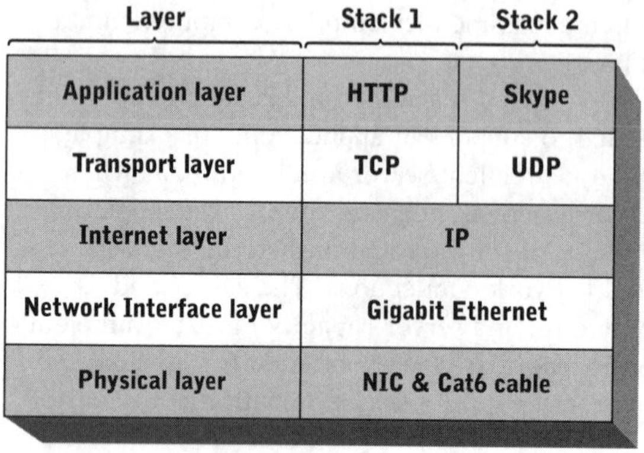

Layer	Stack 1	Stack 2
Application layer	HTTP	Skype
Transport layer	TCP	UDP
Internet layer	IP	
Network Interface layer	Gigabit Ethernet	
Physical layer	NIC & Cat6 cable	

FIGURE 13.4 Two protocol stacks with three shared layers
Courtesy of Course Technology/Cengage Learning

Protocol stacks have several advantages for implementing network I/O and services:

- They divide the task of network interaction into several well-defined pieces that can be implemented, installed, and updated separately.
- They provide the flexibility needed to keep up with rapid changes in protocol standards.
- They insulate application programs and many portions of the OS from details of low-level network communication protocols and physical network implementation, which ensures software portability across a wide range of network protocols and transmission media.

A network protocol stack is a computer's doorway to external resources, such as Web sites and applications, databases, shared files and folders, and shared I/O devices.

Static Resource Connections

Connections to remote resources can be static or dynamic. Before accessing a remote resource, a user or system administrator must know the server and resource names to create a **static connection**. The remote resource can be given a local object, resource, or service name. Figure 13.5 shows the dialog box for creating a static connection between the local S drive and the SharedFiles folder on the Fileserver.mgt.unm.edu server. Similar static connections can be created for printers and other shared resources.

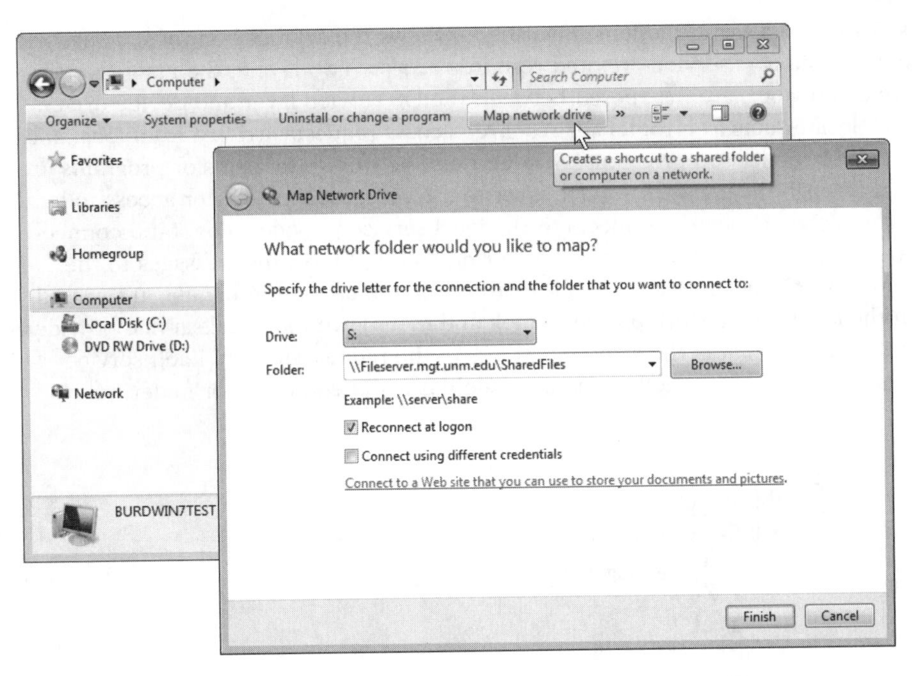

FIGURE 13.5 Displaying and creating static connections in Windows 7
Courtesy of Course Technology/Cengage Learning

Static connections are difficult to initialize and maintain because the OS must be configured to establish a connection each time it starts. In addition, if a remote resource's name or location changes, the configuration of all computers with static connections to this resource must be changed, too.

Remote resource access in current OSs is based on the following premises:

- Operating systems, application programs, and user interfaces are simpler if there's no distinction between local and remote resource access.
- All resources are potentially shared across a network.
- Any computer system is potentially both a client and a server.
- Resources can be moved between computer systems.

Software and user interfaces are simplified by providing a common method of accessing both local and remote resources. For example, a word-processing program running on a workstation should use the same service calls and parameters to access document files stored on local disks *and* on server disks. A Web browser should access resources on remote machines in the same manner it accesses resources on the local machine. This characteristic of software and user interfaces is called **location transparency** or **network transparency**.

The second and third premises go hand in hand. If every local resource might be needed by remote users or processes, every computer is potentially both a client and a server. To provide remote access, all OSs need to incorporate client/server or P2P resource sharing and access functions. Figure 13.6 shows the arrangement of software components that support service-oriented resource access. Two layers, the resource locator and service provider, are interposed between the OS service layer and device drivers. The resource locator acts as a router for resource access requests arriving from local and remote users and application programs. Service requests from local users or programs are passed down through the local OS service layer to the resource locator. For access to local resources, the resource locator connects to the local service provider. For static connections to external resources, the resource locator creates and forwards messages to the external server or P2P node. Service requests from remote users or programs are passed up through the low-level network protocol stack and then down to a local service provider. The service request format is the same regardless of the request's origin. Each service provider is an interface to a specific resource, such as a shared printer or folder.

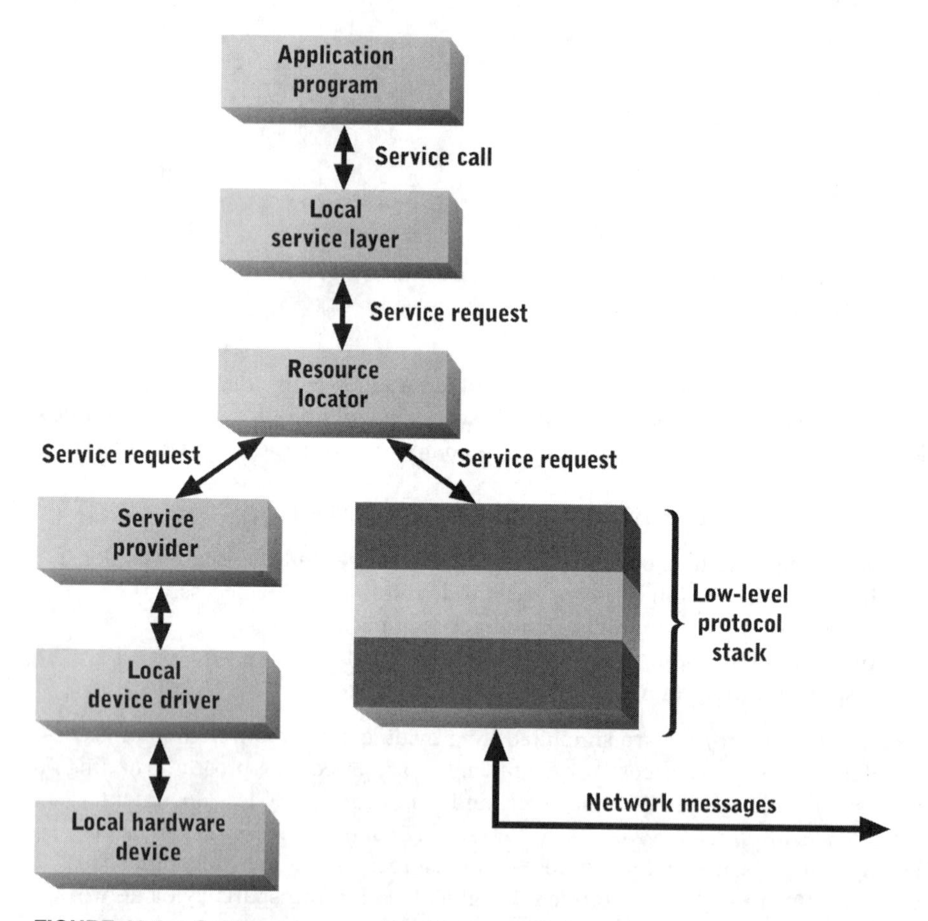

FIGURE 13.6 Software resources used to access local and remote resources
Courtesy of Course Technology/Cengage Learning

The resource locator maintains a local **resource registry** containing the names and locations of known resources and services. When local service provider processes are started, they register themselves with the resource locator, which updates the resource registry accordingly. Resource requests are checked against this registry. If all connections to resources are static, the resource locator generates an error if a requested resource isn't listed in the resource registry.

Dynamic Resource Connections

As mentioned, static connections are difficult to administer. When resources are added, moved, renamed, or deleted, all users' resource registries must be updated to reflect the changes. A more complex, but flexible, approach to remote resource access uses dynamic connections. With **dynamic connections**, a resource user asks for a resource, and if it isn't found in the local resource registry, the resource locator searches for it in external locations. If the resource is found, the resource locator communicates with the host server or P2P node on behalf of the local process or user, who is unaware of whether the resource is local or external.

Domain Name System (DNS), used on the Internet, is one example of dynamic connections. Every IP packet carries the destination node's IP address. Because nearly all remote service accesses use Internet protocols, finding the IP addresses of servers and P2P nodes with needed resources must be the first step to sending access request messages. However, the Internet is a dynamic place, where servers and P2P nodes can be moved from network to network (thus changing their IP addresses), and most users and application programs know servers and P2P nodes by DNS names, such as *www.microsoft.com*, not by their corresponding IP addresses. How does a client or P2P node that wants to connect to another node by DNS name convert the name into an IP address?

Every network attached to the Internet has at least two servers designated as DNS servers. A DNS server maintains a registry of DNS names and corresponding IP addresses for each node on the local network and for DNS servers elsewhere on the Internet. For example, the University of New Mexico (UNM) has two large DNS servers with the IP addresses 129.24.8.1 and 129.24.8.4 that respond to requests for the IP addresses of any DNS name ending with "unm.edu." These servers are registered with Internet naming authorities, and the registrations are stored in root DNS servers that act as DNS master servers for the entire Internet. In the worst case, a client wanting to know the IP address corresponding to a DNS name in UNM takes the following steps:

1. Contact a root DNS server to find the IP address of a UNM DNS server. (IP addresses of root DNS servers are updated periodically via automated downloads.)
2. Send a request to the UNM DNS server's IP address asking for the IP address of a specific node, such as *averia.unm.edu*.

If the request is for the IP address of a node such as *www.unm.edu*, the UNM DNS responds with the IP address. If the request is for an IP address of a node in a UNM subnet (for example, *www.mgt.unm.edu*), the UNM DNS server might respond with the IP address of a DNS server for the subnet, which the client then contacts for the IP address it wants.

As clients send DNS requests and receive responses, the client OS updates a local registry of DNS names and IP addresses. If a client requests an IP address for one DNS name in the unm.edu network and shortly after requests an IP address for the same DNS name, the answer is already in the local DNS registry. If the client later needs the IP address of another DNS name in the unm.edu network, the UNM DNS server's IP address can be retrieved from the local DNS registry, thus avoiding Step 1.

Distributed resource directories and servers, such as those used for DNS queries, offer the flexibility to create dynamic connections to remote resources as needed. When a resource access request is sent to the local resource locator, it searches the local resource registry and, if the resource isn't listed there, initiates a search by using protocols similar to the DNS query protocol. Of course, the flexibility of dynamic connections comes at a high cost—the complexity of the protocol and its software implementation in the client and a potentially large collection of servers. However, in today's world of widely distributed and rapidly changing resources, there's simply no alternative.

NOTE

It should be clear by now that the traditional distinctions between client and server OSs have become blurry. All modern OSs are organized internally as a collection of server processes that can respond to requests from local and remote users and processes. In addition, they maintain one or more resource registries and create dynamic connections to remote resources as needed. There are some distinctions between client and server OSs, such as scalability and security, but they're mainly a matter of different configuration, not fundamentally different architecture.

DIRECTORY SERVICES

When resources are distributed across network nodes, resource users and providers must have some way of finding one another. The term **directory services** describes middleware that does the following:

- Stores the name and network address of distributed resources
- Responds to directory queries
- Accepts directory updates
- Synchronizes replicated or distributed directory copies

Directory services are integral components of all network operating systems. Typically, network OS directories store information about these items:

- Registered users and their permissions to access directory objects
- Shared hardware resources, such as printers
- Shared files, databases, and programs
- Computer systems and specialized hardware devices, such as network storage appliances

Directory services are distributed much like Internet name services. Directories are organized hierarchically to create a single namespace for all network resources and objects. In large networks, responsibility for maintaining directory content and answering queries is distributed throughout the network. Directory content can be replicated in multiple servers in different parts of the network to reduce response time and improve fault tolerance.

Lightweight Directory Access Protocol

In the 1980s, the International Telecommunications Union (ITU) developed the X.500 standard, which defines nonproprietary directory services for e-mail and network addresses. This standard was never fully implemented in a commercial product or OS. However, it became the basis for another standard called **Lightweight Directory Access Protocol (LDAP)**, which the Internet Engineering Task Force (IETF) adopted as a formal Internet standard. It was updated regularly in the late 1990s, and these efforts still are in progress. LDAP is widely used, although not all products have all the features of the most recent updates.

An LDAP directory stores information about LDAP objects. Each object is an instance of an **objectclass**, which defines the attributes common to all member objects. For example, the objectclass Shared_Printer might define attributes named Building, Room, Manufacturer, Model, Color, Duplex, and Pages_Per_Minute. Each directory entry for a shared printer is an object of the type Shared_Printer and contains values for some or all defined attributes.

LDAP objects are organized in a hierarchical directory structure. Objects can be grouped into container objects that can be grouped into other container objects (see Figure 13.7). LDAP defines several standard container object types, including Country (C), Organization (O), and Organizational Unit (OU). All objects in an LDAP schema have a distinguished name (DN) attribute, which uniquely identifies the object in an objectclass. A fully qualified DN, such as the following, specifies a complete path from a directory root node through one or more container objects to a specific object:

```
DN=Stephen Burd,O=Faculty,O=School of Management,
OU=University of New Mexico,C=USA
```


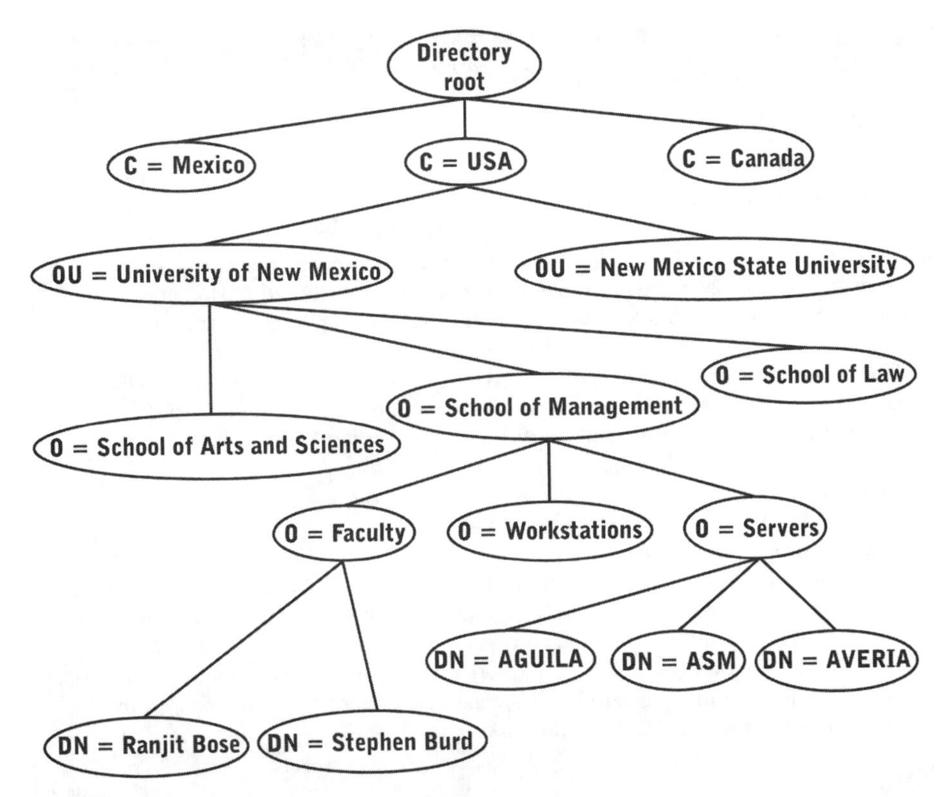

FIGURE 13.7 An LDAP hierarchy of objects and container objects
Courtesy of Course Technology/Cengage Learning

LDAP also defines a standard protocol for updating and querying LDAP directories. LDAP queries contain plaintext characters, although the syntax is complex and awkward. Many client programs, such as e-mail software and Web browsers, have a more user-friendly interface to LDAP services.

Strict adherence to the LDAP standard guarantees interoperability between all LDAP clients and servers. However, an important area of interoperability has yet to be addressed: standard schema names and structure. Currently, there are no standard objectclass or attribute names for entities and resources common to most directories, such as people and shared printers. As a result, there's no standard way for a directory from one vendor or organization to query the directory of another organization or vendor.

For example, one directory might define an objectclass named Employee, and another directory might use the name Worker for a similar objectclass, and the two directories might use different attribute names for identical data items, such as office phone number. One directory can send a query to the other asking it to enumerate its defined objectclasses and attributes, but without a standard naming system, there's no way to determine semantic equivalence of schema classes and attributes automatically.

In the absence of standard class and attribute names, directory vendors and user organizations have developed proprietary directory schemas that are usually incompatible. Even if a set of standard schema classes and attributes is developed, the existing base of incompatible directory schemas will hinder true directory interoperability for years to come.

TECHNOLOGY FOCUS

Microsoft Active Directory

Microsoft **Active Directory** is the directory service and security system built into Windows Server. Active Directory stores information about many network resources, including computers, I/O devices, and users (see Figure 13.8). It incorporates security services that enable system administrators to limit resource access to specific users or user groups. Active Directory also supports large directories and distributed organizations with directory partitioning across Internet domains, directory replication across multiple servers, and automatic synchronization of replicated directories or directory partitions.

Windows client OSs query Active Directory to locate and access network resources. For example, a user can search for shared directories matching specific criteria. Programs can also interact with Active Directory by calling Active Directory-specific service functions. For example, a word-processing program might query Active Directory for all printers in an organizational unit, and then construct and display the result to users when they select the Print function from the File menu.

Windows client and server OSs use Active Directory to store and access security information. Every user, group, and computer object is assigned a unique security identifier, and every Active Directory resource or container object has an **access control list (ACL)** describing rights granted or denied to users, groups, and computers for accessing resources. Figure 13.9 shows the ACL for the Active Directory object named Subnets. Members of the Domain Admins group have special permissions for Subnets and, by default, have these permissions for all child objects.

(*continued*)

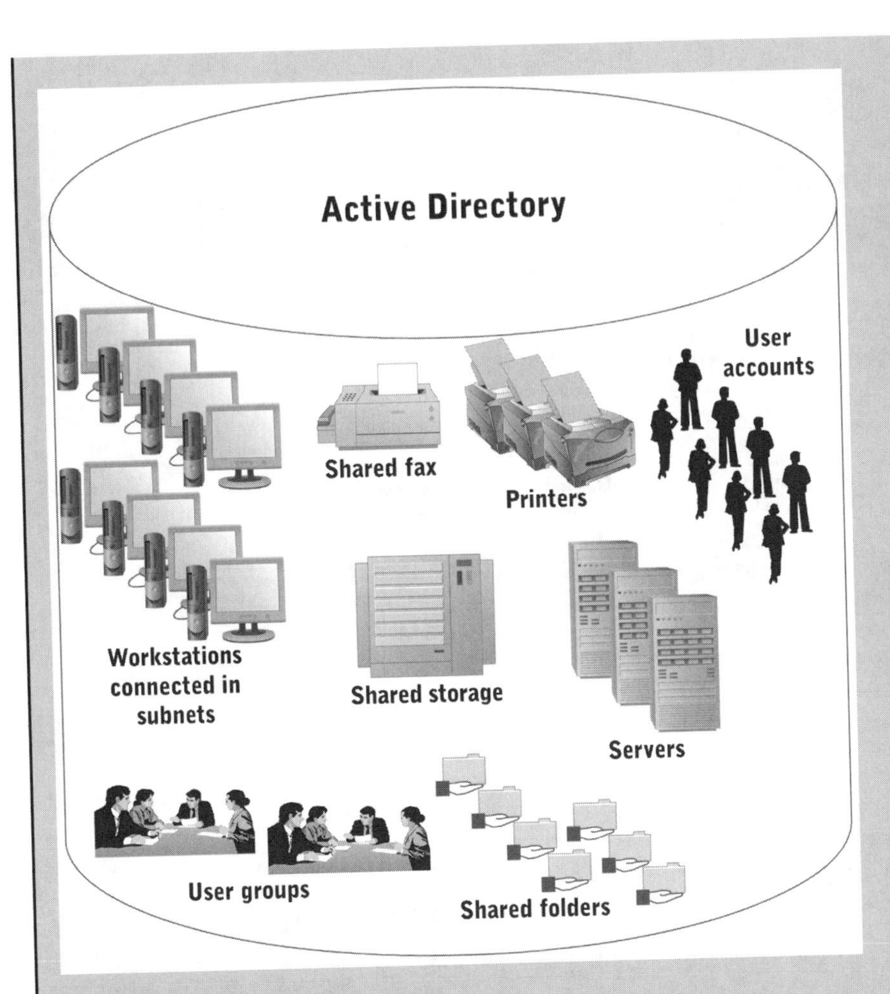

FIGURE 13.8 Active Directory objects
Courtesy of Course Technology/Cengage Learning

Active Directory ACLs and permissions inheritance can be used to distribute system administration tasks across users and organizations. If Active Directory objects are organized hierarchically with container objects, the directory administrator can delegate management functions to other users by assigning rights to container objects. For example, he or she can assign rights such as Create All Child Objects and Modify Permissions to other users for lower-level container objects so that they can administer one part of Active Directory.

Active Directory is based on LDAP and DNS, so it responds to standard LDAP information requests and uses LDAP concepts, such as objectclasses and OUs, to store and organize directory information hierarchically. Active Directory clients rely on a DNS server to locate Active Directory servers, which register their names and services by supplying a service (SRV) record to a DNS server. Active Directory clients query the DNS

(continued)

server for LDAP servers registered in their domain. The DNS server returns IP addresses, which the client uses to send TCP/IP messages to an Active Directory server.

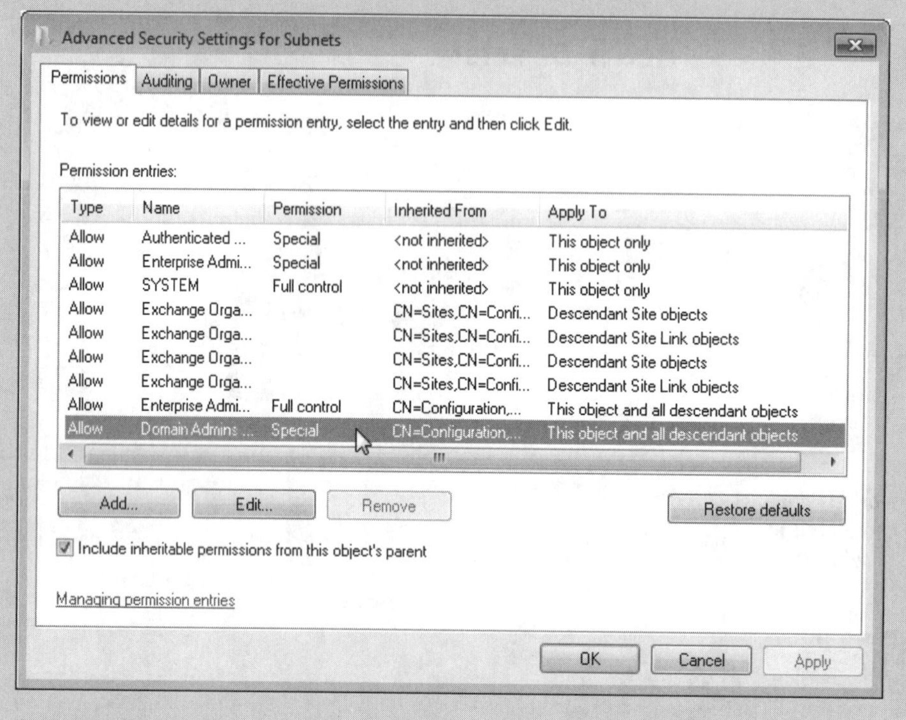

FIGURE 13.9 Viewing an ACL
Courtesy of Course Technology/Cengage Learning

Like many other commercial products, Active Directory doesn't implement LDAP fully. Unlike other directory service software, it relies on DNS and requires DNS servers to accept dynamic updates. Although dynamic updates are an Internet DNS standard, many DNS servers don't accept them by default. DNS administrators are reluctant to enable them because they're a potential security and reliability threat. Microsoft's DNS server software does support dynamic updates, so most Active Directory installations use Microsoft DNS server software.

In addition, Active Directory doesn't support distributed or component-based software directly. Shared files, including executable files, and databases can be registered in Active Directory, which could support locating and accessing programs and data sources. However, Active Directory has no standard way to "glue together" components and external data sources to create complex systems. Windows relies on a parallel directory and registration system to track COM+ components (discussed later in "Components and Distributed Objects") and provide component-related directory and interprocess communication services.

INTERPROCESS COMMUNICATION

When an application is split into multiple processes, these processes must communicate with one another to share data and coordinate their activities. However, when processes are executing on different computers, how do they communicate and coordinate activities? A variety of protocols and standards have been developed that enable process coordination across networks. In this section, you concentrate on lower-level P2P protocols and standards, which enable processes to communicate synchronously across a network (see Figure 13.10). System software often uses these protocols to exchange data and coordinate activities. Distributed applications usually use higher-level protocols, described later in the chapter. However, these higher-level protocols are often layered above interprocess communication protocols.

Application layer	DCE
	RPC
	Named pipes
Transport layer	Sockets
	TCP
Internet layer	IP

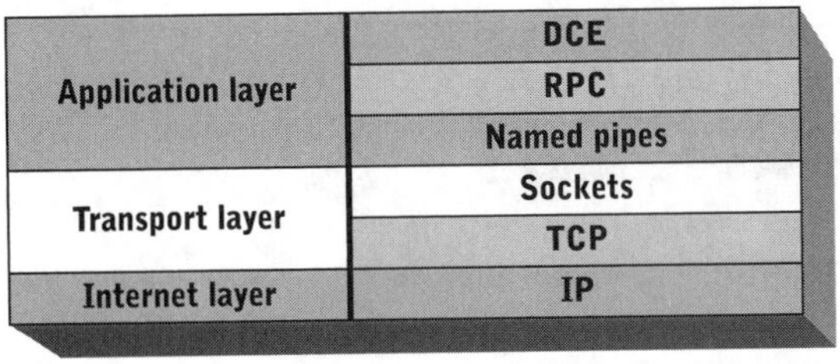

FIGURE 13.10 Interprocess communication protocols layered over TCP/IP
Courtesy of Course Technology/Cengage Learning

Sockets

As described in Chapter 9, a socket is a unique combination of an IP address and a port number, separated by a colon. For example, the socket 129.24.8.1:53 is the network listening address for the primary DNS name server at the University of New Mexico. A port number is an unsigned 16-bit integer, so there are 65,536 possible port numbers. Some port numbers are permanently assigned to standard Internet or vendor-specific services, but many are available for other uses, including client/server or P2P communication between application programs.

All current OSs support sockets and provide system service calls so that programs can initialize sockets, receive messages sent to a socket, and send messages to sockets anywhere on the Internet. Figure 13.11 shows a communication example between client and server processes on two computers. Client processes on Computer A are attached to sockets 129.24.8.212:2 and 129.24.8.212:6 and communicate with two server processes on Computer B attached to sockets 207.46.230.219:1 and 207.46.230.219:6. Each socket uniquely identifies a client or server process on the Internet.

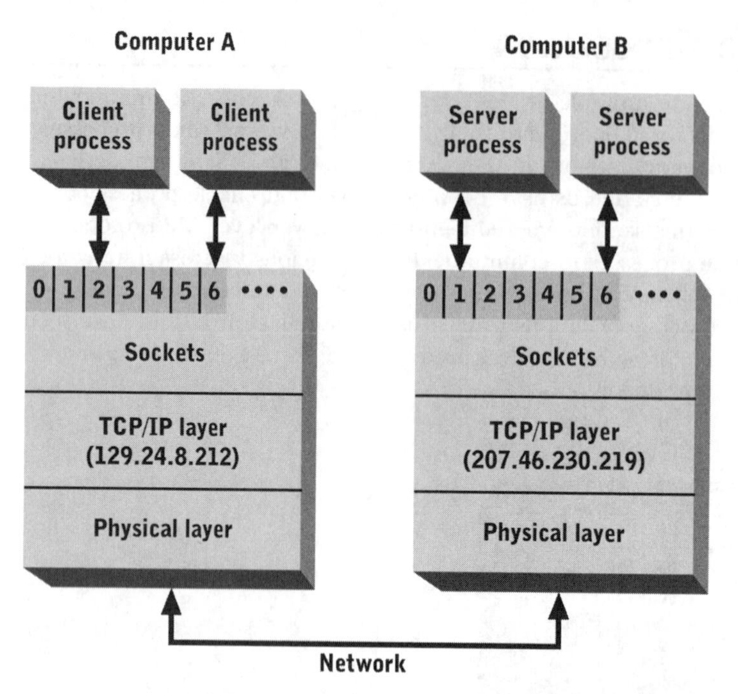

FIGURE 13.11 Multiple processes communicating through sockets
Courtesy of Course Technology/Cengage Learning

Named Pipes

A **pipe** is a region of shared memory through which multiple processes executing on the same machine can exchange data. Pipes are commonly used for communication between OS components, for queuing requests to an OS service (such as a Web server), and for exchanging messages between components in a large program. Processes read or write a pipe as though it were a file, and the OS manages data movement between processes and the shared memory region.

A **named pipe** is a pipe with two additional features:

- A name that's permanently placed in a file system directory
- The capability to communicate between processes on different computers

When a named pipe is created, a directory entry is also created in the local file system. Programs can read or write the named pipe as they would read or write an ordinary file. Typically, the server side of a client/server application creates the named pipe. In a P2P application, either side of the application can create it. A client or peer opens a named pipe as a network resource and reads or writes it as though it were a file on a shared directory. A server or peer on the machine where the named pipe was created reads and writes the pipe as though it were a local file.

The OSs at both ends of the pipe manage communication to and from the pipe. Named pipes are actually a high-level interface to sockets, so the OS assigns a free socket to the named pipe when it's created. The OS also allocates I/O buffers and routes data flowing in and out of the pipe through the low-level network protocol stack, as shown in Figure 13.12. The OS on the client side also allocates a socket each time a remote named pipe is opened.

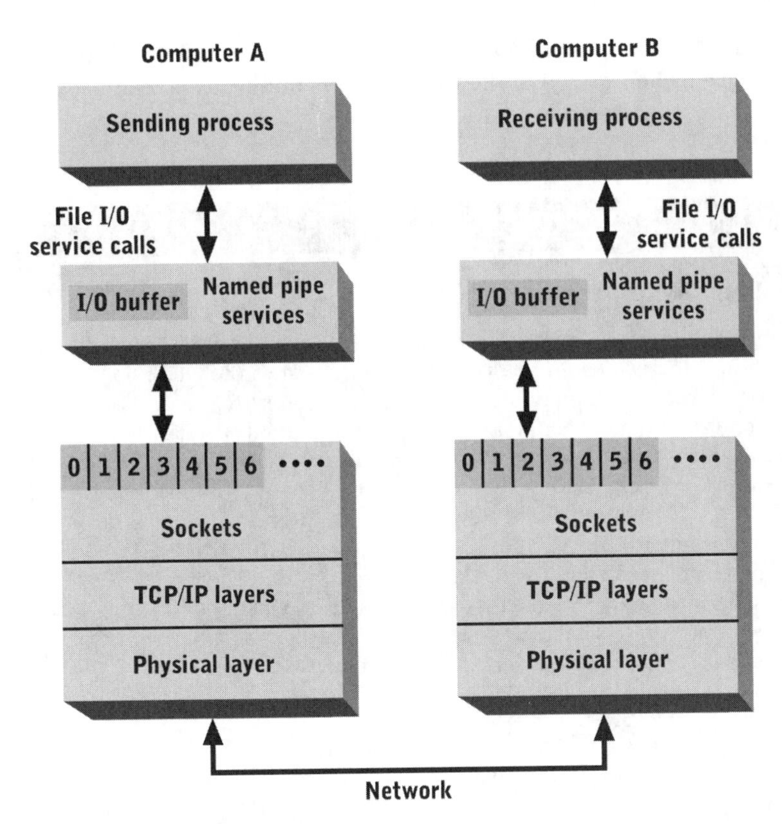

FIGURE 13.12 Two processes communicating through a named pipe
Courtesy of Course Technology/Cengage Learning

Data can flow in either direction through a named pipe, although bidirectional data flow is usually done with two one-way pipes. In client/server applications, multiple clients can send messages to a single server by writing to the same named pipe. The server can tell which client sent which messages or data because each client has a unique socket number.

Remote Procedure Calls

With the **Remote Procedure Call (RPC)** protocol, a process on one machine can call a process on another machine. As with function or procedure calls in a single program, the calling process follows these steps:

1. Pass parameters to the called process.
2. Wait for the called process to complete its task.
3. Accept parameters back from the called process.
4. Resume execution with the instruction following the call.

Parameter passing between machines can potentially cause problems because data representation varies across CPUs and sometimes across operating systems. Common differences include little endian versus big endian memory storage, character coding (ASCII or Unicode), and which IEEE floating-point format is used for real numbers. If the calling and called process execute on machines with different data representation formats, the parameters must be converted when passed to the called process and when passed back to the calling process.

TECHNOLOGY FOCUS

Distributed Computing Environment

Distributed Computing Environment (DCE) is a standard for distributed OS services defined by the Open Group (*www.opengroup.org*), formerly known as the Open Software Foundation. This wide-ranging standard covers network directory services, file-sharing services, RPC, remote thread execution, system security, and distributed resource management. Its main goal is to promote interoperability of distributed software across OSs and middleware products. Many OSs comply partially with DCE, including Windows and many versions of UNIX. IBM and Hewlett-Packard are principal supporters of and contributors to the standard.

DCE defines a subset of OS services and a standard interface to these services. DCE functions are incorporated into an OS or supplied as an optional component (see Figure 13.13). In theory, DCE services can replace their counterparts in an existing OS. In practice, DCE-compliant services usually translate DCE service calls into native OS service calls.

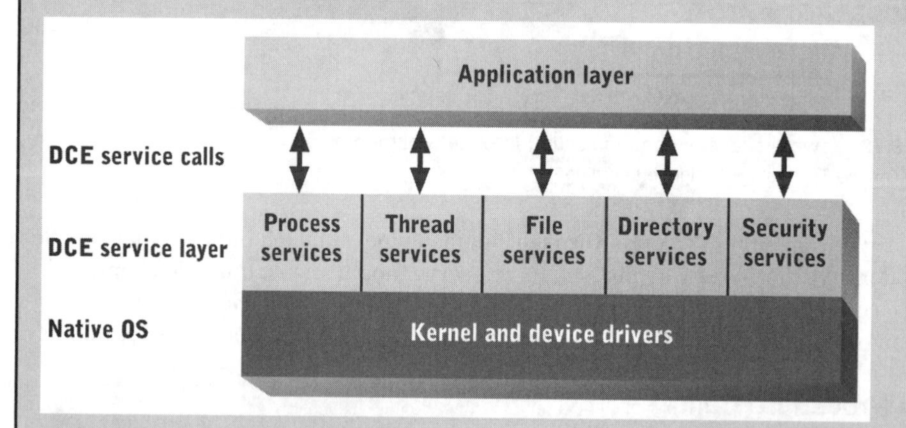

FIGURE 13.13 DCE software layers
Courtesy of Course Technology/Cengage Learning

Security is an integral part of every DCE service. DCE security is based on the **Kerberos** security model, which defines interactions between clients, services, and a trusted security service. Clients and servers authenticate one another by asking the security server to authenticate the other party. The security server issues security "tickets" to each party that are exchanged to verify identities.

The security database also maintains an ACL for each service and resource. After a client has been authenticated, its identity is checked against the ACL for the resource

(*continued*)

it's attempting to access. If the client is part of the ACL, it's issued a ticket that it presents to the server to gain access. Tickets, passwords, and other security mechanisms are encrypted during network transmission. Tickets are also time-stamped and expire in minutes or hours.

DCE file services are based on a fully distributed file and directory structure. Every file in a DCE file system has a unique name consisting of its hostname, a file system resource name in the host, and its pathname in the file system resource. File systems can be distributed or replicated across multiple servers. DCE file services use transaction logging for all updates to ensure rapid recovery in the event of a crash.

Clients, servers, and all DCE services interact through RPC. RPC messages are formatted according to an interface definition language (IDL). DCE-compliant client and server programs are compiled and linked with DCE library routines that implement the IDL. These routines perform the low-level aspects of passing messages, exchanging parameters, converting data formats, and interacting with DCE security services.

The DCE standard is used in many UNIX variants, but it has also penetrated deeply into other OSs. It's widely used, although competing standards, such as CORBA (discussed later in "Components and Distributed Objects"), are more commonly used in new software development projects. However, DCE will probably be around for many years, given its widespread operating support and large installed base.

THE INTERNET

The Internet and World Wide Web are widely used frameworks for implementing and delivering information system applications. However, there's a general lack of agreement on definitions of the Internet, Web, and related terms. The following definitions are used in this chapter:

- The **Internet** is a global collection of networks that are interconnected with TCP/IP.
- The **World Wide Web (WWW)**, also called the Web, is a collection of resources (programs, files, and services) that can be accessed over the Internet by standard protocols, such as Hypertext Transfer Protocol (HTTP).
- An **intranet** is a private network that uses Internet protocols but is accessible only by a limited set of internal users (usually members of the same organization or workgroup). It also describes privately accessible resources that are organized and delivered via one or more Web protocols over a TCP/IP network.

The Internet is the infrastructure on which the Web is based. In other words, Web resources are delivered to users over the Internet. An intranet uses the same protocols as the Internet and Web but restricts access to a limited set of users. Access can be restricted in many ways, including using privately registered resource names, firewalls, and user/group account names and passwords.

Much of the Web is organized by using client/server architecture. Web resources, such as Web pages, are managed by server processes that can execute on dedicated servers or multipurpose computers. Clients are programs that send requests by using standard Web resource request protocols, which define valid resource formats and a standard means of requesting resources. Any program, not just a Web browser, can use Web protocols to access Web resources.

Standard Web Protocols and Services

Web resources are identified by a unique **Uniform Resource Locator (URL)**, such as *http://averia.unm.edu/default.htm*. A URL has four components, as shown in Figure 13.14:

- *Protocol*—An optional header specifying the resource access protocol (http:// is the default value)
- *Host*—The IP address or registered name of an Internet host computer or device
- *Port*—An optional port number that, together with an IP address, specifies a socket (if omitted, a standard port number for the protocol is assumed)
- *Resource*—The complete pathname to a resource on the host (if omitted, the host can return a default resource if configured correctly)

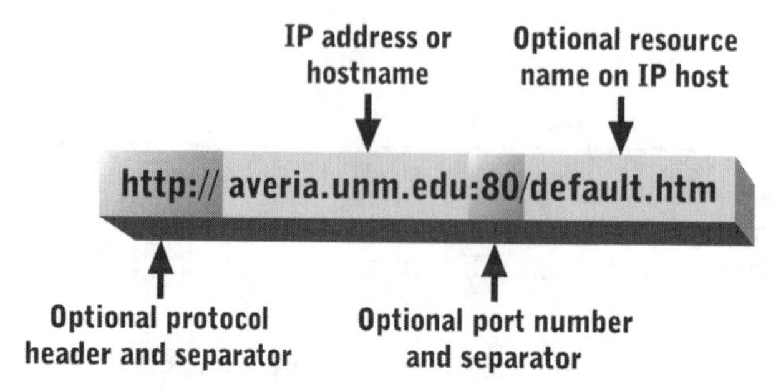

FIGURE 13.14 URL components
Courtesy of Course Technology/Cengage Learning

Web standards define a number of protocols for resource format, content, transfer, and manipulation. The number of standard protocols is growing, and their content changes rapidly. Table 13.1 summarizes major Web protocol categories.

TABLE 13.1 Web protocols

Category	Sample protocols
Formatted and hyperlinked documents	Hypertext Markup Language (HTML) and Extensible Markup Language (XML)
File and document transfer	File Transfer Protocol (FTP) and Hypertext Transfer Protocol (HTTP)
E-mail and messaging	Simple Mail Transfer Protocol (SMTP) and Internet Message Access Protocol (IMAP)

TABLE 13.1 Web protocols (*continued*)

Category	Sample protocols
Videoconferencing	H.323 and Session Initiation Protocol (SIP)
Executable programs	Java, JavaScript, and VBScript
Interprocess communication	Remote Procedure Call (RPC) and named pipes
Web services	Simple Object Access Protocol (SOAP) and Universal Description, Discovery, and Integration (UDDI)
Instant messaging	SIP for Instant Messaging and Presence Leveraging Extensions (SIMPLE), Instant Messaging and Presence Service (IMPS), and Extensible Messaging and Presence Protocol (XMPP)

The Web began with the development of **Hypertext Markup Language (HTML)**. Originally, HTML defined a device-independent document-formatting language in which links to other documents could be embedded. The first Web browser, Mosaic, displayed HTML documents on any computer and output device to which it was ported. Mosaic spawned the current generation of Web browsers, such as Mozilla Firefox and Microsoft Internet Explorer. Both HTML and Web browser software have evolved through several generations to include capabilities such as forms, style sheets, data transfer from client to server, and embedded scripts and programs. HTML will eventually be replaced by **Extensible Markup Language (XML)**, which extends HTML to describe the structure, format, and content of documents.

Hypertext Transfer Protocol (HTTP) is a companion protocol to HTML and XML that specifies the language by which clients request documents and how servers respond to those requests. HTTP is an extension of an older Web protocol, called **File Transfer Protocol (FTP)**, that specifies a client/server request and response language for copying files from one Internet host to another. Because HTTP is an extension of FTP, servers that respond to HTTP requests can also respond to FTP requests. **HTTPS** is a secure version of HTTP that encrypts HTTP requests and responses.

With the **Telnet** protocol, users on one Internet host can interact with the OS command layer of another host (see Figure 13.15). Telnet emulates a VDT and is limited to interacting with command-line interfaces, such as Windows Cmd.exe and the UNIX Bourne shell. **Secure Shell (SSH)** is an improved version of Telnet that encrypts data flowing between client and server to address a major security issue in Telnet.

E-mail standards have been part of the Internet since the 1970s. The earliest e-mail protocol is **Simple Mail Transfer Protocol (SMTP)**, which defines how text messages are forwarded and routed between Internet hosts. E-mail client programs on each host interact with a server process to access forwarded messages and send messages to users on

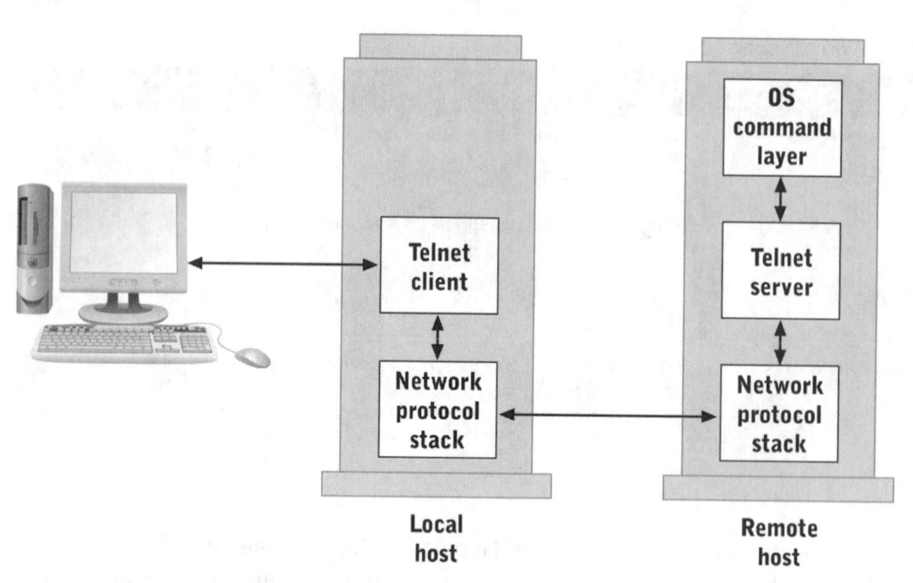

FIGURE 13.15 Telnet connection
Courtesy of Course Technology/Cengage Learning

other hosts. SMTP was later extended by the **Multipurpose Internet Mail Extensions (MIME)** protocol to allow including nontext files in e-mail messages. MIME is also used by HTTP and XML.

Today's e-mail server software is based on **Post Office Protocol 3 (POP3)**, **Internet Message Access Protocol 4 (IMAP4)**, or both. POP3 standardizes the interaction between e-mail clients and servers so that client and server can run on different Internet hosts. Under POP3, e-mail messages are held on the server temporarily, downloaded to the client when a connection is established, and deleted from the server as soon as the download is finished. IMAP4 extends POP3 to permanently store and manage e-mail messages on the server, which enables users to access stored e-mail from any Internet host.

In later HTML versions, program code or scripts can be embedded in HTML documents. Java applets, described in Chapter 10, can be called from an HTML document with parameters passed in either direction. Figure 13.16 shows sample HTML code to call a Java applet named VocabMan.class. The applet is downloaded to a Web browser along with the surrounding HTML document. Figure 13.17 shows the Web page with the embedded Java applets displayed as boxes containing asterisks. The Web browser executes the applet when the cursor is placed over a box, and the applet displays answer text when the mouse button is clicked.

As described in Chapter 10, scripting languages are slimmed-down programming languages. JavaScript and VBScript are two widely used scripting languages. Scripts embedded in HTML pages can perform many functions that full-fledged programs can, without

```
<li>A compiler allocates storage space and makes an entry in the symbol
    table when a(n)
    <applet width="175"
            height="25"
            code="VocabMan.class"
            codebase="http://asm.unm.edu/sa2e_student/Java/">
            <param name="answer"
                   value="data declaration">
              You must have Java enabled for this feature
    </applet>
    is encountered in source code.<br><br>
</li>
```

FIGURE 13.16 A Java applet call embedded in HTML
Courtesy of Course Technology/Cengage Learning

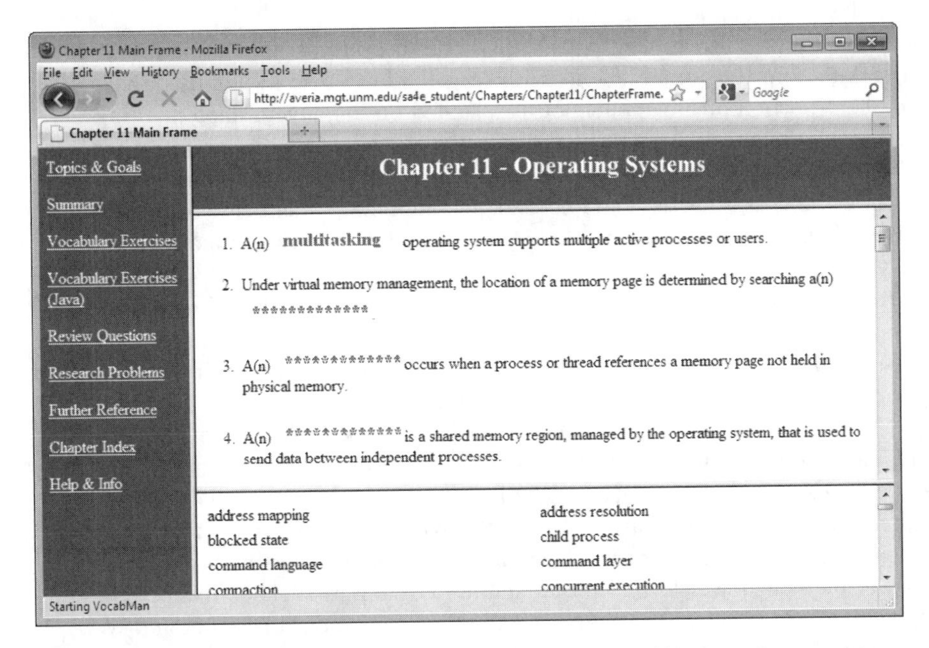

FIGURE 13.17 Vocabulary exercise answers displayed or hidden by a Java applet
Courtesy of Course Technology/Cengage Learning

the need for compiling or link editing. Typical functions include data validation and customizing page layout or content based on Web browser configuration.

Web servers can run programs and scripts on the server in response to requests sent by clients as a URL. Internet search engines are a common example of this approach. The Web browser downloads an HTML form in which the user enters search criteria. When the user clicks a Send or Go button, the search criteria ("vacation" and "caribbean," in this case) are transmitted back to the Web server encoded in a URL, such as this one:

```
http://www.google.com/search?q=vacation+caribbean&ie=utf-8&oe=utf-8
```

This URL contains an encoded function call to a search engine program. The text after the question mark contains input parameters for the search engine program. When the program has found the requested data, it encodes processing results in HTML, and the Web server sends them back to the Web browser for display.

The Internet as an Application Platform

Internet and Web technologies are an attractive alternative for implementing distributed applications. For example, a geographically dispersed computer consulting firm, with dozens of offices and many employees on extended assignment at client locations, uses an automated payroll system. Employees need to update withholding, insurance, and other payroll information and enter billable hours. How can all the firm's employees interact with the payroll system quickly and easily?

One way to address this problem is to build a client/server application that uses a private network to connect remote clients with servers at the firm's administrative offices. The client portion of the application is installed on employees' laptop computers, and employees connect to payroll servers via a modem and private phone number. The disadvantages of this approach include the cost to build and maintain a private network, the cost to develop full-fledged, self-contained client-side software, and the difficulties of installing, configuring, and updating client software on many different computers.

An alternative is building a client/server application that uses a Web browser interface. The application runs on a Web server that can be accessed from any computer with an Internet connection. Employees can access the payroll server with their own laptop computers or computers in a client office or hotel business suite.

Figure 13.18 shows the architecture of a typical Web-based three-layer payroll system. The client interacts with the Web server to download HTML pages that can include embedded scripts or Java applets. The bulk of the application code resides on the server, and program functions are called by the Web server in response to client requests encoded in URLs. The server-side application code interacts with a back-end database server by using lower-level Internet standards, such as sockets, named pipes, or RPC. All parts of the application communicate by using standard Internet and Web protocols. The application can be secured via layered security protocols, such as HTTPS and Kerberos.

Implementing the application via the Internet expands its accessibility and eliminates the need to install custom client software on employees' laptop computers. The application can be updated by updating software only on the Web server. It's also cheaper to develop and deploy because it's built around existing Web standards and relies on Web browser software already installed on clients.

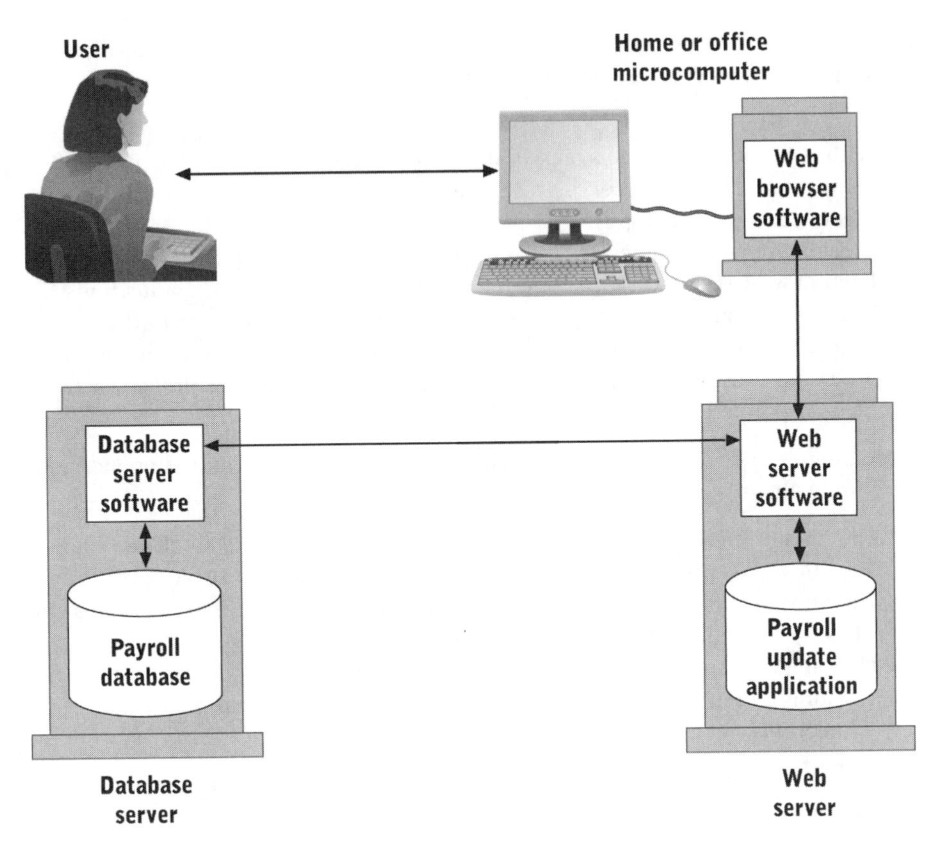

FIGURE 13.18 A distributed Web-based application
Courtesy of Course Technology/Cengage Learning

Each resource participating in the payroll update application can also participate in other applications. For example, the database server hardware and software can host marketing and production databases. The Web server can store product manuals and host an order entry application for customers. The Web browser on an employee's laptop computer can access other Web applications or information resources anywhere on the Internet.

The primary disadvantages of implementing applications via the Internet are security, performance, and reliability. If an employee can access the system via the Web, others might also be able to gain access. System access can be restricted by various means, including user accounts and passwords, but the risk of a security breach is always present. Performance and reliability are limited by employees' Internet connection points and the available Internet capacity between these connection points and the application server. Unreliable or overloaded local Internet connections can render the application unusable.

Although it's possible to implement distributed applications by using Internet standards such as sockets, named pipes, HTTP, and HTML, this approach isn't optimal for the following reasons:

- With lower-level protocols, server addresses are stored in client configuration files or source code. If server resources are moved, the clients must be reconfigured or recompiled.
- Breaking up server-side processes into small manageable pieces is difficult. Each new distributed piece requires a new set of hard-coded connections.
- Developers usually create large complex server processes to avoid the complexity of large numbers of connections between many smaller server processes. However, doing so reduces the chance that server processes can be incorporated into multiple distributed applications.

Newer techniques and standards for deploying and supporting distributed applications address these problems in several ways:

- They make it easier to break up large server processes into small reusable pieces by providing standards for communication between server processes.
- They improve flexibility by providing directory services so that processes can locate one another with location-independent names.
- They make it easier to build systems from small reusable processes by providing the infrastructure to manage large numbers of interprocess connections.

COMPONENTS AND DISTRIBUTED OBJECTS

A **component** is a standardized, interchangeable software module with the following characteristics:

- Is executable
- Has a unique identifier
- Has a well-known interface

Components are ready-to-use software. They're compiled and linked for the OS and CPU on which they'll run, or they're Java programs that can run in an installed Java Virtual Machine (JVM).

Every component has a unique identifier (ID), which is a number or symbolic name. Like Internet names and sockets, component names must be unique. Unique IDs enable one component to find and request services from another component. Component IDs must be registered with a directory service that other components can query.

A component's interface is the set of services it provides or tasks it performs. Each service or task is similar to a function or subroutine in a program, in that it has a name and parameters. When one component asks another to perform a task, it sends it a message containing the task or service name and the required input parameters. The component receiving the message performs the task and, if necessary, returns results.

Component-Based Software

Components are important in software development because complex programs and applications can be constructed from smaller previously developed parts. Other complex products have been built from components for decades. For example, a car contains tens or hundreds of thousands of parts, few of which are manufactured by the companies assembling and selling cars. Automobile manufacturers specify standards for components such as engines, bearings, and tires that can often be used in several models. Other companies manufacture parts based on these specifications. Without component-based construction, cars would be more expensive, less reliable, and more difficult to repair and modify.

Component-based design and construction provide similar benefits to complex software products. For example, grammar-checking in most word-processing programs can be implemented as a function or subroutine that's called by other parts of the program. The grammar-checking function's source code is integrated into the rest of the program's source code during compiling and linking. The executable program is then delivered to users.

Now consider two possible changes to the grammar-checking function:

- The developers of another word-processing program want to incorporate the grammar-checking function into their product.
- The developers of the grammar-checking function modify it to improve speed and accuracy.

For the first change, the developers integrate the grammar-checking function's source code into the new word-processing program. They add the necessary grammar-checking function calls to their program's source code and then compile, link, and distribute this program to users.

For the second change, the developers deliver the grammar-checking function's source code to the developers of both word-processing programs. Both development teams integrate the new source code into their programs, recompile and relink the programs, and deliver a revised program to their users.

What's wrong with the approach? Nothing in theory, but a great deal in practice. The grammar checker developers can provide their function to other developers only as source code, which creates potential problems in intellectual property rights and software piracy. Also, integrating the grammar-checking function is difficult or impossible if the word-processing program is written in a different programming language from the grammar checker.

When the grammar-checking function is updated, the developers of both word-processing programs must recompile and relink their entire product to update the embedded grammar checker. The new executable program must then be delivered to users and installed on their computers—an expensive and time-consuming process.

A component-based approach to software design and construction solves all these problems. Component developers can deliver their product as a ready-to-use executable function, and developers of word-processing programs simply plug in the component. Updating a single component doesn't require recompiling, relinking, and redistributing the entire application. Applications already installed on users' machines can be updated by installing only the new component. This mechanism is the same one many companies use to update installed software, such as Windows Update and Symantec Live Update for its antivirus and other security software.

Components and Objects

Talking about components without using object-oriented (OO) terminology is difficult. Components are similar to objects in an OO program because they send and respond to messages, encapsulate internal data, and interact with other components through a well-defined interface. OO program design concepts can also be applied to component-based applications. In essence, component-based design and development scale up OO programming concepts to the level of application programs and entire information systems.

Components are usually developed with OO software development tools and programming languages, but this method isn't required. A component can be implemented with any programming language, as long as it interacts with other components via messages and a well-defined public interface. Nonetheless, the majority of components are implemented with OO tools because they're naturally suited to component development. A component behaves as a distributed object, regardless of its internal implementation.

Connection Standards and Infrastructure

Interoperability across hardware or software components requires well-defined and widely adopted standards. For example, consider phone connections and the services provided by telecommunication companies (see Figure 13.19). In the United States, all phones are connected to the public telephone grid by a four-conductor wire with an RJ-11 connector. Each wire in an RJ-11 connector carries a specific electrical signal with known voltage and other characteristics. This standard ensures that any phone can be connected easily to any phone interface.

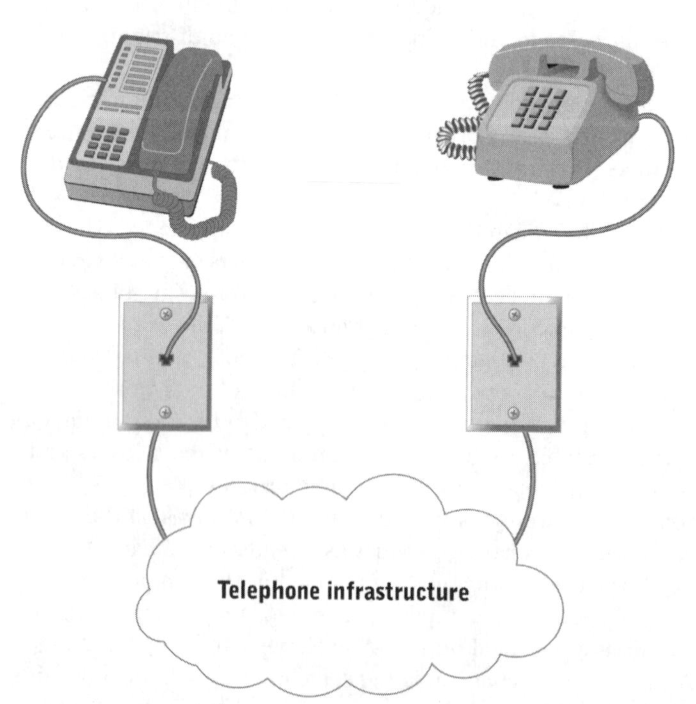

FIGURE 13.19 Standard connectors and infrastructure enable communication between phones
Courtesy of Course Technology/Cengage Learning

Phones rely on basic services supplied by the public telephone infrastructure. A unique phone number identifies every phone or set of phones. The infrastructure provides connection services between phones as well as directory services to match numbers to people, organizations, and physical locations. Switching and routing are handled by the infrastructure with no specific instructions from sending and receiving phones. Additional services, such as forwarding and voicemail, can also be provided. The standards that enable any phone to access any other phone or service have been developed over many years through industry cooperation and occasional governmental intervention.

Software components require similar standards for connections and services. They need the software equivalent of an RJ-11 plug and an infrastructure that provides directory, routing, and forwarding services. In a single computer, OS interprocess communication services can connect components, and configuration files can supply directory information.

Connecting components located on different machines running different OSs requires a standard network protocol, even when all components execute on the same machine because it provides the flexibility needed to distribute components in the future. Standard Internet services and protocols, such as TCP/IP and sockets, provide part of a component connection solution, but they don't address two important issues:

- Format and content of valid messages and responses
- Means of uniquely identifying each component on the Internet and routing messages to and from that component

Addressing these issues requires additional standards, protocols, and services. Many organizations have participated in developing these standards over the past two decades. Currently, four standard families are well developed and widely implemented: CORBA, COM+, SOAP, and Java EE, discussed in the following sections.

CORBA

In the 1980s, many computer hardware and software organizations joined forces to create an industry-wide component interoperability standard known as the **Common Object Request Broker Architecture (CORBA)**. CORBA specifies the middleware objects use to interact across networks and has these two key components:

- **Object Request Broker (ORB)**, a service that maintains a component directory and routes messages between components
- **Internet Inter-ORB Protocol (IIOP)**, a component message-passing protocol

ORB is a server process that can reside anywhere on a network or be distributed across many network nodes. It acts as a component registry, a message router, and a translator. Components must register themselves with ORB before other components can connect to them. ORB assigns a unique identifier to each registered component, so that component's methods can be invoked by other components anywhere on the Internet.

A component that wants to invoke a method in another component sends an IIOP-formatted request to the nearest ORB. ORBs cooperate with one another to locate the component and establish a connection. After a connection has been established, parameters can be passed in both directions. If necessary, ORB can translate parameters from one format to another, which enables incompatible objects, such as those developed with different languages or running on incompatible CPUs, to exchange data.

The CORBA standard is robust, inherently scalable, and independent of programming language, OS, and CPU architecture. These features are the result of deliberate design, a long development history, and participation by many computing vendors and organizations. The standard is implemented widely, with support from the Open Group and major computer and software vendors, such as IBM, Hewlett-Packard, and Oracle.

COM+

Component Object Model Plus (COM+), a Microsoft specification for component interoperability, has its roots in older Microsoft specifications, including Object Linking and Embedding (OLE) and the Component Object Model (COM). In the early and mid-1990s, Microsoft incorporated DCE services into COM and called the resulting specification the Distributed Component Object Model (DCOM). COM+ is the most recent upgrade of this standard.

Like CORBA, COM+ defines component registration, message-routing services, and a component communication protocol. COM+ services and protocol are similar to CORBA, with the following key differences:

- Components aren't assigned a permanent identifier, and their internal states can't be stored permanently. COM+ components can't remember information from one invocation to the next. COM+ components are similar to functions or subroutines, and CORBA components are objects.
- COM+ components are registered in the Windows Registry of the client machine on which they're installed. The Windows Registry stores information other than component registrations, including hardware configuration, software configuration, and user profile information. A CORBA ORB is dedicated to component services.

COM+ is widely used. Most developers of general-purpose application software, such as word-processing and spreadsheet programs, use COM+.

SOAP

Both CORBA and COM+ have some major disadvantages for building distributed component-based software. For CORBA, the main problem is complexity. CORBA infrastructure requirements are substantial, and the programming constructs needed to access these services are complex. This complexity has made companies reluctant to invest heavily in the technology and created a shortage of skilled CORBA-trained personnel.

For COM+, the main problem is dependence on proprietary technology and limited support outside of Microsoft products. A commitment to COM+ entails a commitment to Microsoft OSs and other system software. Although Microsoft dominates the desktop OS market, most organizations have a diverse collection of server OSs and software. Also, few non-Microsoft development tools support COM+.

Simple Object Access Protocol (SOAP) is an open standard, developed by the World Wide Web Consortium (W3C; *www.w3.org*), for distributed object interaction that attempts to address the shortcomings of both CORBA and COM+. Unlike CORBA, SOAP has few infrastructure requirements, and its programming interface is simpler. Perhaps the best evidence of SOAP's long-term potential for success is that Microsoft adopted it as the basis of its .NET distributed software platform.

A key to SOAP's simplicity and minimal infrastructure requirements are its reliance on existing Internet protocols—HTTP and XML. Messages between objects are encoded in XML and transmitted by using HTTP, which enables the objects to be located anywhere on the Internet. Figure 13.20 shows a client sending a service request to a server as a SOAP message. The same transmission method supports server-to-client and peer-to-peer communication. The SOAP encoder/decoder and HTTP connection manager are standard components of a SOAP programmer's toolkit. Applications can also be embedded scripts that use a Web server to provide SOAP message-passing services. SOAP messages can be transmitted by using other protocols, such as FTP and SMTP, but HTTP is the most common transmission protocol.

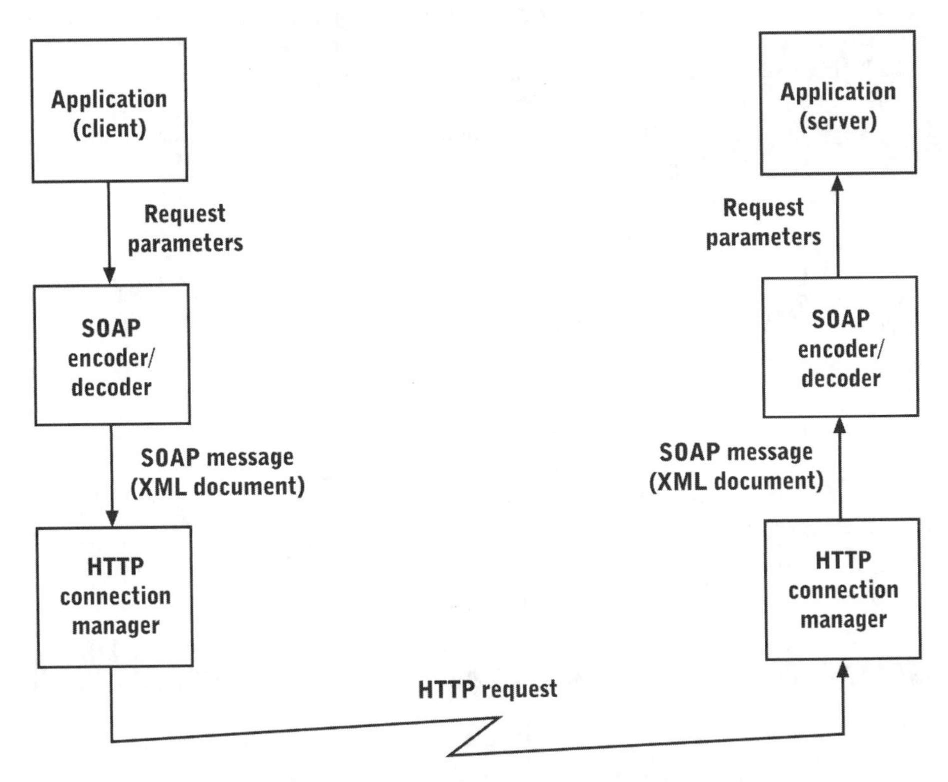

FIGURE 13.20 Client/server communication with a SOAP message
Courtesy of Course Technology/Cengage Learning

Although SOAP has been widely deployed for many types of distributed applications, it has some major limitations. Early SOAP versions left many implementation specifics undefined, thus creating problems of interoperability. SOAP 1.2 filled many gaps in earlier standards but is still an incomplete solution. Building and deploying industrial-strength SOAP-based distributed applications isn't as simple as it might seem because developers must use additional protocols to address issues such as security and message delivery guarantees. Nonetheless, SOAP has been a popular basis for distributed applications for the past decade because it filled a largely unmet need for a powerful yet easy-to-use approach to distributed applications. Recent protocols, such as Java RESTful Web services, will probably supplant SOAP gradually because they're more complete standards and support newer and more powerful Web application architectures.

TECHNOLOGY FOCUS

Java Platform, Enterprise Edition

Java Platform, Enterprise Edition (Java EE) is a family of standards for developing and deploying component-based distributed applications written in Java. Figure 13.21 shows key elements of the Java EE architecture, which follows the three-layer architecture described earlier. The client tier can include Java components and Web browsers displaying HTML pages with embedded scripts. Browser-based client interfaces are sometimes called "thin clients" because they contain little or no program code. "Thick clients" are collections of complete Java objects that run under the control of a Java Virtual Machine (described in Chapter 10) and communicate directly with corresponding components in the Web/business tier.

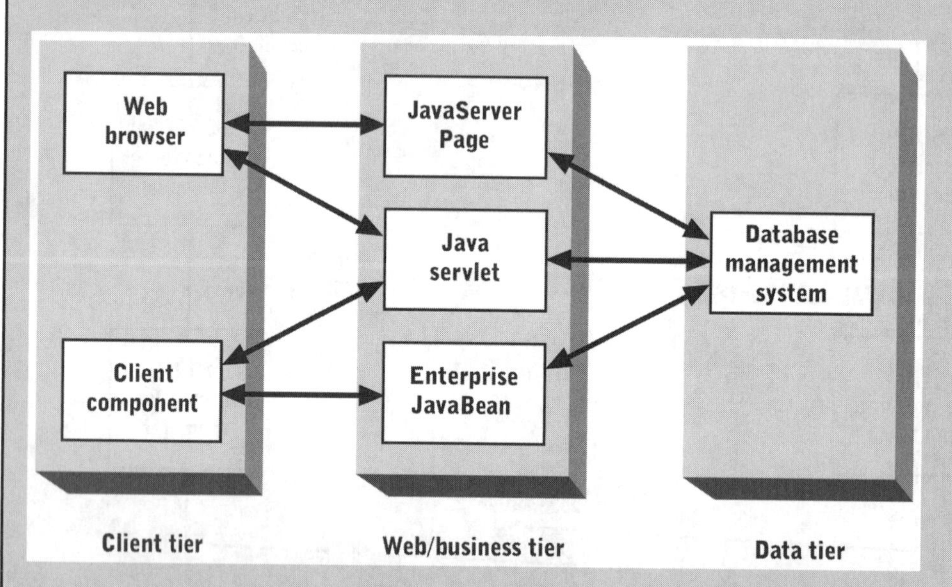

FIGURE 13.21 Java EE architecture
Courtesy of Course Technology/Cengage Learning

In the Web/business tier, components can be **JavaServer Pages (JSP)**; servlets, which run under the control of a Web server; and **Enterprise JavaBeans (EJBs)**, which run in a system software component called a "business container." JSP are components that generate formatted Web pages by using embedded scripts. Servlets are full-fledged

(*continued*)

Java programs that can perform more complex operations, including computations and interactions with other EJBs and databases. JSP and servlets communicate with the client tier by using HTML or XML messages. The latest Java EE standard includes support for **JavaServer Faces (JSF)**, which enable developers to create user interfaces that run on a server but interact with a client Web browser or component.

EJBs are Java programs that perform complex, behind-the-scenes processing. For example, an EJB might implement a distributed object with object attributes stored in the database, perform a complex query against multiple tables in multiple databases, or manage a complex customer order with related financial, inventory, and shipping transactions. EJBs provide services that can be called by components running on the client or on a Web server or by other EJBs.

Component interactions are based on many standards, including the following:

- *Remote Method Invocation (RMI)*—Enables objects executing on different computers to send messages and receive responses
- *Java Naming and Directory Interface (JNDI)*—Enables objects executing on different machines to locate one another and query their available methods
- *Java Authentication and Authorization Service (JAAS)*—Authenticates users and restricts access to components
- *Java Database Connectivity (JDBC)*—Enables objects to interact with relational databases by using SQL statements

Developing and deploying Java EE applications is a complex endeavor because of the elaborate architecture, the number of related standards, and the number of required system software elements. Many vendors, including IBM and Oracle, offer development packages and system software suites to support Java EE applications. Developers typically require months of training to learn the intricacies of the tools and system software.

Java EE application portability is enhanced by a common programming language (Java) and well-defined standards for describing and storing components and system software configuration information in Java source code and XML files. The files have standard internal formats and naming conventions and are placed in a standardized directory structure. This standardization enables developers to move all or part of a Java EE application from one computer or system software suite to another simply by copying the corresponding files and restarting the supporting system software services.

The earliest version of Java EE was developed after CORBA and COM+ but before SOAP. Java EE is incompatible with COM+ and, in its original form, with CORBA, too. However, subsequent revisions to Java EE and CORBA have brought them closer together. The current RMI standard enables CORBA components and Java EE EJBs to interact by using the CORBA IIOP standard. Java components can be defined with the CORBA interface definition language and registered with a CORBA ORB. Java EE also includes limited interoperability with SOAP, which is expected to expand in future Java EE standards.

Java EE has been used widely in enterprise-level information systems. Although its complexity rivals CORBA, it has been more successful, mainly because of support from Sun Microsystems and Oracle, which merged in 2009. The battle for the future of distributed systems seems to have narrowed to Java EE and Microsoft .NET. Microsoft .NET had the advantages of an early adoption of Web services standards, including SOAP, and its considerable resources and market presence. However, Java EE has a well-established reputation as a platform for reliable and scalable industrial-strength Web applications and a strong support base from many vendors and software developers.

EMERGING DISTRIBUTION MODELS

The previous section describes distributed computing in terms of distributed components and objects—a fine-grained view of distributed software that concentrates on how small parts of an application program can be distributed across a network yet cooperate to perform a larger function. A new class of distribution architectures emerged in the 2000s based on a more coarse-grained view of distribution. These distribution architectures deal with larger units of software and the hardware hosting them and address how to distribute entire information systems and deploy them rapidly and flexibly.

By the 2000s, nearly all medium and large businesses had built or purchased information systems to streamline nearly every aspect of the business. Ubiquitous high-speed Internet connections enabled interconnecting these systems with customer and supplier information systems, a setup sometimes described as **business-to-business (B2B)**. Software support for business functions grew more complex, as did the supporting computing and network infrastructure. Managing applications and infrastructure became an increasingly complex task requiring a larger share of organizational resources.

At the same time, higher levels of automation were a barrier to starting new ventures. A small startup company or a larger organization entering a new business area could purchase needed application software, but how could they quickly acquire, assemble, and manage the infrastructure supporting the software? Where could they find the required investment capital and technical expertise to do so? In effect, the increasing complexity of information systems and supporting infrastructure had become a barrier to entry in many industries.

This problem of a barrier to entry has many historical precedents. For example, in the late 1800s and early 1900s, there was no electrical power grid, yet electricity enabled a new generation of manufacturing techniques in many industries that speeded production, reduced costs, and increased profits. Larger factories could afford to build their own electricity-generating stations to power production machinery. Smaller factories couldn't afford the needed infrastructure and, therefore, were in danger of being left behind in this chapter of the Industrial Revolution. In the end, the problem was solved by developing local, regional, and then national power grids. Electricity became a commodity that could be purchased by anyone willing to be hooked to the grid.

In many ways, computing in the late 20th century is similar to the U.S. electrical grid of the late 19th century. Electricity-generating stations and large computing centers existed but were just beginning to be connected. Standards were the exception rather than the norm, and there were substantial gaps in connectivity and compatibility. As with electricity, a decade or two changed the landscape. By the mid-2000s, vast computing power was available from large and small providers over a national and international grid—the Internet.

The dissimilarity lies in what can be delivered over that grid. Electrical power is a simple, uniform commodity delivered in only one direction. In contrast, the Internet enables bidirectional delivery of a bewildering array of computing, software, and information services. Customers can purchase access to components, application software, or entire information systems, or they can "rent" platforms and infrastructure on which to install their own software and services. Standardization and compatibility are more complex in this diverse arena.

The term **cloud computing** has been coined to summarize new approaches to distributing and accessing software and hardware services across the Internet. The term is appropriate in many ways. It embodies a physical reality assumed to exist and work but about which few details are known. For years, network engineers have drawn diagrams with clouds representing parts of the network over which they have no control (for example, the Internet as a whole or a network purchased from a large telecommunications provider). In these diagrams, a cloud describes something that can be seen from the outside but not from the inside. The term "cloud" also embodies the concepts of change and size: The Internet and cloud computing change shape and can grow over time. Three current distribution modes that fit in the cloud computing paradigm, summarized in Figure 13.22, are described in the following sections.

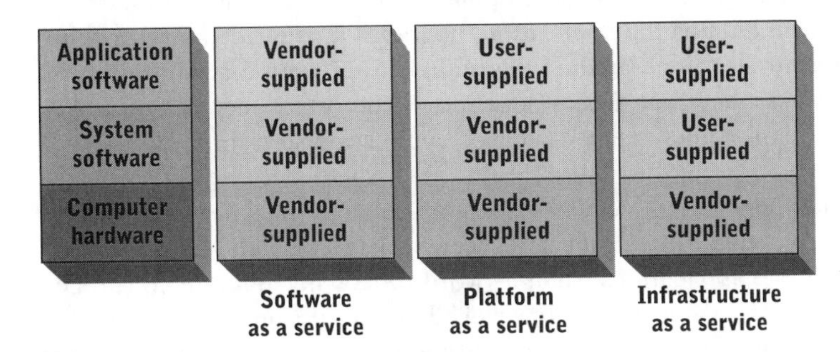

FIGURE 13.22 Cloud distribution models
Courtesy of Course Technology/Cengage Learning

Software as a Service

As described earlier, operating systems and some application software are constructed as a set of services that can be accessed by both internal and external users and software components. In the 1990s, the term **service-oriented architecture (SOA)** was coined to describe this design philosophy, partly reflecting the growing dominance of client/server architecture. As Internet connections became more common and Web browser technology matured, most application software was constructed with the view layer implemented as a Web page that accessed a collection of interconnected back-end services housed on organizational servers.

During this same era, companies such as Yahoo! and Google emerged as leaders in providing services to both businesses and end users via Web browsers. Internet search engines and e-mail applications were the earliest examples of widely used Web-based services, but they were soon followed by many others in application areas such as mapping (Google Maps), social networking (MySpace and Facebook), and communication (Skype).

Software as a service (SaaS) is a Web-based architectural approach in which users interact via a Web browser or other Web-enabled view layer with application software provided by a third party. Instead of installing application software on locally owned and administered computers, application software is installed on computers owned by a software provider, and the user accesses this software over the Internet. It's a straightforward extension of SOA in which services are defined in large chunks, and the resources providing services are no longer local.

From a business perspective, SaaS offers benefits by eliminating the need to purchase and maintain application software, system software, and supporting hardware. A third party can provide an entire information system, and the user organization essentially rents access to it. The only local resources required to implement this architecture are desktop or other computers with a Web browser or another interface and an Internet connection. The user organization no longer needs racks of servers and associated system software and support personnel. If an organization follows SaaS for all its application software, it has essentially outsourced many formerly internal IT functions. From the systems operations and maintenance perspective, all that remains is management of desktop and other end-user computing devices, the local network, and Internet connections.

SaaS also has substantial benefits from a consumer perspective. Consumers are freed from the need to purchase, maintain, and upgrade application software, such as office suites. The service provider maintains the application, upgrades it as needed, and supplies the hardware and labor resources to operate and maintain the software. Users no longer purchase software or upgrades; they rent access per time period or per metered use. User computing devices can be simpler and cheaper because application software no longer resides on them.

To use SaaS successfully, customers need reliable high-capacity Internet connections. If the connection isn't reliable, users might not have access to software and other resources when needed. Simple applications, such as word processing and e-mail, can be supported with ordinary Internet connections. However, data- and graphics-intensive applications, such as online game playing and video editing, require high-capacity Internet connections.

Platform as a Service

Some organizations aren't willing to embrace SaaS fully. Reasons vary but include concerns over ownership and privacy of data, use of custom-developed application software that provides a competitive advantage, and concerns about "locking up" important IT resources with a single vendor.

Platform as a service (PaaS) describes an architectural approach in which an organization rents access to system software and hardware on which it installs its own application software and other services. The simplest example is a hosted Web site. Typically, the user organization develops Web content on a desktop computer or an inexpensive local test server. After the content is developed, it's uploaded to a platform provider that places the site on its own servers and makes it accessible to the intended users. The customer retains ownership of the content and applications and can move them to another platform or service provider if needed. More complex PaaS examples range from Web-based applications, such as e-commerce sites, to entire information systems supporting all facets of an organization's operations.

From the customer/user's perspective, the main advantage of PaaS is avoiding the need to operate its own servers and system software. In the hosted Web site example, the customer bypasses the need to purchase and maintain the server, its OS, the Web server software, and any supporting software and hardware, such as firewalls. The customer also gains the flexibility to scale the Web site up or down in size rapidly. If the Web site is inundated with accesses, for example, the customer can simply rent a more powerful platform. For complex application software, the customer avoids the cost of purchasing, maintaining, and supporting related system software, such as component directories, object brokers, and DBMSs.

Infrastructure as a Service

Infrastructure as a service (IaaS) is similar in many ways to PaaS, but the service provider supplies little or no system software. Hardware virtualization is a key supporting technology in IaaS because it enables a customer/user to configure application and system software for a generic platform as virtual servers and then deploy these servers to a third-party hosting site. The customer is responsible for purchasing and configuring most or all software components, including application, system, and database software, and encapsulating them in virtual servers that are compatible with the hosting infrastructure.

IaaS can also be structured for only part of an application's infrastructure requirements. For example, companies such as Amazon and Google provide back-end data storage services that customers can use to extend locally owned data storage infrastructure. Similarly, many companies provide access to large-scale computing infrastructure that can be used for complex simulations in marketing research, engineering analysis, and other areas.

Although an IaaS customer organization avoids fewer internal costs than it would with SaaS or PaaS, it retains the advantage of being able to respond rapidly to changing demands by scaling capacity up or replicating capacity to new geographic locations. With virtualization, capacity can be increased by upsizing or cloning virtual servers. In specialized cases, such as storage and computation, additional capacity can be purchased as needed.

Risks

SaaS, PaaS, and IaaS share common vendor-related risks. First, vendor reliability is a critical issue. For all three, internal and external functions become dependent on vendor-provided resources. Most organizations can't tolerate major interruptions in system availability, so the reliability and availability of vendor-supplied services must be guaranteed to whatever level the customer requires.

Second, vendor lock-in is a risk, although its level varies across architectures, as shown in Figure 13.23, and with additional specifics. With vendor lock-in, a customer is forced to continue using a vendor because of high cost or other difficulties in switching. In cloud architectures, vendor lock-in arises from lack of compatibility between vendors, making it difficult to move customer-owned resources from one vendor to another. Vendor lock-in isn't a recent phenomenon. It has been a risk through most of computing history and has occurred in hardware platforms, system software, and application software, usually when standards are lacking or just emerging, as with cloud computing.

With cloud architectures, the risk of vendor lock-in rises with the number of hardware and software components the vendor provides. The risk is lower with IaaS because most vendors use similar virtualization environments, which enables customers to move virtual servers between vendors easily.

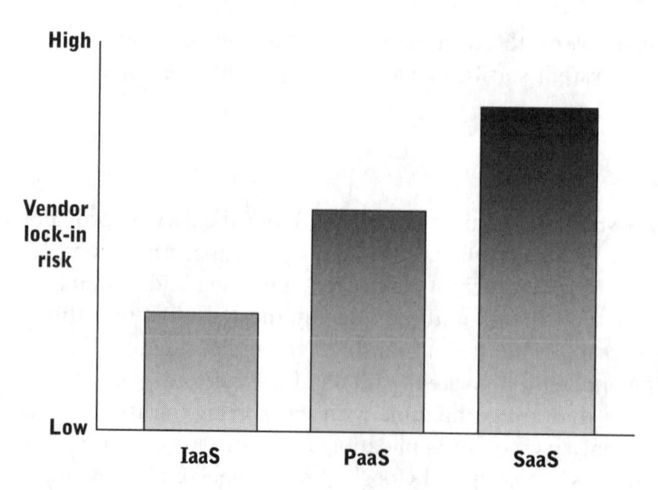

FIGURE 13.23 Vendor lock-in risk is lowest for IaaS and highest for SaaS
Courtesy of Course Technology/Cengage Learning

Lock-in risk rises with PaaS because the vendor provides system software components in addition to hardware. The precise risk level depends on the system software components and the extent to which these components follow widely implemented standards. For example, if a vendor's platform includes Linux and other system software components covered by the GNU public license, customers can move their application software to other vendors with few problems. In contrast, if a vendor uses proprietary system software (for example, Google's BigTable), customers who want to change vendors must reimplement portions of their application software that use the proprietary components.

The highest lock-in risk occurs with SaaS because the vendor controls all key components of the customer's information system. Customers usually negotiate contract terms to reduce the risk, including the right to export data from the system in standardized formats and long-term pricing and support agreements.

Other cloud-related risks include service availability, data security and privacy, and legal ownership of stored data. Most cloud vendors guarantee minimum levels of availability via service-level agreements with financial and other penalties for failure to meet the agreements. However, it's up to each user organization to determine the full impact—financial and otherwise—of a major service outage, such as lost productivity, lost sales, dissatisfied customers, and damage to its reputation. Although these issues concern many managers considering the use of cloud services, they should be just as concerned about them when evaluating the reliability of internal IT services.

Security, privacy, and legal ownership of data are thorny issues in a rapidly shifting legal landscape. Some industries, such as health care, have a legal and regulatory obligation to secure data at rest and in transit and to protect this data from unauthorized use. For many other industries, ensuring data security and privacy is simply good business practice; rapidly changing legislation at all levels of government has increased organizations' obligations in these areas with heavy penalties and legal liability for noncompliance. Meeting these obligations is simpler when most data traverses internal networks, but with cloud computing, most or all data traverses public networks, which are more difficult to protect. However, as with service-level agreements, many internal networks are less secure than managers assume.

BUSINESS FOCUS

Moving Office Applications into the Cloud

Many businesses, large and small, have moved some or all of their office applications into the cloud, and more are considering doing so. Although there are many vendors of cloud computing services, Google stands out as a leader in the market, offering services that cross all three cloud distribution models.

As of mid-2010, Google claims more than two million business customers. Google Apps is perhaps the best-known SaaS offering because it's used by both businesses and home users. It includes the following:

- *Gmail for business*—Combines traditional e-mail, voicemail, chat, and instant messaging into a single application and interacts with cell phones and other mobile devices
- *Google Calendar*—Individual, group, and project-based calendaring; tightly integrated with Gmail
- *Google Docs*—Documents, spreadsheet, and presentation graphics integrated into a single Web-based suite
- *Google Sites*—Web page and Web site creation; intended mainly for document distribution and workgroup coordination
- *Google Video*—Video sharing for a variety of uses, including marketing, training, and recruiting

The Google Apps suite competes with traditional software products from companies such as Microsoft, Lotus, Corel, and OpenOffice.org. Microsoft, in particular, has been feeling pressure because Google Apps competes against some of its most profitable software products, such as the following:

- *Microsoft Office*—Competes with the combination of Gmail for business, Google Calendar, and Google Docs. Office is considered to have more features than Google Apps and typically has better performance for many tasks, especially graphics-intensive tasks, such as image editing.
- *Microsoft Exchange*—Competes with Gmail. Microsoft Exchange is a back-office e-mail, messaging, and calendaring server that provides support services for Microsoft Office in many medium and large organizations.
- *Microsoft SharePoint*—Competes with Google Sites. Enables users to develop Web sites for document distribution and workgroup coordination.

The Microsoft Office suite is typically installed on desktop computers. However, in an organizational setting, it derives much of its power from Microsoft Exchange and SharePoint running on servers. Organizations using all three products often have substantial investments in server hardware, supporting OSs, and related software and

(continued)

provide extensive in-house support for users. Capital costs for servers and related hardware are high, as are annual operating expenses for software licenses and IT staff.

In contrast, organizations using Google Apps can avoid application-related investments in server hardware and high annual operating expenses. Instead, organizations pay an annual per-user license fee to Google, and employees can use applications "anytime, anywhere." Using Google Apps does require a reliable, high-speed network with high-capacity Internet connections. However, building and supporting this type of network is generally less complex and costly than providing the supporting infrastructure and technical support for locally installed software.

Compare the combination of Microsoft Office, Exchange, and SharePoint with the Google Apps suite for the following business settings (the same ones used in the Chapter 11 Business Focus):

- A small electrical supply wholesaler with 25 employees, 15 user workstations, and an application and services suite that includes basic accounting and inventory control functions, file and printer sharing, and a small Web site not used for e-commerce
- An engineering design and consulting partnership with 25 employees, 15 user workstations, and a mix of applications and services, including basic accounting functions, bid preparation, CAD and drafting, construction management, file and printer sharing, and a small Web site not used for e-commerce
- A large catalog seller of musical equipment with hundreds of employees; three warehouses in New Jersey, St. Louis, and Portland; and a mix of applications and services, including all accounting functions, catalog preparation and distribution, phone and Web-based sales, and inventory control and logistics

In your comparison, be sure to pay careful attention to the following issues:

- Differences in application capability and performance and their importance (or lack of importance) to users in each business setting
- Cost of desktop and portable computing devices (locally installed applications require more powerful computers)
- Cost of server and network hardware
- Annual software licensing costs for applications, supporting software (such as Exchange), and operating systems
- Annual costs of IT support personnel

Summary

- Information systems are typically distributed across many computer systems and geographic locations. Client/server, three-layer, and n-layer architecture define frameworks for dividing applications into multiple layers that can execute on different computers. Multilayer architectures require standard methods and services to communicate with one another. System software that implements communication standards and gives clients and servers the capability to interact is called middleware.
- A network protocol enables users and applications to interact with resources and applications on remote computers. The OS service layer accepts resource access requests and routes them to remote or local service processes. Local requests to remote resources are routed through the protocol stack to external servers, and remote requests to access local resources are received through the protocol stack. Connections to remote resources can be static or dynamic. Dynamic connections are more flexible but require a distributed registry of resource names and locations.
- With directory services, users, resources, and components can find one another on the Internet. Directory services must span the range of resource types and Internet protocols. LDAP is a widely used directory service standard that can track users, distributed resources, and objects. Although it defines a standard method for defining and accessing directory structure, it doesn't define standard content templates, which limits interoperability between different LDAP directories.
- Distributed processes must communicate with one another to exchange data and synchronize their activities. P2P interprocess communication protocols include sockets, named pipes, RPC, and DCE. Sockets implement direct process-to-process communication via protocol stacks. Named pipes enable multiple clients to interact with a single server via a shared file system object. RPC allows one process to execute another as a subroutine with parameter passing and format translation. DCE combines all these approaches and adds security and minimal directory services.
- The Internet is a global network based on TCP/IP and many other protocols. The Web is the set of resources accessible over the Internet via standard protocols. Internet protocols define methods for sharing many types of resources, including files, documents, and applications. Multitiered applications can be implemented over the Internet by using a Web browser for the user interface and Internet and Web protocols to connect the layers.
- Component-based applications are divided into many different cooperating processes or distributed objects. Each distributed object implements a public interface to the services it provides. Component-based applications require protocols and infrastructure for component registration, discovery, and communication. CORBA, COM+, Java EE, and SOAP are standards that address the full range of component infrastructure and communication services.
- Cloud computing is an increasingly attractive alternative to managing an entire suite of hardware, system software, and application software internally. Cloud distribution models include software as a service, infrastructure as a service, and platform as a service. SaaS moves application software and all supporting layers into the cloud. PaaS enables users to

install and run their own applications on vendor-supplied hardware and system software. IaaS provides distributed access to vendor-owned hardware with user-owned system and application software. Cloud computing models share varying levels of vendor lock-in, security, and data ownership risks.

Key Terms

access control list (ACL)
Active Directory
business logic layer
business-to-business (B2B)
client
client/server architecture
cloud computing
Common Object Request Broker Architecture (CORBA)
component
Component Object Model Plus (COM+)
data layer
directory services
distributed computing
Distributed Computing Environment (DCE)
distributed processing
Domain Name System (DNS)
dynamic connections
Enterprise JavaBeans (EJBs)
Extensible Markup Language (XML)
File Transfer Protocol (FTP)
HTTPS
Hypertext Markup Language (HTML)
Hypertext Transfer Protocol (HTTP)
infrastructure as a service (IaaS)
Internet
Internet Inter-ORB Protocol (IIOP)
Internet Message Access Protocol 4 (IMAP4)
intranet
Java Platform, Enterprise Edition (Java EE)
JavaServer Faces (JSF)
JavaServer Pages (JSP)
Kerberos
Lightweight Directory Access Protocol (LDAP)
location transparency
middleware
Multipurpose Internet Mail Extensions (MIME)
n-layer architectures
n-tier architectures
named pipe
network transparency
Object Request Broker (ORB)
objectclass
peer-to-peer (P2P) architecture
pipe
platform as a service (PaaS)
Post Office Protocol 3 (POP3)
protocol stack
Remote Procedure Call (RPC)
resource registry
Secure Shell (SSH)
service-oriented architecture (SOA)
Simple Mail Transfer Protocol (SMTP)
Simple Object Access Protocol (SOAP)
software as a service (SaaS)
static connection
Telnet
three-layer architecture
three-tier architecture
Uniform Resource Locator (URL)
view layer
World Wide Web (WWW)

Vocabulary Exercises

1. _______ architecture divides an application into multiple processes, some of which send requests, some of which respond to requests, and others that do both.
2. System software that connects parts of a distributed application or enables users to locate and interact with remote resources is called _______ .
3. A(n) _______ is a process that sends a request. A(n) _______ is a process that responds to requests.
4. A server process creates a(n) _______ that allows clients to send data or messages via a shared filename.
5. With the _______ protocol, a process on one machine can call a process on another machine as a subroutine with parameters passed in either or both directions.
6. Three-layer architecture divides an application into _______ , _______ , and _______ layers.
7. A(n) _______ to a remote resource must be initialized by the user or a system administrator.
8. _______ is the original Internet e-mail standard. More recent e-mail standards include _______ and _______ .
9. The _______ is a global collection of networks that are interconnected with TCP/IP.
10. Web pages are encoded in and delivered from a Web server to a Web browser via _______ .
11. The _______ standard defines methods for embedding graphics and other nontext data in e-mail messages and Web pages.
12. The _______ is a collection of resources that can be accessed over the Internet by standard protocols, such as FTP and HTTP.
13. _______ is a family of component infrastructure and interoperability standards supported by Microsoft.
14. _______ defines a standard for describing and accessing directories of users and distributed resources.
15. A(n) _______ contains a protocol, an Internet host, an optional socket, and a resource path.
16. _______ is a wide-ranging standard covering network directory services, file sharing services, RPC, remote thread execution, system security, and distributed resource management.
17. With the _______ protocol, a client can interact with a remote computer's command layer as though it were a directly connected VDT.
18. A(n) _______ is a standardized, interchangeable, and executable software module that has a unique identifier and a well-known interface.
19. _______ is a family of component infrastructure and interoperability standards supported by a broad consortium of computer companies.
20. The set of software layers for implementing network I/O and services is called a(n) _______ .

21. Web/business tier components of the Java EE distributed application architecture include ________ , ________ , and ________ .
22. ________ is a secure version of HTTP that encrypts HTTP requests and responses.
23. ________ is an improved version of Telnet that encrypts data flowing between client and server.
24. A(n) ________ is a shared memory region that enables multiple processes executing on the same machine to exchange messages.
25. ________ describes a cloud computing model in which users install their own applications on vendor-supplied hardware with vendor-supplied system software.
26. ________ describes a cloud computing model in which users access vendor-owned software, running on vendor-owned hardware and system software, via the Internet.

Review Questions

1. Describe client/server, three-layer, and n-layer architecture. What are the differences between a client and a server? What is the function of each layer in a three-layer application? Why might more than three layers be used?
2. What is middleware?
3. What is a protocol stack? What are the components of a typical protocol in a client computer that can access many Web servers?
4. What are the differences between static and dynamic connections to remote resources? Which connection type requires a resource registry? Where should the resource registry be located?
5. An OS acts as both client and server. How are software components organized to perform both functions at the same time?
6. Describe three low-level P2P interprocess communication standards. What are the advantages and disadvantages of using these standards to implement distributed multilayer applications?
7. Do the terms Internet and Web describe the same thing?
8. What are the components of a URL?
9. Describe at least five standard Internet and Web protocols.
10. How can the Internet be used as a platform to implement distributed multilayer applications? Which Internet and Web protocols are used, and how are they used?
11. What is a component? Component-based design and development have been the norm in manufacturing durable goods for decades. Why has this approach only recently been adopted for designing and deploying information systems?
12. Describe the COM+ and CORBA standards for component infrastructure and communication. Which standard would you choose to support a new large-scale information system? Why?
13. What are directory services? What types of information might be made available through directory services? Describe the LDAP standard.

14. Describe the components of the Java EE architecture. What standards govern the form of the client and Web/business tiers? What standards govern communication between components? Is Java EE compatible with other distributed application standards?
15. Describe the role of DNS in enabling dynamic connections. Could DNS function with a fully centralized directory server architecture? Why or why not?
16. Briefly describe the three most common approaches to cloud computing, and compare the risk levels and potential economic benefits of each approach.

Research Problems

1. The Kerberos security model is a part of both the DCE and CORBA standard families. Investigate the capabilities and limitations of Kerberos. What degree of security does it provide? What infrastructure components are needed to use it? Does it address all security issues in distributed applications and resources?
2. The CORBA (*www.corba.org*), Java EE (*http://java.sun.com*), and SOAP (*www.w3.org*) standards are evolving rapidly. Visit these Web sites, and determine what features have been added by recent standard revisions. What needs are these new features intended to satisfy? Are they being implemented widely? Why or why not?
3. Distributed.net (*www.distributed.net*) is an organization that coordinates research on distributed computing applications. People and organizations can join Distributed.net and contribute idle computing power to ongoing research projects that require or use massively parallel approaches to solving complex computational problems. Investigate some current projects and the methods and protocols that coordinate software on members' computers.
4. Although Google is the clear market leader in cloud computing services, it does have competitors. Identify one or two competitors in each cloud market segment—SaaS, PaaS, and IaaS—and compare their services with Google's. In which market segment is Google most dominant? In which is it least dominant?

CHAPTER 14

SYSTEM ADMINISTRATION

CHAPTER GOALS

- Describe system administration responsibilities and tasks
- Explain the process of acquiring computer hardware and system software
- Describe tools and processes for evaluating application resource requirements and computer system performance
- Summarize measures for ensuring system security
- Describe physical environment factors affecting computer hardware

In previous chapters, you have explored the inner workings of computer hardware and system software. In this chapter, you take a step back from bits, bytes, and instructions to examine hardware and software from a system administrator's point of view. A system administrator's job includes many varied tasks, and describing them in detail could fill an entire book. This chapter concentrates on tasks requiring in-depth knowledge of the topics covered in previous chapters (see Figure 14.1).

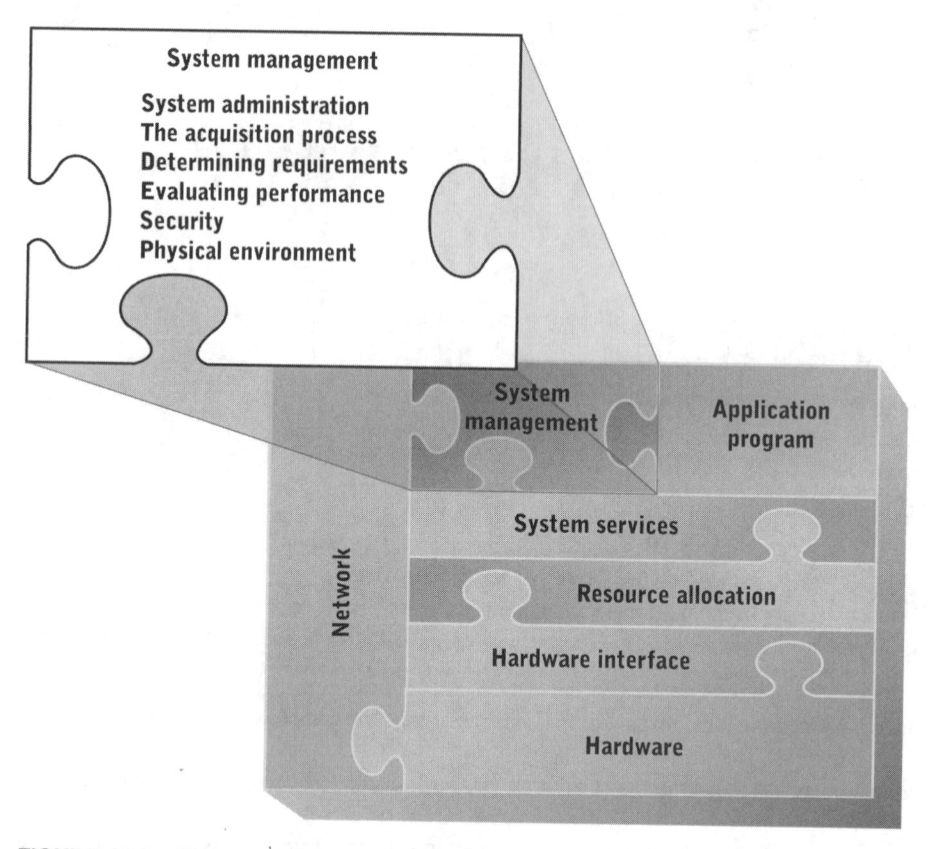

FIGURE 14.1 Topics covered in this chapter
Courtesy of Course Technology/Cengage Learning

SYSTEM ADMINISTRATION

In the context of information systems, **system administration** covers a range of activities and responsibilities. The main responsibility is to ensure efficient and reliable delivery of information services. A typical information system (IS) administrator performs many tasks to meet this responsibility, which can be grouped into the following broad categories:

- Acquiring new IS resources
- Maintaining existing IS resources
- Developing and implementing an IS security policy

The assignment of these responsibilities to specific people varies considerably across organizations. In small organizations, the user or organizational administrator might assume these responsibilities. In large organizations, these responsibilities might be divided among many people. Medium-size organizations might have a single technical specialist who handles all system administration responsibilities.

Each activity area entails a variety of system administration tasks. Some tasks are straightforward and take only a short time, such as maintaining user accounts and performing file system backups. Other tasks require many different skills and take longer, such as strategic planning and resource acquisition.

Strategic Planning

IS resource acquisition and deployment should occur only in the context of a well-defined strategic plan for the organization as a whole. For the purposes of this chapter, a **strategic plan** is defined as a set of long-range goals and a plan to attain these goals. The planning horizon is typically three years and beyond. At a minimum, goals must include the following:

- Services to be provided
- Resources needed to provide these services

The strategic plan addresses the following issues related to achieving stated goals:

- Strategies for developing services and markets for them
- Strategies for acquiring enough resources for operations and growth
- Organizational structure and control

In all cases, strategic plans must address the basic question "How do we get there from here?" An information system is only one part of an organization, and an IS strategic plan is only part of an organization's overall strategic plan. An IS strategic plan must be evaluated with plans of other organizational units. Information systems are normally a support service for other organizational units and functions, such as customer services, accounting, and manufacturing. Therefore, an IS strategic plan tends to follow, rather than lead, the strategic plans of other units in the organization.

Hardware and Software as Infrastructure

The resources devoted to most organizational activities can be classified into two categories: capital expenditures and operating expenditures. **Capital expenditures** purchase **capital resources**, which are assets or resources expected to provide benefits beyond the current fiscal year. **Operating expenditures** provide benefits only in the current fiscal year.

Typical capital expenditures include buildings, land, equipment, and research and development costs. Typical operating expenditures include salaries, rent, lease payments, supplies, and equipment maintenance. Computer and software purchases are capital expenditures, even though the expected useful lifetime of computer hardware and software has decreased in recent years because of rapid technological change.

Many capital resources provide benefits to a wide range of organizational units and functions. An office building, for example, benefits all the units and functions housed in it. These resources are referred to as **infrastructure** and have the following characteristics:

- Service to a large and diverse group of users
- Costs that are difficult to allocate to users separately
- Recurring need for new capital expenditures
- High maintenance costs

The computer hardware and system software that provide IS services are infrastructure. This statement is obvious in organizations that rely extensively on large computer systems used by many units in the organization. It's less obvious but no less true in organizations with highly decentralized computer hardware and system software. IS strategic planning issues are similar to those in many other infrastructure-based service organizations, such as utility companies supplying communications, electrical power, and water. In general, service organizations must address these strategic issues:

- What services will be provided? How will service users be charged?
- What infrastructure is required to provide the services?
- How can the infrastructure be operated, maintained, and improved at minimal cost?

For example, with the communication services from local phone companies, the main strategic question is what types of services should be provided: basic phone only, expanded phone services, Internet access, mobile services, information storage and retrieval, and so on. Answers to this question lead to decisions on the nature of the infrastructure and its capital and operating expenditures.

Standards

Providing infrastructure-based services to a wide variety of users requires adopting **service standards**. However, standardization tends to stifle innovation and produce solutions that are suboptimal for some users. Returning to the local phone company as an example, all users agree on and abide by certain standards for phone service, including acceptable user devices, basic service availability, and interactions with the infrastructure (for example, line voltage and signal encoding). The nature of this infrastructure requires standardization to provide service at a reasonable cost.

However, standardization often causes problems for some users, especially for those demanding services at or near the leading edge of technology. For example, many phones are connected to the nearest switching center with low-capacity copper wiring, which limits the availability of high-speed Internet services and, therefore, the ability of some users to work at home. The provider of an infrastructure-based service must constantly balance the benefits of standardization in reduced costs and simplified service against its costs in stifled innovation and failure to meet the needs of some users.

Standardizing hardware and system software is a particularly complex issue involving many choices and compatibility issues. Standardization issues are further complicated by the diverse components required for information processing in even a small organization.

Competitive Advantage

Discussing hardware and system software as an infrastructure ignores certain opportunities. Infrastructure management tends to concentrate on providing short-term services at minimal cost—a viewpoint that tends to preclude major technical innovations and radical redefinition of services. The term **competitive advantage** describes the way in which an organization uses resources to give it a major edge over its competitors. It can take a number of forms, including the following:

- Providing services that competitors are unable to provide
- Providing services of unusually high quality
- Providing services at an unusually low price
- Generating services at an unusually low cost

Hardware and system software can be applied to achieve a competitive advantage in any or all of these areas. Examples include investment updates to customers via text messages, radio frequency ID (RFID) tags for shipments and inventory control, and use of social networking Web sites to communicate with customers.

Each of these examples was once novel but is now commonplace. Such is the nature of applying technology for a competitive advantage: Rapid technology changes and adoption by competitors severely restrict the useful life of most technology-based competitive advantages. Furthermore, pursuing competitive advantages through new technology involves substantial risks. The costs of technology are usually high for developers and early adopters but then decrease rapidly after introduction, so late adopters might incur much lower costs yet realize most of the benefits. Early adopters also face the inefficiency of starting at the beginning of a learning curve.

BUSINESS FOCUS

A Standard Hardware Platform?

Cooper State University (CSU) is a large school with more than 1000 faculty members and a full-time enrollment of 20,000 students. Academic programs span a wide range of fields, including hard sciences, humanities, engineering, education, management, law, and medicine. CSU offers many graduate degree programs and is highly ranked in terms of externally funded research.

CSU has a campuswide computing organization, Computer Information Services (CIS), which is responsible for supporting the computing and information-processing needs of both academic and administrative users. It supports administrative users by operating a large mainframe computer, several LANs, and many administrative applications, such as payroll, accounts payable, class scheduling and registration, and student academic records. It supports academic users with shared midrange computers and file servers and many microcomputers in classroom labs and general-purpose facilities. CIS also operates the campus network that links most offices, classrooms, and buildings to the Internet.

CIS has been under considerable budgetary pressure for several years. Demand for computing services has grown rapidly, but funding hasn't kept pace. Declining hardware costs have been a budgetary bright spot, but they have been more than offset by increases in support costs for CSU's wide variety of hardware platforms, software packages, and administrative applications. Integrating this disparate collection of hardware and software with the growing campus network has been a difficult and costly task.

Recently, CIS proposed standardizing the hardware platforms and OSs for microcomputers and midrange computers to minimize support costs, contracting with a single vendor for each computer class, and negotiating volume pricing on hardware, software, and maintenance. All CSU workstation and midrange computer purchases, including those made with non-CIS funds, would be part of the contract.

Reactions to the proposal have ranged from indifference to near revolt. Many administrative and academic departments that rely on CIS for most computing and information-processing needs are supportive, provided CIS can continue meeting their needs. Some academic departments that fund their own computer purchases are concerned with loss of control over the acquisition process.

The computer science, engineering, information systems, and physics departments are vehemently opposed to the proposal. They argue that their computing needs are

(continued)

unique and require hardware and software at the cutting edge of technology. They claim that standardized platforms would lag behind technology and interfere with their teaching and research missions. They point out that most of their computing needs are met with funds from external sources, such as research grants and contracts, and oppose restrictions on how this money is spent.

Questions:

- Are the benefits that CIS anticipates from a standardized platform likely to be realized?
- Are the concerns of the four opposing departments valid? Should any other departments share these concerns?

THE ACQUISITION PROCESS

The process of acquiring hardware and system software is ongoing in most organizations. New hardware and software can be acquired to do the following:

- Support new applications.
- Increase support for existing applications.
- Reduce the cost of supporting existing applications.

The nature of this process depends on which of these goals (or combination of goals) motivates the new acquisition. It also depends on other factors, such as the following:

- The mix of applications the hardware and software will support
- Existing plans for upgrades or changes in these applications
- Compatibility requirements with existing hardware and software
- Existing technical capabilities

Hardware and software exist to support current and future applications. Planning for acquisition is little more than guesswork if you don't have a thorough understanding of present and anticipated application needs.

The acquisition process consists of these steps:

1. Determine the applications the hardware and software will support.
2. Specify hardware and software capability and capacity requirements.
3. Draft a request for proposal and circulate it to potential vendors.
4. Evaluate responses to the request for proposal.
5. Contract with a vendor or vendors for purchase, installation, and/or maintenance.

The acquisition process occurs whenever a new system is implemented or an existing system is upgraded. In either case, hardware and system software requirements are determined by activities in the design discipline of the Unified Process (UP). Activities in the requirements discipline produce detailed estimates of system activity, such as transaction volume and file or database content. Decisions about what processes will be automated, type of user interfaces, OSs and other system software, and selection of development tools are made during early design discipline activities. These decisions, combined with activity estimates, are translated into detailed hardware requirements.

Determining and Stating Requirements

Computer performance is measured in terms of application tasks that can be performed in a given time frame. Performance can be measured in terms of throughput (execution speed), response time, or some combination of the two. The first step in stating hardware and software requirements is to state the application tasks to be performed and their performance requirements.

Different techniques are used to determine requirements, depending on the motivation for the proposed acquisition. For existing applications, hardware and system software requirements might be based on measurements of existing performance and resource consumption. Requirements for new applications are more difficult to derive. Techniques for finding this information are discussed later in "Determining Requirements and Evaluating Performance."

Although application requirements are the primary basis for hardware and system software requirements, other factors must also be considered, including the following:

- Integration with existing hardware and software
- Availability of maintenance services
- Availability of training
- Physical parameters, such as size, cooling requirements, and disk space for system software
- Availability of upgrades

These factors are integral to the overall requirements statement. Some requirements, such as physical parameters, are essential; others are less important bases for differentiating between potential vendors.

Request for Proposal

A **request for proposal (RFP)** is a formal document sent to vendors that states requirements and solicits proposals to meet these requirements. It's often a legal document as well, particularly in governmental purchasing. Vendors rely on information and procedures specified in the RFP and invest resources in the response process with the expectation that stated procedures will be followed consistently and completely. An RFP is generally considered a contract offer, and a vendor's response is an acceptance of this offer. Problems such as erroneous or incomplete information, failure to enforce deadlines, failure to treat all respondents equitably, and failure to state all relevant requirements and procedures can lead to litigation.

The general outline of an RFP is as follows:

- Identification of requestor
- Format, content, and timing requirements for responses
- Requirements
- Evaluation criteria

The identification describes the organization requesting proposals. It should include the name of a person to whom questions can be addressed as well as postal and e-mail addresses, phone numbers, and fax numbers.

The RFP should state procedural requirements for submitting a valid proposal and, when possible, include an outline of a valid proposal describing each section's required content. In addition, it should clearly state deadlines for questions, proposal delivery, and other important events.

The requirements statement composes the majority of the RFP. Requirements should be categorized by type and listed completely. Relevant categories include the following:

- Hardware and software capability
- Related services, such as installation and maintenance
- Warranties and guaranties
- Deadline dates for delivery or implementation of requested products or services
- Financial considerations—for example, acquisition cost, maintenance cost, lease cost, and payment options

Requirements should be separated into those that are essential and those that are optional or subject to negotiation. For example, minimum hardware capacity is generally stated as an absolute requirement, whereas some related services, such as 24/7 technical support, might be described only as desirable.

Evaluation criteria are stated as specifically as possible. A point system or weighting scheme is often used to evaluate optional or desirable requirements. Weight might also be given to factors that aren't stated as part of the hardware or software requirements, such as a vendor's financial stability and good or bad previous experiences with a vendor.

Evaluating Proposals

Proposal evaluation is a multistep process, with the following usual steps:

1. Determine the acceptability of each proposal.
2. Rank acceptable proposals.
3. Validate high-ranking proposals.

Each proposal is evaluated to determine whether it meets the basic criteria, including essential requirements, financial requirements, and deadlines. Proposals that fail to satisfy minimal criteria in any category are eliminated from further consideration.

The remaining proposals are ranked by evaluating the extent to which they exceed minimal requirements. This ranking includes providing excess capability or capacity, satisfying optional requirements, and other factors. Measurements of subjective criteria, such as compatibility, technical competence, and vendor stability, are also considered at this stage.

A subset of highly ranked proposals is then chosen for validation. This subset should be small because of the time and expense involved in evaluating proposals. To validate a proposal, the evaluator determines the correctness of vendor claims and the vendor's ability to meet commitments in the proposal. The ranking process relies primarily on what vendors assert in the proposal; the validation stage is where the accuracy of these assertions is determined.

Various methods and sources of information can be used to validate proposals. The most reliable is a benchmark of the proposed system with actual applications. (Benchmarks were introduced in Chapter 4 and are discussed more in the following section.) For all but the largest systems, vendors usually deliver and install hardware and system software for customer evaluation. If purchased or already developed application software is available, the test system can be benchmarked with the application suite.

Alternatives to on-site benchmarking with actual application software include the following:

- Benchmarking at alternate sites with actual applications
- Benchmarking test applications
- Validating with published evaluations

With systems that are difficult to install, testing applications at another site, such as on the premises of the vendor or another customer, might be possible.

DETERMINING REQUIREMENTS AND EVALUATING PERFORMANCE

Information systems are a combination of hardware, system software, and application software. One of the most difficult system administration tasks is determining hardware requirements for a specific set of application software. This task is difficult for a few reasons:

- Computers are complex combinations of interdependent components.
- OS and other system software configuration can affect raw hardware performance substantially.
- The hardware and system software resources that applications require can't always be predicted precisely.

The starting point for determining hardware requirements is the application software that will run on the hardware platform. The nature and volume of application software inputs must be described to determine what application code will be executed and how often. This information determines the demands that application software will place on system software, such as number and type of service calls, and on hardware, such as number of CPU instructions.

If the application software has already been developed, its resource consumption in terms of hardware and system software can be measured. Software and hardware tools can be used to monitor resource utilization while application programs are running, which generates precise measurements of resource requirements.

Determining resource requirements is more complex when application software hasn't been developed yet. Detailed requirements and design details typically aren't available for the entire system until several UP iterations have been completed. Estimates of hardware and system software requirements are only as precise as the specifications of application software requirements. General statements of user and software requirements can generate only rough estimates of hardware requirements. More accurate hardware requirements can't be determined until all design discipline activities are completed.

Unfortunately, application developers seldom have the luxury of waiting until all design discipline activities are finished to acquire hardware, especially when implementing large systems. To accomplish the acquisition process described previously, enough lead time is required. The acquisition process is typically started in an early UP iteration, after the automated system's scope has been determined.

Informal methods of estimating requirements based on requirements models can sometimes be used. These methods typically use comparisons to similar application software and work volumes. For example, a developer of an online transaction-processing system might attempt to identify a similar system inside or outside the organization and examine it to determine the adequacy of its existing hardware and system software. If current hardware resources match current application demand, they can be used as a baseline for estimating the hardware requirements of the system being developed. Requirements can be adjusted to account for differences in transaction volume and application-processing characteristics.

This basic approach to estimating requirements is used widely and successfully, particularly for small hardware platforms running typical application software. The risk of inaccurate estimates is minimized for smaller hardware platforms because they can be modified quickly and inexpensively.

Large hardware platforms and large-scale one-of-a-kind application software require more formal methods of estimating requirements. Lead times for hardware acquisition and configuration are longer for mainframes and large midrange computers than for smaller hardware classes. Hardware and software costs are high, which increases the economic risks of inaccurate estimates. Estimating requirements in this environment typically follows formal methods based on two types of mathematical models:

- *Application demand model*—Represents units of work application software performs in terms of its demand for low-level hardware services. For example, an online transaction update might be described in terms of required number of CPU instructions, disk I/O bytes, and bytes input from or output to I/O devices. A complete model includes numerical representations of each type of work the application performs.
- *Resource availability model*—Describes a computer's capability to deliver resources to application software. The resources described in this model must match those used in the application demand model. Modeled resources can be low-level, such as CPU instructions and disk I/O bytes, or high-level, such as OS service calls or HTTP requests to a Web server.

After both models are specified, evaluators can test them by using a variety of analysis methods, including integer programming and simulation. Simple models can be constructed and evaluated by IS professionals with the assistance of decision support software. However, building and analyzing complex models require personnel with an extensive background in mathematics and experience in developing and analyzing computer performance models.

Benchmarks

As described in Chapter 4's Technology Focus on benchmarking, a benchmark is a measure of computer system performance while executing one or more processing tasks. There are many types of benchmarks testing many different aspects of performance. Benchmarks can test low-level aspects of hardware performance—such as CPU instruction execution speed or write speed to a RAID storage device—as well as measure performance when performing complex tasks, such as processing database queries and updates.

Using published benchmarks to evaluate new computers for an application or application suite is a complex matter because evaluators must do the following:

- Select the benchmark tests most relevant to the intended applications.
- Determine the relationship between benchmark tests and the actual work the new computer will perform.

To address the first issue, evaluators examine benchmarks for similarities to and differences from their own application software. Testing organizations supply detailed descriptions and source code to help evaluators select benchmarks.

To determine the relationship between a benchmark test and application software, evaluators test their own application software on a system with known performance on the most relevant benchmarks. The purpose of this testing is to establish a mathematical relationship between units of work in the application software and units of work in the benchmark.

For example, evaluators testing a computer with a known performance rating from the Transaction Processing Performance Council (TPC)—for example, 500,000 transactions per minute—run their own application software on the same computer and generate a result of 300,000 transactions per minute. The ratio of the two results (300,000 ÷ 500,000 = 0.6) enables evaluators to adapt the TPC-C benchmark for any computer to their software. For example, if evaluators anticipate a transaction volume of 150,000 transactions per minute, they need a computer meeting the TPC-C benchmark of at least 150,000 ÷ 0.6 = 250,000.

It might not be possible to test computer systems with actual applications if they're large, difficult to install, or not yet developed. In these cases, evaluators must examine the performance benchmark tests closely and estimate the ratio computed previously.

CAUTION

Care must be taken when using benchmarks, especially if only one benchmark is used at a time to test only one type of performance. Overall system performance under mixed application demand can vary substantially from separate benchmark results.

Measuring Resource Demand and Utilization

Accurate performance analysis requires accurate performance data. This data is needed as input to numerical performance models and when evaluating the performance of existing systems to determine the need for reconfiguration or hardware upgrades. Automated tools to measure resource demand and utilization include the following:

- Hardware monitors
- Software monitors
- Program profilers

The information these monitors generate describes the behavior of devices, resources, or subsystems over some period of time.

Monitors are programs or hardware devices that detect and report processing or I/O activity. A **hardware monitor** is a device attached directly to the communication link between two hardware devices, often used to monitor the use of communication channels, disk drives, and network traffic. It monitors communication activity between the two devices and stores communication statistics or summaries that can be retrieved and printed in a report.

A **software monitor** is a program typically embedded in OS service routines that detects and reports processing activity or requests. Software monitors are usually disabled by default but can be enabled to monitor high-level processing requests, such as file open, read, write, and close operations, or low-level kernel routines, such as flushing file or I/O buffers or virtual memory paging. When it's enabled, a software monitor generates statistics of service utilization or processing activity that can be displayed in real time or stored in a file for later analysis.

Software monitors can alter the activity being measured. For example, one that measures systemwide CPU activity generates inflated measurements because the software monitor itself requires CPU cycles to execute. Resources the monitor consumes must be estimated or measured and subtracted from the overall measurements to determine correct measurements for processes other than the monitor.

Monitors can operate continuously or intermittently. A continuous monitor records all activity as it occurs; it provides complete information on activity, but its operation can consume excessive system resources. An intermittent, or sampling, monitor checks for activity periodically. The advantage of a sampling monitor is that fewer resources are expended in running the monitor. Another advantage is that less data is accumulated in output files, which can become very large. Continuous utilization statistics can be estimated based on the sampled activity.

Monitors help identify performance bottlenecks as a precursor to configuring hardware and/or system software for maximal performance. For example, monitoring I/O activity on all secondary storage channels can indicate that some disks are used continuously, and others aren't used as much. Based on this information, a system administrator might decide to move frequently accessed files from heavily used disks to less used disks or to reallocate disks among controllers or I/O channels. In addition, an entire system can be tested at full load to determine its maximum sustainable resource delivery.

A **program profiler** describes the resource or service utilization of an application program during execution. Typically, monitor subroutines are added to the program's executable file during link editing. As the program runs, these subroutines write data to a file each time the application requests a system service. They can also record other statistics, such as elapsed (wall clock) time to complete each service request, CPU time consumed by service calls, and CPU time consumed by program subroutines. This information can be used to derive an application demand model as well as identify segments of the program that might be implemented inefficiently.

TECHNOLOGY FOCUS

Windows Performance Monitoring

Windows includes the Performance Monitor utility. System administrators can use it to monitor hardware and software resource use in real time. It defines a number of system objects about which performance and utilization data can be captured. Figure 14.2 shows some of these objects. Monitored objects can be a single hardware device, such as the CPU; groups of hardware devices, such as all physical disks; OS services, such as a network name service; and resource management data structures, such as a service queue or virtual memory paging file.

(*continued*)

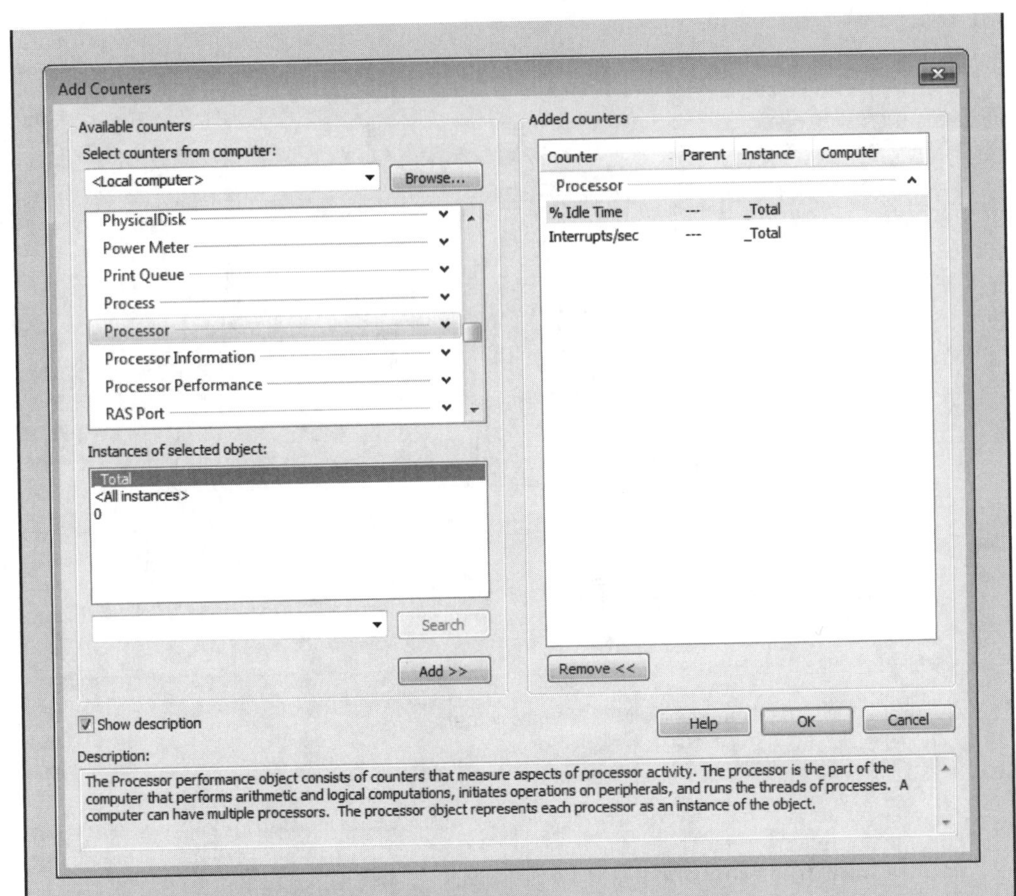

FIGURE 14.2 Objects and counters in Performance Monitor

Courtesy of Course Technology/Cengage Learning

Each object has data items called "counters." A counter monitors a specific performance or resource utilization aspect of an object. For example, there are 19 counters for the PhysicalDisk object, including current queue length, average seconds per disk read or write operation, number of disk read or write operations per second, and number of bytes read or written per second. Using different counters, a system administrator can monitor aggregate or highly specific aspects of performance.

Resource utilization and other system activity can be displayed in various formats. Figure 14.3 shows a line graph displaying total CPU utilization (upper line) and network interface bytes per second (lower line). Statistics such as mean, minimum, and maximum are shown for the currently selected counter. The display can be configured for any number of counters, and each counter is assigned a color and type of line. Similar display options are available for other output formats, such as bar graphs and text reports. Performance data can also be captured in a file for later analysis.

A system administrator needs accurate data to make configuration decisions. For example, decisions about changes to the size and/or location of virtual memory page files

(continued)

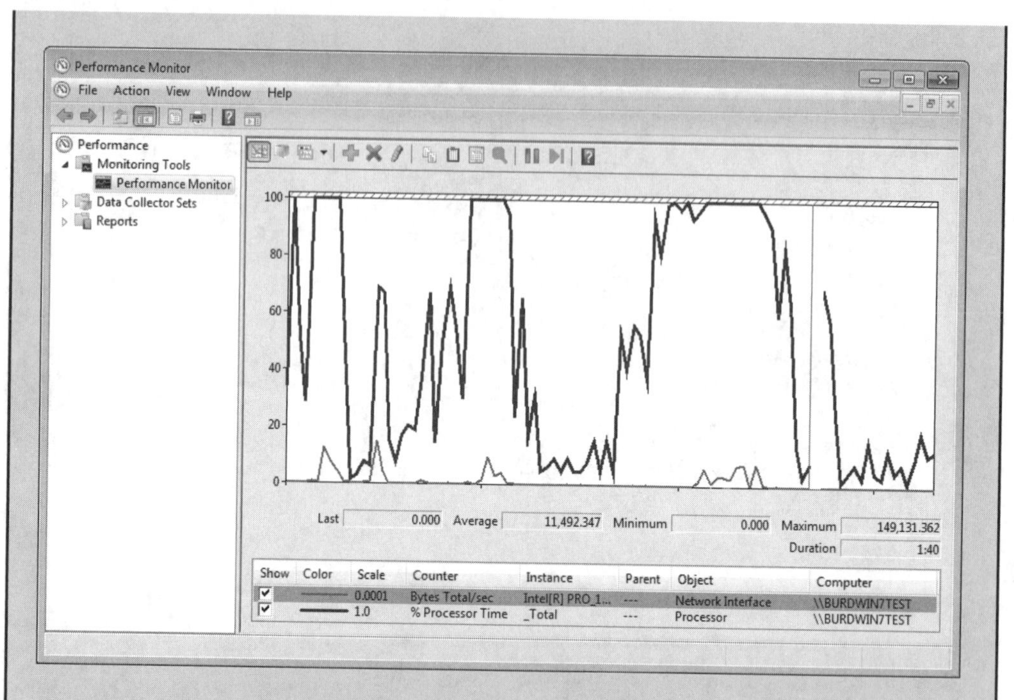

FIGURE 14.3 A real-time display of performance data
Courtesy of Course Technology/Cengage Learning

require accurate data on current virtual memory utilization. Decisions about upgrading CPU capacity require data describing current CPU utilization and utilization of related resources, such as memory and bus I/O. Performance Monitor is a flexible and powerful means of acquiring this data.

Accurate data is not the only requirement for good configuration decisions, however. Interpreting data requires a thorough understanding of the OS and hardware components, which enables system administrators to associate cause and effect and understand the tradeoffs between types of system performance. Without this understanding, performance assessment is little more than guesswork, and configuration decisions can produce suboptimal or detrimental effects on overall system performance.

SECURITY

For most organizations, information-processing resources are a sizable investment. Some resources, such as specific items of hardware and software, are tangible and have well-defined dollar values. Others, such as databases, user skill, and reliable operating procedures, are less tangible but also of considerable value. As applied to information systems, the term "security" describes all measures for protecting the value of these investments, including physical protection against equipment loss or damage and economic protection against loss of information's value through unauthorized disclosure.

A well-integrated approach to system security includes measures for the following:

- Protect physical resources against accidental loss or damage.
- Protect data and software resources against accidental loss or damage.
- Protect all resources against malicious tampering.
- Protect sensitive software and data resources against unauthorized access and accidental disclosure.

A complete discussion of all these measures could fill an entire book. This section gives you an overview of some commonly used security measures.

Physical Security

Access to computers and related equipment should be restricted to prevent theft, tampering, and unauthorized access. Locked doors and limited distribution of keys, key cards, and other lock control mechanisms are the most direct way to protect equipment. Additional protective measures for rooms containing servers and other dedicated equipment include architectural details, such as reinforced doors, reinforced walls, and barriers above drop ceilings.

Other physical controls are required for computers in public and semipublic areas, such as offices and campus computer labs. Most computer cases can be locked to prevent removal of key system components. Computers and peripheral equipment can also be locked to desks or other furniture by cable locks.

Access Controls

All operating systems incorporate access control features that enable restricting access to resources such as data files, programs, and hardware devices. Access control is based on two key processes:

- **Authentication**, which is the process of determining or verifying the identity of a user or process owner
- **Authorization**, which is the process of determining whether an authenticated user or process has enough rights to access a resource

A challenge-response dialogue using a username and password is the most common means of authentication. A user enters a name or other identifier and a password to prove his or her identity. The OS verifies the username and password by searching a local security database or interacting with a security server, such as a Kerberos server.

Although password-based authentication is most common, other methods are often used as supplements or alternatives for improved security. ID cards with bar codes or embedded ROM chips can supplement passwords. **Biometric authentication** methods are sometimes used instead of password-based authentication. These methods identify a person by using physical characteristics, such as fingerprints, facial features, or retinas.

Until the early 1990s, authentication was normally handled by the OS. Each OS maintained its own security database and enforced its own access controls. As organizations became more tightly integrated through information technology, handling authentication and access control exclusively through OSs became a problem. If a user needed access to resources on dozens of different servers, a user account and authorizations had to be created and maintained on each server. This approach is too complex for both users and administrators.

Operating systems now cooperate with one another to perform authentication and authorization. Typically, each OS interacts with a directory or security service, such as Kerberos or Microsoft Active Directory, and relies on this service to handle authentication and perhaps some or all authorization functions. Because all security functions are centralized, security procedures are simpler for both users and administrators.

After being authenticated, a user or process is given a digital ID. OSs and security systems have different names for the ID, including user identification (UID), security identification (SID), and security ticket. (The term "ticket" is used for the rest of this chapter.) Tickets are the basis for determining users' authorization to access resources the OS manages and can govern their ability to do the following:

- Access the OS or resources from specific locations or at specific times.
- Read, write, create, and delete directories and files.
- Run programs.
- Access hardware resources, such as printers and communication devices.

For each restricted action and resource, the OS or security service maintains a list of users or groups with access authority, called an authorization list or access control list. The list can include different authority levels for different users—for example, read authority for all users who access a file but write authority only for system administrators.

When a user or process attempts an action or requests a resource, the OS or security service looks up the user's ticket in the access control list to determine whether the user has enough authority. Typically, authority checks are an integral part of the OS service routines that application programs and administration utilities use to access resources. For example, a service routine to open a file typically asks for a ticket and passes it to a security service for authentication before processing the request (see Figure 14.4).

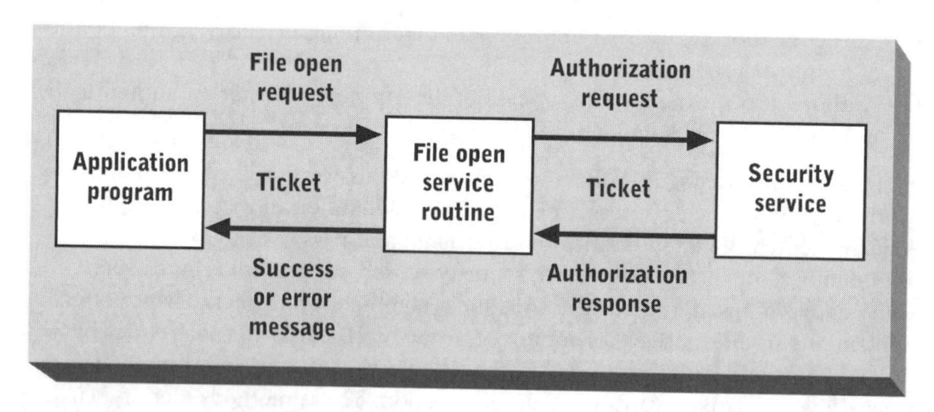

FIGURE 14.4 Authorization in a file open service call
Courtesy of Course Technology/Cengage Learning

Password Controls and Security

Because password-based authentication is so common, operating systems and security services use methods such as the following to enhance it:

- Restrictions on the length and composition of valid passwords
- Requirements that passwords be changed periodically

- Analysis of password content to identify passwords that can be guessed easily
- Encryption of passwords in files and during transmission over a network

With most OSs, the system administrator can create and enforce password policies on a per-user, per-group, or per-system basis. Figure 14.5 shows password controls that can be set in Windows, including restrictions on password length, age, complexity, and uniqueness. OS password policies should be supplemented with organizational policies, including rules against sharing passwords, creating passwords based on personal information (such as birth dates or family member names), and writing passwords down in easily accessed locations. Some password controls, such as password age restrictions, can be overridden for certain accounts.

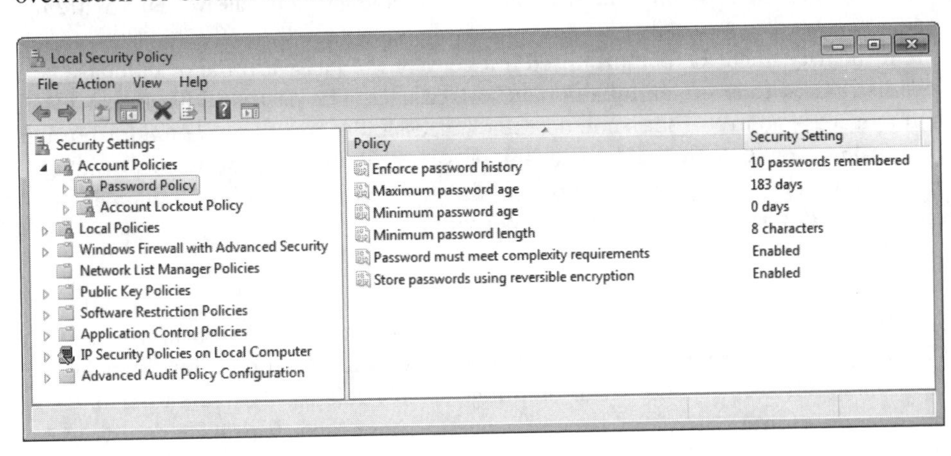

FIGURE 14.5 Windows password policies
Courtesy of Course Technology/Cengage Learning

A related group of policies, shown in Figure 14.6, can be set to deal with failed attempts to log on. For example, accounts can be deactivated after a specified number of failed logons and reactivated manually by the system administrator or automatically after a specified time interval.

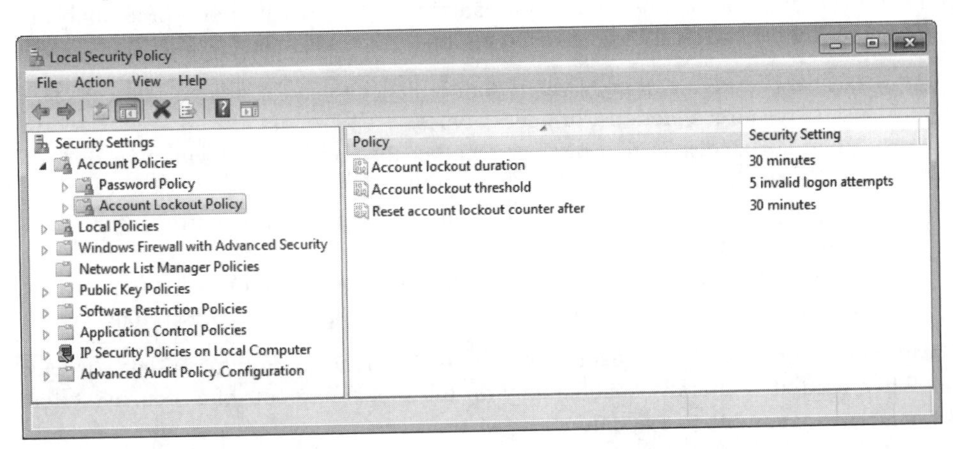

FIGURE 14.6 Windows account lockout policies
Courtesy of Course Technology/Cengage Learning

Locking out accounts after a specified number of failed logon attempts prevents unauthorized users from repeatedly attempting to guess correct passwords for valid user accounts. Password-guessing programs are sometimes used to attempt remote logon with words and/or names taken from a dictionary. Because many users choose common words or names as their passwords, it's possible for these programs to find a correct password by trying every word in their dictionaries. Automatic account deactivation coupled with delayed reactivation limits the number of guesses that can be made in any time period.

Some OSs and security services use similar password-guessing programs against local password files periodically or continuously. When passwords are guessed successfully, the OS forces the user to change his or her password at the next logon. Similar capabilities can be embedded in the program that enables users to change their passwords. An attempt to change a password to an easily guessed value is denied.

System security might be compromised if a copy of the password file or database is somehow distributed to or stolen by an unauthorized user. The password file is usually protected by access controls, but some degree of accessibility must be granted for normal logon processing. Most OSs further protect the password file by encrypting part or all of its content. However, a stolen password file can be decrypted successfully if enough computing resources are used to crack the encryption key.

Auditing

In accounting, **auditing** is the process of examining records to determine whether generally accepted accounting principles were applied correctly in preparing financial reports. In computer security, the term "auditing" usually refers only to creating and managing records of user activity or resource access. These records provide data to determine whether the security policy has been implemented correctly or whether resources or the system itself have been compromised.

Most OSs and security services include an audit function that's disabled by default but can be enabled for specific users, resources, actions, or access types. When auditing is enabled, the OS or security service writes an entry to a log file each time an audited action is performed. This log entry includes information such as which ticket was presented to gain access and the access date and time.

Although auditing can be a useful tool for examining the security policy and analyzing security breaches, it has several limitations, including the following:

- Log files can grow quickly when auditing is enabled for a large number of users, resources, actions, or access types.
- Auditing reduces system performance because of the overhead of writing log file entries.
- Auditing examines historical data, so it's incapable of preventing future security breaches.
- Extracting useful information from large auditing logs requires automated search tools and a consistently implemented program of log file analysis.

Because of its limitations and effect on system performance, auditing is used sparingly, if at all. When used, it's generally enabled only for the specific aspects of a security system that an administrator wants to evaluate.

Virus Protection

In biology, a virus is a DNA fragment that infects a living host at the cellular level and uses the host's cellular machinery to reproduce itself and perform actions detrimental to the host. In computers, a **virus** is a program or program fragment that does the following:

- Infects a computer by installing itself permanently, usually in a hard-to-find location
- Performs malicious acts on the infected computer
- Replicates and spreads itself by using services on the infected computer

Viruses come in many types, including the following:

- *Boot virus*—Attaches itself to code that runs when the system boots, such as a BIOS or OS startup routine
- *Macro virus*—Embedded in a macro stored in a desktop application file, such as a spreadsheet or word-processing document
- *Worm*—Stored in a stand-alone executable program and usually sent as an e-mail attachment; runs automatically when the attachment is opened

Viruses are commonplace and can perform many malicious acts, including damaging or destroying important files, opening backdoors for potential hackers, and sending sensitive information to others. No information system can be considered secure unless it includes active virus protection.

Many companies market antivirus software. Features vary, but common capabilities include the following:

- Scanning e-mail messages and attachments for known viruses and disabling or deleting them
- Monitoring access to important system files and data structures and logging or denying access when needed
- Scanning removable media for known viruses whenever they're inserted
- Scanning the file system and important data structures periodically for viruses that might have escaped other scans and monitoring activities
- Monitoring Web page accesses and disabling malicious software that might be embedded

The most important aspect of antivirus software configuration is ensuring that it's enabled and updated regularly. Antivirus software should be installed automatically with the OS and enabled by default. Users are sometimes tempted to disable antivirus software because its continuous monitoring and scanning activities impose a performance penalty.

Antivirus software uses data files, sometimes called "signature files," containing information about known viruses. Because new viruses appear constantly, these files must be updated regularly. Most antiviral software includes a subscription to updated signature files, which the vendor makes available as new viruses are discovered. The best approach to updating these files is based on push technology, meaning one of the organization's servers monitors the vendor's download site for new signature files. When they're released, the server downloads the files and copies them to all the organization's computers.

Software Updates

Operating systems and application software are a complex collection of interconnected components. A typical OS or desktop application suite includes tens of millions of lines of source code. Given software's size and complexity, errors, bugs, and security holes are a certainty. Hackers and viruses often attempt to exploit these problems to perform malicious acts or gain access to secure information or resources.

Software developers are in a constant race to fix bugs, errors, and security holes as they're discovered. They do so by developing new software versions or software patches, sometimes called service packs, to apply to existing installations. A key part of any security system is updating system and application software.

Until recently, applying patches and upgrades was a manual process. Users or system administrators needed to monitor information sources about software upgrades and patches constantly and take actions to apply them. However, software now includes features to streamline or automate the update process. Typically, software examines its own internal configuration whenever it runs and then sends a query to a server to determine whether its configuration is current. If the response indicates an upgrade is needed, the software can inform the user or system administrator or download and install required updates automatically. Figure 14.7 shows the configuration dialog box for Windows automatic updates. Similar update options exist for many other OSs and application software.

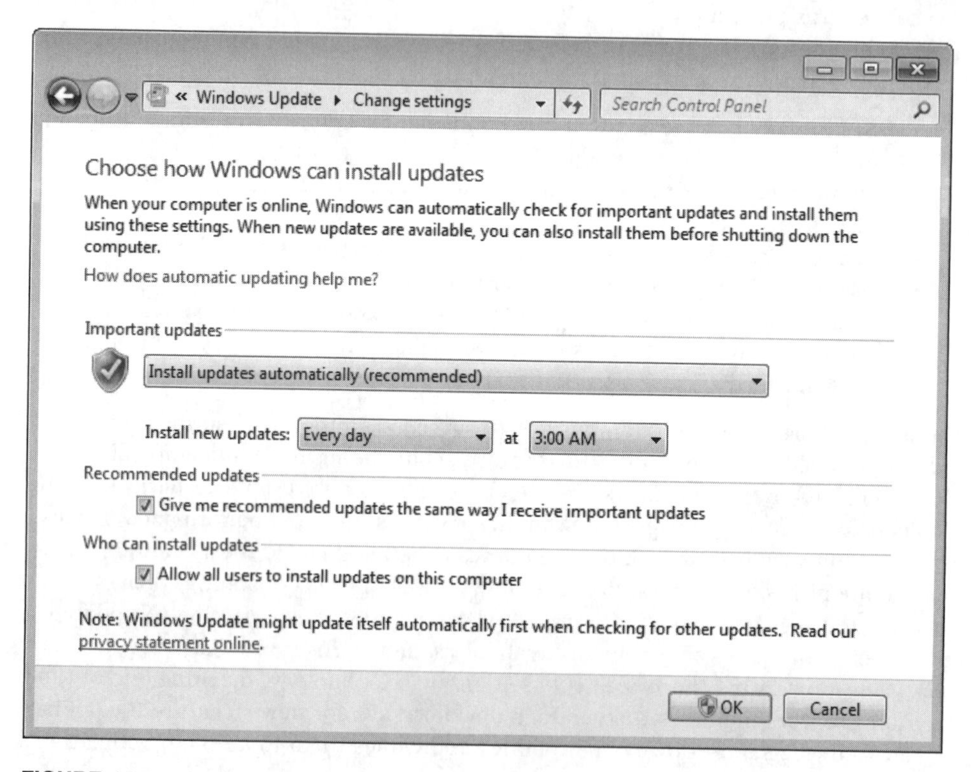

FIGURE 14.7 Configuring automatic updates in Windows
Courtesy of Course Technology/Cengage Learning

A system administrator can also control software updates via some directory services and network OSs. Figure 14.8 shows installing an application program via a Microsoft Active Directory group policy object (GPO). In this example, the GPO is applied to workstations in computer classrooms. Each time a workstation boots, it checks whether Intel Processor ID Utility is installed and performs the installation automatically, if necessary. Automatic software installation improves security by ensuring that the latest software versions and security patches are always installed on all an organization's computers.

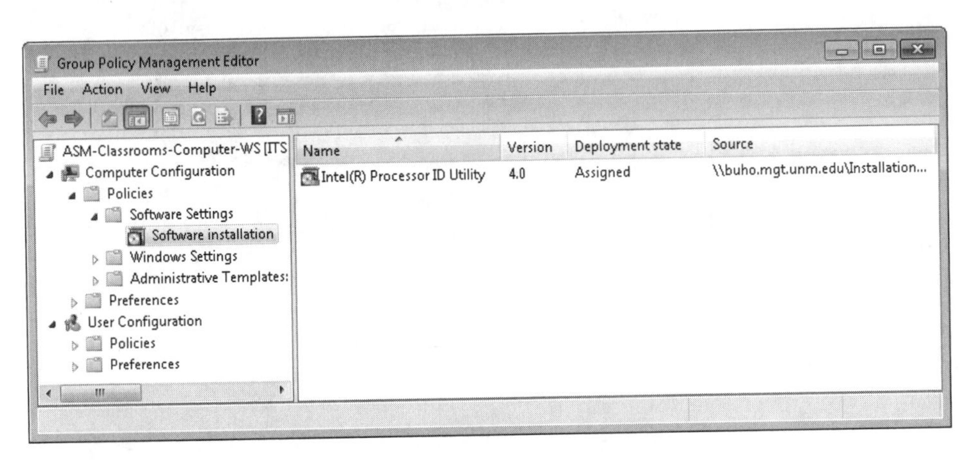

FIGURE 14.8 Viewing a software installation policy
Courtesy of Course Technology/Cengage Learning

Firewalls

In architecture and machinery, a firewall is a physical barrier that prevents fire from spreading between two structures or compartments that share a common wall or divider. In computer security, a **firewall** is a hardware device, software, or a combination of hardware and software that prevents unauthorized users in one network from accessing resources on another network. Typically, a firewall is a stand-alone device with embedded software that physically separates a private network from a public network, such as the Internet (see Figure 14.9). Firewalls are widely deployed in information systems to protect servers and information resources from unauthorized access over the Internet.

The simplest firewall type is a **packet-filtering firewall**, which examines each packet and matches header content to a list of allowed or denied packet types. Recall from Chapter 9 that TCP/IP packets include source and target sockets (combinations of IP addresses and port numbers). A packet-filtering firewall consults a list of valid source

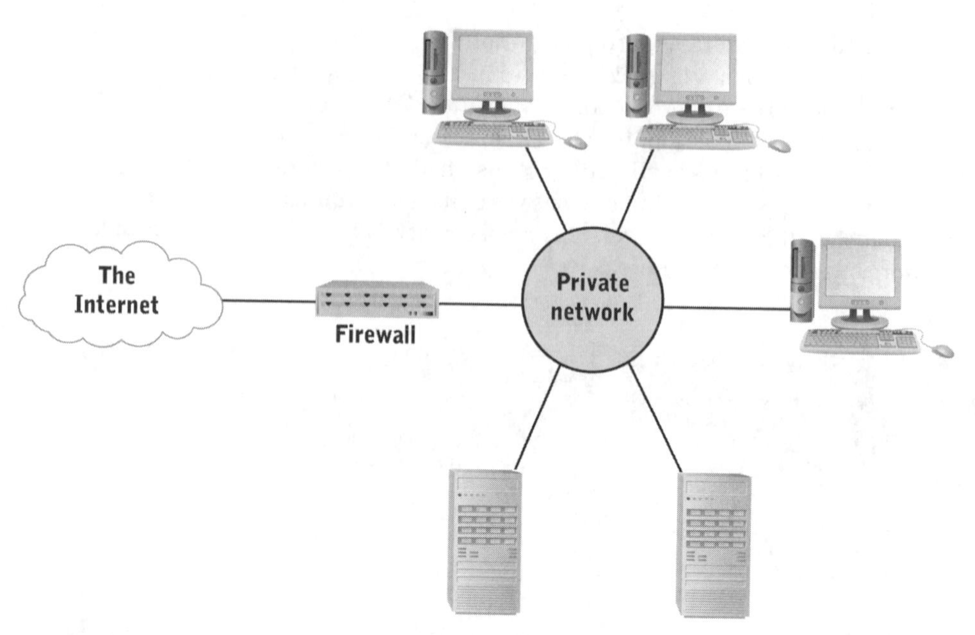

FIGURE 14.9 A firewall between the Internet and a private network
Courtesy of Course Technology/Cengage Learning

and target IP addresses or address ranges and valid port numbers for these addresses. For example, if a public Web server is located in the private network, the firewall is configured to allow incoming packets addressed to the Web server's IP address and port 80, the standard HTTP port.

Configuring the firewall becomes more complex as the volume and variety of legitimate traffic between networks increases. For example, an organization enables strategic partners, such as suppliers, to interface directly with its production databases. An application that supports these interactions has components running both inside and outside the private network. The firewall must be configured to allow transmitting packets between servers and enable software to support the application. Specific entries must be added to the firewall database for the IP addresses strategic partners use and the port numbers applications use.

More complex approaches to firewall use and configuration include application firewalls and stateful firewalls. An **application firewall**, sometimes called a **proxy server**, is a server that handles service requests from external users of applications. It accepts service requests from clients on the Internet or other untrusted networks on behalf of servers in the internal network, relays requests to the correct servers, and relays responses back to clients (see Figure 14.10). Application firewalls improve security by shielding internal servers and resources from direct access by outside users. To gain access to secure resources, a hacker or other intruder must compromise two servers, usually with two different security systems.

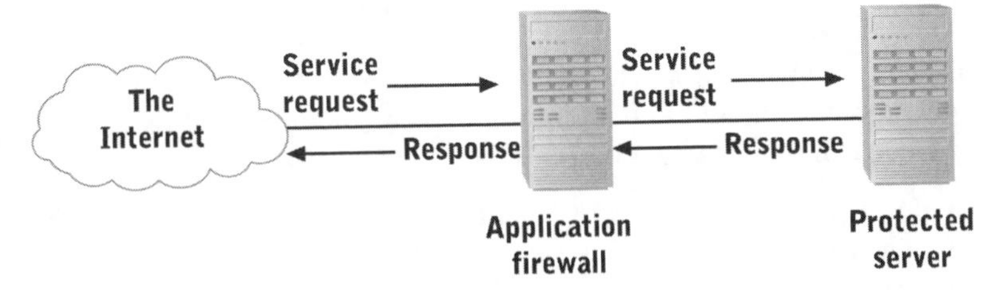

FIGURE 14.10 An application firewall
Courtesy of Course Technology/Cengage Learning

A **stateful firewall** tracks the progress of complex client/server interactions. It's most useful in client/server transaction-processing environments, where each transaction has a predictable processing pattern. The firewall is configured with the sequence of events composing valid client/server interactions, including packet types and transmission sequence, port and server access order, and maximum time durations between valid requests and responses. Packets that don't match specified patterns of normal activity are blocked automatically.

Installing and configuring any firewall correctly requires a detailed understanding of network protocols, OS security procedures, and the mechanisms by which client, server, and peer processes interact with one another. The process starts with an inventory of all existing services and the packet types, IP addresses, and ports that support client/server or P2P interaction for each service. Note that a typical information system includes many services that aren't obvious, such as Remote Procedure Calls, authentication and authorization, and background services (for example, time synchronization). System administrators sometimes use a port scanner or network monitor to determine what IP addresses and ports existing services are using.

After the inventory is complete, the administrator determines which services need to be accessible from outside the organization's private network. Based on this information, the administrator can configure firewalls to block packets associated with unimplemented services and packets from unauthorized IP addresses. Because configuration is complex and error prone, any changes to firewall configuration should be tested thoroughly to ensure that unauthorized external access is blocked and all authorized internal and external activity is enabled.

PHYSICAL ENVIRONMENT

Installing hardware requires special attention to many aspects of the physical environment. Issues to be considered when choosing or preparing a location for hardware include the following; each is discussed in the next sections:

- Electrical power
- Heat dissipation
- Moisture
- Cable routing
- Fire protection

Electrical Power

Hardware is sensitive to fluctuations in power levels. Computer circuitry is designed to operate at a constant low power level, and fluctuations can cause momentary loss of operation or damage to electrical circuits. Fluctuations can be of several types:

- Momentary power surges
- Momentary power sags
- Long-term voltage sags
- Total loss of power

Power surges or spikes can be caused by a number of events. Lightning strikes in power generation or transmission facilities tend to cause the most dangerous types of power surges. Dangerous spikes can also be caused by failure of power transformers or other transmission equipment, which leads to brief power surges of very high intensity. Because the surges are brief, they might not engage standard protection devices, such as fuses or circuit breakers, before major damage has occurred. Standard fuses or breakers for a floor or building don't provide adequate protection for computer equipment.

Power sags normally occur when a device that requires a large amount of power is started. You can see this in the home when air conditioners, refrigerators, and electric dryers are started and cause a momentary dimming of lights. Small power sags are almost always present when multiple devices share a single electrical circuit. Large power sags are a symptom of overloaded circuits. Longer-term power sags are often caused by the power provider. The common term for this event is a **brownout**. Brownouts occur when the demand for electricity exceeds the provider's generation and transmission capabilities during peak demand periods, such as hot summer days. To avoid a complete service interruption for some users, the power provider reduces the voltage level temporarily on a systemwide basis to spread available power evenly to all users.

Most computer equipment is designed to operate reliably over a range of voltage levels. In the United States, transformers in small computer systems are designed for 110-volt alternating current (AC) power input. Large computer systems in the United States can use 220- or 440-volt power inputs. Computer systems in other countries use different voltage levels. Computer system power transformers can usually deliver consistent voltage output with input voltage variations of up to 10%. This characteristic provides some protection against surges and sags caused by the startup or shutdown of other equipment as well as brownouts.

Auxiliary power-conditioning equipment is needed to deal with powerful surges and total power loss. Equipment can be protected against high power surges by a **surge protector**, which detects incoming power surges and diverts them to ground in nanoseconds. Surge protectors differ in the speed at which they react, the intensity of the surge they can suppress, and whether they can be reused after a surge. Surge protectors can be purchased as separate devices or integrated into other power-conditioning devices.

NOTE

By itself, total power loss rarely causes damage to hardware. However, a tripped circuit breaker or blackout is often accompanied by power surges. The main problem with power loss lies in the loss of data. Data held in RAM, including process data areas, secondary storage buffers, and communication buffers, is lost when power is interrupted.

Protection against power loss requires an auxiliary power source. It can take several forms, including a combination of redundant power circuits, auxiliary generators, and battery backup. An **uninterruptible power supply (UPS)** is a device that provides power to attached devices in the event of external power failure. UPSs vary in their power delivery capacity, switching time, and battery life. Surge protection is normally incorporated into a UPS. This feature is almost essential because switching between external and internal power supplies might introduce surges.

Some UPSs, particularly those with short delivery times, are designed to work with a computer's OS. They include a standard (for example, USB) communication port and cable that can be connected to the computer. When a power failure is detected, the UPS sends a signal to the computer to indicate that it has switched over to battery power. This signal enables the OS to initiate protective actions before a total loss of power. Typically, the OS starts a normal or emergency shutdown procedure when an interrupt is detected. More complex UPSs support two-way communication, which can give the OS additional information, such as error conditions and battery discharge rate.

Heat Dissipation

All electrical devices generate heat as a byproduct of normal operation. As you learned in Chapter 4, excessive heat can cause intermittent or total failure of electrical circuits, so all computer equipment needs some means of **heat dissipation**. In equipment that generates little heat, vents in the equipment cabinet are normally enough to dissipate heat. Care must be taken to ensure that vents don't become blocked, which impedes free movement of air through the cabinet.

Many computer hardware devices use fans to move air through the unit. Fans force cool exterior air into the cabinet or draw hot interior air out. Either method requires at least two vents and a clear pathway for air movement. Vents are normally positioned at opposite corners of the cabinet to ensure that all components receive adequate cooling. Fan-based cooling also requires some filtering to remove dust and other contaminants.

When heat is dissipated from an equipment cabinet, it collects in the room where the cabinet is located. Heat must be dissipated from the room, especially when many hardware devices are situated in a small space, as in a server closet. Normal room or building air-conditioning might be adequate; otherwise, supplemental cooling capacity is necessary. Supplemental cooling counteracts heat buildup more effectively and adds extra protection if the primary cooling system fails.

In extreme cases, auxiliary cooling can be provided in an equipment cabinet in the form of a refrigerant-based heat exchanger, a liquid cooling system, or even a liquid nitrogen system. These measures are often used with supercomputers, in which CPUs and other components operate at extremely high clock rates or access speeds.

Moisture

Excessive moisture is an enemy of electrical circuitry because of the danger of short circuits, which form when water conducts electricity between two otherwise unconnected conductors. Short circuits can cause circuit damage and fire when computer equipment is operating. In addition, water can damage computer equipment even when power is turned off because impurities in the water remain on computer components as the water evaporates. These impurities can corrode exposed electrical contacts or other hardware components, such as power connectors, printed circuit boards, and metal cabinets. They can also form a short circuit.

Well-designed cabinets are one defense against the dangers of moisture, but they protect only against spills and leaks. Another protective measure is to mount hardware cabinets and devices above floor level, which minimizes the danger from roof leaks, broken pipes, and similar problems that can lead to standing water.

Hardware must also be protected against condensation resulting from excessive humidity. Low humidity is a problem, too. Low humidity levels increase the buildup of static electricity, which increases the likelihood of circuit damage caused by inadvertent static discharges. In general, the humidity level of a room containing computer equipment should be near 50%.

Humidity can generally be controlled automatically by optional components of heating, ventilation, and air-conditioning systems. A household dehumidifier can be used in small rooms, but be aware that when the unit powers on, it might cause a brief power sag if it's on the same circuit as computer equipment. Also, household dehumidifiers collect airborne water in a pan, which must be emptied regularly. An overflowing pan can be a safety or electrical hazard.

Cable Routing

Computer facilities must provide protection for data communication lines. Because the configuration of these lines changes frequently, ease of access is also important. Computer facilities usually deal with this problem in two ways—raised floors and dedicated cabling conduits.

A raised floor is often used in a room containing multiple hardware cabinets and serves several purposes. The main purpose is to have an accessible location for cables connecting different devices. The floor is made up of load-bearing supports on which a grid of flooring panels is laid. The flooring panels can be installed or removed from the grid easily. Cables are routed under walkway areas.

Other advantages of raised floors include protection from standing water and ensuring movement of chilled air. Several inches of water can accumulate without reaching the level of equipment cabinets. Moisture sensors placed below the floor panels can detect the buildup of standing water. The space between the actual floor and the floor panels can also be used as a conduit for chilled air. When used in this manner, equipment cabinets are vented at the bottom and top. Chilled air is forced through the floor into the bottom of the cabinets, and heat is dissipated through the top.

Dedicated cabling conduits normally provide cable access between rooms or floors. To prevent electromagnetic interference, these conduits shouldn't be used to route both electrical power and electrical data communication lines. In addition, access panels should be installed at regular intervals for adding, removing, or rerouting cables. Conduits should be shielded to limit external electromagnetic interference.

Fire Protection

Fire protection is an important consideration both for ensuring safety of personnel and protecting expensive computer hardware. As with cooling, the normal fire protection mechanisms incorporated in buildings are inadequate for rooms housing computer hardware. In fact, these measures actually increase the danger to personnel and equipment because they usually rely on water—for example, automatic sprinklers.

Carbon dioxide, fire retardant foams and powders, and various gaseous compounds are alternate methods of fire protection. Carbon dioxide is generally unacceptable because it's a suffocation hazard and promotes condensation in computer equipment. Fire-retardant foams and powders are unacceptable because of their moisture content, and powders generally have corrosive properties.

Until the 2000s, many large computer facilities used halon 1301 gas, which doesn't promote condensation and doesn't displace oxygen to the extent that carbon dioxide does. It also gives personnel adequate time to evacuate a room or floor. Unfortunately, halon gas is a chlorofluorocarbon (CFC), a class of gases with ozone-depleting properties. Most industrialized countries have signed the Montreal Protocol, which has phased out CFC production and bans importation of newly produced CFCs. Other alternatives are currently being used, including halocarbon compounds and inert gas mixtures. None has yet emerged as a clear successor to halon, which is still in wide use as a result of industry-wide recycling efforts.

Fire detection is also a special problem in computer facilities. Electrical fires often don't generate heat or smoke as quickly as conventional fires do, and normal detection equipment might be slow to react. Fast detection is an economic necessity. A fire in one item of computer equipment can spread quickly and cause damage to attached equipment through power surges. Normal building fire-detection equipment is typically supplemented in a computer room with additional smoke detectors placed near large concentrations of equipment and below raised floors to detect fires in cabling.

Disaster Planning and Recovery

Because disasters such as fire, flood, and earthquakes can't be avoided, plans must be made to recover from them. Disaster planning is particularly critical in online systems and systems in which extended downtime causes extreme economic impact. A number of measures are normally taken, including the following:

- Periodic data backup and storage of backups at alternate sites
- Backup and storage of critical software at alternate sites
- Duplicate or supplementary equipment installed at alternate sites
- Arrangements for leasing existing equipment at alternate sites, such as another company or a service bureau

The measures that are suitable for a specific installation depend heavily on local characteristics, such as size and distribution of computing resources, available budget, available service vendors, and available network capacity to and from those vendors.

Summary

- A system administrator's main responsibility is to ensure efficient, reliable delivery of IS services. Broad categories of system administration tasks include acquiring new IS resources, maintaining existing IS resources, and designing and implementing an IS security policy. The assignment of tasks varies considerably across organizations.
- IS resource acquisition and deployment should occur only in the context of a well-defined strategic plan for the organization. IS strategic plans tend to follow, rather than lead, the strategic plans of other organizational units. IS resources can be considered organizational infrastructure. Many IS resources are also capital assets because they're used for multiple years. Strategic issues relevant to information systems include services to be provided, charges for services, infrastructure composition, and infrastructure improvement and maintenance.
- Providing infrastructure-based services to a wide variety of users requires service standards. However, standardization tends to stifle innovation and produce solutions that are suboptimal for some users. Standardization often causes problems for users who need services at or near the leading edge of technology. Managing IS resources as an infrastructure can result in ignoring opportunities to use IS resources for a competitive advantage.

- Planning for IS resource acquisition requires a thorough understanding of current and anticipated application needs. The acquisition process includes determining the applications that the hardware and software will support, specifying detailed hardware and software requirements, drafting and circulating a request for proposal (RFP), evaluating RFP responses, and negotiating a purchase and/or support contract.
- Hardware requirements depend on the hardware and system software resources required by application software. If application software has already been developed, its hardware and system software resource consumption can be measured. Determining resource requirements is more complex when application software hasn't been developed yet. In this case, requirements must be determined by benchmark testing.
- A well-integrated approach to system security protects an organization's hardware, software, and data resources against accidental loss or damage, malicious tampering, unauthorized access, and accidental disclosure. Authentication and authorization enforced through the OS or security service are the first line of defense. Other defensive measures include password control, auditing, virus protection, regular software updates, and firewalls.
- Hardware installation and operation require special attention to many aspects of the physical environment, particularly electrical power, heat dissipation, moisture, cable routing, and fire protection.

Key Terms

application firewall
auditing
authentication
authorization
biometric authentication
brownout

capital expenditures
capital resources
competitive advantage
firewall
hardware monitor
heat dissipation
infrastructure
monitors
operating expenditures
packet-filtering firewall
power sags
power surges
program profiler
proxy server
request for proposal (RFP)
service standards
software monitor
stateful firewall
strategic plan
surge protector
system administration
uninterruptible power supply (UPS)
virus

Vocabulary Exercises

1. ________ are expected to provide service over a period of years.
2. ________ is the process of determining or verifying the identity of a user or process owner.
3. ________ is the process of determining whether an authenticated user or process has enough rights to access a resource.
4. The term ________ usually refers only to creating and managing records of user activity or resource access.
5. A(n) ________ accepts service requests from an untrusted network and relays the requests to the appropriate servers.
6. Because IS resources can be considered ________, service standards and costs for operation, maintenance, and improvement are important components of the IS strategic plan.
7. A(n) ________ is a formal (legal) document that solicits bids from hardware and software vendors.
8. A(n) ________ is a program or program fragment that infects a computer by installing itself permanently, performing one or more malicious acts, and replicating and spreading itself by using services of the infected computer.
9. A(n) ________ detects and reports processing or I/O activity.
10. Opportunities to use IS resources for a ________ might be missed if these resources are managed only as infrastructure.
11. A(n) ________ provides auxiliary power during blackouts and can notify the OS when it's activated.
12. Computer hardware must be protected against ________ and ________ in electrical power.
13. A(n) ________ tracks the progress of complex client-server interactions and blocks packets that don't conform to normal activity patterns.

14. A(n) ________ is a hardware device, software, or a combination of hardware and software that prevents unauthorized users in one network from accessing resources in another network.
15. Long-range acquisition of hardware and software should be made in the context of an overall ________ for the organization.
16. The resource demands of an existing application program can be measured with a(n) ________.

Review Questions

1. What is infrastructure? In what ways do hardware and system software qualify as infrastructure?
2. What basic strategic planning questions should be addressed for infrastructure?
3. What are the advantages and disadvantages of standardization in hardware and system software?
4. What is a request for proposal (RFP)? How are responses to an RFP evaluated?
5. What problems are encountered when attempting to determine hardware and system software requirements for application software that hasn't been developed yet?
6. What is a monitor? List types of monitors and the information they provide.
7. Describe authentication and authorization. Which depends on the other? How and why are these processes more complex in a networked organization than in an organization that supports all information processing with a single mainframe?
8. What password-protection measures are normally implemented by system administrators, operating systems, and security services?
9. Describe the pros and cons of enabling auditing of resource accesses.
10. What is a virus? How can users and system administrators prevent virus infections?
11. Why is it important to install OS and application software updates in a timely manner? How can users and system administrators ensure that they're installed in a timely manner?
12. Describe the main firewall types and how each can improve system security. What information does a system administrator need to configure a firewall correctly?
13. Why are conventional methods of fire protection inadequate or dangerous for computer equipment?
14. What problems associated with electrical power must be considered in planning the physical environment for computer hardware?

Research Problems

1. Some manufacturers and vendors of large midrange computers and mainframes have developed documentation and software to help engineers, account representatives, and purchasers configure computer hardware and system software. Some software can help users match software demands against the capabilities of particular hardware

configurations. Investigate the online offerings of a few large computer vendors, such as *www.ibm.com*, *www.hp.com*, and *www.fujitsu.com*.

2. Investigate firewall products from a major computer vendor, such as *www.checkpoint.com*, *www.cisco.com*, or *www.netgear.com*. Which products are best suited to a small LAN with no publicly accessible resources? Which are best suited to a small LAN containing one publicly accessible Web site? Which are best suited to large organizations with e-commerce Web sites and internal resources accessed by strategic partners?
3. Locate one or more benchmarking programs for PCs on the Web. Download and install the benchmarks and test your own computer's performance. What do the results tell you about your computer? Do they offer any guidance for improving system performance? What types of application tasks do the benchmark programs simulate? Would similar benchmark programs be useful with midrange computers or mainframes?

APPENDIX

MEASUREMENT UNITS

Many measurement units describe the capabilities of computer system components, such as storage devices, I/O devices, device controllers, and networks. This appendix describes common measurement units and abbreviations as well as common mistakes in use and interpretation.

TIME UNITS

Time intervals, such as CPU cycle time or disk access time, are expressed in fractions of a second, as shown in Table A.1.

TABLE A.1 Time measurements

Abbreviation	Term	Fraction
ms	Millisecond	10^{-3} (one thousandth)
μs	Microsecond	10^{-6} (one millionth)
ns	Nanosecond	10^{-9} (one billionth)
ps	Picosecond	10^{-12} (one trillionth)

Note that microseconds are sometimes abbreviated as "ms," too. However, microseconds are rarely used when describing computer hardware, so you can generally assume that "ms" refers to milliseconds.

CAPACITY UNITS

Bits and bytes are the smallest data storage and transmission units. A bit is a binary digit containing the value 0 or 1, and a byte contains 8 bits. The terms "bit" and "byte" are often quantified with a prefix to indicate larger magnitudes. These prefixes are based on the value 2^{10} (1024_{10}) because it approximates traditional units based on powers of 10, such as thousands, millions, and billions. Table A.2 shows the abbreviations and prefixes used in capacity measurements.

TABLE A.2 Capacity measurements

Abbreviation	Prefix	Value
K or k	Kilo	1024
M or m	Mega	1024^2 (1,048,596)
G or g	Giga	1024^3 (1,073,741,824)
T or t	Tera	1024^4 (1,099,511,627,776)
P or p	Peta	1024^5 (1,125,899,906,842,624)
E or e	Exa	1024^6 (1,152,921,504,606,846,976)

Both uppercase and lowercase letters are used, although uppercase is more common. The values these prefixes represent are slightly larger than their decimal counterparts, which can cause confusion when an abbreviation based on 1024 is used to describe a quantity based on 1000.

For example, you're considering the capacity of a typical 500 GB disk for a laptop or desktop PC. Buried in the fine print of a typical ad or technical specification sheet, you see a statement such as "GB means 1 billion bytes and TB equals 1 trillion bytes." In other words, the stated capacity in GB is based on units of 1000^3, not 1024^3. So the correct capacity of the 500 GB disk is as follows:

$$1000^3 \div 1024^3 \times 500 \approx 465.66 \text{ GB}$$

Another source of confusion is representing the terms "bit" and "byte" with the letter b or B. For example, both "megabit" and "megabyte" can be abbreviated as MB. A common method of distinguishing between these terms is to abbreviate "bit" with the lowercase b and "byte" with the uppercase B. For example, the abbreviation GB is interpreted as gigabytes and the abbreviation Gb as gigabits. Unfortunately, this convention isn't followed universally. Any abbreviation containing the letters b and B deserves careful scrutiny to determine whether "bit" or "byte" is intended.

DATA TRANSFER RATES

Data transfer rates describe the capacity of communication channels and the input speed, output speed, or throughput of I/O devices and device controllers. A data transfer rate is always expressed as a quantity of data per time interval. For example, LAN data transmission capacity might be described as 1 gigabit per second, and printer output speed might be described as four pages per minute.

Communication channel capacities are usually expressed in bits or bytes per second. Magnitude prefixes, such as "kilo" and "mega," can be used (such as 500 kilobits per second and 100 megabytes per second). Bit and byte capacity measures are abbreviated as described previously, and the phrase "per second" (abbreviated as ps) is appended to the measurement unit. For example, "gigabits per second" is abbreviated as Gbps, and "megabytes per second" is abbreviated as MBps. As with capacity measures, examine any abbreviation containing the letter b or B to determine whether it means "bit" or "byte."

Printer output rates are generally expressed in pages per minute, abbreviated as ppm. The number of bytes in a page varies widely, depending on page content and data-encoding method. Video output rates can be expressed in frames per second, abbreviated as fps. The number of bytes per frame depends on frame size, image content, and data-encoding method.

GLOSSARY

10 Gigabit Ethernet An Ethernet standard, based on the IEEE 802.11ae and 802.11ak standards, for 10 Gbps transmission. *See also* Ethernet *and* IEEE 802 standards.

24-bit color A color display scheme that represents each pixel's color as three 8-bit numbers, each representing the intensity of an additive or a subtractive color; results in chromatic depth of around 16 million colors. *See also* chromatic depth.

802.11 An early IEEE wireless networking standard that defines a 2 Mbps maximum raw data transfer rate in the 2.4 GHz band.

802.11a An IEEE wireless networking standard that divides frequency bands in the 5.2, 5.7, and 5.8 GHz ranges into 12 channels; uses orthogonal frequency division multiplexing (OFDM) to achieve standard transmission speeds of 6, 9, 12, 18, 24, 36, 48, and 54 Mbps.

802.11b An IEEE wireless networking standard that divides the 2.4 GHz band into 14 channels, each with 22 MHz of bandwidth; uses direct sequence spread spectrum (DSSS) to yield raw data transfer rates of 22, 44, and 88 Mbps.

802.11g An IEEE wireless networking standard that combines the frequencies and bit-encoding methods of 802.11b with the OFDM transmission method of 802.11a; raw transmission speeds are the same as 802.11a.

802.11n An IEEE wireless networking standard that expands on 802.11g; capable of broadcasting or receiving on up to four frequencies in the 2.4 or 5 GHz bands.

80x86 processors A family of processors that enhanced the capabilities of the original 8088/8086 processors and enabled Microsoft OSs to develop beyond MS-DOS.

absolute addressing In programming, using memory address operands that refer to actual physical memory locations; this method requires knowing process offsets. *See also* process offset.

access arm In magnetic disks, the device where read/write heads are mounted; it's attached to a positioning servo for placing read/write heads on specific tracks.

access control list (ACL) A list describing rights granted or denied to users, groups, and computers for accessing network resources.

access point (AP) A device that connects a wireless network to a wired network and manages media access, performs error detection, and implements security protocols for the wireless network.

access time The time required to perform one complete read or write operation; a measure of storage device time.

acknowledge (ACK) An ASCII control character sent by a receiver if no data errors are detected.

Active Directory A directory service and security system built into Windows Server. *See also* directory services.

active matrix display An LCD that uses one or more transistors for every pixel.

ADD An instruction that accepts two numeric inputs and produces their arithmetic sum.

additive colors The primary colors for video display (red, green, and blue).

address The location of a data element in a storage device; often used in data structures.

address bus The portion of the system bus that transmits a memory address when primary storage is the sending or receiving device. *See also* system bus.

address mapping The process of the CPU determining the physical memory address that corresponds to a memory reference.

address resolution *See* address mapping.

addressable memory The highest numbered storage byte that can be represented in a CPU or computer; usually determined by the number of bits used to represent an address.

Advanced Intelligent Tape (AIT) A magnetic tape standard developed by Sony based on Digital Audio Tape; uses helical scanning and an improved tape drive technology to pack more data onto a single tape. AIT includes a small RAM cache in cartridges, which stores directory information to speed searching and data access. *See also* Digital Audio Tape (DAT) *and* helical scanning.

algorithm A program in which different instructions are applied to different data input values, depending on the outcome of decisions the program makes. This term also applies to processing steps that describe the solution to a problem.

allocation unit The smallest number of secondary storage bytes that can be allocated to a file; can't be smaller than the unit of data transfer between the storage device and device controller.

American Standard Code for Information Interchange (ASCII) A standard 7-bit coding method for character data and some device control codes.

amplifier A device that increases a signal's amplitude and can extend a signal's range by boosting signal power to overcome attenuation; a drawback is that any existing noise or distortion in the signal is amplified as well. *See also* attenuation *and* repeater.

amplitude A measure of wave height or power; the maximum distance between a wave's peak and its zero value.

amplitude modulation (AM) A modulating method that represents bit values as specific wave amplitudes.

amplitude-shift keying (ASK) *See* amplitude modulation.

analog signal A signal that uses the full range of a carrier wave characteristic to encode continuous data values; because it's continuous in nature, it can represent any data value within a range of values.

analog-to-digital converter (ADC) An audio device that accepts a continuous electrical signal representing sound, samples it at regular intervals, and outputs a bitstream representing the samples. *See also* sampling.

AND An instruction that generates the result true only if both its data inputs are true.

applet A Java program that runs inside another program, such as a Web browser, and performs functions such as accepting user input and displaying forms and images. *See also* Java.

application development software Programs used to develop other programs, including application software, system software, and other application development programs; encompasses compilers and interpreters for programming languages and integrated software development packages.

application firewall A server that handles service requests from external users of applications; improves security by shielding internal servers and resources from direct access by outside users.

Application layer The OSI layer that includes communication protocols used by programs that make and respond to high-level requests for network services. *See also* Open Systems Interconnection (OSI) model.

application software A stored set of instructions for responding to a specific request or performing a specific task.

architectural design The first set of activities in the UP's design discipline; involves selecting and describing the configuration of all hardware, network, systems software, and application development tools to support system development and operations. *See also* design discipline *and* Unified Process (UP).

areal density The surface area allocated to a bit on a storage medium, typically measured in bits, bytes, or tracks per inch; also called "recording density" or "bit density."

arithmetic logic unit (ALU) One of the main CPU components, containing circuitry for performing computation, comparison, and logic instructions. *See also* central processing unit (CPU).

arithmetic SHIFT A SHIFT instruction that performs division or multiplication. *See also* SHIFT.

array An ordered list of data elements, in which each element can be referenced by an index to its position; array elements are normally referenced by the array name and the index value.

assembler A program that translates an assembly-language program into binary CPU instructions; the earliest example of automated program development tools. *See also* assembly language.

assembly language A programming language that uses mnemonics to represent CPU instructions and memory addresses; also called a second-generation language (2GL).

Association for Computing Machinery (ACM) A professional society for computer scientists, programmers, and engineers.

Association for Information Technology Professionals (AITP) A professional society for information system managers and application developers.

asynchronous transmission A method in which messages are sent on an as-needed basis, so sender and receiver don't synchronize their clocks continuously.

attenuation A reduction in signal amplitude caused by interactions between the signal's energy and the transmission medium; proportional to the medium's length.

audio response unit A device that generates spoken messages based on text input; commonly used for automated phone bank tellers and automated call routing.

auditing The process of creating and managing records of user activity or resource access.

authentication The process of determining or verifying the identity of a user or process owner.

authorization The process of determining whether an authenticated user or process has enough rights to access a resource.

average access time Typically expressed as an average of access times for all storage locations. *See also* access time.

back-end CASE tool An application development tool that generates source code instructions based on analysis and design models; also called a code generator. *See also* computer-assisted software engineering (CASE) tool.

bandwidth The difference between maximum and minimum frequencies that can be transmitted through a transmission medium.

bar code A series of vertical bars of equal length but varied thickness and spacing, used to encode numeric data.

bar-code scanner An optical input device that detects specific patterns of bars or boxes representing numeric data.

bare-metal hypervisors Hypervisors that are installed much like an OS but provide only minimal OS-type functions, such as the capability to start and stop virtual machines (in contrast to hypervisors installed over a traditional OS). *See also* hypervisor *and* virtualization environments.

base A multiplier that describes the difference between one position and the next in a numbering system.

benchmark A measure of CPU or computer performance when carrying out one or more specific tasks; used to compare the performance of multiple computers, measure a computer's performance and determine how to improve it, and determine the computer or system configuration that meets an application's requirements.

benchmark program A program that performs specific tasks that can be counted or measured. *See also* benchmark.

benchmark suite A collection of benchmark programs for evaluating computer systems. *See also* benchmark.

big endian A CPU or memory architecture in which the most significant byte is stored at the lowest memory address. *See also* most significant byte.

billions of floating-point operations per second (gigaflops or GFLOPS) A measurement of the rate at which floating-point operations are performed; used to measure CPU performance.

binary number A number in which each digit can have only one of two possible values (0 or 1).

binary signals Digital signals in which one of two values is encoded by modulating a wave characteristic. *See also* digital signal.

biometric authentication A method of verifying identity based on physical characteristics, such as fingerprints and facial features. *See also* authentication.

bit Derived from the term "binary digit," it represents one digit of a binary number and can have the value 0 or 1.

bitmap A stored set of numbers describing the content of all pixels in an image.

bit string A group of bits describing a single data value.

bit time The duration of each bit in a carrier signal.

blade A circuit board containing most of a server but lacks secondary storage, external I/O connections, and a power supply; concentrates more computing power in less space and with lower power requirements than a typical cluster.

block A generic term for describing secondary storage data transfer units. Also refers to a set of logical records grouped on a storage device for efficient processing, storage, or transport as well as a portion of a program that's always executed as a unit.

block check character (BCC) The combined parity bits from each position in a group of characters or bytes; added to the end of the block before transmitting. *See also* parity bit.

block checking An error-checking method for groups of characters or bytes in which the sender combines parity bits for each position into a block check character (BCC) and adds it to the end of the block. *See also* parity bit.

blocked state The state of an active thread that's been suspended by the OS and is waiting on the stack until interrupt processing has been completed. *See also* interrupt *and* thread.

blocking The grouping of logical records in physical records. *See also* logical record *and* physical record.

blocking factor A numeric ratio of logical records to physical records. *See also* logical record *and* physical record.

Blu-ray disc (BD) An update to DVD-ROM, originally designed for high-definition video discs but has been adapted to data storage. *See also* DVD read-only memory (DVD-ROM).

Boolean data type A data type that can store only the value true or false; requires only a single bit for storage.

Boolean logic A formal logic system in which statements can be evaluated only as true or false; well suited to the binary numbers used in computer processing.

BRANCH An instruction that causes the processor to depart from sequential instruction order; its operand is loaded into the register that the control unit uses to fetch the next instruction.

branch prediction An approach to dealing with conditional BRANCHes in which the CPU guesses whether a branch condition will be true or false based on past experience; a form of parallel processing.

broadband A high-bandwidth communication channel.

broadcast mode A communication mode in which the same message is transmitted to all devices on a network simultaneously. *See also* simplex mode.

brownout A temporary reduction in voltage level by a power provider, usually because of demand for electricity exceeding the provider's generation and transmission capabilities during peak demand periods.

buffer A small reserved area of main memory (usually DRAM or SRAM) that holds data in transit from one device to another and is used to resolve differences in data transfer unit size.

buffer overflow An error condition that results when receiving more data than can be stored in a buffer. *See also* buffer.

bus A shared electrical or optical communication channel that connects two or more devices in a computer or network.

bus arbitration unit A simple processor attached to a peer-to-peer bus that decides which devices must wait when multiple devices want to become a bus master. *See also* bus master *and* peer-to-peer bus.

bus clock A clock circuit that generates timing pulses, which are transmitted to all devices attached to the system bus to coordinate their activities.

bus cycle The time interval from one bus clock pulse to the next; also the time required to perform one data transfer operation on a bus.

bus master A device attached to a bus that can initiate a data transfer operation or send a command to another device; it controls all access to the bus.

bus protocol A communication protocol, used by all devices attached to a bus, that governs the format, content, and timing of data, memory addresses, and control messages sent across the bus.

bus slaves Devices that must go through the bus master for access to the bus. *See also* bus master.

bus topology A network topology in which every node is directly connected to a single shared transmission line. *See also* bus *and* network topology.

business logic layer The software layer that carries out the rules and procedures of business processing. *See also* three-layer architecture.

business modeling discipline Activities in the Unified Process for developing models of an organization and the system environment. *See also* Unified Process (UP).

business-to-business (B2B) The interconnection of a company's information systems with customer and supplier information systems to improve efficiency.

byte A string of 8 bits; generally the smallest unit of data that can be read from or written to a storage device. *See also* bit.

cache An area of high-speed memory (usually RAM) for storage device accesses that improves the performance of read and write operations.

cache controller A special-purposes processor or software that manages cache content; it guesses what data will be requested in the near future and loads this data from the storage device into the cache before it's actually requested.

cache hit An access to data already contained in the cache.

cache miss An access to data that isn't stored in the cache.

cache swap An operation performed after a cache miss. The cache controller guesses which data is least likely to be needed in the near future, writes it to the storage device, and purges it from the cache. The requested data is then read from the storage device and placed in the cache. *See also* cache miss.

call instruction In programming, an instruction that transfers control to the first instruction in a function. *See also* function.

capital expenditures Funds an organization uses on capital resources. *See also* capital resources.

capital resources Assets or resources expected to provide benefits beyond the current fiscal year.

Carrier Sense Multiple Access/Collision Avoidance (CSMA/CA) A MAC protocol, used in wireless networks, that uses a three-step carrier sense and transmission sequence to try to prevent collisions. *See also* collision *and* Media Access Control (MAC).

Carrier Sense Multiple Access/Collision Detection (CSMA/CD) A MAC protocol that allows collisions to occur but has methods for detecting and recovering from them.

See also collision *and* Media Access Control (MAC).

carrier wave A wave with encoded bits in a communication channel.

Category 5 Similar to Category 6 but can't achieve the same transmission speeds reliably.

Category 6 The most widely used twisted-pair wiring standard; consists of four twisted pairs that transmit at speeds up to 1 Gbps.

cathode ray tube (CRT) An older video display device that's an enclosed glass vacuum tube with an electron gun generating a narrow beam of electrons toward the tube's front surface, which is coated with colored phosphors that emit light when struck by electrons.

CD digital audio (CD-DA) A read-only format for storing and distributing music on a CD.

CD read-only memory (CD-ROM) A standard 120-mm read-only optical disc; compatible with CD-DA but includes additional formatting information to store directory and file information. *See also* CD digital audio (CD-DA).

central processing unit (CPU) A general-purpose processor that executes all instructions and controls all data movement in a computer system. *See also* general-purpose processor.

channel *See* I/O channel *or* communication channel.

character A symbol in a written language, including letters, numerals, and punctuation marks.

character-framing methods Approaches to clock synchronization when messages consist of ASCII or Unicode characters.

chief information officer (CIO) The person who's responsible for planning, maintaining, and operating all information-processing resources in an organization; managers such as the database administrator, network administrator, and computer operations manager often report to the CIO.

child The file version that's been updated with changes to the parent file. *See also* parent.

child process A process created and controlled by the parent process that spawned it. *See also* process *and* spawn.

chromatic depth The number of distinct colors or gray shades that can be displayed in a grayscale image. *See also* grayscale.

chromatic resolution *See* chromatic depth.

circuit switching A channel-sharing strategy that grants exclusive use of a communication channel for the duration of the session.

class A data structure containing both traditional (static) data elements and programs that manipulate data; it combines related data items in much the same way a record does, but it extends a record to include methods for manipulating data items.

clear-to-send (CTS) signal A CSMA/CA signal that a wireless access point transmits after it detects no potential collision. *See also* Carrier Sense Multiple Access/Collision Avoidance (CSMA/CA).

client A program or computer that requests services from another program or computer.

client/server architecture A method of organizing software to provide and access distributed information and computing resources; divides software into two classes—client and server.

clock cycle The time interval between two clock timing pulses. *See also* system clock.

clock rate The frequency (expressed in Hz) at which the system clock generates timing pulses. *See also* system clock.

cloud A specific way of organizing computing resources for maximum availability and accessibility and minimum complexity in the interface; includes front-end Web-based interfaces and a large collection of computing and data resources (collectively called "back-end resources").

cloud computing A cloud-based approach to distributing and accessing software and hardware services across the Internet. *See also* cloud, infrastructure as a service (IaaS), platform as a service (PaaS), *and* software as a service (SaaS).

cluster A group of similar or identical computers, connected by a high-speed network, that cooperate to provide services or run a common application; has the advantages of scalability and fault tolerance but can be complex to configure and manage.

CMY Cyan, magenta, and yellow; the primary colors in printing. *See also* subtractive colors.

CMYK Cyan, magenta, yellow, and black; represents the primary colors in printing plus a separate ink for black.

coaxial cable A transmission medium that contains a single copper conductor surrounded by a thick plastic insulator, a metallic shield, and a tough plastic outer wrapping.

code Program instructions for performing a task.

coercivity The capability of a substance or magnetic storage medium to accept and hold a magnetic charge; directly proportional to mass.

collating sequence The specific order for assigning numeric codes to characters or symbols.

collision The noise or interference produced when multiple nodes attempt to transmit across the same medium at the same time, and their messages mix.

colon hexadecimal notation The written format of 128-bit IPv6 addresses; written in the form *hhhh:hhhh:hhhh:hhhh:hhhh:hhhh:hhhh:hhhh*, with *hhhh* representing a sequence of four hexadecimal digits. *See also* Internet Protocol version 6 (IPv6).

command language A set of commands and syntax requirements for implementing an operating system's command layer via a text interface.

command layer The operating system layer that serves as the user interface; via this layer, users can run applications and OS utilities and manage system resources, such as files, folders, and I/O devices.

Common Object Request Broker Architecture (CORBA) An industry-wide interoperability standard specifying the middleware that objects use to interact across networks.

communication channel A combination of a sending device, a receiving device, the transmission medium connecting them, and a communication protocol. *See also* communication protocol *and* transmission medium.

communication protocol A set of rules and conventions for representing the content of data and commands, encoding and transmitting data (bits), channel organization, and communication coordination (clock synchronization and error detection and correction).

compact disc (CD) A technology developed by Sony and Phillips for storing and distributing music in the CD-DA format on a 120-mm optical disc. *See also* CD digital audio (CD-DA).

compaction The process of reallocating all programs in memory so that free partitions form a contiguous block in upper memory; used to address the problem of fragmentation. *See also* contiguous *and* fragmentation.

competitive advantage The way in which an organization uses resources to offer better or cheaper services so that it has a major edge over its competitors.

compiler A program that translates some source code instructions into executable code and others into library calls; a compiler's output is called object code. *See also* object code.

compiler library A file containing related executable functions and an index of the library contents. *See also* compiler *and* link editor.

complete path The access path that begins at the root directory and proceeds through all directories along a path to the file being accessed.

complex instructions Instructions that combine primitive processing operations.

complex instruction set computing (CISC) A computer and processor design approach, using complex instructions that do more work per instruction; it reduces the extra memory required for program storage and execution in RISC CPUs. *See also* reduced instruction set computing (RISC).

component A standardized, interchangeable software module that's executable, has a unique identifier, and has a well-known interface.

Component Object Model Plus (COM+) A Microsoft specification for component interoperability; defines component registration, message-routing services, and a component communication protocol.

composite signal A complex signal created by combining multiple simple signals.

compression A technique that reduces the number of bits used to encode data.

compression algorithm A mathematical compression technique implemented as a program for translating data inputs into equivalent but smaller data outputs. *See also* compression.

compression ratio The ratio of data size in bits or bytes before and after compression. *See also* compression.

computer-assisted software engineering (CASE) tool An application development tool suite that supports the Unified Process requirements and design disciplines. *See also* design discipline *and* requirements discipline.

computer network A collection of hardware and software components that enable users and computer systems to share information, software, and hardware resources and make it possible to use many types of communication methods.

computer operations manager The person who oversees the operation and maintenance of a large information-processing facility or an information system.

computer science The study of implementing, organizing, and applying computer software and hardware resources.

concurrent execution A method of sharing CPU control among threads by using time slices. *See also* thread.

condition A comparison or other logical operation that produces a Boolean (true or false) result.

conditional BRANCH A BRANCH instruction that occurs only if a specified condition is met. The condition is evaluated, and the Boolean result is stored in a register; the register's contents are checked, and the BRANCH is performed only if the Boolean result is true.

conductivity The capability of an element or a substance to enable electron flow.

conductor A substance that exhibits conductivity, allowing electrons to flow through it.

connectionless protocol A communication protocol that doesn't require sender and receiver to establish a connection before transmitting any data.

connection-oriented protocol A communication protocol that requires sender and receiver to establish a connection before transmitting any data.

contiguous The condition of all portions of a program or the OS being loaded into sequential physical locations in memory. *See also* memory allocation.

control bus The portion of the system bus that carries commands, command responses, status codes, and similar messages; computer components coordinate their activities by sending signals over this bus. *See also* system bus.

control structure A source code instruction that controls the execution of other source code instructions; includes unconditional BRANCHes, such as a `goto` statement; conditional BRANCHes, such as an `if-then-else` statement; and loops, such as `while-do`. *See also* source code.

control unit One of the main CPU components, responsible for moving data, accessing instructions, and controlling the arithmetic logic unit. *See also* central processing unit (CPU).

core A term describing the logic, computation, and control circuitry of a single CPU. *See also* multicore architecture.

core memory In early computers, a technology for implementing primary storage as rings of ferrous materials embedded in a two-dimensional wire mesh.

crosstalk In parallel transmission channels, noise added to the signal in a wire from EMI generated by adjacent wires. *See also* parallel transmission.

current directory The directory that's being accessed. *See also* directory.

cursor A symbol on a video display that indicates the current position; also called a "pointer."

cycle In communication, the full range of a sine wave, from zero to positive peak, back to zero, to negative peak, and back to zero again.

cycle time The inverse of the clock rate; in most CPUs, it's the time required to fetch and execute the simplest instruction in the instruction set. *See also* clock rate.

cyclic redundancy checking (CRC) The most widely used error-detection method; uses a complex algorithm to generate CRC bit strings for groups of characters or bytes.

cylinder In magnetic disks, consists of all tracks at an equivalent distance from the edge or spindle on all platter surfaces. *See also* platters *and* track.

data bus The portion of the system bus that transmits data between computer components. *See also* system bus.

data declaration A source code instruction type that defines the name and data type of program variables. *See also* source code.

data layer The software layer that manages stored data, usually in databases. *See also* three-layer architecture.

Data Link layer The OSI layer serving as the interface between network software and hardware; responsible for media access control and converting messages and addresses from one format to another. *See also* Open Systems Interconnection (OSI) model.

data operation A source code instruction type that updates or computes a data value, such as an assignment statement or a computation. *See also* source code.

data striping A fault-tolerance technique that breaks a unit of data into smaller segments and stores these segments on multiple disks. *See also* fault tolerance.

data structure A related group of primitive data elements organized for some type of common processing; it's defined and manipulated in software because the CPU can't manipulate data structures directly.

data transfer rate The rate at which data is transmitted through a medium or communication channel; measured in data units per time interval; essentially, it's a measure of communication capacity. For a storage device, it's computed by dividing 1 by the access time and multiplying the result by the unit of data transfer.

database administrator The person responsible for overseeing an organization's database and ensuring data integrity, reliability, security, and availability.

datagrams Messages accepted from Transport-layer protocols and forwarded to their destinations. *See also* Transport layer.

debugging tools Utilities that simulate program execution and help programmers trace errors. *See also* application development software.

debugging version A program version containing symbol table entries and debugging checkpoints to help locate and correct errors. *See also* production version *and* symbol table.

decimal point The period or comma in the decimal numbering system that separates the whole and fractional parts of a numeric value. *See also* radix point.

decoding The process the control unit performs of extracting an instruction's op code and operands, loading data inputs, and signaling the ALU.

decompression algorithm An algorithm that restores compressed data to its original or nearly original state. *See also* compression.

deployment discipline Activities in the Unified Process for installing and configuring infrastructure and application software components and bringing them into operation. *See also* Unified Process (UP).

design discipline Activities in the Unified Process for determining the structure of a specific information system that fulfills the system requirements. *See also* Unified Process (UP).

design models Models that specify detailed blueprints for software component construction and the interaction between software components and users.

detailed design Design activities that specify system details, including databases, application software, user and system interfaces, and backup and recovery mechanisms.

device controller A processor that controls the physical actions of storage and I/O devices; connects these devices to the system bus or a subsidiary bus.

differential backup A type of backup that's a variation on an incremental backup, in which backup times aren't reset as files are copied. *See also* incremental backup.

Digital Audio Tape (DAT) An early magnetic tape technology on which Digital Data Storage standards are based. *See also* Digital Data Storage (DDS).

Digital Data Storage (DDS) Magnetic tape standards developed by Hewlett-Packard and Sony and based on Digital Audio Tape; DDS drives use helical scanning. *See also* Digital Audio Tape (DAT) *and* helical scanning.

digital signal A signal that can contain one of a finite (countable) number of possible values.

digital signal processor (DSP) A microprocessor specialized for processing continuous streams of audio or graphical data; commonly embedded in audio and video hardware.

digital-to-analog converter (DAC) An audio device that accepts a bitstream representing sound samples and generates a continuous analog signal that can be amplified and routed to a speaker.

digitizer A device consisting of a digitizing tablet and a pen, stylus, or both that captures a pointing device's position as input data.

Direct3D A video controller image description language that's part of the Microsoft DirectX suite embedded in Windows OSs. *See also* image description language (IDL).

direct access *See* random access.

direct-attached storage (DAS) A storage access model in which software running on a CPU accesses secondary storage devices in the same computer.

direct memory access (DMA) A method of data transfer that enables the CPU to execute instructions while another device (the DAM controller) manages all transfers between memory and other storage or I/O devices. *See also* DMA controller.

directory A data structure containing information about files and other directories.

directory services Middleware that stores the name and network address of distributed resources, responds to directory queries, accepts directory updates, and synchronizes directory copies. *See also* middleware.

disciplines Related groups of system development activities in the Unified Process. *See also* Unified Process (UP).

discrete signal *See* digital signal.

disk defragmentation Reorganizing data on a disk drive so that a file's contents are stored in sequential sectors, tracks, and platters; an OS utility is used to perform this task.

disk mirroring A fault-tolerance technique in which all disk write operations are made simultaneously or concurrently to two storage devices. *See also* fault tolerance.

dispatching Giving CPU control to a thread in the ready state. *See also* ready state *and* thread.

distortion Changes to the data signal caused by interaction with the communication channel; can include echoes, resonance, and selective attenuation.

distributed computing *See* distributed processing.

Distributed Computing Environment (DCE) A standard for distributed OS services defined by the Open Group Standard; covers network directory services, file-sharing services, RPC, remote thread execution, system security, and distributed resource management. *See also* distributed processing *and* Remote Procedure Call (RPC).

distributed processing Spreading parts of an information system across many computer systems and locations.

dithering A process that generates continuous color approximations by placing small dots of different colors in an interlocking pattern.

diversity As specified in the 802.11n standard, using antenna pairs for redundant data transmission across different frequencies to increase signal reliability. *See also* 802.11n.

DMA controller A device attached to the system bus and main memory that manages data transfers, freeing the CPU to execute instructions. *See also* direct memory access (DMA).

Domain Name System (DNS) A name-resolution protocol used on the Internet; makes use of dynamic connections to find requested IP addresses. *See also* dynamic connection.

dot matrix printer An impact printer that moves a print head containing a matrix of pins over the paper, and a pattern of pins matching the character or symbol to be printed is forced out of the print head.

dots per inch (dpi) A measure of print or display resolution (pixel density); a smaller pixel size represents a higher dpi and, therefore, higher image or print quality. *See also* resolution.

dotted decimal notation The written format of IPv4 addresses; written in the form *ddd.ddd.ddd.ddd*, with *ddd* representing a decimal number between 0 and 255. *See also* Internet Protocol version 4 (IPv4).

double data rate (DDR) A series of technologies, each doubling the data transfer rate of the previous synchronous DRAM version. *See also* synchronous DRAM (SDRAM).

double inline memory module (DIMM) A small printed circuit board that's essentially a SIMM with independent electrical contacts on both sides of the module. *See also* single inline memory module (SIMM).

double-precision A data format that combines two adjacent fixed-length data items to hold a single value; increases accuracy or numeric range.

doubly linked list A data structure in which each list element contains pointers to both the previous and next list elements.

drive array An arrangement of hard drives enclosed in a storage cabinet and accessed as though they're a single storage device.

dual inline packages (DIPs) An early form of packaging for RAM or ROM circuits; had two rows of electrical contact pins.

dual-porting The simultaneous read/write capability in video RAM.

DVD An optical disc format for distributing movies and other audiovisual content; stands for both "digital video disc" and "digital versatile disc."

DVD read-only memory (DVD-ROM) A format for general-purpose read-only data storage on DVD.

dynamic connection A more complex, but flexible, approach to remote resource access, in which connections between a client and a server or remote resource aren't established until the time of the access request. *See also* static connection.

dynamic link libraries (DLLs) Repositories of reusable software modules organized for dynamic linking; also refers to a Windows file format for storing reusable software modules.

dynamic linking A linking process performed during program loading or execution. *See also* link editor.

dynamic RAM (DRAM) A type of RAM that stores each bit by using a single transistor and capacitor.

early binding *See* static linking.

effective data transfer rate The data-transmission capacity actually achieved with a communication protocol; always less than the raw data transfer rate. *See also* raw data transfer rate.

electromagnetic interference (EMI) An alteration of wave characteristics caused by external electrical or magnetic phenomena, such as radio equipment and nearby power lines.

electronically erasable programmable ROM (EEPROM) A type of nonvolatile memory that can be programmed, erased, and reprogrammed by signals sent from a CPU; the only type of ROM that's currently used.

encapsulation A message-translation process that embeds all or part of a datagram in the message format of a physical network.

Enterprise JavaBeans (EJBs) Java components that run in a business container on a server; capable of performing complex, behind-the-scenes processing.

erasable programmable ROM (EPROM) A type of nonvolatile memory that's manufactured blank, written (programmed) with a special EPROM writer, and erased by exposure to ultraviolet light.

Ethernet A widely used LAN technology, developed by Xerox in the early 1970s, that's closely related to the IEEE 802.3 standard. *See also* IEEE 802 standards *and* local area network (LAN).

even parity An error-detection method in which the sender sets the parity bit to 0 if the count of 1-valued data bits is even or to 1 if the count of 1-valued data bits is odd. *See also* parity bit.

excess notation A format that can be used to represent signed integers with a fixed number of bits; essentially, it divides a range of ordinary binary numbers in half and uses the lower half for negative values and the upper half for nonnegative values. *See also* signed integer.

exclusive OR (XOR) An instruction that generates the value true if either (but not both) data input is true.

executable code A program consisting entirely of CPU instructions that are ready to be loaded and run.

executing The act of a processor performing a function in response to an instruction. *See also* instruction *and* processor.

execution cycle The CPU cycle in which instructions are retrieved from registers, the specified data transformation is performed, and data outputs are stored in registers.

explicit priority A priority-based scheduling method that assigns a priority level to each thread and can dispatch the highest-priority threads first or assign larger time slices to high-priority threads. *See also* dispatching, priority-based scheduling, *and* thread.

Extended Binary Coded Decimal Interchange Code (EBCDIC) An IBM mainframe coding method for representing character data in an 8-bit format.

Extensible Markup Language (XML) An extension of HTML that describes the structure, format, and content of documents. *See also* Hypertext Markup Language (HTML).

external function call A placeholder in object code that the compiler generates for missing executable code; contains the name and type of the called function as well as the memory addresses and types of function parameters. *See also* compiler *and* function.

external I/O buses Subsidiary buses that connect external devices to the system bus; they provide the connection points for external devices and aggregate their capacity to better match a system bus connection's capacity.

fault tolerance In an FMS, methods of securing file content against hardware failure. *See also* file management system (FMS).

fetch cycle The CPU cycle (also called the "instruction cycle") in which data inputs are prepared for transformation into data outputs; instructions are fetched from primary storage and stored in registers.

fiber-optic cable A guided transmission medium for optical signals, generally consisting of plastic or glass fibers sheathed in a protective plastic coating.

fields In a file's logical structure, the components of a record; usually contains information about a single person, a thing, or an event. *See also* record.

fifth-generation language (5GL) A nonprocedural programming language used to develop software that mimics human intelligence.

file A sequence of records on secondary storage; the common organization schemes for files are sequential and indexed. *See also* record.

File Allocation Table (FAT) The file system used in DOS and early Windows versions to record the allocation of storage device locations to files and directories.

file association The relationship between file types and the programs or OS utilities that manipulate them.

file close operation The act of severing the relationship between a file and a process by flushing file I/O buffers, deallocating buffer memory, updating the file's directory entries, and updating the open file table.

file control layer The FMS layer that provides service functions for manipulating files and directories, processes service calls from the command layer, issues commands to the storage I/O control layer to interact with hardware, and maintains the directory and storage allocation data structures. *See also* file management system (FMS).

file management system (FMS) The collection of system software that performs file and secondary storage management and access functions; usually part of the OS.

file migration A file management technique that balances each file version's storage cost with anticipated user demand for this version; older versions of files are moved automatically to less costly storage media or devices.

file open operation The process of associating a file with an active process by allocating buffers and updating internal tables.

File Transfer Protocol (FTP) An older Web protocol that specifies a client/server request and response language for copying files from one Internet host to another.

firewall A hardware device or software (or a combination) that prevents unauthorized users in one network from accessing resources on another network.

firmware Software, such as system BIOS, stored in nonvolatile memory; can be loaded into main memory at high speeds. *See also* nonvolatile memory (NVM).

first come, first served (FCFS) A priority-based scheduling method in which the scheduler always dispatches the ready thread that has been waiting the longest. *See also* dispatching, priority-based scheduling, *and* thread.

first-generation languages (1GLs) The earliest programming languages, consisting of binary CPU instructions.

fixed-length instructions In this type of instruction format, the amount by which the instruction pointer must be incremented after each fetch is a constant; this increment is the length of an instruction.

flag A Boolean variable representing each bit in a PSW register's bit string; generally used to store the result of a comparison operation, control conditional BRANCH execution, or indicate actual or potential error conditions. *See also* program status word (PSW).

flash RAM The most common type of nonvolatile memory; typically used to store firmware and in portable secondary storage systems, such as USB flash drives.

flat memory model An approach to assigning memory addresses in which memory locations are described by single unsigned integers corresponding to linear positions.

flat panel displays Newer video display devices that are thinner, generate higher quality images, and consume less power than CRTs.

floating-point notation A method for representing real numbers that consists of two parts: a mantissa and an exponent; the mantissa holds the bits that are interpreted to derive the real number's digits, and the exponent value indicates the radix point's position. *See also* real number.

font A collection of characters of similar style and appearance.

formula A sequence of computation and data movement instructions that a processor executes to solve a processing problem.

forwarding table A table of node addresses and transmission lines or connection ports that each central network node maintains to make message-forwarding decisions.

fourth-generation languages (4GLs) High-level programming languages that have higher instruction explosion and support nonprocedural programming, database manipulation, and advanced I/O capabilities. *See also* instruction explosion.

fragmentation The scattering of storage locations allocated to a single process or purpose throughout noncontiguous locations in physical memory or a secondary storage device. *See also* contiguous *and* memory allocation.

fragmented The condition of a hard disk (or other storage drive) with many programs and files scattered across it in noncontiguous storage locations. *See also* contiguous.

frequency The number of wave cycles occurring in 1 second; measured in hertz. *See also* cycle.

frequency-division multiplexing (FDM) A channel-sharing technique that partitions a single broadband channel into multiple narrowband subchannels, each representing a different frequency band.

frequency modulation (FM) A modulation method that represents bit values by varying carrier wave frequency while holding amplitude constant.

frequency-shift keying (FSK) *See* frequency modulation.

front-end CASE tool An application development tool that primarily supports developing system models. *See also* computer-assisted software engineering (CASE) tool.

full backup A type of backup in which the FMS copies all files and directories for an entire storage volume; can include storage allocation tables, partition tables, and other important disk management data structures.

full-duplex mode A communication mode in which two transmission lines are used; allows simultaneous communication in both directions.

fully qualified reference *See* complete path.

function A named instruction sequence in a high-level programming language that's always executed as a unit; also called a subroutine or a procedure.

gate A circuit that can perform a processing function on a single binary electrical signal, or bit. *See also* switch.

gateways Nodes connecting two or more networks or network segments that might be physically implemented as workstations, servers, or routers.

general-purpose processor A processor that can be instructed to perform a wide variety of tasks. *See also* processor.

general-purpose registers Registers used only by the currently running program; they typically hold intermediate results or frequently used data values. *See also* registers.

germanium, antimony, and tellurium (GST) A glasslike compound, used in phase-change memory, that can change between amorphous and crystalline states. *See also* phase-change memory (PCM).

Gigabit Ethernet An Ethernet standard, based on the IEEE 802.3z and 802.3ab standards, for 1 Gbps transmission. *See also* Ethernet *and* IEEE 802 standards.

gigahertz (GHz) A measurement of wave or system clock frequency; one billion cycles per second.

Google File System (GFS) A file system Google developed to meet rapidly increasing storage requirements; offers scalability to petabyte storage, the capability to handle extremely large files, data storage on distributed commodity servers, and simultaneous file access by multiple distributed applications.

grandparent The file version before the parent version. *See also* parent.

graph directory structure A directory structure in which files and subdirectories can be contained in multiple directories, and directory links can form a cycle. *See also* links.

Graphics Interchange Format (GIF) A common bitmap compression format for still images.

grayscale A display device or encoding method that can display black, white, and shades of gray but no other colors.

grid A group of dissimilar computers, connected by a high-speed network, that cooperate to provide services or run a shared application; unlike a cluster, computers in a grid work cooperatively at some times and independently at others and can be located far away from each other.

Grosch's Law An outdated mathematical relationship between computer size and cost per unit of instruction execution, which states that computing power, measured by millions of instructions per second, is proportional to the square of hardware cost. In other words, large and powerful computers are always more cost effective than smaller ones.

guided transmission A transmission medium that routes signals between two locations through a physical connection, such as copper wire or optical fiber; also called "wired transmission."

H.323 The oldest and most widely deployed VoIP protocol suite; also addresses video and data conferencing. *See also* Voice over IP (VoIP).

half-duplex mode A communication mode that uses a single shared channel, and each node takes turns using the transmission line to transmit and receive.

half-toning Simulating shades of gray by dithering black and white dots. *See also* dithering.

HALT An instruction that suspends the normal flow of instruction execution in the current program; in some CPUs, it causes the CPU to cease all operations, and in others, it causes a BRANCH to a predetermined memory address. *See also* BRANCH.

hard disk A magnetic disk medium with a rigid metal base (substrate) where data is recorded as patterns of magnetic charge.

hardware independence Embedding hardware's physical details into system software so that users and application programmers don't need to know them to interact with hardware.

hardware monitor A device attached directly to the communication link between two hardware devices, often used to monitor the use of communication channels, disk drives, and network traffic; monitors communication activity between the two devices and stores communication statistics that can be retrieved and printed in a report.

head-to-head (HTH) switching time The time needed to switch a hard drive's read/write circuitry to the correct read/write head before accessing a sector.

heat dissipation Conducting excessive heat away from a device, thus reducing its temperature.

heat sink An object designed to absorb heat and rapidly dissipate it via air or water movement; it's placed in direct physical contact with an electrical device.

helical scanning A geometric approach to recording data on a tape surface in which data is read and written by rotating the read/write head at an angle and moving from tape edge to tape edge. *See also* linear recording.

hertz (Hz) In computers, a unit of measure for the frequency of system clock timing pulses; one Hz corresponds to one clock cycle per second. *See also* clock rate *and* system clock.

hexadecimal notation A numbering system with a base value of 16; uses digits from 0 to 9 and letters from A to F, which represent the decimal values 10 to 15.

hierarchical directory structure A multilevel system of directories in which directories can contain other directories, but a directory can't be contained in more than one parent.

high-order bit *See* most significant digit.

hit ratio The ratio of cache hits to read accesses. *See also* cache hit.

home directory The main directory associated with and owned by a single user.

host channel adapter (HCA) An interface that connects a device to an InfiniBand switch; used by devices, such as general-purpose servers, that can initiate and respond to data transfer requests.

HTTPS A secure version of HTTP that encrypts HTTP requests and responses. *See also* Hypertext Transfer Protocol (HTTP).

hub A central connection point for nodes in a LAN; provides separate point-to-point connections between nodes and the hub by using cabling in a physical star topology and attaching these connections to its internal shared bus. *See also* local area network (LAN).

Hypertext Markup Language (HTML) A device-independent formatting language that describes Web documents; links to other documents can be embedded in it.

Hypertext Transfer Protocol (HTTP) A Web protocol that specifies the language by which clients request documents and how servers respond to those requests.

hypervisor An OS that enables dividing a single physical computer or cluster into multiple virtual machines. *See also* virtual machine (VM).

IEEE 802 standards A collection of IEEE network standards describing physical network hardware, transmission media, transmission methods, and protocols.

image description language (IDL) A language (usually device independent) that uses compact bit strings or ordinary ASCII or Unicode text to describe primitive image components, such as straight lines and simple shapes; reduces storage space requirements because a description of a simple image component is usually much smaller than a bitmap.

implementation discipline Activities in the Unified Process for building, acquiring, and integrating application software components. *See also* Unified Process (UP).

inclusive OR An instruction that generates the value true if either or both data inputs are true; usually called just an "OR instruction."

incremental backup A type of backup in which the FMS archives only files that have been modified since the previous incremental or full backup. *See also* full backup.

index In file organization, an array of pointers to records. *See also* pointer *and* record.

indirect addressing A method of computing physical memory addresses automatically; the CPU adds the process offset to all memory address operands before accessing memory. *See also* process offset.

InfiniBand A data connection standard for high-speed interconnection of network switches, servers, and secondary storage devices; based on a switched fabric architecture. *See also* switched fabric.

information architecture The requirements and constraints that define important characteristics of information-processing resources and how these resources interact with one another.

infrastructure Capital resources that provide benefits to a wide range of organizational units and functions; typically, they serve a large, diverse group of users; have high maintenance costs; involve costs difficult to allocate to users separately; and require recurring capital expenditures. *See also* capital resources.

infrastructure as a service (IaaS) A cloud-based architectural approach similar to PaaS, in which customers can configure application and system software for a generic platform as virtual servers and then deploy these servers to a third-party hosting site; often used to provide back-end storage services and large-scale computing infrastructures for running complex simulations. *See also* cloud computing *and* platform as a service (PaaS).

inkjet printer A printer that produces printed images by placing small drops of liquid ink onto paper; ink is forced out of the nozzle by mechanical movement or by heat.

input/output (I/O) units Devices that perform external communication functions.

input pads A general class of input devices that convert pressure into input, such as for capturing signatures or drawings; these devices typically use infrared detectors, photosensors, pressure-sensitive pads, or magnetic fields.

Institute for Electrical and Electronics Engineers (IEEE) Computer Society A subgroup of the IEEE that specializes in computer and data communication technologies.

instruction A signal or command to a processor to perform one of its functions. *See also* processor.

instruction cycle *See* fetch cycle.

instruction explosion The one-to-many (1:N) relationship between later-generation programming statements and the CPU actions implementing them.

instruction format A template that specifies the number of operands and the position and length of the op code and operands in instructions.

instruction pointer (IP) A special-purpose register that stores the address of the next instruction the control unit should fetch from memory. *See also* registers.

instruction register A special-purpose register that holds an instruction the control unit has fetched from memory. *See also* registers.

instruction set The collection of instructions that a CPU can process. *See also* instruction.

integer A whole number—that is, a value that doesn't have a fractional part.

integrated circuit (IC) A semiconductor device that incorporates several transistors and their interconnections on a single chip.

integrated development environment (IDE) A collection of automated support tools to speed program development and testing; typically includes tools such as program editors, interpreters, compilers, debuggers, prototyping tools, function and class libraries, and so forth.

interleaved execution *See* concurrent execution.

International Alphabet 5 (IA5) The international equivalent of the ASCII coding method for character data. *See also* American Standard Code for Information Interchange (ASCII).

International Organization for Standardization (ISO) An international group with functions similar to those of the American National Standards Institute.

Internet A global collection of networks interconnected with TCP/IP.

Internet Inter-ORB Protocol (IIOP) A component message-passing protocol in CORBA. *See also* Common Object Request Broker Architecture (CORBA).

Internet Message Access Protocol 4 (IMAP4) A protocol that extends POP3 to permanently store and manage e-mail messages on the server, which enables users to access stored e-mail from any Internet host. *See also* Post Office Protocol 3 (POP3).

Internet Protocol (IP) A core protocol for packet switching and routing on the Internet.

Internet Protocol version 4 (IPv4) The original version of IP; uses 32-bit addresses.

Internet Protocol version 6 (IPv6) An update to IPv4 that uses 128-bit addresses; developed to address the limited number of node addresses, handle streaming multimedia, and multicast more efficiently.

interpretation A process for source code instructions that interleaves source code translation, link editing, and execution. *See also* link editor *and* source code.

interpreter A program that reads a single source code instruction, translates it into CPU instructions or a DLL call, and executes the instructions or DLL call immediately. *See also* dynamic link libraries (DLLs).

interrupt A signal sent to the CPU over the control bus that some event requires it to execute a specific program or process; used to prevent inefficiency caused by I/O wait states. *See also* I/O wait states.

interrupt code A numerical value of an interrupt, indicating the type of event that has occurred; usually equivalent to the bus port number of the peripheral device sending the interrupt. *See also* interrupt.

interrupt handler An OS service routine that processes interrupts; each interrupt handler is a separate program stored in a separate part of primary storage.

interrupt register A register in the CPU's control unit that stores interrupt codes received over the bus or generated by the CPU. *See also* interrupt code.

intranet An internal private network that uses Internet protocols but is accessible only by a limited set of internal users; also describes privately accessible resources that are organized and delivered via one or more Web protocols over a TCP/IP network.

I/O channel A device controller dedicated to a mainframe bus port that enables many devices to share access to, and the capacity of, the port; originally used to describe a specific hardware component of IBM's 7000 series mainframe computers.

I/O port A communication pathway from the CPU to a peripheral device; in most computers, it's a memory address, or set of contiguous memory addresses, that can be read or written by the CPU and a single peripheral device.

I/O wait states Idle processor cycles consumed while waiting for secondary storage or I/O devices to complete access requests.

iterations Repeated steps in an SDLC process; for example, in the UP, each iteration, consisting of specific activities, is 4 to 6 weeks. *See also* systems development life cycle (SDLC) *and* Unified Process (UP).

Java An object-oriented programming language that supports almost any combination of hardware platform and OS.

Java Platform, Enterprise Edition (Java EE) A family of standards for developing and deploying component-based distributed applications written in Java; follows a three-layer architecture, with the client, Web/business, and data tiers.

JavaServer Faces (JSF) Java components that enable developers to create user interfaces that run on a server but interact with a client Web browser or component.

JavaServer Pages (JSP) Server-side Java components that generate formatted Web pages by using embedded scripts.

Java Virtual Machine (JVM) A hypothetical computer and operating system serving as the target machine for Java interpreters and compilers; avoids the need to translate source code instructions for a specific platform and OS.

job control language (JCL) *See* command language.

Joint Photographic Experts Group (JPEG) A common bitmap compression format for still images.

journaling *See* transaction logging.

JUMP *See* BRANCH.

keyboard controller A microprocessor integrated into a keyboard that generates a bitstream output of scan codes according to an internal program or lookup table. *See also* scan code.

Kerberos A security model that defines standard interactions between clients, services, and a trusted security service.

kernel The OS layer that manages resources and interacts with hardware; includes a resource allocation layer and interface programs called device drivers.

label A mnemonic representing a program instruction's memory address.

large-format printer A more current term for plotters. *See also* plotter.

laptop computer A full-featured, portable microcomputer with an integrated display and a battery; rivals traditional microcomputers in power and cost. *See also* microcomputer.

laser printer A printer that operates by charging areas of a photoconductive drum;

toner is attracted to charged areas of the drum and then to paper.

late binding *See* dynamic linking.

Latin-1 An ISO standard character-coding table containing the ASCII-7 characters in the lower 128 table entries and most of the characters used by Western European languages in the upper 128 table entries. *See also* multinational characters.

law of diminishing returns The economic principle stating that when multiple resources are required to produce something useful, adding more of a single resource produces fewer benefits; can be applied to buffer and cache sizes as well as many other computer components.

least significant byte In storage bytes, the rightmost byte in a multiple-byte data item containing digits of the lowest weight.

least significant digit The rightmost digit in a bit string that represents the lowest weight.

level one (L1) cache An SRAM cache between the CPU and SDRAM primary storage, used to limit wait states; the L1 cache is closest to the CPU.

level two (L2) cache An SRAM cache between the CPU and SDRAM primary storage, used to limit wait states; the L2 cache is the next level away from the CPU.

level three (L3) cache An SRAM cache between the CPU and SDRAM primary storage, used to limit wait states; the L3 cache is the farthest level away from the CPU.

light-emitting diodes (LEDs) A video display technology that uses phosphorescent compounds to produce red, green, and blue light.

Lightweight Directory Access Protocol (LDAP) An Internet standard for directory services, based on the X.500 standard and adopted by the Internet Engineering Task Force.

linear address space The set of sequentially numbered storage locations in a peripheral device; these locations must be converted into a disk's corresponding platter, sector, and track for the CPU to physically access the correct sector. *See also* logical access.

linear recording A geometric approach to recording data on a tape surface in which bits are placed along parallel tracks that run along the tape's entire length. *See also* helical scanning.

Linear Tape Open (LTO) A magnetic tape standard developed by Hewlett-Packard, IBM, and Seagate; uses linear recording and has technology improvements in tape cartridges, coercible materials, read/write heads, and tape control. *See also* linear recording.

line turnaround A control message that's sent when one node in a half-duplex channel has stopped sending; the receiver then assumes the role of sender.

links The UNIX term for pointers from one directory to another in a graph directory structure; also refers to pointers connecting two data items in a data structure or external function calls in an object code file. *See also* graph directory structure.

link editor A program that combines object code files into an integrated set of executable code with a consistent scheme of memory addresses and references.

linked list A data structure that uses pointers so that list elements can be scattered among nonsequential storage locations. *See also* doubly linked list *and* singly linked list.

liquid crystal display (LCD) A video display device containing liquid crystals sandwiched between two polarizing filter panels; the crystals change from opaque to transparent when an electrical charge is applied.

little endian A CPU or memory architecture in which the least significant byte is stored at the lowest memory address. *See also* least significant byte.

load A data transfer from main memory into a register.

local area network (LAN) A network that spans a limited area, such as a single building or office floor.

location transparency A characteristic of software and user interfaces meaning that local and remote resources are accessed in the same way; also referred to as "network transparency."

logic instructions Instructions that implement Boolean operations, such as ADD, AND, OR, NOR, and XOR.

logical access A read or write operation from the hypothetical storage device representing a peripheral device. *See also* linear address space.

logical record A collection of data items, or fields, that an application program accesses as a single unit.

logical SHIFT A SHIFT instruction used to extract a single bit from a bit string. *See also* SHIFT.

logical topology The path messages traverse as they travel between end and central network nodes. *See also* network topology.

long integers Double-precision representations of integers. *See also* double-precision.

longitudinal redundancy checking (LRC) *See* block checking.

lossless compression A compression algorithm in which data content is unchanged when compressed and then decompressed.

lossy compression A compression algorithm in which data content is altered or lost when compressed and then decompressed; usually applied only to audio and video data.

low-order bit *See* least significant digit.

machine data types *See* primitive data types.

machine independence *See* hardware independence.

machine languages *See* first-generation languages (1GLs).

machine state The saved register values of interrupted processes or programs that represent their state before an interrupt. *See also* interrupt.

magnetic decay The tendency of magnetically charged particles to lose their charge over time; it's constant over time and proportional to the power of the charge.

magnetic leakage The reduction in strength of a stored magnetic charge because of interference from adjacent magnetic charges of opposite polarity.

magnetic tape A ribbon of plastic with a coercible (usually metallic oxide) coating, used to store data.

magneto-optical (MO) drive A secondary storage device that uses a laser and reflected light to sense magnetically recorded bit values; data reading is based on the polarity of the reflected laser light, which is determined by the polarity of the magnetic charge.

magnetoresistive RAM (MRAM) A type of nonvolatile memory under development that stores bit values by using two magnetic elements, one with fixed polarity and the other with polarity that changes when a bit is written; has better longevity than conventional flash RAM.

main memory *See* primary storage.

mainframe A computer system designed to handle the information-processing needs of a large number of users and applications and optimized to store large quantities of data and move it from one place to another quickly and efficiently.

Mammoth A magnetic tape standard, developed by Exabyte, based on Digital Audio Tape; uses helical scanning and an improved tape drive technology to pack more data onto a single tape. *See also* Digital Audio Tape (DAT) *and* helical scanning.

manipulation In computer processing, refers to working with data by executing processor instructions, such as addition, subtraction, and equality comparisons.

mark sensor An optical input device that scans for light or dark marks at specific locations on a page.

Media Access Control (MAC) A protocol for determining how to share a transmission medium efficiently.

megahertz (MHz) A measurement of wave or system clock frequency; one million cycles per second.

memory allocation The assignment of specific memory addresses to system software, application programs, threads and processes, and data.

memory bus A subsidiary bus that connects only the CPU and memory; because its data transfer rate is higher than that of the system bus, it improves overall computer performance.

memory map A list of the memory location of every function and program variable; produced by the link editor based on the symbol table's contents. *See also* link editor *and* symbol table.

memory protection A procedure for protecting memory allocated to one program from unauthorized access by another program; adds overhead to each write operation.

mesh topology A network topology in which every node pair is connected by a point-to-point link; requires many transmission lines if the number of end nodes is large. *See also* network topology.

message In network communication, a unit of data or information transmitted from a sender to a recipient. In components and object-oriented programs, a request sent from one object or component to another.

methods Programs for manipulating data items in a class. *See also* class.

metropolitan area networks (MANs) Networks that typically cover a town or city; the market for WiMAX networks. *See also* Worldwide Interoperability for Microwave Access (WiMAX).

microchip A semiconductor device capable of integrating hundreds, thousands, and even billions of electrical devices on a single chip.

microcomputer Also called a PC or a workstation, a computer system designed to meet a single user's information-processing needs; can include portable computers, such as laptops and handheld computers.

microprocessor A microchip containing all the circuits and components of a CPU.

middleware System software that that "glues" together parts of a client/server or multitier application and enables clients and servers or distributed components to locate and communicate with one another.

midrange computer A computer system designed to provide information processing for multiple users and run many application programs simultaneously; sometimes called a "minicomputer."

millions of floating-point operations per second (megaflops or MFLOPS) A measurement of the rate at which floating-point operations are performed; used to measure CPU performance.

millions of instructions per second (MIPS) A measurement of the rate at which instructions are executed; assumed to measure CPU performance when manipulating single-precision integers.

modulator-demodulator (modem) A device that translates analog signals into digital signals and vice versa, enabling computer hardware to communicate over voice-grade phone lines.

monitors Video display panels. The term also refers to hardware or software that tracks and reports processing or I/O activity. *See also* hardware monitor *and* software monitor.

monochrome A display device or encoding method that can display only one of two colors (usually black and white), so it requires only 1 bit per pixel.

monophonic Capable of generating only one frequency (note) at a time.

Moore's Law Gordon Moore's observation that the rate of increase in transistor density on microchips doubles every 18 to 24 months, with no increase in unit cost.

most significant byte In storage bytes, the leftmost byte in a multiple-byte data item containing digits of the highest weight.

most significant digit The leftmost digit in a bit string that represents the greatest weight.

MOVE An instruction that copies data bits to storage locations and can copy data between any combination of registers and primary storage locations.

Moving Pictures Experts Group (MPEG) An organization that creates standards for motion picture recording and encoding technology; each standard is divided into layers numbered 1 (systems), 2 (video), and 3 (audio).

MP3 The audio-encoding standard that's layer 3 of the MPEG-1 standard. *See also* Moving Pictures Experts Group (MPEG).

multicasting Transmission situations involving multiple senders and receivers.

multicomputer configuration Any arrangement of multiple computers used to support specific services or applications; includes clusters, blades, and grids.

multicore architecture A microprocessor architecture that embeds multiple CPUs and cache memory on a single chip.

multilevel coding A technique for embedding multiple bit values in a single wave characteristic, such as frequency or amplitude; treats groups of bits as a single unit for the purpose of signal encoding.

multimode graded-index cable A multimode fiber-optic cable in which fibers vary in density from the center to the edge, which reduces the number of light reflections.

multimode step-index cable A multimode fiber-optic cable in which both the optical fiber and cladding have different but uniform densities throughout the cable, resulting in many light reflections.

multinational characters Modified Latin characters, such as ç and á, used by Western European languages other than English.

multiple-core CPUs Processors that improve performance by integrating multiple processing cores and memory caches on a single chip and by increasing raw CPU speed; the processors share primary storage and a single system bus.

multiple-processor architecture A more traditional approach to multiprocessing that uses two or more processors on a single motherboard or set of interconnected motherboards; slower than multicore architecture. *See also* multicore architecture *and* multiprocessing.

multiprocessing Any CPU architecture in which duplicate CPUs or processor stages can execute in parallel; a form of parallel processing.

Multipurpose Internet Mail Extensions (MIME) A protocol that's an extension of SMTP; enables including nontext content in e-mail messages. *See also* Simple Mail Transfer Protocol (SMTP).

multitasking An operating system's support for running multiple programs simultaneously.

multithreaded A process or program divided into two or more threads, each of which can be scheduled and executed independently. *See also* thread.

Musical Instrument Digital Interface (MIDI) A standard for storing and transporting control information between computers and electronic musical instruments and synthesizers.

n-layer architecture A client/server architecture with more than three layers; used when processing requirements or data resources are complex.

n-tier architecture *See* n-layer architecture.

named pipe A pipe with a name that's placed permanently in a file system directory; can communicate between processes on different computers. *See also* pipe.

narrowband A low-bandwidth communication channel; typically a subchannel of a broadband channel.

native applications Programs that are compiled and linked for a particular CPU and OS.

negative acknowledge (NAK) An ASCII control character sent by a receiver if data errors are detected.

netbook computer A laptop computer that emphasizes small size, reduced weight, low cost, and wireless networking; capable of performing only light-duty tasks. *See also* laptop computer.

network adapter *See* network interface card (NIC).

network administrator The person who's responsible for managing an organization's network infrastructure; can also apply to the manager of a local area network.

network-attached storage (NAS) A storage architecture with a dedicated storage server attached to a general-purpose network to handle storage access requests from other servers.

network interface card (NIC) A device that connects a node, such as a computer or network printer, to a network transmission cable.

Network layer The OSI layer that forwards messages to their correct destinations. *See also* Open Systems Interconnection (OSI) model.

network topology The spatial organization of network devices, physical routing of network cabling, and flow of messages from one network node to another.

network transparency *See* location transparency.

New Technology File System (NTFS) The Microsoft file system introduced with Windows NT; targeted to applications requiring high-speed directory and file operations, reliability and fault tolerance, secure file and disk content, and the capability to handle large disks, files, and directories.

noise In a communication channel, unwanted signal components added to the data signal that might be interpreted incorrectly as data; can be introduced by factors such as electromagnetic interference and distortion.

noncontiguous memory allocation A memory allocation scheme in which portions of a process can be allocated to free partitions anywhere in memory; uses small fixed-size partitions. *See also* contiguous *and* memory allocation.

nonprocedural language A programming language that describes a processing requirement without describing a specific procedure for satisfying the requirement.

nonvolatile A term describing storage devices that hold data without loss for long periods; secondary storage is usually nonvolatile.

nonvolatile memory (NVM) A generic term for memory devices with long-term or permanent data retention.

NOT An instruction that transforms the Boolean value true (1) into false (0) and the value false into true.

numeric range The set of all data values that can be represented by a data-encoding method.

object One instance, or variable, of a class. *See also* class.

object code The output of an assembler or a compiler; contains a mixture of CPU instructions, library calls, and other information the link editor needs. *See also* link editor.

object-oriented programming (OOP) A programming paradigm that views data and programs as two parts of an integrated whole; better addresses program reuse and long-term software maintenance.

Object Request Broker (ORB) A CORBA service that maintains a component directory and routes messages between components. *See also* Common Object Request Broker Architecture (CORBA).

objectclass An LDAP concept that defines attributes common to all members of a class. *See also* directory services *and* Lightweight Directory Access Protocol (LDAP).

octal notation A base-8 numbering system that uses digits from 0 to 7.

odd parity An error-detection method in which the sender sets the parity bit to 0 if the count of 1-valued data bits in the character is odd and to 1 if the count of 1-valued data bits is even. *See also* parity bit.

offset register A register containing the process offset value; used in indirect addressing. *See also* indirect addressing *and* process offset.

on-off keying (OOK) A signal-coding method that generates square waves by rapidly switching (pulsing) an electrical or optical power source to represent bit values;

essentially the digital equivalent of amplitude modulation. *See also* amplitude modulation (AM).

op code In an instruction's bit string, it's the first group of bits and represents the instruction's unique binary number.

OpenGL A video controller image description language developed by Silicon Graphics but now maintained by Khronos Group as an open standard. *See also* image description language (IDL).

Open Systems Interconnection (OSI) model A conceptual model for network hardware and software that organizes network functions into seven layers; useful as a general model of networks, a framework for comparing networks, and an architectural roadmap that enhances interoperability between network architectures and products.

operands In an instruction's bit string, they're groups of bits after the op code that hold the instruction's input values; they can contain a data item or the location of a data item. *See also* op code.

operating expenditures Funds expended during the current fiscal year to support normal business activities.

operating system (OS) A collection of utility programs for supporting users and application programs, allocating resources to multiple users and application programs, and controlling access to hardware.

optical character recognition (OCR) A technology that combines optical-scanning technology with a special-purpose processor or software to interpret bitmap content. *See also* optical scanner.

optical scanner A device that generates bitmap representations of printed images; detects light reflected from the page with an array of photosensors.

organic LED (OLED) A newer LED, manufactured with TFT technology, that achieves high-quality color display with organic compounds. *See also* light-emitting diodes (LEDs) *and* thin film transistor (TFT).

overflow An error that occurs when the result of a processing operation exceeds the format's numeric range. *See also* numeric range.

packets Basic units of data communication in a network.

packet-filtering firewall The simplest type of firewall; examines each packet and matches header content to a list of allowed or denied packet types.

packet switching The most common type of TDM, in which messages are divided into packets and then transmitted to their destination as channel capacity becomes available. *See also* time-division multiplexing (TDM).

page A small fixed-size portion of a program, normally between 1 and 4 KB, swapped between primary and secondary storage. *See also* virtual memory management.

page fault A reference to a page held in secondary storage. *See also* page *and* virtual memory management.

page file *See* swap space.

page frame A memory page used in virtual memory management. *See also* page *and* virtual memory management.

page hit A reference to a page held in memory. *See also* page *and* virtual memory management.

page tables Tables that store information about page locations, allocated page frames, and secondary storage space. *See also* page *and* virtual memory management.

palette A table of colors used to represent pixel color; the number of bits used to represent each pixel determines the table size.

parallel access An access method that can access multiple storage locations simultaneously; can also be achieved by subdividing data items and storing the component pieces on multiple storage devices.

parallel transmission Sending each bit position of a message over a separate transmission line simultaneously.

parent The original version of a file, after updates have been applied to generate a new version (the child). *See also* child.

parent process The original process that initiates and controls execution of a child process. *See also* process *and* spawn.

parity bit A bit appended to a character that stores redundant information used for error checking; its value is a count of other bit values in the character.

parity checking Validating character data by recomputing the value of a parity bit. *See also* even parity *and* odd parity.

passive matrix display An LCD that shares transistors among rows and columns of pixels.

peer-to-peer (P2P) architecture A software architecture in which the roles of client and server are combined into a single application or group of related applications.

peer-to-peer bus In this arrangement, any device can assume control of the bus or act as a bus master for transfers to any other device. *See also* bus arbitration unit *and* bus master.

Pentium processors A family of processors that improved memory access and raw CPU speeds and added features such as support for higher-speed system buses, pipelined instruction execution, and multimedia processing instructions.

Peripheral Component Interconnect (PCI) A family of bus standards, developed by Intel in the early 1990s, found in nearly all small and midrange computers of that era; since the late 2000s, largely replaced by PCI Express.

peripheral devices Storage and I/O devices in a computer, other than the CPU and primary storage.

personal computer (PC) *See* microcomputer.

personal digital assistant (PDA) A handheld computer, usually integrated with a cell phone, that supports light-duty tasks.

petaflops (PFLOPS) A measurement of the rate (10^{15} per second) at which floating-point operations are performed; used to measure CPU performance.

phase In communication, a specific time point in a sine wave's cycle; measured in degrees. *See also* cycle.

phase-change memory (PCM) A type of nonvolatile memory under development that uses a GST compound capable of switching between amorphous and crystalline states; has fast write times and high longevity. *See also* germanium, antimony, and tellurium (GST).

phase-shift keying (PSK) *See* phase-shift modulation.

phase-shift modulation A modulation method that represents bits as sudden shifts in wave phase.

phonemes Vocal sounds that are basic components of human speech; they correspond roughly to the sounds of each letter of the alphabet.

photosensor A device that converts incoming light energy into outgoing electrical energy.

Physical layer The OSI layer where communication between devices actually takes place; includes hardware devices that encode and decode bitstreams and the transmission lines that transport them. *See also* Open Systems Interconnection (OSI) model.

physical memory The actual number of memory bytes that are physically installed in a computer system; can be smaller than addressable memory but never larger. *See also* addressable memory.

physical record The unit of storage transferred between the device controller and memory in a single operation.

physical topology The physical placement of cables and device connections in a network topology. *See also* network topology.

pipe A region of shared memory through which multiple processes executing on the same computer can exchange data; used for communicating between OS components, queuing requests to an OS service, and exchanging messages between program components.

pipelining A method of organizing CPU circuitry so that multiple instructions can be in different stages of execution at the same time; a form of parallel processing.

pixel An abbreviation of "picture element," it's a single unit of data in an image; also refers to a single point on a display surface.

plasma display A video display device that uses an active matrix display and generates light by applying an electrical charge to neon gas.

platform as a service (PaaS) A cloud-based architectural approach in which an organization rents access to system software and hardware on which it installs its own application software and other services. *See also* cloud computing.

platters In magnetic disk media, they're flat, circular disks with metallic coatings that are rotated beneath read/write heads; data is normally recorded on both sides.

plotter A printer that generates line drawings on sheets or rolls of paper up to 64 inches wide. *See also* large-format printer.

point A standard measurement unit for font size; equals 1/72 of an inch.

pointer A data element containing the address of another data element; typically used in data structures; also a synonym for "cursor" in a display device.

polyphonic Capable of generating many frequencies (notes) simultaneously.

pop The process of removing register values from the top of a stack and loading them back into the correct registers. *See also* stack.

port A TCP connection identified by a unique integer number; many ports are standardized to specific Internet services. *See also* Transmission Control Protocol (TCP).

Portable Document Format (PDF) An Adobe image description language developed to generate and manage documents as an integrated whole rather than a collection of independent images and pages.

Post Office Protocol 3 (POP3) A protocol that standardizes the interaction between e-mail clients and servers so that client and server can run on different Internet hosts; e-mails are held on the server temporarily, downloaded to the client when a connection is established, and then deleted from the server.

PostScript An Adobe image description language designed mainly for printed documents, although it's also a programming language for generating video display output.

power sags Momentary reductions in the voltage or amperage of electrical power.

power surges Momentary increases (spikes) in the voltage or amperage of electrical power.

preemptive scheduling A scheduling method that enables a higher-priority thread to interrupt and suspend a lower-priority thread. *See also* scheduling *and* thread.

Presentation layer The OSI layer that makes sure data transmitted by one network node is interpreted correctly by the other network node; used mainly by applications that format data for user display. *See also* Open Systems Interconnection (OSI) model.

primary storage High-speed storage in a computer system that holds currently running programs and data immediately needed by these programs.

primitive data types The integer, real number, character, Boolean, and memory address data types that CPUs can manipulate directly.

priority-based scheduling A scheduling method that determines which ready thread should be dispatched to the CPU based on user or thread priority. *See also* scheduling *and* thread.

procedure *See* function.

process A unit of executing software that's managed independently by the OS and can request and receive hardware resources and OS services. Also refers to transforming input data by applying calculations, manipulations, and other operations.

process control block (PCB) A data structure containing information about an active process; used by the OS to keep track of each process and allocate resources, secure resource access, and protect active processes from interference by other active processes.

process family The collective name for a parent process and all its descendants. *See also* process *and* spawn.

process list *See* process queue.

process offset In memory allocation, the difference between the first address in physical memory and the address of the first process instruction. *See also* memory allocation *and* process.

processor A device capable of performing data manipulation and transformation operations.

process queue A data structure containing a list of PCBs for all active processes; can be searched by OS components. *See also* process control block (PCB).

production version A program version that omits the symbol table and debugging checkpoints to reduce program size and increase execution speed. *See also* debugging version *and* symbol table.

program A stored set of instructions for performing a specific task.

program editors Writing tools similar to word-processing applications but customized for writing programs instead of documents. *See also* application development software.

program profiler A software utility that describes the resource or service utilization of an application program during execution.

program status word (PSW) A special-purpose register containing a bit string describing the CPU's status and the currently running program. *See also* flag *and* registers.

program translator A program that translates instructions in a programming language into CPU instructions. *See also* application development software.

programmer A software developer who builds and tests software; might also perform tasks in the requirements and design disciplines. *See also* software developers.

programming language A language for expressing computer-processing functions or instructions.

protocol stack A complex set of software layers that an OS uses to implement network I/O and services.

proxy server *See* application firewall.

push The process of copying register values to the top of a stack. *See also* stack.

Quarter Inch Committee (QIC) A committee that develops open standards for magnetic tape drives on smaller computers.

qubit An atom or any other matter that stores data in multiple simultaneous quantum states.

radio frequency (RF) Electromagnetic radiation propagated through space; describes transmissions using frequencies between 50 Hz and 1 THz.

radix *See* base.

radix point In numbering systems other than decimal, the period or comma that separates the whole and fractional parts of a numeric value. *See also* decimal point.

random access An access method that can access any storage location directly and in any order; primary storage devices and disk storage devices use random access.

random access memory (RAM) Semiconductor devices used to implement primary storage; they don't provide permanent storage because RAM's contents are erased when the system power is turned off.

raw data transfer rate The maximum number of bits or bytes per second a communication channel can carry; ignores the communication protocol and assumes error-free transmission.

read-only memory (ROM) The earliest type of nonvolatile memory, with data content written permanently during manufacture; this primary storage device can be read, but no further data can be written.

read/write head A mechanism in a storage device that reads and writes data to and from the storage medium; also referred to as a "read/write mechanism."

ready state The state of an active thread that's idle, pending availability of a CPU. *See also* thread.

ready-to-send (RTS) signal A CSMA/CA signal sent by a node waiting to transmit. *See also* Carrier Sense Multiple Access/Collision Avoidance (CSMA/CA).

real number A number that can contain both whole and fractional components; the fractional portion is represented by digits to the right of the radix point. *See also* radix point.

real resources A computer's hardware devices and associated system software that physically exist. *See also* virtual resources.

real-time scheduling A scheduling method that guarantees a thread a minimum amount of CPU time and enough resources to complete its function in a specified time if the thread makes an explicit request when it's created. *See also* scheduling *and* thread.

record A data structure composed of other data structures or primitive data elements; commonly used as a unit of input and output to and from files or databases.

reduced instruction set computing (RISC) A computer and processor design approach that typically includes fixed-length instructions, short instruction length, and a large number of general-purpose registers; the main feature is the absence of some complex instructions from the instruction set. *See also* complex instruction set computing (CISC).

redundant array of independent disks (RAID) A disk storage technique that improves performance and fault tolerance. *See also* fault tolerance.

refresh cycle In dynamic RAM, the period during which circuitry supplies fresh infusions of power automatically; read and write operations can't be performed during this cycle. Also refers to the transfer of a full screen of data from the display generator to the monitor.

refresh rate The number of refresh cycles per second on a video display device; normally stated in hertz. *See also* refresh cycle.

registers Internal storage locations in a CPU; each is capable of holding a single instruction or data item. *See also* central processing unit (CPU).

relative addressing *See* indirect addressing.

relative path The access path that begins at the current directory's level and extends downward to a specific file. *See also* current directory.

release version *See* production version.

Remote Procedure Call (RPC) A protocol that enables a process on one computer to call a process on another computer.

repeater A device that functions much like an amplifier but extracts data embedded in the signal it receives and retransmits a new signal containing the same data; therefore, noise or distortion aren't retransmitted. *See also* amplifier.

request for proposal (RFP) A formal document stating hardware or software requirements and soliciting proposals from vendors to meet these requirements.

requirements discipline Activities in the Unified Process for developing models of system and user requirements. *See also* Unified Process (UP).

resistance The loss of electrical power that occurs as electrons pass through a conductor; low resistance means little power is lost.

resolution The number of pixels displayed per linear measurement unit.

resource registry A database, maintained by the resource locator, containing the names and locations of known resources and services on a network.

return instruction In programming, an instruction executed at the end of a function to return control to the calling function. *See also* function.

return wire In electrical transmission through wires, the channel component that completes an electrical circuit between sending and receiving devices.

RGB Red, green, and blue; the primary colors for video display. *See also* additive colors.

ring topology A network topology in which every node is connected directly to two other nodes, with a set of links forming a closed ring. *See also* network topology.

Rock's Law Arthur Rock's addendum to Moore's Law, stating that the cost of fabrication facilities for the latest chip generation doubles every four years. *See also* Moore's Law.

rotational delay The time a hard disk controller must wait for the right sector to rotate beneath read/write heads.

router A device that intelligently forwards messages between networks; it stores messages in a buffer, examines their contents, and applies decision rules to determine where to forward them.

routing tables Internal maps of a network containing information routers use to forward messages and choose from multiple possible paths to a recipient; routers periodically exchange this information with other routers to learn about networks beyond those to which they're connected. *See also* router.

run queue A data structure listing all active TCBs. *See also* thread *and* thread control block (TCB).

running state The state of an active thread that's been dispatched and has CPU control. *See also* dispatching *and* thread.

sampling The process of converting analog sound waves to digital representation; it analyzes the content of the audio sound spectrum many times per second and converts it to a numeric representation.

sandbox The protected area in which Java applets and servlets run; provides extensive security controls to prevent these programs from accessing unauthorized resources or damaging the hardware, OS, or file system. *See also* Java.

scaling out An approach to increasing processing and other computer system power by partitioning processing and other tasks among multiple computer systems; examples include clusters and grids. *See also* cluster *and* grid.

scaling up An approach to increasing processing and other computer system power by using larger and more powerful computers; examples include multicore and multiple-processor architectures. *See also* multicore architecture *and* multiple-processor architecture.

scan code A 1- or 2-byte data element generated by a keyboard controller; represents a specific keyboard event. *See also* keyboard controller.

scanning lasers Devices that sweep a narrow laser beam back and forth across bar codes. *See also* bar-code scanner.

scheduler The portion of the OS that makes scheduling decisions for threads. *See also* thread.

scheduling The decision-making process the OS uses to determine which ready thread moves to the running state. *See also* thread.

scripting language A simple programming language that enables programmers to assemble software quickly by "gluing" together the capabilities of many other programs, such as Web servers and database management systems; scripts can be embedded in HTML pages and many other programs.

second-generation language (2GL) *See* assembly language.

secondary storage System devices that provide large-capacity and long-term data storage.

sector The data transfer unit for magnetic disk and optical disc drives; the size is generally stated in bytes and can vary from one device to another. Also refers to a fractional portion of a track on magnetic disk media.

Secure Shell (SSH) An improved version of Telnet that encrypts data between client and server to address a major security issue in Telnet. *See also* Telnet.

segmented memory model An approach to assigning memory addresses in which primary storage is divided into equal-sized segments called pages, identified by sequential nonnegative integers; each byte of memory has a two-part address: The first part identifies the page, and the second part identifies the byte in the page.

semiconductors Materials with conductivity that varies in response to the electrical inputs applied; they have resistance properties that can be modified between those of a conductor and an insulator by adding chemical impurities.

sequential access time The time required to read the second of two adjacent sectors on the same track and platter of a hard disk.

serial access An access method that stores and retrieves data items in a linear (sequential) order; mainly used to hold backup copies of data stored on other storage devices.

Serial Advanced Technology Attachment (SATA) A storage device and cabling standard commonly used in PCs; compatible with older parallel ATA standards but uses serial transmission.

Serial Attached SCSI (SAS) A storage device and cabling standard commonly used in servers; compatible with older parallel SCSI standards but uses serial transmission.

serial transmission Sending bits sequentially over a single transmission line; the receiver reassembles the bits into larger data units, such as bytes.

server A mode of use rather than a class of computer system; manages shared resources and enables users to access these resources over a network.

server consolidation Using virtual machines as small servers hosted by a hypervisor running on a larger machine; reduces total hardware requirements because not every server has to be installed on a separate computer. *See also* hypervisor *and* virtual machine (VM).

service call A request to execute an OS service-layer function.

service layer The OS layer containing reusable components packaged as functions that can be called from other programs; also acts as an intermediary between programs, which request and use resources, and the kernel, which manages and provides access to resources. *See also* kernel.

service-oriented architecture (SOA) A design philosophy under which operating systems and some application software are constructed as a set of services that can be accessed by both internal and external users and software components.

service standards Standards for providing infrastructure-based services to a wide variety of users. *See also* infrastructure.

servlet A full-fledged Java program that runs in a Web server and performs functions such as calculations, database access, and creation of Web pages; runs in a protected area called the sandbox. *See also* Java Platform, Enterprise Edition (Java EE).

Session layer The OSI layer that establishes and manages communication sessions. *See also* Open Systems Interconnection (OSI) model.

shell *See* command layer.

SHIFT An instruction that moves all bit values right or left, according to the number of positions specified by the operand; after shifting, empty positions are filled with 0s, and bit values that shift beyond the bit string's bounds are discarded.

shortcuts The Windows term for links in a graph directory structure. *See also* graph directory structure *and* links.

shortest time remaining (STR) A priority-based scheduling method that chooses the next thread to be dispatched based on the expected amount of CPU time needed to complete the process. *See also* dispatching, priority-based scheduling, *and* thread.

sibling processes The child processes of a single parent process. *See also* process *and* spawn.

signal A data transmission event or group of events representing a bit or group of bits; a message sent from one active process to another.

signal-to-noise (S/N) ratio A mathematical relationship between the power of a carrier signal and the power of noise in the communication channel; measured in decibels.

signal wire In electrical transmission through wires, the channel component used to carry data.

signed integer An integer that uses a sign bit to indicate whether the value is negative or positive.

Simple Mail Transfer Protocol (SMTP) The earliest e-mail protocol; defines how text messages are forwarded and routed between Internet hosts.

Simple Object Access Protocol (SOAP) An open standard, developed by the World Wide Web Consortium; has a simple programming interface and few infrastructure requirements.

simplex mode A communication mode in which messages flow in only one direction.

sine wave A waveform that varies continuously between positive and negative states.

single inline memory module (SIMM) A small printed circuit board that incorporates multiple DIPs and has a row of electrical contacts on the edge; the entire package is designed to lock into a SIMM slot on a motherboard. *See also* dual inline packages (DIPs).

single-mode cable A fiber-optic cable with fibers that are much thinner in diameter than multimode fibers and vary continuously in density from center to edge to eliminate light reflections.

singly linked list A data structure in which each list element contains a pointer to the next list element.

skew The timing difference between the arrival of bits sent in a parallel transmission channel; skew increases with distance and transmission rate. *See also* parallel transmission.

socket The combination of an IP address with a port number, such as 129.24.8.4:53; used to establish connections.

software as a service (SaaS) A Web-based or cloud-based architectural approach in which users interact via a Web browser or other Web-enabled view layer with application software provided by a third party and installed on the provider's hardware. *See also* cloud computing.

software developers People who create application software for specific processing needs; can include many job titles, with each role contributing to a different part of the SDLC. *See also* systems development life cycle (SDLC).

software monitor A program typically embedded in OS service routines that detects and reports processing activity or requests; can also generate statistics of service utilization or processing activity that can be displayed in real time or stored in a file for later analysis.

solid-state drive (SSD) A storage device that mimics the behavior of a magnetic disk but uses flash RAM or other nonvolatile memory devices as the storage medium and read/write mechanism; expected to replace magnetic disks gradually.

sound card An expansion card connected to the system bus that contains components for sound input and output.

source code Instructions or statements in a high-level programming language; normally stored in a file that's named to indicate both its function and programming language.

spawn The act of a parent process creating a child process. *See also* process.

speaker dependent Requiring training to recognize the sounds of human speakers; a characteristic of speech-recognition systems. *See also* speech recognition.

special-purpose processor A processor that's designed to perform only one specific task; essentially, a processor with a single internal program. *See also* processor.

special-purpose registers Registers used by the CPU for specific tasks; they include the instruction register, instruction pointer, and program status word. *See also* registers.

speculative execution An approach to dealing with condition BRANCHes in which the CPU executes instructions after a branch prediction but before the final branch condition value is known with certainty; a form of parallel processing. *See also* branch prediction.

speech recognition The process of recognizing and responding to the meaning embedded in spoken words, phrases, or sentences.

speech synthesis A complex process for generating human speech based on character or text input.

stack A reserved area of primary storage accessed on a last-in, first-out (LIFO) basis; this mechanism enables a program suspended by an interrupt to resume execution in exactly the same state as before an interruption.

stack overflow An error condition that occurs when attempting to push register values onto a stack that's already at its maximum capacity. *See also* stack.

stack pointer A special-purpose register that points to the next empty address in the stack and is incremented or decremented automatically each time the stack is pushed or popped. *See also* stack.

star topology A network topology in which every node is connected directly to a shared hub, switch, or router. *See also* network topology.

start bits Bits added to the beginning of messages in asynchronous transmission to alert the receiver to synchronize its clock.

stateful firewall A firewall that tracks the progress of complex client/server interactions. *See also* firewall.

static connection A mapping between a local resource name and a remote resource that must be initialized before use.

static linking A linking process in which library calls and other functions can't be changed after they're inserted into executable code. *See also* link editor.

static RAM (SRAM) A type of RAM that's implemented entirely with transistors; the basic storage unit is a flip-flop circuit.

storage allocation table A data structure that records which allocation units are free and which belong to files. *See also* allocation unit.

storage area network (SAN) A high-speed interconnection between general-purpose servers and a separate storage server; storage accesses are at the level of disk sectors in a logical address space.

storage bus A subsidiary bus that connects secondary storage devices to the system bus; reduces the length and number of physical connections to the system bus and aggregates the lower data transfer capacity of secondary storage devices to better match the higher capacity of a single system bus connection.

storage I/O control layer An FMS layer; the part of the kernel that accesses storage locations and manages data movement between storage devices and memory. Includes device drivers, interrupt handlers, and buffers and cache managers. *See also* file management system (FMS).

storage medium A device or substance in a storage device that actually holds data.

store A data transfer from a register into primary storage.

store and forward An interconnected system of end nodes and transfer points used to route data between source and destination nodes.

strategic plan A set of long-range goals for services to be provided and the resources needed to provide these services.

string A sequence of characters forming a meaningful word, phrase, or other useful group.

subroutine *See* function.

subtractive colors In printing, the primary colors are generated by using the inverse of the primary video display colors, so cyan is the absence of red, magenta is the absence of green, and yellow is the absence of blue.

supercomputer A computer system designed for rapid mathematical computations and used for computation-intensive applications, such as simulations, 3D modeling, weather prediction, computer animation, and real-time analysis of large databases.

Super Digital Linear Tape (SDLT) A magnetic tape standard developed by Quantum Corporation; the cartridge has only one reel, and the device records in parallel linear tracks in an end-to-end format.

supervisor The master interrupt handler program; it examines the interrupt code stored in the interrupt register, uses it as an index to the interrupt table, extracts the corresponding memory address, and transfers control to the interrupt handler at that address. *See also* interrupt code *and* interrupt handler.

surge protector A hardware device that detects incoming power surges and diverts them to ground. *See also* power surges.

sustained data transfer rate The maximum data transfer rate a storage device or communication channel can sustain during lengthy data transfer operations.

swap file *See* swap space.

swap space A secondary storage region reserved for the task of holding pages not held in memory; it's divided into page frames in the same manner as memory. *See also* page *and* virtual memory management.

switch A central connection point for nodes in a LAN; examines incoming destination addresses and temporarily connects the sending transmission line to the receiving transmission line. Also refers to a building block of processing circuitry; it controls electrical current flow in a circuit and is implemented as a transistor.

switched fabric An architecture for interconnecting devices with multiple data transmission pathways and a mesh of switches resembling the interwoven threads of fabric.

symbol table An internal table updated by the compiler that keeps track of data names, types, and assigned memory addresses in programs. *See also* compiler.

symbolic debugger An automated tool for testing executable programs; includes features for tracing calls to source code statements or functions, tracing changes to variables' contents, and detecting runtime errors.

synchronous DRAM (SDRAM) A read-ahead RAM that uses the same clock pulse as the system bus; read and write operations are broken into simple steps that can be completed in one bus clock cycle.

synchronous idle characters Control messages consisting of a predetermined pattern of signal transitions designed for easy clock synchronization.

synchronous transmission A method that ensures sender and receiver clocks are always synchronized by sending data in continuous streams of fixed-size byte groups called "blocks."

system administration A wide range of managerial activities for ensuring efficient and reliable delivery of information services.

system bus The internal communication channel connecting all hardware devices.

system clock A digital circuit that generates timing pulses (signals) and transmits them to other devices in the computer; all actions, especially a CPU's fetch and execution cycles, are timed according to this clock.

system development tools Tools that enable systems analysts and designers to develop models of information systems that are then used as the starting point for developing application programs. *See also* application development software.

system overhead The resources consumed by resource allocation procedures.

system requirements models Models that provide the detail needed to develop a system that meets users' needs.

system software A program for handling resource allocation to application software, performing utility functions needed by application software, or managing computer resources; includes operating systems, database management systems, antivirus software, and network security software, for example.

systems analyst A software developer who contributes to the business modeling and requirements disciplines. *See also* software developers.

systems architecture The structure, interaction, and technology of computer components.

systems designer A software developer who contributes to the design discipline. *See also* software developers.

systems development life cycle (SDLC) The process for developing an information system; follows a series of steps or activities. *See also* Unified Process (UP).

systems programmers People who develop or maintain system software and might also perform tasks such as hardware troubleshooting and software installation and configuration. *See also* system software.

tape drive A slow serial access device containing motors that wind and unwind tapes and read/write heads to access tape content.

target channel adapter (TCA) An interface that connects a device to an InfiniBand switch; used by simpler devices, such as network switches and storage appliances.

TCP/IP *See* Internet Protocol (IP) *and* Transmission Control Protocol (TCP).

TCP/IP model A layered protocol model describing current Internet standards and technology. *See also* Internet Protocol (IP) *and* Transmission Control Protocol (TCP).

Telnet A Web protocol in which users on one Internet host can interact with another host's OS command layer; it emulates a VDT and is limited to interacting with command-line interfaces.

testing discipline Activities in the Unified Process for verifying the correct functioning of infrastructure and application software components and ensuring that they satisfy system requirements. *See also* Unified Process (UP).

thin film transistor (TFT) A technology for manufacturing active matrix displays, in which wiring and transistors are added in thin layers to a glass substrate; similar to semiconductor fabrication technology.

thread A subdivision of a process that can be scheduled and executed independently; shares all resources allocated to its parent process. *See also* process.

thread control block (TCB) A data structure containing information that an OS uses to keep track of and manage threads. *See also* thread.

thread cycle The amount of CPU time needed to complete a thread's function. *See also* thread.

thread list *See* run queue.

three-layer architecture A variation of client/server architecture that divides software into three client or server processes called layers: the view layer, the business logic layer, and the data layer.

three-tier architecture *See* three-layer architecture.

third-generation language (3GL) A high-level programming language that uses mnemonics to represent instructions, variables, and labels; is machine independent; and has instruction explosion higher than 1:1. It has no advanced capabilities for interactive I/O, database processing, or nonprocedural programming. *See also* instruction explosion.

time-division multiplexing (TDM) A technique that divides a channel's data transfer capacity into time slices and allocates them to multiple users.

timer interrupt An interrupt generated at regular intervals (between several dozen and several thousand CPU cycles) to give the scheduler an opportunity to suspend the currently executing thread. *See also* scheduling *and* thread.

traces Arrangements of conductive molecules (usually straight lines) that enable electrons to flow from one place or device to another.

track One concentric circle of a platter; the surface area that passes under a read/write head when its position is fixed. *See also* platters.

track-to-track (TTT) seek time The average time needed to move a read/write head between two adjacent tracks; typically measured in milliseconds.

transaction To an FMS, any change to file contents or attributes, such as an added record, a modified field, or changed access controls. *See also* file management system (FMS).

transaction logging A form of automated backup in which all changes to file content and attributes are recorded automatically in a log file in addition to being written to the file's I/O buffer; provides a high degree of protection against data loss caused by program or hardware failure but imposes a performance penalty.

transistors Electrical switches made of semiconductor material that has been treated with chemical impurities to enhance the semiconducting effects; they're combined to implement gates. *See also* gate.

Transmission Control Protocol (TCP) A core Internet protocol for translating messages into packets and guaranteeing their delivery.

transmission medium A communication path that carries signals.

Transport layer The OSI layer responsible for formatting messages into packets suitable for transmission over the network; adds required header and trailer information, including network addresses, error-detection data, and packet-sequencing data. *See also* Open Systems Interconnection (OSI) model.

tree directory structure *See* hierarchical directory structure.

trillions of floating-point operations per second (teraflops or TFLOPS) A measurement of the rate at which floating-point operations are performed; used to measure CPU performance.

truncation The act of deleting bits that don't fit in a storage location.

twin-axial cable A transmission medium that's similar to coaxial cable but is thinner and contains two internal conductors. *See also* coaxial cable.

twisted-pair cable A transmission medium consisting of two copper wires twisted around one another and encased in nonconductive material, usually plastic.

twos complement notation A notation system that represents positive integers as ordinary binary values and negative integers by adding 1 to the complement of the positive value.

Type I error In data transmission, the probability of not detecting a real error.

Type II error In data transmission, the probability of incorrectly identifying good data as an error.

unblocked The term describing a physical record containing just one logical record. *See also* logical record *and* physical record.

unconditional BRANCH An instruction in which the processor always departs from the normal execution sequence. *See also* BRANCH.

undelete operation The act of restoring a record or file by re-creating its index information (directory entry) and recovering its previously allocated storage locations.

underflow A condition that occurs when a value is too small to represent in floating-point notation; also refers to overflow of a negative exponent in floating-point notation. *See also* floating-point notation.

unguided transmission A transmission medium that uses the atmosphere or space to carry messages encoded in radio frequency or light signals; also called "wireless transmission."

Unicode A standard 16-bit or 32-bit character-coding method that assigns nonnegative integers to represent printable characters; includes alphabets, ideographs, and characters for most of the world's languages.

Unified Process (UP) A systems development life cycle based on object-oriented techniques; follows a series of repeated steps. *See also* iterations.

Uniform Resource Locator (URL) A unique identification for a Web resource, composed of a protocol, host, port, and resource.

uninterruptible power supply (UPS) A device that provides power to attached devices in the event of external power failure;

UPSs vary in their power delivery capacity, switching time, and battery life.

unresolved reference *See* external function call.

unsigned integer A data type that stores positive integer values as ordinary binary numbers; its value is always assumed to be positive.

User Datagram Protocol (UDP) A connectionless protocol that lacks the connection management features of TCP but can support communication between multiple hosts and senders; used mainly for communication that requires low processing overhead and doesn't require a guarantee of reliable delivery. *See also* connectionless protocol.

variable A mnemonic representing a data item's memory address in assembly language or a name representing a data item's memory address in a high-level programming language. *See also* assembly language.

variable-length instruction In this type of instruction format, the amount by which the instruction pointer is incremented after a fetch is the length of the most recently fetched instruction; this format complicates instruction fetching because the number of bytes to be fetched isn't known in advance.

vector In graphics, a line segment with a specific angle and length in relation to a point of origin; also refers to a one-dimensional array.

vector list A series of concatenated or linked vectors that can be used to construct complex shapes; images constructed from a vector list resemble connect-the-dots drawings.

versioning A process in which a file's original version is archived automatically whenever the file is modified.

vertical redundancy checking *See* parity checking.

victim If all page frames are allocated, the page currently in memory that must be written to the swap space before the reference page is loaded into a page frame. *See also* page, swap space, *and* virtual memory management.

video bus A subsidiary bus that connects only memory and the video display device; improves performance by removing display update traffic from the system bus and providing a high-capacity one-way communication channel optimized for video data.

video controller A device connected to the system bus (or a dedicated video bus) that accepts commands and data from the CPU and generates analog or digital video signals, which are transmitted to the monitor.

video display terminal (VDT) An early I/O device containing an integrated keyboard and TV screen; capable of displaying only text and primitive graphics.

video RAM (VRAM) A type of RAM used in a video controller; differs from ordinary RAM because it can be written by the bus interface circuitry or video processor while being read by display generator circuitry. *See also* dual-porting.

view layer The software layer that accepts user input and formats and displays processing results. *See also* three-layer architecture.

virtualization environments Hypervisors that are installed as applications in other OSs. *See also* hypervisor.

virtual machine (VM) A collection of files on a physical computer that define the virtual machine's configuration and the contents of its virtual disk drives; creates an environment separate from the physical computer in which different OSs can run, application software can be tested, and so forth.

virtual memory management A memory management method in which the OS divides memory and programs into partitions called pages, which are held in secondary storage until needed. *See also* page.

virtual resources Resources (hardware and system software) that are apparent to a user or program as being available but don't necessarily exist physically.

virus A program or program fragment that infects a computer by installing itself permanently, performs malicious acts on the infected computer, and replicates and spreads itself by using services on the infected computer.

Voice over IP (VoIP) A family of technologies and standards for carrying voice messages and data over a single packet-switched network.

volatile A term describing storage devices that can't hold data for long periods; primary storage is usually volatile.

volume Part of the logical view of secondary storage; consists of an entire physical disk, a partition of the disk, or a removable storage medium and on larger computers, can span multiple physical disks.

wait state An idle clock cycle during which the CPU is waiting for a response from another device. *See also* clock cycle.

wavelength-division multiplexing (WDM) In long-distance telecommunication, using FDM to multiplex single-mode optical fibers. *See also* frequency division multiplexing (FDM).

wide area network (WAN) A network that spans large physical distances, such as multiple buildings, cities, regions, or continents.

wired transmission *See* guided transmission.

wireless transmission *See* unguided transmission.

wires *See* traces.

word A unit of data containing a fixed number of bytes or bits, loosely defined as the amount of data a CPU processes at one time; word size normally matches the size of general-purpose registers. *See also* general-purpose registers.

working directory *See* current directory.

workstation A type of microcomputer that's typically more powerful than a PC to support demanding numeric or graphics processing tasks. *See also* microcomputer.

Worldwide Interoperability for Microwave Access (WiMAX) A group of wireless networking standards codified in the IEEE 802.16 standard; targeted to applications involving fixed and mobile Internet access spanning distances up to 50 kilometers (about 30 miles). *See also* metropolitan area networks (MANs).

World Wide Web (WWW) Also referred to as just "the Web," a collection of resources that can be accessed over the Internet by standard protocols, such as HTTP.

INDEX

B

C

D

E

F

G

H

J

K

L

M

O

P

Q

R

S

T

U

V

W

X